3G WIRELESS NETWORKS

McGRAW-HILL TELECOMMUNICATIONS

Ali	*Digital Switching Systems*
Ash	*Dynamic Routing in Telecommunications Networks*
Azzam	*High-Speed Cable Modems*
Azzam/Ransom	*Broadband Access Technologies*
Bartlett	*Cable Communications*
Bates	*Broadband Telecommunications Handbook*
Bates	*Optical Switching and Networking Handbook*
Bayer	*Computer Telephony Demystified*
Bedell	*Wireless Crash Course*
Clayton	*McGraw-Hill Illustrated Telecom Dictionary,* 3/e
Collins	*Carrier Grade Voice over IP*
Davis	*ATM for Public Networks*
Gallagher	*Mobile Telecommunications Networking with IS-41*
Harte	*CDMA IS-95*
Harte	*Cellular and PCS: The Big Picture*
Harte	*Delivering xDSL*
Harte	*GMS Superphones*
Heldman	*Competitive Telecommunications*
Lachs	*Fiber Optics Communications*
Lee	*Lee's Essentials of Wireless Communications*
Lee	*Mobile Cellular Telecommunications,* 2/e
Lee	*Mobile Communications Engineering,* 2/e
Louis	*Telecommunications Internetworking*
Macario	*Cellular Radio,* 2/e
Muller	*Bluetooth Demystified*
Muller	*Desktop Encyclopedia of Telecommunications*
Muller	*Desktop Encyclopedia of Voice and Data Networking*
Muller	*Mobile Telecommunications Factbook*
Pattan	*Satellite-Based Cellular Communications*
Pecar	*Telecommunications Factbook,* 2/e
Richharia	*Satellite Communications Systems,* 2/e
Roddy	*Satellite Communications,* 3/e
Rohde/Whitaker	*Communications Receivers,* 3/e
Russell	*Signaling System #7,* 3/e
Russell	*Telecommunications Pocket Reference*
Russell	*Telecommunications Protocols,* 2/e
Shepard	*Optical Networking Demystified*
Shepard	*Telecommunications Convergence*
Simon	*Spread Spectrum Communications Handbook*
Smith	*LMDS*
Smith	*Practical Cellular and PCS Design*
Smith	*Wireless Telecom FAQs*
Smith/Gervelis	*Cellular System Design and Optimization*
Turin	*Digital Transmission Systems*
Winch	*Telecommunications Transmission Systems,* 2/e

3G Wireless Networks

Clint Smith, P.E.

Daniel Collins

McGraw-Hill
New York Chicago San Francisco Lisbon
London Madrid Mexico City Milan New Delhi
San Juan Seoul Singapore Sydney Toronto

Cataloging-in-Publication Data is on file with the Library of Congress

McGraw-Hill

A Division of The ***McGraw-Hill*** *Companies*

1 2 3 4 5 6 7 8 9 0 AGM/AGM 0 7 6 5 4 3 2 1

ISBN 0-07-136381-5

The sponsoring editor for this book was Stephen S. Chapman and the production supervisor was Pamela A. Pelton. It was set in Century Schoolbook by MacAllister Publishing Services, LLC.

Printed and bound by Quebecor/Martinsburg.

This book is printed on recycled, acid-free paper containing a minimum of 50 percent recycled, de-inked fiber.

This book would not have been possible without the continued assistance from Sam, Rose, and Mary whom always support my many efforts. Therefore, this book is dedicated to them as a small token of thanks and appreciation for the numerous hours they surrendered.

I also want to thank the many people who assisted me in putting together this important effort. Although there are few names on the book, there are many who helped in its creation.

—Clint Smith, P.E.

To my wonderful wife, Ann, who is the inspiration, strength and support behind everything I do.

—Daniel Collins

CONTENTS

PREFACE

The wireless industry continues to provide new opportunities and challenges. The proliferation of wireless devices, plus the ever-increasing bandwidth requirements envisioned for future packet data applications, is creating a vast array of effort to support that end goal. As with any new technology or platform being introduced for use in the market, no true benchmark can be followed because there is no real legacy or current information from which to pattern after.

This book is meant to help reduce the potential confusion with the multitude of issues associated with mobile data implementation into an existing or new wireless system, which uses either UMTS or CDMA2000. The present operators are currently in the process of building a 2.5G platform in anticipation of high-speed packet usage or they are still contemplating if and when to make the plunge.

As always, when migrating from a 2G platform to a 2.5G or even 3G, numerous decisions and alterations to the existing network must take place. Because the IMT2000 specification has several platforms all called 3G, and since they meet particular data throughput requirements, the decisions are vast and critical for an operator to make. Whichever platform is chosen will fundamentally determine the success or failure of the wireless operator in any given market. It is also interesting to note that depending on the market and services desired, plus the legacy system, several methods are available for migrating from a 2G to a 3G platform.

For example, various vendor implementations of CDMA2000 exist for 1XRTT and 3XRTT (future), and just picking a particular platform type is the first step. The next, of course, is understanding how the vendor will try and realize the system. This book will cover the common functions that exist among the various implementations, and it also will address the design rules to follow. The specific vendor card requirements and methods for how to implement the standards are not covered here.

The book is meant to establish design guidelines and a fundamental understanding of UMTS and CDMA2000 in a concise area for the stated purpose of helping orientate the engineer into focusing with each vendor on the specific areas that are most relevant for the design and implementation. This is not to say that the nuances associated with each vendor's implementation are not important; however, in a multi-vendor environment that most operators are now working in, the fundamental commonality among overall design principals is the common thread that binds them together from an engineering perspective.

Clint Smith, P.E.
Daniel Collins

ACKNOWLEDGMENTS

I want to thank the members of Qualcomm for their technical assistance and guidance on CDMA2000.

—Clint Smith, P.E.

Grateful thanks to Lars Nilsson of Ericsson for very useful input and to Elleni Letta of TeleworX for reviewing the GSM and UMTS portions of the book and providing very helpful feedback.

—Daniel Collins

CHAPTER 1

Wireless Communications

1.1 The Amazing Growth of Mobile Communications

Over recent years, telecommunications has been a fast-growing industry. This growth can be seen in the increasing revenues of major telecommunications carriers and the continued entry into the marketplace of new competitive carriers. No segment of the industry, however, has seen growth to match that experienced in mobile communications. From relatively humble beginnings, the last 15 years have seen an explosion in the number of mobile communications subscribers and it appears that growth is likely to continue well into the future.

The growth in the number of mobile subscribers is expected to continue for some years, with the number of mobile subscribers surpassing the number of fixed network subscribers at some point in the near future. Although it may appear that such predictions are optimistic, it is worth pointing out that in the past, most predictions for the penetration of mobile communications have been far lower than what actually occurred. In fact, in several countries, the number of mobile subscribers already exceeds the number of fixed subscribers, which suggests that predictions of strong growth are well founded. It is clear that the future is bright for mobile communications. For the next few years at least, that future means third-generation systems, the subject of this book.

Before delving into the details of third-generation systems, however, it is appropriate to review mobile communications in general, as well as first- and second-generation systems. Like most technologies, advances in wireless communications occur mainly through a process of steady evolution (although there is the occasional quantum-leap forward). Therefore, a good understanding of third-generation systems requires an understanding of what has come before. In order to place everything in the correct perspective, the following sections of this chapter provide a history and a brief overview of mobile communications in general. Chapter 2, "First Generation (1G)," and Chapter 3, "Second Generation (2G)," provide some technical detail on first- and second-generation systems, with the remaining chapters of the book dedicated to the technologies involved in third-generation systems.

1.2 A Little History

Mobile telephony dates back to 1920s, when several police departments in the United States began to use radiotelephony, albeit on an experimental basis. Although the technology at the time had had some success with maritime vessels, it was not particularly suited to on-land communication. The equipment was extremely bulky and the radio technology did not deal very well with buildings and other obstacles found in cities. Therefore, the experiment remained just an experiment.

Further progress was made in the 1930s with the development of *frequency modulation* (FM), which helped in battlefield communications during the Second World War. These developments were carried over to peacetime, and limited mobile telephony service became available in the 1940s in some large cities. Such systems were of limited capacity, however, and it took many years for mobile telephone to become a viable commercial product.

1.2.1 History of First-Generation Systems

Mobile communications as we know it today really started in the late 1970s, with the implementation of a trial system in Chicago in 1978. The system used a technology known as *Advanced Mobile Phone Service* (AMPS), operating in the 800-MHz band. For numerous reasons, however, including the break-up of AT&T, it took a few years before a commercial system was launched in the United States. That launch occurred in Chicago in 1983, with other cities following rapidly.

Meanwhile, however, other countries were making progress, and a commercial AMPS system was launched in Japan in 1979. The Europeans also were active in mobile communications technology, and the first European system was launched in 1981 in Sweden, Norway, Denmark, and Finland. The European system used a technology known as *Nordic Mobile Telephony* (NMT), operating in the 450-MHz band. Later, a version of NMT was developed to operate in the 900-MHz band and was known (not surprisingly) as NMT900. Not to be left out, the British introduced yet another technology in 1985. This technology is known as the *Total Access Communications System* (TACS) and operates in the 900-MHz band. TACS is basically a modified version of AMPS.

Many other countries followed along, and soon mobile communications services spread across the globe. Although several other technologies were developed, particularly in Europe, AMPS, NMT (both variants), and TACS were certainly the most successful technologies. These are the main first-generation systems and they are still in service today.

First-generation systems experienced success far greater than anyone had expected. In fact, this success exposed one of the weaknesses in the technologies—limited capacity. Of course, the systems were able to handle large numbers of subscribers, but when the subscribers started to number in the millions, cracks started to appear, particularly since subscribers tend to be densely clustered in metropolitan areas. Limited capacity was not the only problem, however, and other problems such as fraud became a major concern. Consequently, significant effort was dedicated to the development of second-generation systems.

1.2.2 History of Second-Generation Systems

Unlike first-generation systems, which are analog, second-generation systems are digital. The use of digital technology has a number of advantages, including increased capacity, greater security against fraud, and more advanced services.

Like first-generation systems, various types of second-generation technology have been developed. The three most successful variants of second-generation technology are *Interim Standard 136* (IS-136) TDMA, IS-95 CDMA, and the *Global System for Mobile communications* (GSM). Each of these came about in very different ways.

1.2.2.1 IS-54B and IS-136 IS-136 came about through a two-stage evolution from analog AMPS. As described in more detail later, AMPS is a *frequency division multiple access* (FDMA) system, with each channel occupying 30 KHz. Some of the channels, known as control channels, are dedicated to control signaling and some, known as voice channels, are dedicated to carrying the actual voice conversation.

The first step in digitizing this system was the introduction of digital voice channels. This step involved the application of *time division multiplexing* (TDM) to the voice channels such that each voice channel was divided into time slots, enabling up to three simultaneous conversations on the same RF channel. This stage in the evolution was known as IS-54 B (also known as Digital AMPS or D-AMPS) and it obviously gives a significant capacity boost compared to analog AMPS. IS-54 B was introduced in 1990.

Note that IS-54 B involves digital voice channels only, and still uses analog control channels. Thus, although it may offer increased capacity and some other advantages, the fact that the control channel is analog does limit the number of services that can be offered. For that reason, among others, the next obvious step was to make the control channels also digital. That step took place in 1994 with the development of IS-136, a system that includes digital control channels and digital voice channels.

Today AMPS, IS-54B, and IS-136 are all in service. AMPS and IS-54 operate only in the 800-MHz band, whereas IS-136 can be found both in the 800-MHz band and in the 1900-MHz band, at least in North America. The 1900-MHz band in North America is allocated to *Personal Communications Service* (PCS), which can be described as a family of second-generation mobile communications services.

1.2.2.2 GSM Although NMT had been introduced in Europe as recently as 1981, the Europeans soon recognized the need for a pan-European digital system. There were many reasons for this, but a major reason was the fact that multiple incompatible analog systems were being deployed across Europe. It was understood that a single Europe-wide digital system could enable seamless roaming between countries as well as features and capabilities not possible with analog systems. Consequently, in 1982, the *Conference on European Posts and Telecommunications* (CEPT) embarked on developing such a system. The organization established a group called (in French) *Group Spéciale Mobile* (GSM). This group was assigned the necessary technical work involved in developing this new digital standard. Much work was done over several years before the newly created *European Telecommunications Standards Institute* (ETSI) took over the effort in 1989. Under ETSI, the first set of technical specifications was finalized, and the technology was given the same name as the group that had originally begun the work on its development—GSM.

The first GSM network was launched in 1991, with several more launched in 1992. International roaming between the various networks quickly followed. GSM was hugely successful and soon, most countries in Europe had launched GSM service. Furthermore, GSM began to spread outside Europe to countries as far away as Australia. It was clear that GSM was going to be more than just a European system; it was going to be global. Consequently, the letters GSM have taken on a new meaning—Global System for Mobile communications.

Initially, GSM was specified to operate only in the 900-MHz band, and most of the GSM networks in service use this band. There are, however, other frequency bands used by GSM technology. The first implementation of GSM at a different frequency happened in the United Kingdom in 1993. That service was initially known as DCS1800 since it operates in the 1800-MHz band. These days, however, it is known as GSM1800. After all, it really is just GSM operating at 1800 MHz.

Subsequently, GSM was introduced to North America as one of the technologies to be used for PCS—that is, at 1900 MHz. In fact, the very first PCS network to be launched in North America used GSM technology.

1.2.2.3 IS-95 CDMA Although they have significant differences, both IS-136 and GSM use *Time Division Multiple Access* (TDMA). This means that individual radio channels are divided into timeslots, enabling a number of users to share a single RF channel on a time-sharing basis. For several reasons, this technique offers an increase in capacity compared to an analog system where each radio channel is dedicated to a single conversation. TDMA is not the only system that enables multiple users to share a given radio frequency, however. A number of other options exist—most notably *Code Division Multiple Access* (CDMA).

CDMA is a technique whereby all users share the same frequency at the same time. Obviously, since all users share the same frequency simultaneously, they all interfere with each other. The challenge is to pick out the signal of one user from all of the other signals on the same frequency. This can be done if the signal from each user is modulated with a unique code sequence, where the code bit rate is far higher than the bit rate of the information being sent. At the receiving end, knowledge of the code sequence being used for a given signal allows the signal to be extracted.

Although CDMA had been considered for commercial mobile communications services by several bodies, it was never considered a viable technology until 1989 when a CDMA system was demonstrated by Qualcomm in San Diego, California. At the time, great claims were made about the potential capacity improvement compared to AMPS, as well as the potential improved voice quality and simplified system planning. Many people were impressed with these claims and the Qualcomm CDMA system was standardized as IS-95 in 1993 by the U.S. *Telecommunications Industry Associ-*

ation (TIA). Since then, many IS-95 CDMA systems have been deployed, particularly in North America and Korea. Although some of the initial claims regarding capacity improvements were perhaps a little overstated, IS-95 CDMA is certainly a significant improvement over AMPS and has had significant success. In North America, IS-95 CDMA has been deployed in the 800-MHz band and a variation known as J-STD-008 has been deployed in the 1900-MHz band.

CDMA is unique to wireless mobility in that it spreads the energy of the RF carrier as a direct function of the chip rate that the system operates at. The CDMA system utilizing the Qualcomm technology utilizes a chip rate of 1.228 MHz. The chip rate is the rate at which the initial data stream, the original information, is encoded and then modulated. The chip rate is the data rate output of the PN generator of the CDMA system. A chip is simply a portion of the initial data or message that is encoded through use of a XOR process.

The receiving system also must despread the signal utilizing the exact same PN code sent through an XOR gate that the transmitter utilized in order to properly decode the initial signal. If the PN generator utilized by the receiver is different or is not in synchronization with the transmitter's PN generator, then the information being transmitted will never be properly received and will be unintelligible. Figure 1-1 represents a series of data that is encoded, transmitted, and then decoded back to the original data stream for the receiver to utilize.

The chip rate also has a direct effect on the spreading of the CDMA signal. Figure 1-2 shows a brief summary of the effects on spreading the original signal that the chosen chip rate has on the original signal. The heart of CDMA lies in the point that the spreading of the initial information distributes the initial energy over a wide bandwidth. At the receiver, the signal is despread through reversing the initial spreading process where the original signal is reconstructed for utilization. When the CDMA signal experiences interference in the band, the despreading process despreads the initial signal for use but at the same time spreads the interference so it minimizes its negative impact on the received information.

The number of PN chips per data bit is referred to as the processing gain and is best represented by the following equation. Another way of referencing processing gain is the amount of jamming, or interference, power that is reduced going through the despreading process. Processor gain is the improvement in the signal-to-noise ratio of a spread spectrum system and is depicted in Figure 1-3.

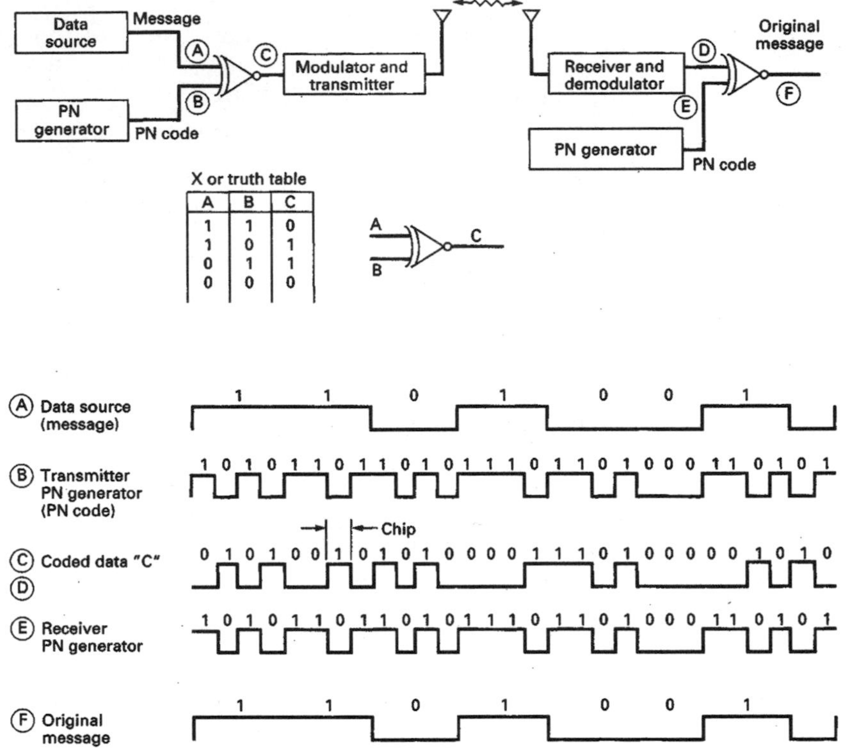

Figure 1-1 CDMA PN coding.

1.2.3 The Path to Third-Generation Technology

In many ways, second-generation systems have come about because of fundamental weaknesses in first-generation technologies. First-generation technologies have limited system capacity, they have very little protection against fraud, they are subject to easy eavesdropping, and they have little to offer in terms of advanced features. Second-generation systems are designed to address all of these issues, and they have done a very successful job.

Systems like IS-95, GSM, and IS-136 are much more secure; they also offer higher capacity and more calling features. They are, however, still optimized for voice service and they are not well suited to data communications.

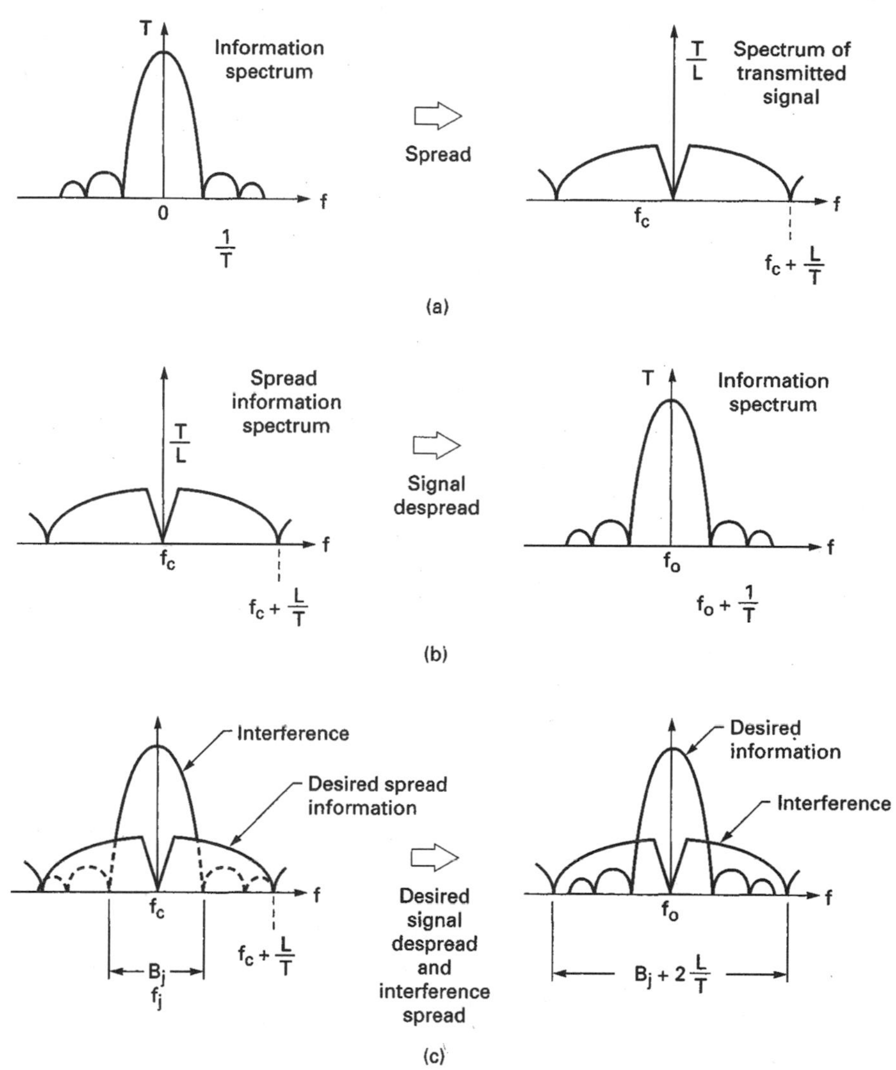

Figure 1-2 Summary of spread spectrum. (a) Using PN sequence and transmitter with chip (**PN**) duration of **T/L**. (b) using correlation and a synchronized replica of the pn sequence at the receiver. (c) When interface is present. **L/T** = chip duration; $\mathbf{f_j}$ = jamming frequency; **Bj** = jammer's bandwidth.

In the current environment of the Internet, electronic commerce, and multimedia communications, limited support for data communications is a serious drawback. Although subscribers want to talk as much as ever, they now want to communicate in a myriad of new ways, such as e-mail, instant messaging, the World Wide Web, and so on. Not only do subscribers want these services, they want mobility too. To provide all of these capabilities means that new advanced technology is required—third-generation technology.

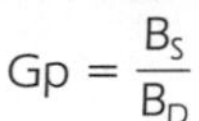

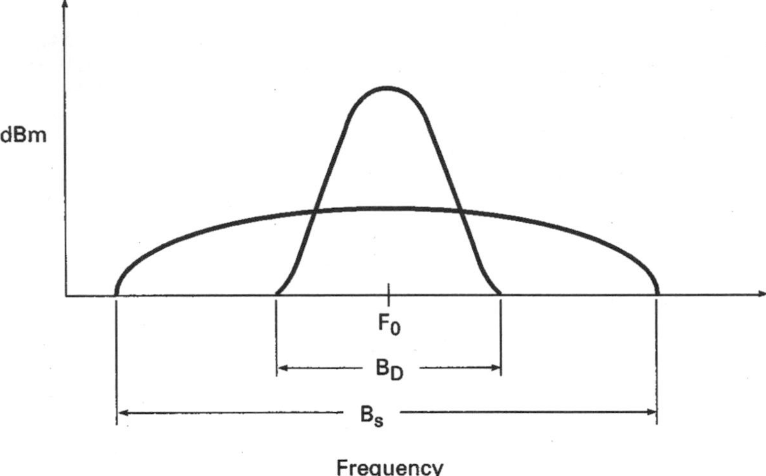

Figure 1-3
Processor gain:

$$Gp = \frac{B_S}{B_D}$$

B_D = bandwidth of initial signal
B_S = bandwidth of initial signal spread

The need for third-generation mobile communications technology was recognized on many different fronts, and various organizations began to the address the issue as far back as the 1980s. The *International Telecommunications Union* (ITU) was heavily involved and the work within the ITU was originally known as *Future Public Land Mobile Telecommunications Systems* (FPLMTS). Given the fact, however, that this acronym is difficult to pronounce, it was subsequently renamed *International Mobile Telecommunications—2000* (IMT-2000).

The IMT-2000 effort within the ITU has led to a number of recommendations. These recommendations address areas such as user bandwidth (144 Kbps for mobile service, and up to 2 Mbps for fixed service), richness of service offerings (multimedia services), and flexibility (networks that can support small or large numbers of subscribers). The recommendations also specify that IMT-2000 should operate in the 2-GHz band. In general, however, the ITU recommendations are mainly a set of requirements and do not specify the detailed technical solutions to meet the requirements. To address the technical solutions, the ITU has solicited technical proposals from interested organizations, and then selected/approved some of those proposals. In 1998, numerous air interface technical proposals were submitted. These were reviewed by the ITU, which in 1999 selected five technologies for terrestrial service (non-satellite based). The five technologies are

- Wideband CDMA (WCDMA)
- CDMA 2000 (an evolution of IS-95 CDMA)
- TD-SCDMA (time division-synchronous CDMA)
- UWC-136 (an evolution of IS-136)
- DECT

These technologies represent the foundation for a suite of advanced mobile multimedia communications services and are starting to be deployed across the globe. Of these technologies, this book deals with four —WCDMA, CDMA2000, TD-SCDMA, and UWC-136.

1.3 Mobile Communications Fundamentals

Even though the term "cellular" is often used in North America to denote analog AMPS systems, most, though not all, mobile communications systems are cellular in nature. Cellular simply means that the network is divided into a number of cells, or geographical coverage areas, as shown in Figure 1-4. Within each cell is a base station, which contains the radio transmission and reception equipment. It is the base station that provides the radio communication for those mobile phones that happen to be within the cell. The coverage area of a given cell is dependent upon a number of factors such as the transmit power of the base station, the transmit power of mobile, the height of the base station antennas, and the topology of the landscape. The coverage of a cell can range from as little as about 100 yards to tens of miles.

Specific radio frequencies are allocated within each cell in a manner that depends on the technology in question. In most systems, a number of individual frequencies are allocated to a given cell and those same frequencies are reused in other cells that are sufficiently far away to avoid interference. With CDMA, however, the same frequency can be reused in every cell. Although the scheme shown in Figure 1-4 is certainly feasible and is sometimes implemented, it is common to sectorize the cells, as shown in Figure 1-5. In this approach, the base station equipment for a number of cells is co-located at the edge of those cells, and directional antennas are used to provide coverage over the area of each cell (as opposed to omnidirectional antennas in the case where the base station is located at the center of a

Figure 1-4
Cellular System.

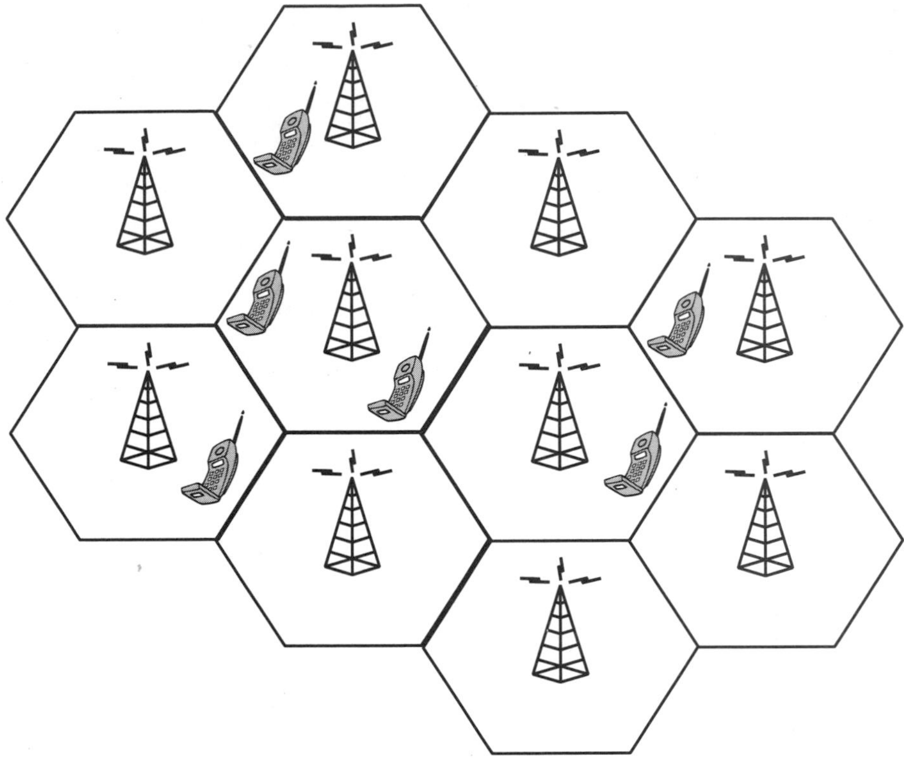

cell). Sectorized arrangements with up to six sectors are known, but the most common configuration is three sectors per base station in urban areas, with two sectors per base station along highways.

Of course, it is necessary that the base stations be connected to a switching network and for that network to be connected to other networks, such as the *Public Switched Telephone Network* (PSTN) in order for calls to be made to and from mobile subscribers. Furthermore, it is necessary for information about the mobile subscribers to be stored in a particular place on the network. Given that different subscribers may have different services and features, the network must know which services and features apply to each subscriber in order to handle calls appropriately. For example, a given subscriber may be prohibited from making international calls. Should the subscriber attempt to make an international call, the network must disallow that call based upon the subscriber's service profile.

Figure 1-5
Typical Sectorized Cell Sites
(a) Three-sector configuration
(b) Two-sector configuration

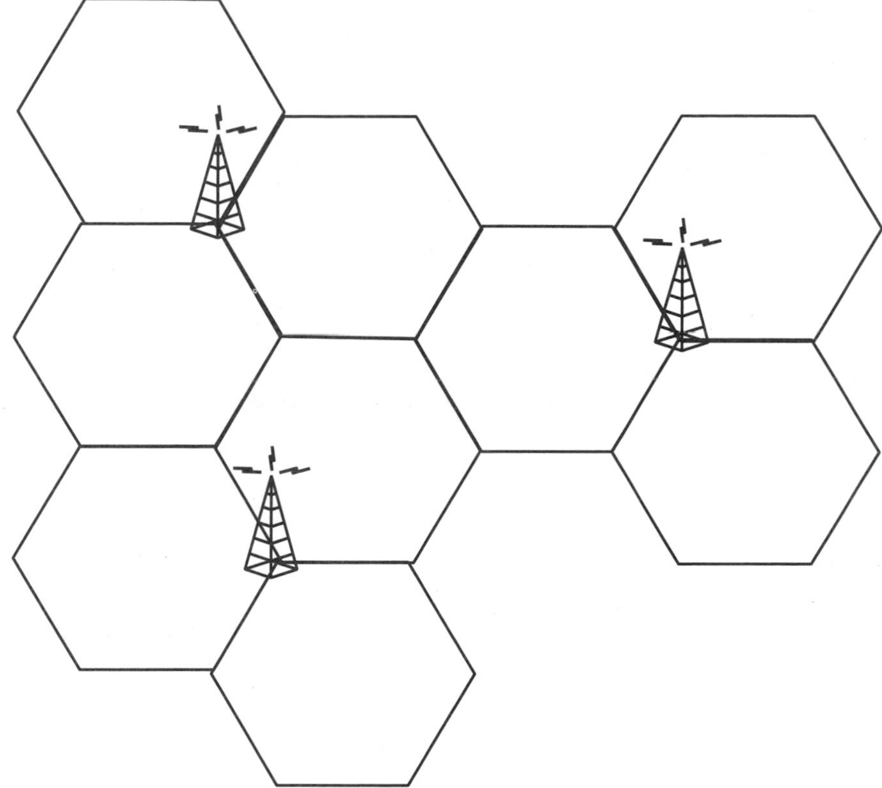

Three-sector configuration

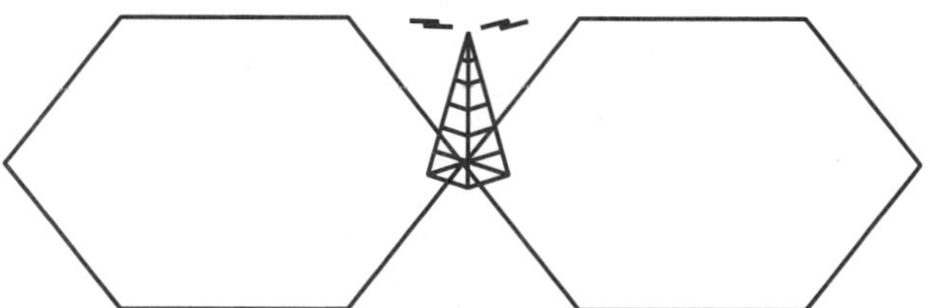

Two-sector configuration

1.3.1 Basic Network Architecture

Figure 1-6 shows a typical (although very basic) mobile communications network. A number of base stations are connected to a *Base Station Controller* (BSC). The BSC contains logic to control each of the base stations. Among other tasks, the BSC manages the handoff of calls from one base station to another as subscribers move from cell to cell. Note that in certain implementations, the BSC may be physically and logically combined with the MSC.

Connected to the BSC is the *Mobile Switching Center* (MSC). The MSC, also known in some circles as the *Mobile Telephone Switching Office* (MTSO), is the switch that manages the setup and teardown of calls to and

Figure 1-6 Basic Network Architecture.

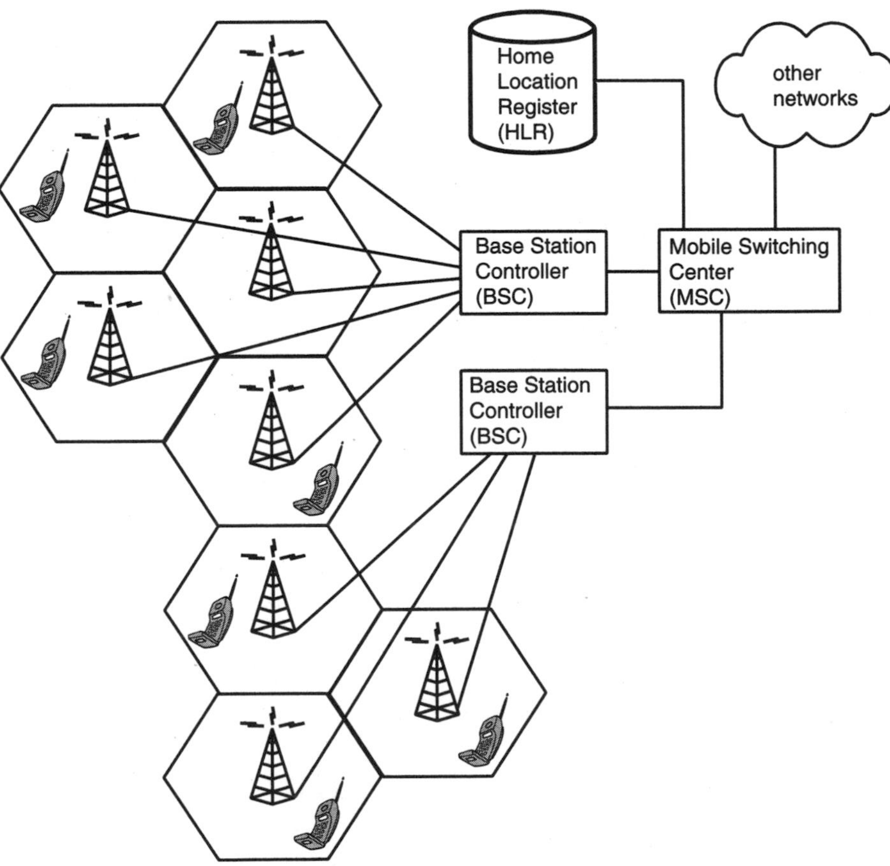

from mobile subscribers. The MSC contains many of the features and functions found in a standard PSTN switch. It also contains, however, a number of functions that are specific to mobile communications. For example, the BSC functionality may be contained with the MSC in certain systems, particularly in first-generation systems. Even if the BSC functionality is not contained within the MSC, the MSC must still interact with a number of BSCs over an interface that is not found in other types of networks. Furthermore, the MSC must contain a logic of its own to deal with the fact that the subscribers are mobile. Part of this logic involves an interface to one or more HLRs, where subscriber-specific data is held.

The HLR contains subscription information related to a number of subscribers. It is effectively a subscriber database and is usually depicted in diagrams as a database. The HLR does, however, do more that just hold subscriber data; it also plays a critical role in mobility management—that is, the tracking of a subscriber as he or she moves around the network. In particular, as a subscriber moves from one MSC to another, each MSC in turn notifies the HLR. When a call is received from the PSTN, the MSC that receives the call queries the HLR for the latest information regarding the subscriber's location so that the call can be correctly routed to the subscriber. Note that, in some implementations, HLR functionality is incorporated within the MSC, which leads to the concept of a "home MSC" for a given subscriber.

The network depicted in Figure 1-6 can be considered to represent the bare minimum needed to provide a mobile telephony service. These days, a range of different features' services are offered in addition to just the capability to make and receive calls. Therefore, most of today's mobile communications networks are much more sophisticated than the network depicted in Figure 1-6. As we progress through this book, we will introduce many other network elements and interfaces as we build from the fundamentals to the sophisticated technologies of third-generation networks.

1.3.2 Air Interface Access Techniques

Radio spectrum is a precious and finite resource. Unlike other transmission media such as copper or fiber facilities, it is not possible to simply add radio spectrum when needed. Only a certain amount of spectrum is available and it is critical that it be used efficiently, and be reused as much as possible. Such requirements are at the heart of the radio access techniques used in mobile communications.

1.3.2.1 Frequency Division Multiple Access (FDMA) Of the common multiple access techniques used in mobile communications systems, FDMA is the simplest. With FDMA, the available spectrum is divided into a number of radio channels of a specified bandwidth, and a selection of these channels is used within a given cell. In analog AMPS, for example, the available spectrum is divided into blocks of 30 kHz. A number of 30-kHz channels are allocated to each cell, depending on the expected traffic load for the cell. When a subscriber wants to place a call, one of the 30-kHz channels is allocated exclusively to the subscriber for that call.

In most FDMA systems, separate channels are used in each direction—from network to subscriber (downlink) and from subscriber to network (uplink). For example, in analog AMPS, when we talk about 30-kHz channels, we are actually talking about two 30-kHz channels, one in each direction. Such an approach is known as *Frequency Division Duplex* (FDD) and normally a fixed separation exists between the frequency used in the uplink and that used in the downlink. This fixed separation is known as the duplex distance. For example, in many systems in North America, the duplex distance is 45 MHz. Thus, in such a system, channel 1 corresponds to two channels (uplink and downlink) with a separation of 45 MHz between them. An FDD FDMA technique can be represented as shown in Figure 1-7.

FDD is not the only duplexing scheme, however. Another technique known as Time Division Duplex is also used. In such a system, only one channel is used for both uplink and downlink transmissions. With TDD, the channel is used very briefly for uplink, then very briefly for downlink, then very briefly again for uplink, and so on. TDD is not very common in North America, but it is widely used in systems deployed in Asia.

1.3.2.2 Time Division Multiple Access (TDMA) With *Time Division Multiple Access* (TDMA), radio channels are divided into a number of time slots, with each user assigned a given timeslot. For example, on a given radio frequency, user A might be assigned timeslot number 1 and user B might be assigned time slot number 3. The allocation is performed by the network as part of the call establishment procedure. Thus, the user's device knows exactly which timeslot to use for the remainder of the call, and the device times its transmissions exactly to correspond with the allocated time slot. This technique is depicted in Figure 1-8.

Typically, a TDMA system is also an FDD system, as shown in Figure 1-8, although TDD is used in some implementations. Furthermore, TDMA systems normally also use FDMA. Thus, the available bandwidth is divided into a number of smaller channels as in FDMA and it is these channels that

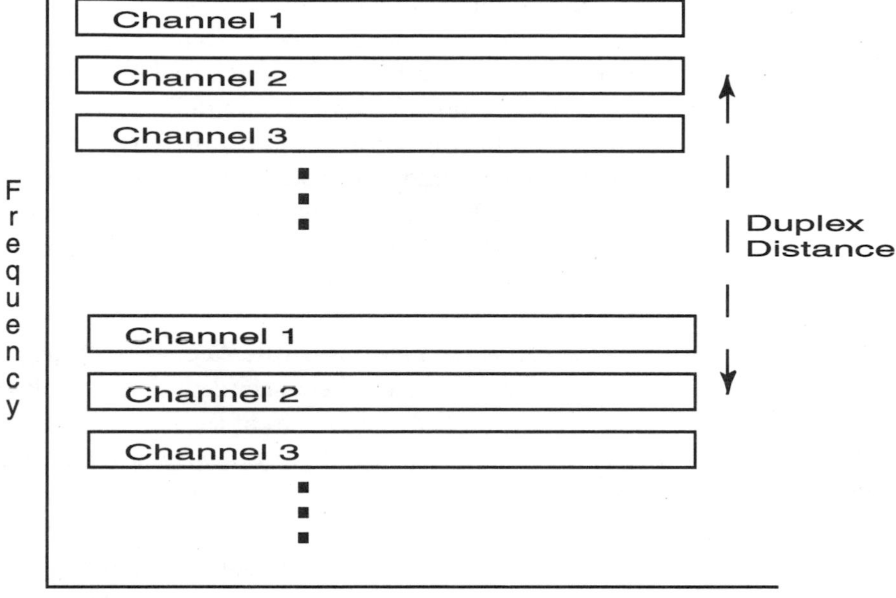

Figure 1-7
FDMA.

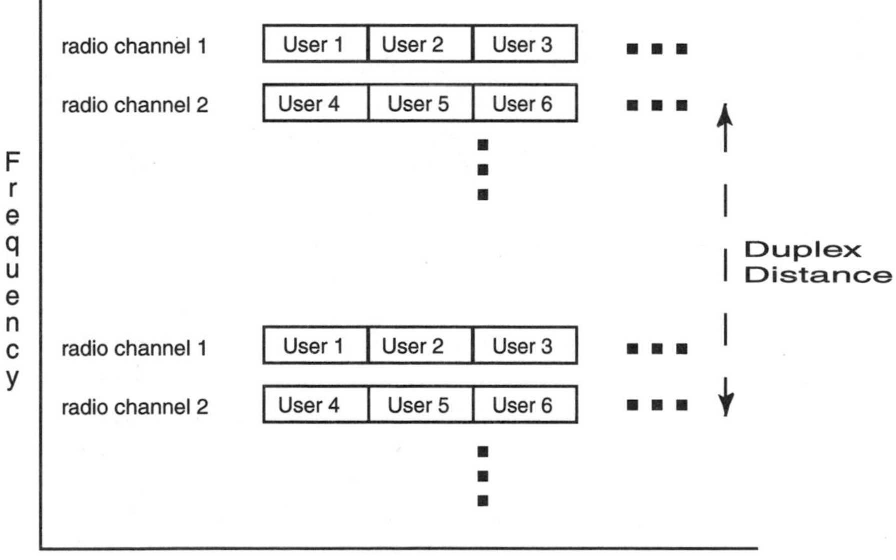

Figure 1-8
TDMA.

are divided into timeslots. The difference between a pure FDMA system and a TDMA system that also uses FDMA is that, with the TDMA system, a given user does not have exclusive access to the radio channel.

Implementing a TDMA system can be done in many ways. For example, different TDMA systems may have different numbers of time slots per radio channel and/or different time slot durations, and/or different radio channel bandwidths. Although, in the United States, the term TDMA is often used to refer to IS-136, such a usage of the term is incorrect because IS-136 is just one example of a TDMA system. In fact, GSM is also a TDMA system.

1.3.2.3 Code Division Multiple Access (CDMA) With CDMA, neither the time domain nor the frequency domain are subdivided. Rather, all users share the same radio frequency at the same time. This approach obviously means that all users interfere with each other. Such interference would be intolerable if the radio frequency bandwidth were limited to just the bandwidth that would be needed to support a single user. To overcome this difficulty, CDMA systems use a technique called spread spectrum, which involves spreading the signal over a wide bandwidth. Each user is allocated a code or sequence and the bit rate of the sequence is much greater than the bit rate of the information being transmitted by the user. The information signal from the user is modulated with the sequence assigned to the user and, at the far end, the receiver looks for the sequence in question. Having isolated the sequence from all of the other signals (which appear as noise), the original user's signal can be extracted.

TDMA systems have a very well-defined capacity limit. A set number of channels and a set number of time slots exist per channel. Once all time slots are occupied, the system has reached capacity. CDMA is somewhat different. With CDMA, the capacity is limited by the amount of noise in the system. As each additional user is added, the total interference increases and it becomes harder and harder to extract a given user's unique sequence from the sequences of all the other users. Eventually, the noise floor reaches a level where the inclusion of additional users would significantly impede the system's capability to filter out the transmission of each user. At this point, the system has reached capacity. Although it is possible to mathematically model this capacity limit, exact modeling can prove a little difficult, since the noise in the system depends on factors such as the transmission power of each individual mobile, thermal noise, and the use of discontinuous transmission (only transmitting when something is being said). By making certain reasonable assumptions in the design phase, however, it is possible to design a

CDMA system that provides relatively high capacity without significant quality degradation.

IS-95/J-STD-008 is the only widely deployed CDMA system for mobile communications. This system uses a channel bandwidth of 1.23 MHz and is an FDD system. The fact that the bandwidth is 1.23 MHz means that the total system bandwidth (typically, 10 MHz, 20 MHz, or 30 MHz) can accommodate several CDMA *radio frequency* (RF) channels. Therefore, like TDMA, IS-95 CDMA also uses FDMA to some degree. In other words, within a given cell, more than one RF channel may be available to system users.

A significant advantage of CDMA is the fact that it practically eliminates frequency planning. Other systems are very sensitive to interference, meaning that a given frequency can be reused only in another cell that is sufficiently far away to avoid interference. In a commercial mobile communications network, cells are constantly being added, or capacity is being added to existing cells, and each such change must be done without causing undue interference. If interference is likely to be introduced, then retuning of part of the network is required. Such retuning is needed frequently and can be an expensive effort. CDMA, however, is designed to deal with interference and, in fact, it allows a given RF carrier to be reused in every cell. Therefore, there is no need to worry about retuning the network when a new cell is added.

1.3.3 Roaming

The discussion so far has focused largely on the methods used to access the network over the air interface. The air interface access is, of course, extremely important. Other aspects, however, are necessary in order to make a wireless communications network a mobile communications network.

Mobility implies that subscribers be able to move freely around the network and from one network to another. This requires that the network tracks the location of a subscriber to a certain accuracy so that calls destined for the subscriber may be delivered. Furthermore, a subscriber should be able to do so while engaged in a call.

The basic approach is as follows. First, when a subscriber initially switches on his or her mobile phone, the device itself sends a registration message to the local MSC. This message includes a unique identification for the subscriber. Based on this identification, the MSC is able to identify the HLR to which the subscriber belongs, and the MSC sends a registration message to the HLR to inform the HLR of the MSC that now serves the subscriber. The HLR then sends a registration cancellation message to the MSC that previously served the subscriber (if any) and then sends a confirmation to the new serving MSC.

When mobile communications networks were initially introduced, only the air interface specification was standardized. The exact protocol used between the visited MSC and the HLR (or home MSC) was vendor-specific. The immediate drawback was that the home system and visited system had to be from the same vendor if roaming was to be supported. Therefore, a given network operator needed to have a complete network from only one vendor. Moreover, roaming between networks worked only if the two networks used equipment from the same vendor. These limitations severely curtailed roaming.

This problem was addressed in different ways on either side of the Atlantic. In North America, the problem was recognized fairly early, and an effort was undertaken to establish a standard protocol between home and visited systems. The result of that effort was a standard known as IS-41. This standard has been enhanced significantly over the years and the current revision of the standard is revision D. IS-41 is used for roaming in AMPS systems, IS-136 systems, and IS-95 systems.

Meanwhile, in Europe, nothing was done to address the roaming issue for first-generation systems, but a major effort was applied to ensuring that the problem was addressed in second-generation technology—specifically GSM. Consequently, when GSM specifications were created, they addressed far more than just the air interface. In fact, most aspects of the network were specified in great detail, including the signaling interface between home and visited systems. The protocol specified for GSM is known as the GSM *Mobile Application Part* (MAP). Like IS-41, GSM MAP has also been enhanced over the years.

Strictly speaking, the term MAP is not specific to GSM. In fact, the term refers to any mobility-specific protocol that operates at layer 7 of the *Open Systems Interconnection* (OSI) seven-layer stack. Given that IS-41 also operates layer 7, the term MAP is also applicable to IS-41.

1.3.4 Handoff/Handover

Handoff (also known as handover) is the ability of a subscriber to maintain a call while moving within the network. The term handoff is typically used with AMPS, IS-136, and IS-95, while handover is used in GSM. The two terms are synonymous.

Handoff usually means that a subscriber travels from one cell to another while engaged in a call, and that call is maintained during the transition (ideally without the subscriber noticing any change). In general, handoff means that the subscriber is transitioned from one radio channel (and/or timeslot) to another. Depending on the two cells in question, the handoff can be between two sectors on the same base station, between two BSCs, between two MSCs belonging to the same operator, or even between two networks. (Note that inter-network handoff is not supported in some systems, often mainly for billing reasons.)

It is also possible to handoff a call between two channels in the same cell. This could occur when a given channel in a cell is experiencing interference that is affecting the communication quality. In such a case, the subscriber would be moved to another frequency that is subject to less interference. A handoff scenario is depicted in Figure 1-9.

How does the system determine that a handoff needs to occur? Basically, two main approaches are used. In first-generation technologies, a handoff is

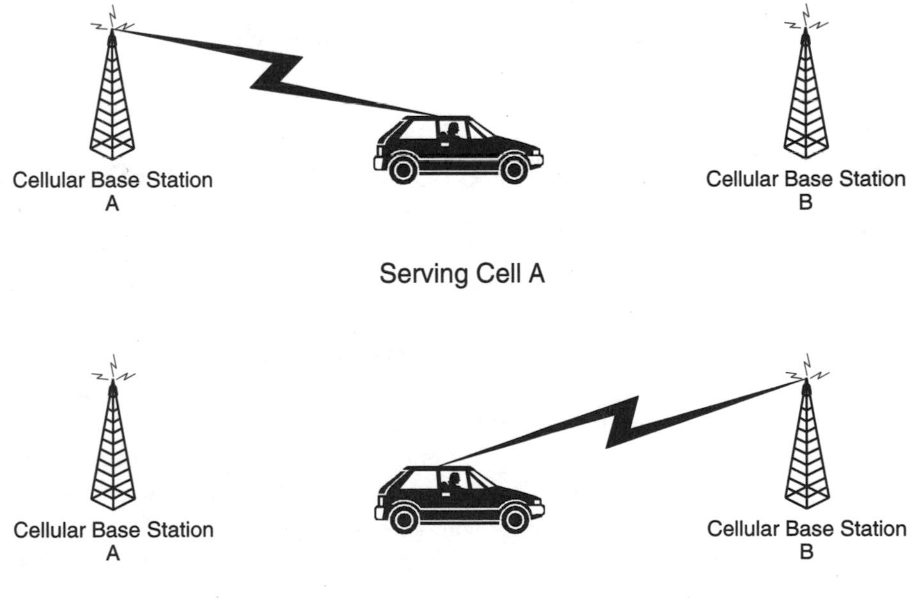

Figure 1-9
Handover.
(a) Pre-handoff
(b) Post-handoff

generally controlled by the network. The network measures the signal strength from a mobile as received at the serving cell. If it begins to fall below a certain threshold, then nearby cells are requested to perform signal strength measurements. If a nearby cell records a better signal strength, then it is highly likely that the subscriber has moved to the coverage of that cell. The new cell is instructed by the BSC or MSC (typically just the MSC, since first-generation systems do not have BSCs) to allocate a channel for the subscriber. Once that allocation is performed, the network instructs the mobile to swap to the new channel. This is known as a network-controlled handoff, because the network determines when and how a handoff is to occur.

In more recent technologies, a technique known as *mobile assisted handover* (MAHO) is the most common. In the approach, the network provides the mobile with a list of base station frequencies (those of nearby base stations). The mobile makes periodic measurements of the signals received from those base stations (as well as the serving base station), including signal strength and signal quality (usually determined from bit error rates), and it sends the corresponding measurement reports to the network. The network analyzes the reports and makes a determination of if and how a handoff should occur. Assuming that a handoff is required, then the network reserves a channel on the new cell and sends an instruction to the mobile to move to that channel, which it does.

1.4 Wireless Migration

In the previous sections of this chapter, some of the various technology platforms were discussed. The existing wireless operators today, regardless of the frequency band or existing technology deployed have or are making very fundamental decisions as to which direction in the 3G evolution they will take. The decision on 3G technology will define a company's position in the marketplace for years to come.

Some existing operators and new entrants are letting the technology platform be defined by the local regulator, thereby eliminating the platform decision. However, the majority of the operators need to determine which platform they must utilize. Since the platforms to pick from utilize different access technologies, they are by default not directly compatible. The utilization of different access technologies for the realization of 3G also introduces several interesting issues related to the migration from 2G to 3G. The migration path from 2G to 3G is referred to as 2.5G and involves an interim position for data services that are more advanced than 2G, but not as robust as the 3G envisioned data services.

Some of the migration strategies for an existing operator involve

- Overlay
- Spectrum segmentation

The overlay approach typically involves implementing the 2.5 technology over the existing 2G system and then implementing 3G as either an overlay or in a separate part of the radio frequency spectrum they are allocated, spectrum segmentation.

The choice of whether to use an overlay or spectrum segmentation is naturally dependant upon the technology platform that is currently being used, 2G, the spectrum available, the existing capacity constraints, and marketing. Marketing is involved with the decision because of the impact to the existing subscriber base and services that are envisioned to be offered.

Some of the decisions are rather straightforward involving upgrading portions of the existing technology platforms that are currently deployed. Other operators have to make a decision as to which technology to utilize since they either are building a new system or have not migrated to a 2G platform, using only 1G.

In later chapters various migration strategies are discussed relative to the underlying technology platform that exists.

1.5 Harmonization Process

Harmonization refers to the vision and objective of the IMT2000 specification that enables the various technology platforms that are defined in that specification to interact with each other. True harmonization relative to the capability of a CDMA2000 and WCDMA system is based on having subscriber units that operate in both technologies. The access infrastructure being able to support both is a goal, but not one that is in the near future.

1.6 Overview of Following Chapters

This chapter has served as a brief introduction to mobile communications systems. The brief overview that has been given, however, is certainly not a sufficient background to enable a good understanding of third-generation technology. Therefore, before tackling the details of third-generation systems, it is necessary to better describe first- and second-generation systems. Chapter 2, "First Generation (1G)," addresses first-generation technology and Chapter 3, "Second Generation (2G)," delves into the second-generation systems. The remaining chapters focus on third-generation systems and some of the migration paths to obtainment of the IMT2000 vision.

References

AT&T. "*Engineering and Operations in the Bell System*," 2nd Ed., AT&T Bell Laboratories, Murry Hill, N.J., 1983.

Barron, Tim. "*Wireless Links for PCS and Cellular Networks*," Cellular Integration, Sept. 1995, pgs. 20–23.

Brewster. "*Telecommunications Technology*," John Wiley & Sons, New York, NY, 1986.

Brodsky, Ira. "*3G Business Model*," Wireless Review, June 15, 1999, pg. 42.

Daniels, Guy. "*A Brief History of 3G*," Mobile Communications International, Issue 65, Oct. 99, pg. 106.

Gull, Dennis. "*Spread-Spectrum Fool's Gold?*" Wireless Review, Jan. 1, 1999 pg. 37.

Homa, Harri, and Antti Toskala. "*WCDMA for UMTS*," John Wiley & Sons, 2000.

Smith, Clint. "*Practical Cellular and PCS Design*," McGraw-Hill, 1997.

Smith, Gervelis. "*Cellular System Design and Optimization*," McGraw-Hill, 1996.

CHAPTER 2

First Generation (1G)

2.1 First Generation (1G)

Although the advancement of technology (any technology) certainly involves quantum leaps forward from time to time, it is common for major progress to also occur as a result of incremental improvements. For mobile communications technology, advancement has come about in both ways—through occasional revolution and almost certain evolution. Therefore, although the book deals primarily with the technology of *third-generation* (3G) wireless networks, an understanding of earlier systems is important. This understanding provides the appropriate perspective from which to view 3G systems and helps us understand how solutions for 3G systems have been developed. In other words, it is easier to understand where we are going if we understand where we have been. To help in that understanding, this chapter provides an overview of *first-generation* (1G) systems.

Cellular communication, referred to as 1G, is one of the most prolific voice communication platforms that has been deployed within the last two decades. Overall, cellular communication is the form of wireless communication that enables several key concepts to be employed, such as the following:

- Frequency reuse
- Mobility of the subscriber
- Handoffs

The cellular concept is employed in many different forms. Typically, when referencing cellular communication, it is usually associated with either the *Advanced Mobile Phone System* (AMPS) or *Total Access Communication Services* (TACS) technology. AMPS, operates in the 800-MHz band (821 to 849 MHz) for base station receiving and (869 to 894 MHz) for base station transmitting. For TACS, the frequency range is 890 MHz to 915 MHz for base receiving and 935 MHz to 960 MHz for base station transmitting.

Many other technologies also fall within the category of cellular communication and those involve the *Personal Communications Service* (PCS) bands, including both the domestic U.S. and international bands. In addition, the same concept is applied to several technology platforms that are currently used in the *specialized mobile radio* (SMR) band (IS-136 and iDEN). However, cellular communication is really utilized by both the AMPS and TACS bands but is sometimes interchanged with the PCS and SMR bands because of the similarities. However, AMPS and TACS systems are an analog-based system and not a digital system.

The concept of cellular radio was initially developed by AT&T at their Bell Laboratories to provide additional radio capacity for a geographic customer service area. The initial mobile systems that cellular evolved from were called *mobile telephone systems* (MTSs). Later improvements to these systems occurred and the systems were referred to as *improved mobile telephone systems* (IMTSs). One of the main problems with these systems was that a mobile call could not be transferred from one radio station to another without loss of communication. This problem was resolved by implementing the concepts of reusing the allocated frequencies of the system. Reusing the frequencies in cellular systems enables a market to offer higher radio traffic capacity. The increased radio traffic enables more users in a geographic service area than with the MTS or IMTS systems.

Cellular radio was a logical progression in the quest to provide additional radio capacity for a geographic area. The cellular system, as it is known today, has its primary roots in the MTS and the IMTS. Both MTS and IMTS are similar to cellular with the exception that no handoff takes place with these networks.

Cellular systems operate on the principal of frequency reuse. Frequency reuse in a cellular market enables a cellular operator the ability to offer higher radio traffic capacity. The higher radio traffic capacity enables many more users in a geographic area to utilize radio communication than are available with a MTS or IMTS system.

The cellular systems in the United States are broken into the *Metropolitan Statistical Area* (MSA) and *Rural Statistical Areas* (RSAs). Each MSA and RSA have two different cellular operations that offer service. The two cellular operations are referred to as A-band and B-band systems. The A-band system is the non-wireline system and the B-band is the wireline system for the MSA or RSA.

2.2 1G Systems

Numerous mobile wireless systems have been deployed throughout the world. Each of the various 1G wireless systems has its own unique advantage and disadvantages, depending on the spectrum available and the services envisioned for delivery.

1G mobility systems are defined as analog systems and are typically referred to as an AMPS or TACS system. It is important to note that analog systems utilize digital signaling in many aspects of their network, including the air interface. However, the analog reference applies to the method that the information content is transported over; that is, no CODEC is involved.

Table 2-1 represents the popular 1G wireless mobility service offerings that have been deployed. As mentioned previously, the two most prolific 1G systems deployed in the world are AMPS and TACS.

All of the 1G systems shown in the table utilize a *Frequency Division Multiple Access* (FDMA) scheme for radio system access. However, the specific channel bandwidth that each use is slightly different, as is the typical spectrum allocations for each of the services. The channel bandwidths are as follows:

- AMPS is the cellular standard that was developed for use in North America. This type of system operates in the 800-MHz frequency band. AMPS systems have also been deployed in South America, Asia, and Russia.
- *Narrow Band AMPS* (NAMPS) is a product that is used in part of the United States, Latin America, and other parts of the world. NAMPS is a cellular standard that was developed as an interim platform between 1G and 2G systems and was developed by Motorola. Specifically,

Table 2-1

1G Technology Platforms

	AMPS	NAMPS	TACS	NMT450	NMT900	C450
Base Tx MHz	869–894	869–894	935–960	463–468	935–960	461–466
Base Rx MHz	824–849	824–849	890–915	453–458	890–915	451–456
Multiple Access Method	FDMA	FDMA	FDMA	FDMA	FDMA	FDMA
Modulation	FM	FM	FM	FM	FM	FM
Radio Channel Spacing	30 kHz	10 kHz	25 kHz	25kHz	12.5kHz	20kHz (b) 10kHz (m)
Number Channels	832	2496	1000	200	1999	222(b) 444(m)
CODEC	NA	NA	NA	NA	NA	NA
Spectrum Allocation	50MHz	50MHz	50MHz	10MHz	50MHz	10MHz

NAMPS is an analog radio system that is very similar to AMPS, with the exception that it utilized 10-kHz-wide voice channels instead of the standard 30-kHz channels. The obvious advantage with this technology is the capability to deliver, under ideal conditions, three times more capacity of a system over that of regular AMPS.

NAMPS is able to achieve this smaller bandwidth through changing the format and methodology for *Supervisory Audio Tone* (SAT) and control communications from the cell site to the subscriber unit. In particular in NAMPS, they use a subcarrier method and use a digital color code in place of SAT. These two methods make it possible to use less spectrum while communicating the same amount of, or even more, information all at the same time and increasing the capacity of the system with the same spectrum.

However, this advantage in capacity, of course, requires a separate transmitter, either a *Power Amplifier* (PA) or transceiver, for each NAMPS channel deployed. However, the control channel that is used for the cell site is the standard control channel, 30 kHz, which is used by AMPS and other technology platforms used for cellular communication. Additionally, the *Carrier-to-Interferer* (C/I) requirements due to the narrower bandwidth channels are different than that of a regular AMPS system, which has a direct impact on the capacity of the system.

- TACS is a cellular band that was derived from the AMPS technology. TACS systems operate in both the 800-MHz band and the 900-MHz band. The first system of this kind was implemented in England. Later these systems were installed in Europe, Hong Kong, Singapore, and the Middle East. A variation of this standard was implemented in Japan, JTACS.
- *Nordic Mobile Telephone* (NMT) is the cellular standard that was developed by the Nordic countries of Sweden, Denmark, Finland, and Norway in 1981. This type of system was designed to operate in the 450-MHz and in the 900-MHz frequency bands. These are noted as NMT 450 and NMT 900. NMT systems have also be deployed throughout Europe, Asia, and Australia.

The basic service offering for 1G systems is and was voice communication. These systems have been extremely successful and many of them are still in service offering 1G services only.

1G systems, however, suffer from a number of difficulties. Some of those difficulties were addressed by additional technology added to the network and some of the difficulties have required the implementation of 2G technology. The biggest problem that led to the introduction of 2G technology was the fact that the 1G systems had limited system capacity. This became a serious issue as the popularity of mobile communications grew to a level that far exceeded anyone's expectations. Other problems included the fact that the technologies in question addressed only the air interface, and other interfaces in the network were not specified (at least not initially), which meant limited roaming, particularly between networks that were supplied by different vendors. The technologies did not initially include security mechanisms, which allowed for fraud. Finally, some limitation in the technologies led to the problem of "lost mobiles," where a subscriber is located at one MSC and the network thinks that the subscriber is elsewhere.

Nevertheless, it is worth emphasizing the popularity of these technologies and the fact that, in some cases, they have been the foundation upon which 2G and 3G technologies have been built.

2.3. General 1G System Architecture

A generic 1G cellular system configuration is shown in Figure 2-1. The configuration involves all the high-level system blocks of a cellular network. Many components comprise each of the blocks shown in Figure 2-1. The individual system components of a cellular network will be covered in later chapters of this book.

Referring to Figure 2-1, the mobile communicates to the cell site through the use of radio transmissions. The radio transmissions utilize a full-duplex configuration, which involves separate transmit and receive frequencies used by the mobile and cell sites. The cell site transmits on the frequency that the mobile unit is tuned to, while the mobile unit transmits on the radio frequency the cell site receiver is tuned to.

The cell site acts as a conduit for the information transfer converting the radio energy into another medium. The cell site sends and receives information from the mobile and the *mobile telephone system office* (MTSO). The MTSO is connected to the cell site either by leased T1/E1 lines or through a microwave system. The cellular system is made up of many cell sites that all interconnect back to the MTSO.

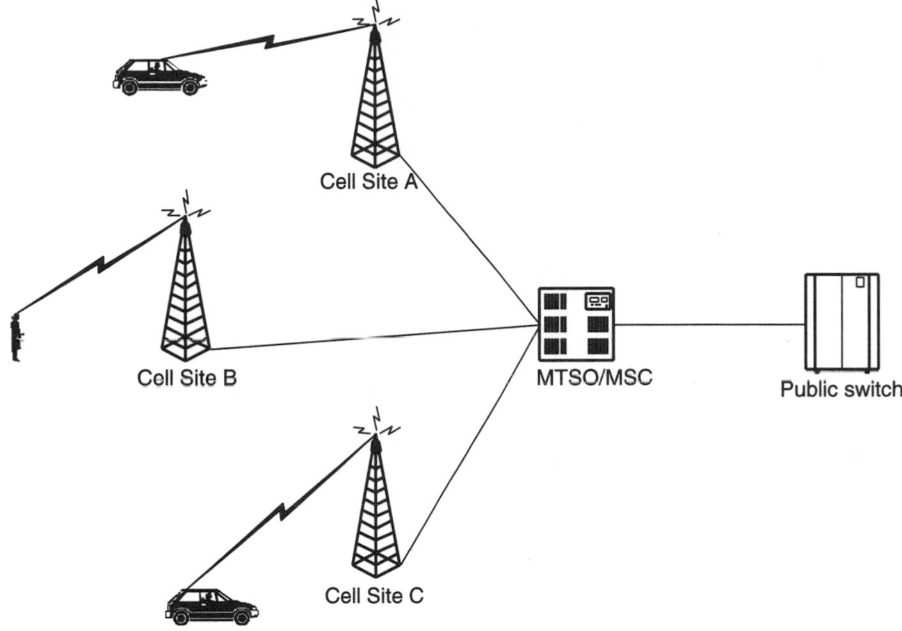

Figure 2-1
General 1G system.

The MTSO processes the call and connects the cell site radio link to the *Public Service Telephone Network* (PSTN). The MTSO performs a variety of functions involved with call processing and is effectively the brains of the network. The MTSO maintains the individual subscriber records, the current status of the subscribers, call routing, and billing information to mention a few items.

2.4 Generic MTSO Configuration

Figure 2-2 is a generic MTSO configuration. The MTSO is the portion of the network that interfaces the radio world with the public telephone network, PSTN. Mature systems often have multiple MTSO locations, and each MTSO can have several cellular switches located within each building.

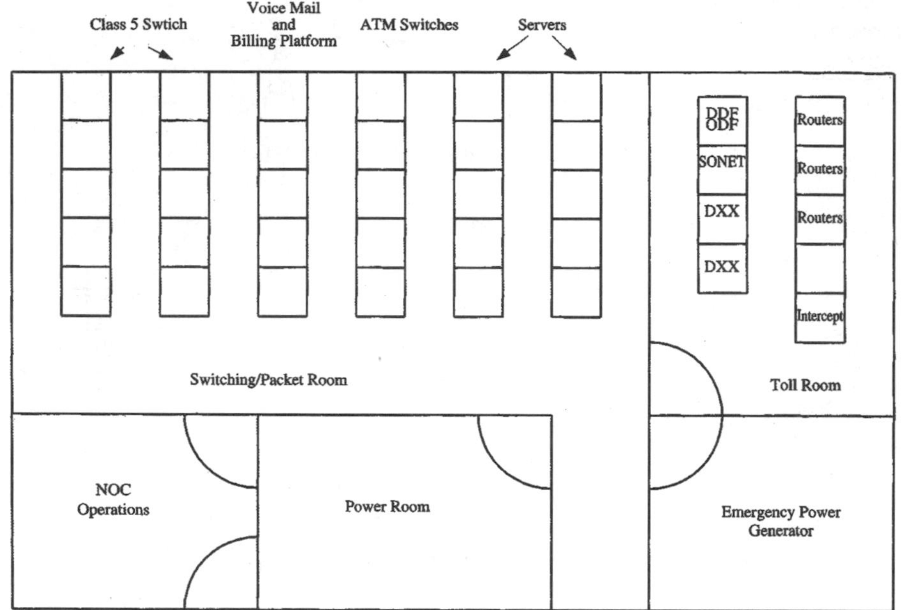

Figure 2-2 General MTSO configuration.

2.5 Generic Cell Site Configuration

Figure 2-3 is an example of a generic cell site configuration, which is a monopole cell site. The site has an equipment hut associated with it that houses the radio transmission equipment. The monopole, which is next to the equipment hut, supports the antennas used for the cell site at the very top of the monopole. The cable tray, which is between the equipment hut and the monopole, supports the coaxial cables that connect the antennas to the radio transmission equipment.

The radio transmission equipment used for a cellular base state, located in the equipment room, is shown in Figure 2-4. The equipment room layout is a typical arrangement in a cell site. The cell site radio equipment consists of a *base site controller* (BSC), a radio bay, and the amplifier, TX, bay. The cell site radio equipment is connected to the *Antenna Interface Frame* (AIF), which provides the receiver and transmit filtering. The AIF is then con-

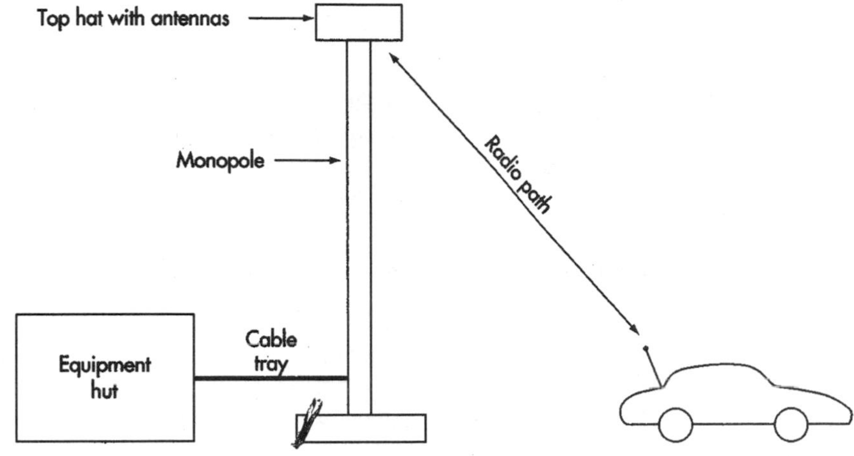

Figure 2-3 General cell site configuration.

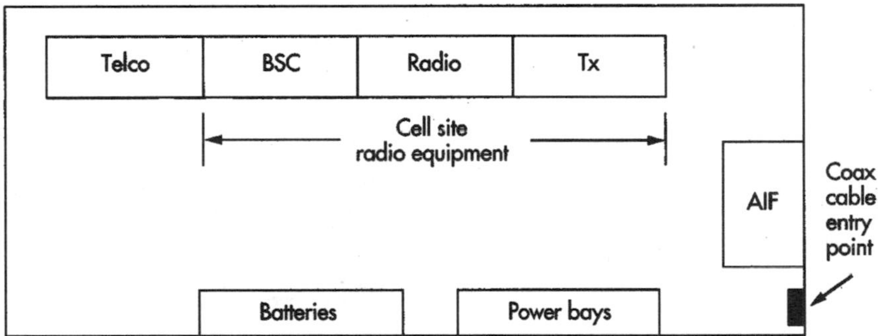

Figure 2-4 Radio transmission equipment for a cellular base station.

nected to the antennas on the monopole through use of the coaxial cables, which are located next to the AIF bay.

The cell site is also connected to the MTSO through the Telco bay. The Telco bay either provides the T1/E1 leased line or the microwave radio link connection. The power for the cell site is secured through the use of power bays and rectifiers, which convert AC electricity to DC. Batteries are used in the cell site in the event of a power disruption to ensure that the cell site continues to operate, until power is restored, or the batteries are exhausted.

2.6 Call Setup Scenarios

Several general call scenarios can occur and they pertain to all cellular systems. A few perturbations of the call scenarios are discussed here that are driven largely by fraud-prevention techniques employed by individual operators. Numerous algorithms are utilized throughout the call setup and processing scenarios, which are not included in Figures 2-5, 2-6, and 2-7. However, the call scenarios presented here provide the fundamental building blocks for all call scenarios utilized in cellular.

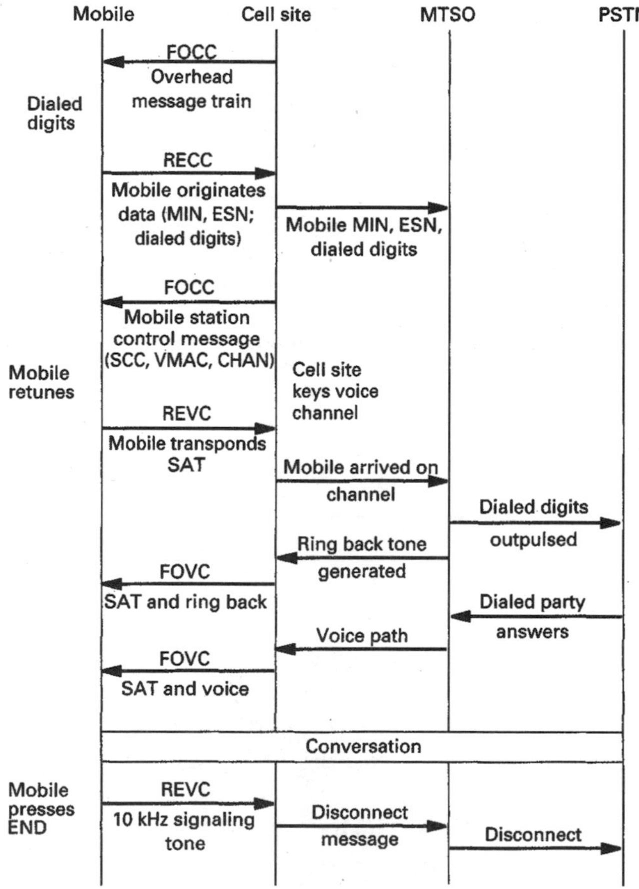

Figure 2-5 Mobile-to-land call setup.

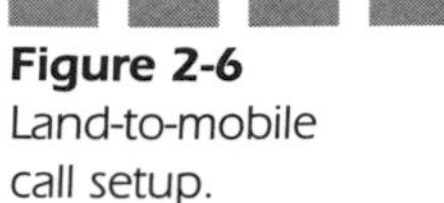

Figure 2-6
Land-to-mobile call setup.

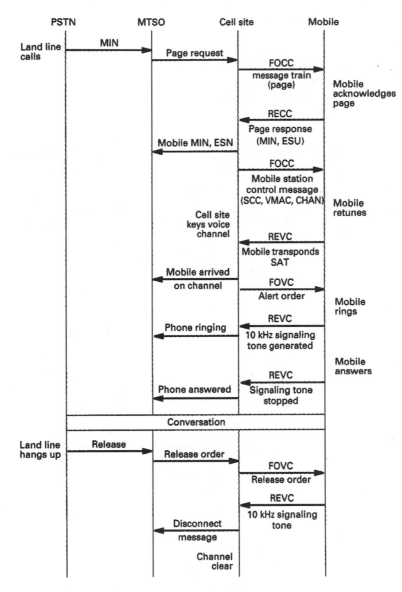

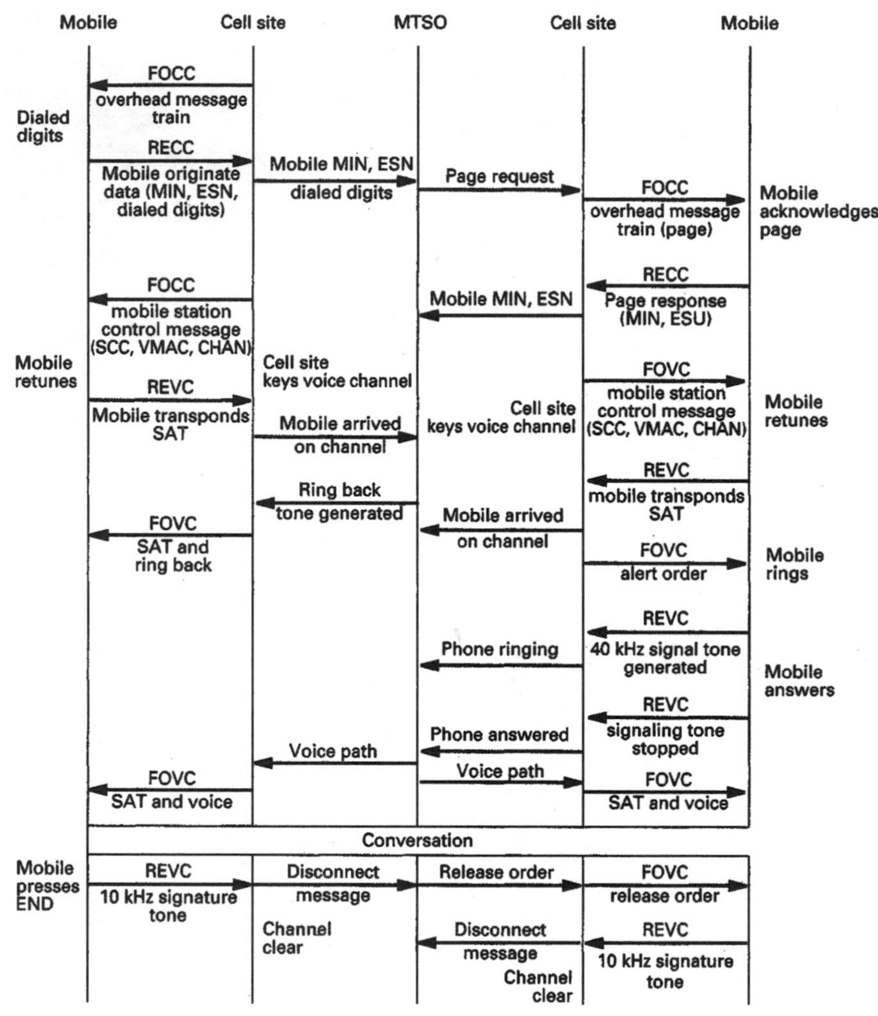

Figure 2-7 Mobile-to-mobile call setup.

2.7 Handoff

The *handoff* concept is one of the fundamental principals of this technology. Handoffs enable cellular to operate at lower power levels and provide high capacity. The handoff scenario presented in Figure 2-8 uses a simplified process. A multitude of algorithms are invoked for the generation and processing of a handoff request and an eventual handoff order. The individual

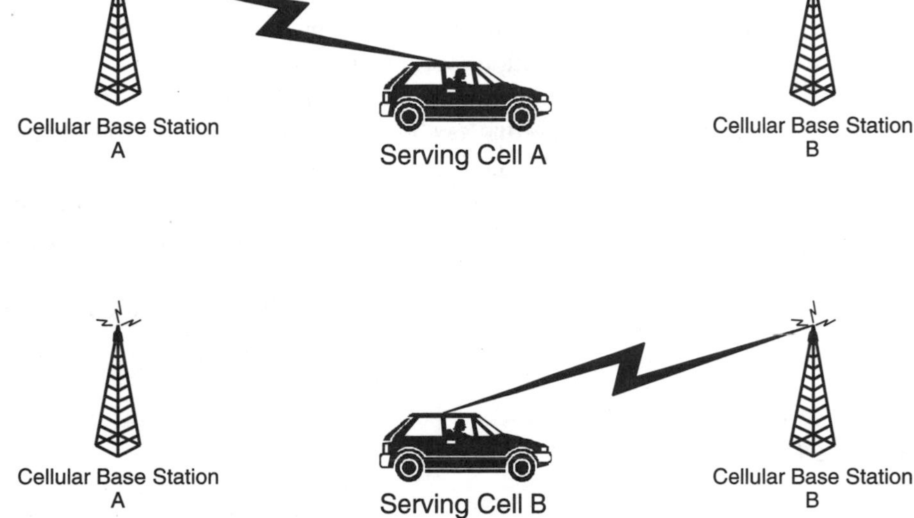

Figure 2-8
Analog handoff.

algorithms are dependent upon the individual vendor for the network infrastructure and the software loads utilized.

Handing off from cell to cell is fundamentally the process of transferring the mobile unit that has a call in progress on a particular voice channel to another voice channel, all without interrupting the call. Handoffs can occur between adjacent cells or sectors of the same cell site. The actual need for a handoff is determined by the actual quality of the RF signal received from the mobile into the cell site.

As the mobile transverses the cellular network, it is handed off from one cell site to another sell site, ensuring a quality call is maintained for the duration of the conversation.

2.8 Frequency Reuse

The concept and implementation of *frequency reuse* was an essential element in the quest for cellular systems to have a higher capacity per geographic area than an MTS or IMTS system. Frequency reuse is the core concept defining a cellular system and involves reusing the same frequency in a system many times over. The capability to reuse the same radio frequency many times in a system is the result of managing the C/I signal levels for an analog system. Typically, the minimum C/I level designed for in a cellular analog system is 17 dB C/I.

In order to improve the C/I ratio, the reusing channel should be as far away from the serving site as possible so as to reduce the interferer component of C/I. The distance between reusing base stations is defined by the D/R ratio, which is a parameter used to define the reuse factor for a wireless system. The D/R ratio, shown in Figure 2-9, is the relationship between the reusing cell site and the radius of the serving cell sites. Table 2-2 illustrates standard D/R ratios for different frequency reuse patterns, N.

Figure 2-9
D/R ratio.

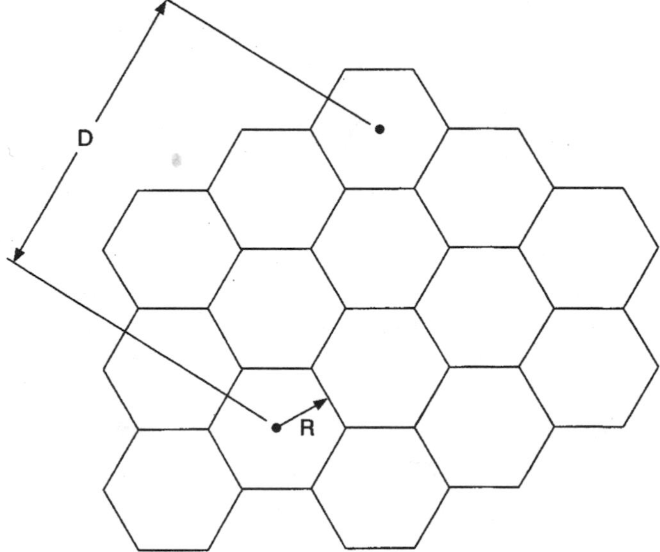

Table 2-2
D/R

D	N (reuse pattern)
3.46	4
4.6 R	7
6R	12
7.55R	19

As the D/R table implies, several frequency reuse patterns are currently in use throughout the cellular industry. Each of the different frequency reuse patterns has its advantages and disadvantages. The most common frequency reuse pattern employed in cellular is the N=7 pattern, which is shown in Figure 2-10.

The frequency repeat pattern ultimately defines the maximum amount of radios that can be assigned to an individual cell site. The N=7 pattern can assign a maximum of 56 channels that are deployed using a three-sector design.

2.9 Spectrum Allocation

The cellular systems have been allocated a designated frequency spectrum to operate within. Both the A-band and B-band operators are allowed to utilize a total of 25 MHz of radio spectrum for their systems. The 25 MHz is divided into 12.5 MHz of transmit frequencies and 12.5 MHz of receive frequencies for each operator. The cellular spectrum is shown in Figure 2-11.

The spectrum chart shown in Figure 2-11 has the location of the A-band and B-band cell sites transmit and receive the frequencies indicated. Currently, a total of 832 individual FCC channels are available in the United States. Each of the radio channels utilized in cellular are spaced at 30-kHz

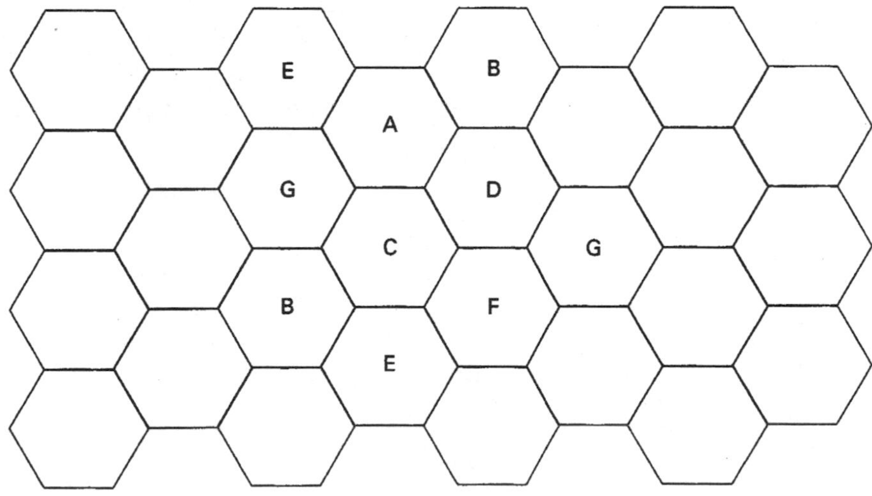

Figure 2-10 N=7 frequency reuse pattern.

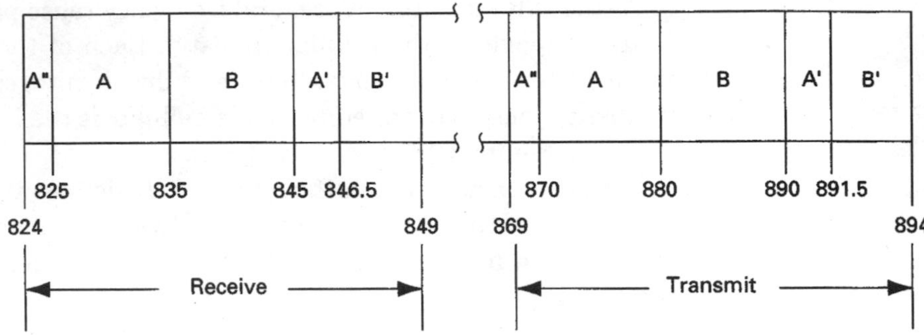

Figure 2-11
AMPS spectrum.

intervals with the transmit frequency operating at 45 MHz above the receive frequency. Both the A-band and B-band operators have available to them a total of 416 radio channels: 21 setup and 395 voice channels.

2.10 Channel Band Plan

The *channel band plan* is essential in any wireless system, especially one that reuses the spectrum at defined intervals. The channel band plan is a method of assigning channels, or fixed bandwidth, to a given amount of a radio frequency spectrum that is then grouped in a local fashion.

An example of a channel band plan is shown in Table 2-3. The channel band plan is for an AMPS (B-band) system utilizing an N=7 frequency reuse pattern. The channels are all by definition 30 kHz in size. Therefore, if one was to count the individual channels listed in the chart, 12.5 Hz of spectrum would be accounted for. Since the cellular system is a duplexed system, 12.5 MHz is used for both transmit and receive, while the total spectrum utilized is 25 MHz per operator.

Table 2-3

FCC Channel Chart for N = 7

Wireline B-Band Channels

Channel group:	A1	B1	C1	D1	E1	F1	G1	A2	B2	C2
Control channel:	334	335	336	337	338	339	340	341	324	343
	355	356	357	358	359	360	361	362	345	364
	376	377	378	379	380	381	382	383	366	385
	397	398	399	400	401	402	403	404	387	406
	418	419	420	421	422	423	424	425	408	427
	439	440	441	442	443	444	445	446	429	448
	460	461	462	463	464	465	466	467	450	469
	481	482	483	484	485	486	487	488	471	490
	502	503	504	505	506	507	508	509	492	511
	523	524	525	526	527	528	529	530	513	532
	544	545	546	547	548	549	550	551	534	553
	565	566	567	568	569	570	571	572	555	574
	586	587	588	589	590	591	592	593	576	595
	607	608	609	610	611	612	613	614	597	616
	628	629	630	631	632	633	634	635	618	637
	649	650	651	652	653	654	655	656	639	658
	717	718	719	720	721	722	723	724	725	726
	738	739	740	741	742	743	744	745	746	747
	759	760	761	762	763	764	765	766	767	768
	780	781	782	783	784	785	786	787	788	789

Nonwireline A-Band Channels

Channel group:	A1	B1	C1	D1	E1	F1	G1	A2	B2	C2
Control channel:	333	332	331	330	329	328	327	326	325	324
	312	311	310	309	308	307	306	305	304	303
	291	290	289	288	287	286	285	284	283	282
	270	269	268	267	266	265	264	263	262	261
	249	248	247	246	245	244	243	242	241	240
	228	227	226	225	224	223	222	221	220	219
	207	206	205	204	203	202	201	200	199	198
	186	185	184	183	182	181	180	179	178	177
	165	164	163	162	161	160	159	158	157	156
	144	143	142	141	140	139	138	137	136	135
	123	122	121	120	119	118	117	116	115	114
	102	101	100	99	98	97	96	95	94	93
	81	80	79	78	77	76	75	74	73	72
	60	59	58	57	56	55	54	53	52	51
	39	38	37	36	35	34	33	32	31	30
	18	17	16	15	14	13	12	11	10	9
	1020	1019	1018	1017	1016	1015	1014	1013	1012	1011
	999	998	997	996	995	994	993	992	991	716
	704	703	702	701	700	699	698	697	696	695
	683	682	681	680	679	678	677	676	675	674

Table 2-3 (cont.)

FCC Channel Chart for N = 7

Wireline B-Band Channels										
D2	E2	F2	G2	A3	B3	C3	D3	E3	F3	G3
344	345	346	347	348	349	350	351	352	353	354
365	366	367	368	369	370	371	372	373	374	375
386	387	388	389	390	391	392	393	394	395	396
407	408	409	410	411	412	413	414	415	416	417
428	429	430	431	432	433	434	435	436	437	438
449	450	451	452	453	454	455	456	457	458	459
470	471	472	473	474	475	476	477	478	479	480
491	492	493	494	495	496	497	498	499	500	501
512	513	514	515	516	517	518	519	520	521	522
533	534	535	536	537	538	539	540	541	542	543
554	555	556	557	558	559	560	561	562	563	564
575	576	577	578	579	580	581	582	583	584	585
596	597	598	599	600	601	602	603	604	605	606
617	618	619	620	621	622	623	624	625	626	627
638	639	640	641	642	643	644	645	646	647	648
659	660	661	662	663	664	665	666			
727	728	729	730	731	732	733	734	735	736	737
748	749	750	751	752	7537	754	755	756	757	758
769	770	771	772	773	774	775	776	777	778	779
790	791	792	793	794	795	796	797	798	799	

Nonwireline A-Band Channels										
D2	E2	F2	G2	A3	B3	C3	D3	E3	F3	G3
323	322	321	320	319	318	317	316	315	314	313
302	301	300	299	298	297	296	295	294	293	292
281	280	279	278	277	276	275	274	273	272	271
260	259	258	257	256	255	254	253	252	251	250
239	238	237	236	235	234	233	232	231	230	229
218	217	216	215	214	213	212	211	210	209	208
197	296	195	194	193	192	191	190	189	188	187
176	175	174	173	172	171	170	169	168	167	166
155	154	153	152	151	150	149	148	147	146	145
134	133	132	131	130	129	128	127	126	125	124
113	112	111	110	109	108	107	106	105	104	103
92	91	90	89	88	87	86	85	84	83	82
71	70	69	68	67	66	65	64	63	62	61
50	49	48	47	46	45	44	43	42	41	40
29	28	27	26	25	24	23	22	21	20	19
8	7	6	5	4	3	2	1			
								1023	1022	1021
1010	1009	1008	1007	1006	1005	1004	1003	1002	1001	1000
715	714	713	712	711	710	709	708	707	706	705
694	693	692	691	690	689	688	687	686	685	684
673	672	671	670	669	668	667				

2.11 1G Systems

The introduction of 1G systems began the wireless revolution toward mobility being an accepted and expected method of communication. However, as implicated in the 1G discussions, the overwhelming demand for mobility services has resulted in the need to improve the wireless system's overall capacity. The capacity increase was needed, but it needs to be provided in a more cost-effective method of increasing capacity without introducing more cell sites into the system.

References

MacDonald. "*The Cellular Concept*," Bell Systems Technical Journal, Vol. 58, No. 1, 1979.

Smith, Clint. "*Practical Cellular and PCS Design*," McGraw-Hill, 1997.

Smith, Gervelis. "*Cellular System Design and Optimization*," McGraw-Hill, 1996.

CHAPTER 3

Second Generation (2G)

3.1 Overview

To better understand the issues with third-generation (3G) and the interim 2.5G radio and network access platforms, it is essential to know the fundamentals of second-generation (2G) systems. This chapter will attempt to cover a vast array of topics with reasonable depth and breath related to some of the more prevalent 2G wireless mobility systems that have been deployed.

Second-generation is the generalization used to describe the advent of digital mobile communication for cellular mobile systems. When cellular systems were being upgraded to 2G capabilities, the description at that time was digital and there was little if any indication of 2G since voice was the service to deliver, not data. Personal communication systems at the time of their entrance were considered the next generation of communication systems and boasted about new services that the subscriber would want and could be readily provided by this new system or systems. However, *Personal Communication Services* (PCS) took on the same look and feel as those originating from the cellular bands.

Second-generation mobility involves a variety of technology platforms as well as frequency bands. The issues regarding 2G deployment are as follows:

- Capacity
- Spectrum utilization
- Infrastructure changes
- Subscriber unit upgrades
- Subscriber upgrade penetration rates

The fundamental binding issue with 2G is the utilization of digital radio technology for transporting the information content.

It is important to note that while 2G systems utilized digital techniques to enhance their capacity over analog, its primary service was voice communication. At the time 2G systems were being deployed, 9.6 Kbps was more than sufficient for existing data services, usually mobile fax. A separate mobile data system was deployed in the United States called *Cellular Data Packet Data* (CDPD), which was supposed to meet the mobile data requirements. In essence, 2G systems were deployed to improve the voice traffic throughput compared to an existing analog system.

Digital radio technology was deployed in cellular systems using different modulation formats with the attempt to increase the quality and capacity of the existing cellular systems. As a quick point of reference in an analog cellular system, the voice communication is digitized within the cell site itself for transport over the fixed facilities to the MTSO. The voice representation and information transfer utilized in *Advanced Mobile Phone Service* (AMPS) cellular was analog and it is this part in the communication link that digital transition is focusing on.

The digital effort is meant to take advantage of many features and techniques that are not obtainable for analog cellular communication. Several competing digital techniques are being deployed in the cellular arena. The digital techniques for cellular communication fall into two primary categories: AMPS and the TACS spectrum. For markets employing the TACS spectrum allocation, the *Global System for Mobile communications* (GSM) is the preferred digital modulation technique. However, for AMPS markets, the choice is between *Time Division Multiple Access* (TDMA) and *Code Division Multiple Access* (CDMA) radio access platforms. In addition to the AMPS/TACS spectrum decision, the IDEN radio access platform is available and it operates in the *specialized mobile radio* (SMR) band, which is neither cellular or PCS. With the introduction of the PCS licenses, three fundamental competing technologies exist, which are CDMA, GSM, and TDMA. Which technology platform is best depends on the application desired, and at present, each platform has its pros and cons, including if it is a regulatory requirement to utilize one particular platform or not.

Table 3-1 represents some of the different technology platforms in the cellular, SMR, and PCS bands.

PCS was described at the time the frequency bands were made available as the next generation of wireless communications. PCS by default has similarities and differences with its counterparts in the cellular band. The similarities between PCS and cellular lie in the mobility of the user of the service. The differences between PCS and cellular fall into the applications and spectrum available for PCS operators to provide to the subscribers.

The PCS spectrum in the United States was made available through an action process set up by the *Federal Communications Commission* (FCC). The license breakdown is shown in Figure 3-1.

The geographic boundaries for PCS licenses are different that those imposed on cellular operators in the United States. Specifically, PCS licenses are defined as MTAs and BTAs. The MTA has several BTAs within its geographic region. A total of 93 MTAs and 487 BTAs are defined in the United States. Therefore, a total of 186 MTA licenses were awarded for the construction of a PCS network, and each license has a total of 30 MHz of spectrum to utilize. In addition, a total of 1,948 BTA licenses were awarded in the United States. Of the BTA licenses, the C band has

Table 3-1
Cellular and SMR 2G Technology Platforms

Cellular and SMR Bands	IS-136	IS-136*	IS-95	GSM	IDEN
Base Tx MHz	869–894	851–866	869–894	925–960	851–866
Base Rx MHz	824–849	806–821	869–894	880–915	806–821
Multiple Access Method	TDMA/FDMA	TDMA	CDMA/FDMA	TDMA/FDMA	TDMA
Modulation	Pi/4DPSK	Pi/4DPSK	QPSK	0.3 GMSK	16QAM
Radio Channel Spacing	30kHz	30kHz	1.25MHz	200kHz	25kHz
Users/Channel	3	3	64	8	3/6
Number Channels	832	600	9 (A) 10 (B)	124	600
CODEC	ACELP/VCELP	ACELP	CELP	RELP-LTP/ ACELP	
Spectrum Allocation	50MHz	30MHz	50MHz	50MHz	30MHz

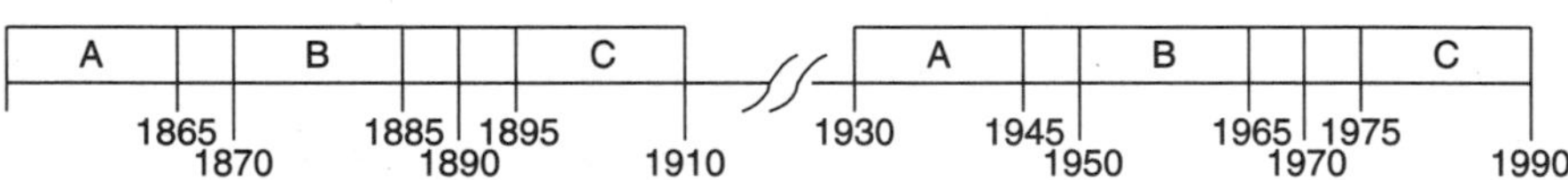

Figure 3-1
U.S. PCS spectrum chart.

30 MHz of spectrum, while the D, E, and F blocks will only have 10 MHz available.

Currently, PCS operators do not have a standard to utilize for picking a technology platform for their networks. The choice of PCS standards is daunting and each has its advantages and disadvantages. The current philosophy in the United States is to let the market decide which standard or standards is the best. This is significantly different than that used for cellular where every operator has one set interface for the analog system from which to operate.

Table 3-2 represents various PCS systems that are used throughout the world, particularly in the United States. The major standards utilized so far for PCS are DCS-1900, IS-95, IS-661, and IS-136. DCS-1900 utilizes a GSM format and is an upbanded DCS-1800 system. IS-95 is the CDMA standard that is utilized by cellular operators, except it is upbanded to the PCS spectrum. The IS-136 standard is an upbanded cellular TDMA system that is used by cellular operators. IS-661 is a Time Division Duplex system offered by Omnipoint Communications with the one notable exception that it was supposed to be deployed in the New York market as part of the pioneer preference license issued by the FCC.

Presently digital or digital modulation is now prevalent throughout the entire wireless industry. Digital communication references any communication that utilizes a modulation format that relies on sending the information in any type of data format. More specifically, digital communication

Table 3-2

2G Technology Platforms

	IS-136	IS-95	DCS1800 (GSM)	DCS1900 (GSM)	IS661
Base Tx MHz	1930–1990	1930–1990	1805–1880	1930–1990	1930–1990
Base Rx MHz	1850–1910	1850–1910	1710–1785	1850–1910	1850–1910
Multiple Access Method	TDMA/FDMA	CDMA/FDMA	TDMA/FDMA	TDMA/FDMA	TDD
Modulation	Pi/4DPSK	QPSK	0.3 GMSK	0.3 GMSK	QPSK
Radio Channel Spacing	30kHz	1.25MHz	200kHz	200kHz	5MHz
Users/Channel	3	64	8	8	64
Number Channels	166/332/498	4–12	325	25/50/75	2–6
CODEC	ACELP/VCELP	CELP	RELP-LTP/ACELP	RELP-LTP/ACELP	CELP
Spectrum Allocation	10/20/30Mhz	10/20/30Mhz	150MHz	10/20/30Mhz	10/20/30Mhz

is where the sending location digitizes the voice communication and then modulates it. At the receiver, the exact opposite is done.

Data is digital, but it needs to be converted into another medium in order to transport it from point A to point B, and more specifically between the base station and the host terminal. The data between the base station and the host terminal is converted from a digital signal into RF energy. Its modulation is a representation of the digital information that enables the receiving device, base station, or host terminal to properly replicate the data.

Digital radio technology is deployed in a cellular/PCS/SMR system primarily to increase the quality and capacity of the wireless system over its analog counterpart. The use of digital modulation techniques enables the wireless system to transport more bit/Hz than would be possible with analog signaling utilizing the same bandwidth. However, the service offering for 2G is mainly a voice offering.

Figure 3-2 is a block diagram representation of the differences between an analog and a digital radio. Reviewing the digital radio portion of the diagram, the initial information content, usually voice, is input into the microphone of the transmission section. The speech then is processed in a vocoder, which converts the audio information into a data stream utilizing a coding scheme to minimize the amount of data bits required to represent the audio. The digitized data then goes to a channel coder that takes the vocoder data and encodes the information even more, so it will be possible for the receiver to reconstruct the desired message. The channel-coded information is then modulated onto an RF carrier utilizing one of several modulation formats covered previously in this chapter. The modulated RF carrier is then amplified, passes through a filter, and is transmitted out an antenna.

The receiver, at some distance away from the transmitter, receives the modulated RF carrier though use of the antenna, which then passes the information though a filter and into a preamp. The modulated RF carrier is then down-converted in the digital demodulator section of the receiver to an appropriate intermediate frequency. The demodulated information is then sent to a channel decoder that performs the inverse of the channel coder in the transmitter. The digital information is then sent to a vocoder for voice information reconstruction. The vocoder converts the digital format into an analog format, which is passed to an audio amplifier connected to a speaker for the user at the other end of the communication path in order to listen to the message sent.

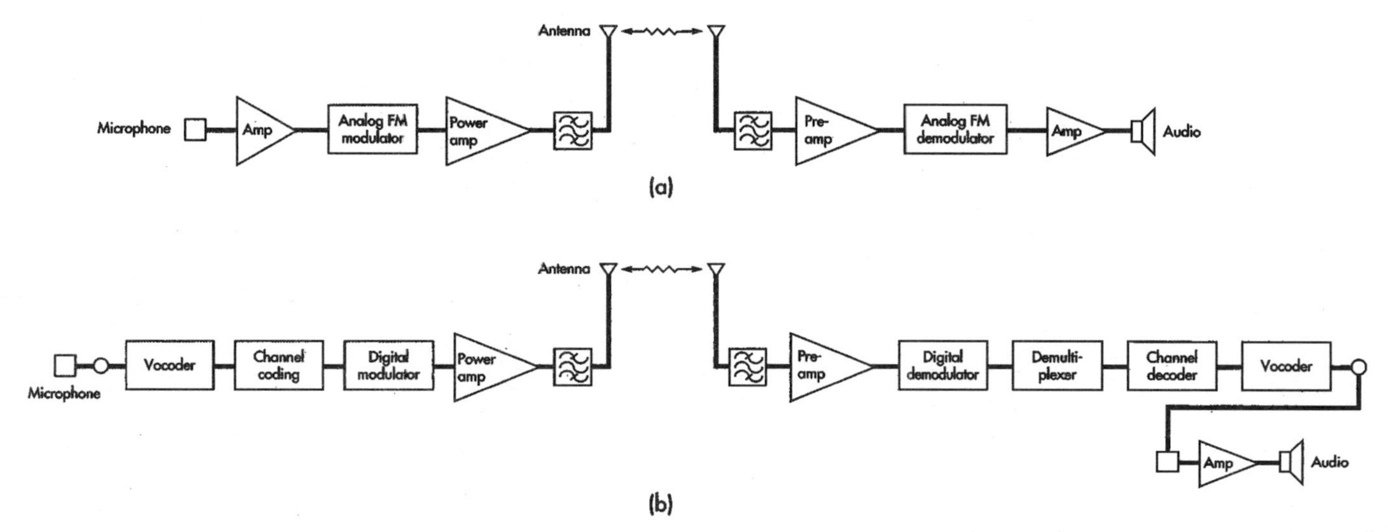

Figure 3-2 Analog and digital radio.

3.2 Enhancements over 1G Systems

The introduction of 2G mobility systems, whereas focused on voice transport, brought about numerous improvements or enhancements for the mobile wireless operators and their customers. The major benefits associated with the introduction of a 2G system are listed here:

- Increased capacity over analog
- Reduced capital infrastructure costs
- Reduced the capital per subscriber cost
- Reduced cellular fraud
- Improved features
- Encryption

The benefits, when looking at this list, were geared toward the operator of the wireless system. The implementation of 2G was a reduction in operating costs for the mobile operators either through improved capital equipment and spectrum utilization to a reduction in cellular fraud. The improved features were centered around SMS services, which the subscriber benefited from. The onslaught of 2G systems, however, primarily benefited the customer in that the overall cost to the subscriber was significantly reduced.

3.3 Integration with Existing 1G Systems

The advent of 2G digital systems brought about several implementation issues that the existing operators and infrastructure vendors needed to solve. At heart of the issue was how to cost-effectively implement 2G into an existing analog network. The problems involved the available spectrum, existing infrastructure, and subscriber equipment. Most of the staff that went through this period of time can remember the issues.

For the cellular operators, several options or rather decisions needed to be made for how to integrate the new system into the existing analog network. However, for PCS operators, the integration with legacy systems did not present a problem since there was no legacy system. The PCS operators in the United States had one other obstacle to overcome and that dealt with microwave clearance issues because the radio frequency spectrum auctioned for use for the PCS operators was currently being used by 2-GHz point-to-point microwave systems.

The integration with the existing 1G, legacy, systems was therefore an issue that only affected the analog systems operating in the 800/900-MHz bands. The 2G technologies that were applicable involved GSM,TDMA, and CDMA radio access systems. Several options were available for the 1G operators to follow, and they are listed in this section relative to each access platform since the actual implementation also is technology-dependent.

3.3.1 GSM

Global System for Mobile Communications (GSM) is the European standard for digital cellular systems operating in the 900-MHz band. This technology was developed out of the need for increased service capacity due to the analog systems' limited growth. This technology offers international roaming, high-speech quality, increased security, and the capability to develop advanced systems features. The development of this technology was completed by a consortium of pan-European countries working together to provide integrated cellular systems across different borders and cultures.

GSM is a European standard that has achieved worldwide success. GSM has many unique features and attributes that make it an excellent digital radio standard to utilize. GSM has the unique advantage of being the most widely accepted radio communication standard at this time. GSM was developed as a communication standard that would be utilized throughout all of Europe in response to the problem of multiple and incompatible standards that still exist there today.

GSM consists of the following major building blocks: the *Switching System* (SS), the *Base Station System* (BSS), and the *Operations and Support System*(OSS). The BSS is comprised of both the *Base Station Controller* (BSC) and the *Base Transceiver Stations* (BTS). In an ordinary configuration, several BTSs are connected to a BSC and then several BSCs are connected to the *Mobile Switching Center* (MSC).

The GSM radio channel is 200 kHz wide. GSM has been deployed in several frequency bands, namely the 900-, 1800-, and 1900-MHz bands. Both the 1800- and 1900-MHz bands required some level of spectrum clearing before the GSM channel could be utilized. However, the 900-MHz spectrum was used by an analog system *Enhanced Total Access Communiction System* (ETACS), which occupied 25-kHz channels. The introduction of

GSM into this band required the reallocation of traffic or rather channels to accommodate GSM.

3.3.2 TDMA (IS-54/IS-136)

IS-136, an enhancement of IS-54, is the digital cellular standard developed in the United States using TDMA technology. Systems of this type operate in the same band as the AMPS systems and are used in the PCS spectrum also. IS-136 therefore applies to both the cellular and PCS bands, as well as in some unique situations to down-banded IS-136, which operates in the SMR band.

TDMA technology enables multiple users to occupy the same channel through the use of time division. The TDMA format utilized in the United States follows the IS-54 and IS-136 standards and is referred to as the *North American Dual Mode Cellular* (NADC). IS-136 is an evolution to the IS-54 standard and enables a feature-rich technology platform to be utilized by the current cellular operators.

TDMA, utilizing the IS-136 standard, is currently deployed by several cellular operators in the United States. IS-136 utilizes the same channel bandwidth, as does analog cellular: 30 kHz per physical radio channel. However, IS-136 enables three and possibly six users to operate on the same physical radio channel at the same time. The IS-136 channel presents a total of six time slots in the forward and reverse direction. IS-136 at present utilizes two time slots per subscriber at this time with the potential to go to half-rate vocoders that require the use of only one time slot per subscriber.

IS-136 has many advantages in its deployment in a cellular system:

- Increased system capacity, up to three times over analog
- Improved protection for adjacent channel interference
- Authentication
- Voice privacy
- Reduced infrastructure capital to deploy
- Short message services

Integrating IS-136 into an existing cellular system can be done more easily than for the deployment of CDMA. The use of IS-136 in a network requires the use of a guardband to protect the analog system from the IS-136 signal. However, the guardband required consists of only a single channel on either side of the spectrum block allocated for IS-136 use. Depending on the actual location of the IS-136 channels in the operator's spectrum, it is possible to only require one or no guardband channel.

The IS-136 has the unique advantage of affording the implementation of digital technology into a network without elaborate engineering requirements. The implementation advantages mentioned for IS-136 also facilitate the rapid deployment of this technology into an existing network.

The implementation of IS-136 is further augmented by requiring only one channel per frequency group as part of the initial system offering. The advantage with only requiring one channel per sector in the initial deployment is the minimization of capacity reduction for the existing analog network. Another advantage with deploying one IS-136 channel per sector initially eliminates the need to preload the subscriber base with dual mode, IS-136 handsets.

3.3.3 CDMA

The operators who chose to deploy CDMA systems had basically two methods to use in deploying CDMA, IS-95 systems. The first method is to deploy CDMA in every cell site for the defined service areas on a 1:1 basis. The other method available is to deploy CDMA on a N:1 basis. Both the 1:1 and the N:1 deployment strategies had their advantages and disadvantages. Of course, a third method involved a hybrid approach to both the 1:1 and N:1 methods (see Table 3-3).

Table 3-3

CDMA Deployment Strategies

Layout	Advantages	Disadvantages
1:1	Consistent Coverage	Cost
	Facilitates gradual growth	Guard Zone requirements
	Integrates into existing 1G System	Digital to Analog boundry handoff
	Large initial capacity gain	Slower deployment then N:1
N:1	Lower Capital cost over 1:1	Engineering complexity
	Faster to implement over 1:1	Lower capacity gain

Figures 3-3 and 3-4 illustrate at a high level the concept of both 1:1 and N:1 deployment scenarios for integrating a CDMA system into an existing 1G analog network.

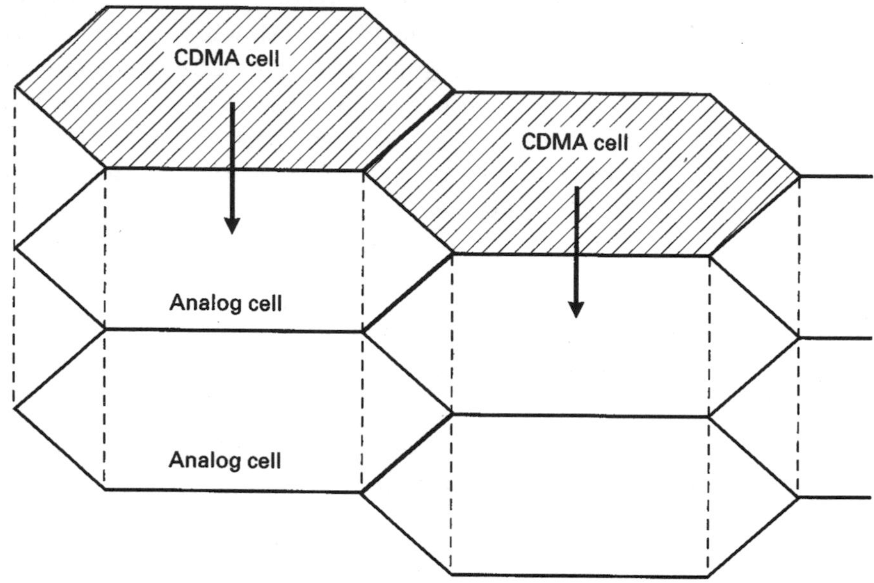

Figure 3-3 1:1 CDMA deployment.

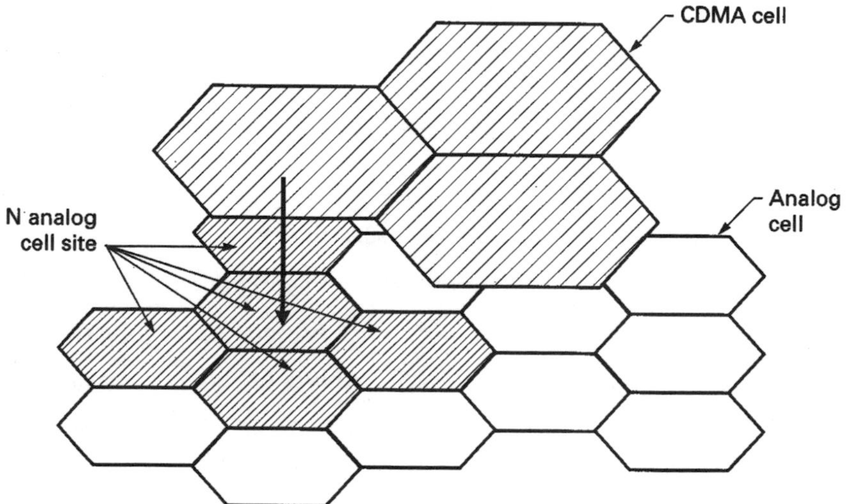

Figure 3-4 N:1 CDMA deployment.

The introduction of CDMA into an existing AMPS system also required the establishment of a guard band and guard zone. The guard band and guard zone are required for CDMA to ensure that the interference received from the AMPS system does not negatively impact the ability for CDMA to perform well.

The specific location that the CDMA channel or channels occupy in a cellular system is dependent upon a multitude of issues. The first issue is how much spectrum will be dedicated to the use of CDMA for the network. The spectrum issue ties into the fact that one CDMA channel occupies 1.77 MHz of spectrum, 1.23 MHz per CDMA channel, and 0.27 MHz of guard-band on each side of the CDMA channel. With a total of 1.77 MHz per CDMA, the physical location in the operator's band that CDMA will operate in needs to be defined. For the B-band carrier, wireline operators, two predominant locations were utilized. The first location in the spectrum is the band next to the control channels, and the other section is in the lower portion of the extended AMPS band. The upper end of the AMPS band is not as viable due to the potential of AGT interference because AGT transmit frequencies have no guard band between AMPS receiving and AGT transmitting. The lower portion of the AMPS band has the disadvantage of receiving A band mobile to base interference, which will limit the size of the CDMA cell site.

The other issue with the guard band ties into the actual amount of spectrum that will be unavailable for use by AMPS subscribers in the cellular market. With the expansive growth of cellular, assigning 1.77 MHz of spectrum to CDMA reduces the spectrum available for AMPS usage by 15 percent or to 59 radio channels from the channel assignment chart. The reduction in the available amount of channels for regular AMPS requires the addition of more cell sites to compensate for the amount of radio channels no longer available in the AMPS system. Utilizing a linear evaluation, the reduction in usable spectrum by 15 percent involves a reduction on the traffic-handling capacity by the AMPS system by a maximum of 21 percent at an Erlang B 2-percent *Grade of Service* (GOS) with a maximum of 16 channels per sector verse 19. The reduction of 21 percent in the initial AMPS traffic-handling capacity requires the need to build more analog cell sites to compensate for this reduction in traffic-handling capabilities. The only way to offset the reduction in the traffic-handling capacity experienced by partitioning the spectrum is to preload the CDMA subscriber utilizing dual mode phones or to build more analog cell sites.

The guard zone is the physical area outside the CDMA coverage area that can no longer utilize the AMPS channels now occupied by the CDMA

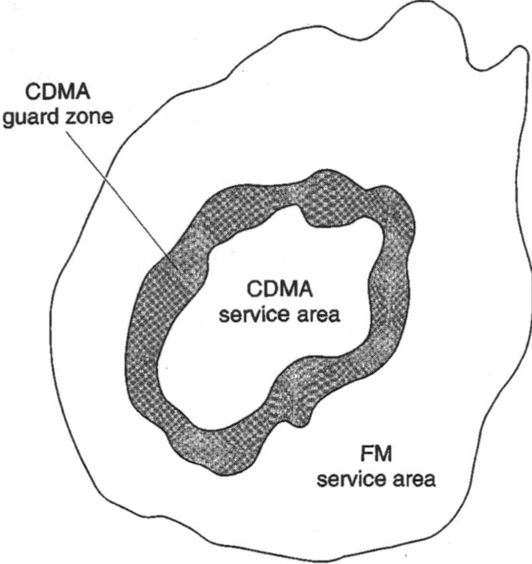

Figure 3-5
Guard zone.

system. Figure 3-5 shows an example of a guard zone versus a CDMA system coverage area. The establishment and size of the guardzone is dependent upon the traffic load expected by the CDMA system. The guard zone is usually defined in terms of a signal strength level from which analog cell sites operating with the CDMA channel sets cannot contribute to the overall interference level of the system. The interesting point about the guard zone is when the operator on one system wants to utilize CDMA and must require the adjacent system operator to reduce his or her channel utilization in the network to accommodate his or her neighbor's introduction of this new technology platform.

However, regardless of the method chosen for the implementation of the CDMA into an existing 1G analog system, part of the radio frequency spectrum needed to be cleared of existing analog radio usage. The impact to this situation, as discussed, was the need to build more cell sites with lower traffic-carrying capacity due to the spectrum reduction, or to increase the blocking percentage that the system would be allowed to operate at. Obviously, the mix of both increased blocking as well as additional cell sites was the method that was followed by the wireless operators.

In addition to the reduction in spectrum that is used for existing 1G subscribers for base stations which had CDMA installed there are also numerous sites which did not have CDMA installed but still had to surrender the use of spectrum to accommodate the introduction of CDMA in the system. Naturally, if the system was a complete 1:1 system, then there would be no need for the implementation of a guard zone and only a guard band. But when the CDMA system butted up to another radio access system like TDMA or even analog where another operator decided not to implement 2G systems, the need for a guard zone and guard band was required.

3.4 GSM

Unlike IS-136 or IS-95, GSM was designed from scratch as a complete system, including air interface, network architecture, interfaces, and services. In addition, the design of GSM included no compatibility with existing analog systems. The reasons for this included the fact that multiple analog systems were used in Europe and it would have taken great effort to design a system that would provide backward compatibility with each of them. The lack of compatibility also meant that carriers had a greater impetus to build GSM coverage as extensively and as quickly as possible.

In the following sections, we spend some time describing the GSM architecture and functionality. The main reason is because GSM is the foundation of a number of more advanced technologies such as the *General Packet Radio Service* (GPRS) and the *Universal Mobile Telecommunications Service* (UMTS). An understanding of GSM is necessary in order to understand those technologies.

3.4.1 GSM Network Architecture

Figure 3-6 shows the basic architecture of a GSM network. Working our way from the left, we see that the handset, known in GSM as the *Mobile Station* (MS), communicates over the air interface with a *Base Transceiver Station* (BTS). Strictly speaking, the MS is composed of two parts—the handset itself, known as the *Mobile Equipment* (ME), and the *Subscriber Identity Module* (SIM), a small card containing an integrated circuit. The SIM contains user-specific information, including the identity of the subscriber, subscriber authentication information, and some subscriber service information. It is only when a given subscriber's SIM is inserted into a handset that the handset acts in accordance with the services the

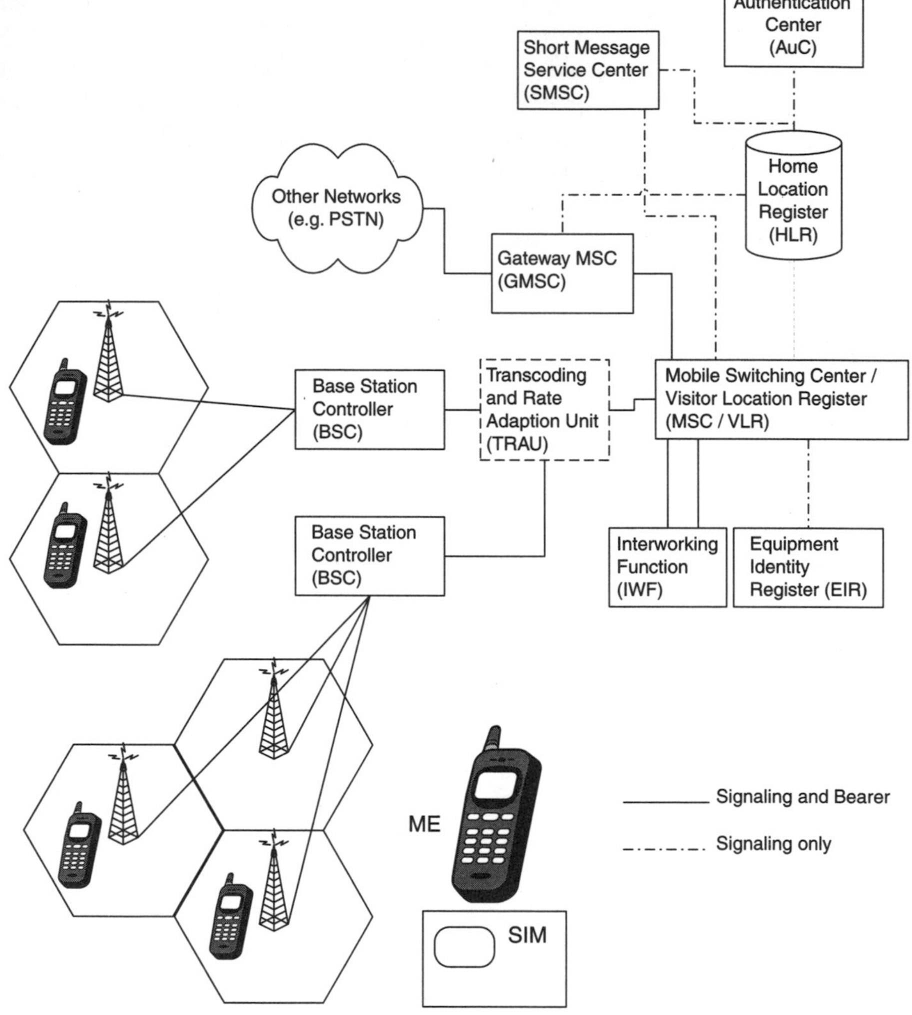

Figure 3-6 GSM System Architecture.

subscriber has subscriber to. In other words, my handset only acts as my handset when my SIM is inserted.

The BTS contains the radio transceivers that provide the radio interface with mobile stations. One or more BTSs is connected to a *Base Station Controller* (BSC). The BSC provides a number of functions related to *radio resource* (RR) management, some functions related to *mobility management* (MM) for subscribers in the coverage area of the BTSs, and a number of operation and maintenance functions for the overall radio network. Together, BTSs and BSCs are known as the *Base Station Subsystem* (BSS).

The interface between the BTS and the BSC is known as the Abis interface. Many aspects of that interface are standardized. One aspect, however, is proprietary to the BTS and BSC vendor, which is the part of the interface that deals with configuration, operation, and maintenance of the BTSs. This is known as the *Operation and Maintenance Link* (OML). Because the internal design of a BTS is proprietary to the BTS vendor, and because the OML needs to have functions that are specific to that internal design, the OML is also proprietary to the BTS vendor. The result is that a given BTS must be connected to a BSC of the same vendor.

One or more BSCs are connected to a Mobile Switching Center (MSC). The MSC is the switch—the node that controls call setup, call routing, and many of the functions provided by a standard telecommunications switch. The MSC is no ordinary PSTN switch, however. Because of the fact that the subscribers are mobile, the MSC needs to provide a number of MM functions. It also needs to provide a number of interfaces that are unique to the GSM architecture.

When we speak of an MSC, a *Visitor Location Register* (VLR) is also usually implied. The VLR is a database that contains subscriber-related information for the duration that a subscriber is in the coverage area of an MSC. A logical split exists between an MSC and a VLR, and the interface between them has been defined in standards. No equipment vendor, however, has ever developed a stand-alone MSC or VLR. The MSC and VLR are always contained on the same platform and the interface between them is proprietary to the equipment vendor. Although early versions of GSM standards defined the MSC-VLR interface (known as the B-interface) in great detail, later versions of the standards recognized that no vendor complies with the standardized interface. Therefore, any "standardized" specification for the B-interface should be considered informational.

The interface between the BSC and the MSC is known as the A-interface. This is an SS7-based interface using the *Signaling Connection Control Part* (SCCP), as depicted in Figure 3-7. Above Layer 3 in the signaling stack, we find the *BSS Application Part* (BSSAP), which is the protocol used for communication between the MSC and the BSC, and also between the MSC and the MS. Since the MSC communicates separately with both the BSC and the MS, the BSSAP is divided into two parts—the *BSS Management Application Part* (BSSMAP) and the *Direct Transfer Application Part* (DTAP). BSSMAP contains those messages that are either originated by the BSS or need to be acted upon by the BSS. DTAP contains

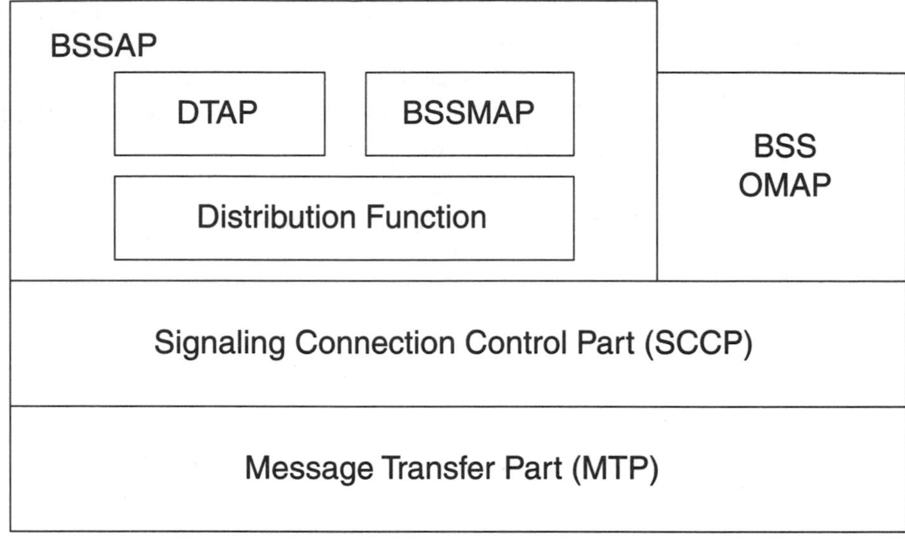

Figure 3-7 BSSAP Protocol Layers.

those messages that are passed transparently through the BSS from the MSC to the MS or vice versa. Note that there is also a *BSS Operation and Maintenance Application Part* (BSSOMAP). Although this is defined in standards, it is normal for the BSC to be managed through a vendor-proprietary management protocol.

In Figure 3-6, we find (in the dashed outline) the *Transcoding and Rate Adaptation Unit* (TRAU). In GSM, the speech from the subscriber is usually coded at either 13 Kbps (full rate, FR) or 12.2 Kbps (enhanced full rate, EFR). In some cases, we also find half-rate coding at a rate of 5.6 Kbps, but that is rare in commercial networks. In any case, it is clear that the speech to and from the MS is very different from the standard 64 Kbps *Pulse Code Modulation* (PCM) used in switching networks.

Since the MSC interfaces with the PSTN network, it needs to send and receive speech at 64 Kbps. The function of the TRAU is to convert the coded speech to or from standard 64 Kbps. Strictly speaking, the TRAU is a part of the BSS. As far as the MSC is concerned, voice to and from the BSS is passed at 64 Kbps and the BSS takes care of the transcoding. In practice, however, it is common for the TRAU to be physically separate from the BSC and placed near the MSC. This reduces the bandwidth required between the MSC location and the BSC location and can mean significant savings in transport cost, particularly if the BSC and MSC are separated by a significant distance. In cases where the BSC and TRAU are separated, the interface between them is known as the Ater interface. This interface is proprietary to the BSS equipment vendor. Hence, the BSC and TRAU must be from the same vendor.

In Figure 3-6, we find also find a *Home Location Register* (HLR)—a node found in most, if not all, mobile networks. The HLR contains subscriber data, such as the details of the services to which a user has subscribed. Associated with the HLR, we find the *Authentication Center* (AuC). This is a network element that contains subscriber-specific authentication data, such as a secret authentication key called the Ki. The AuC also contains one or more sophisticated authentication algorithms. For a given subscriber, the algorithm in the AuC and the Ki are also found on the SIM card. Using a random number assigned by the AuC and passed down to the SIM via the HLR, MSC, and ME, the SIM performs a calculation using the Ki and authentication algorithm. If the result of the calculation on the SIM matches that in the AuC, then the subscriber has been authenticated. The interface between the HLR and AuC is not standardized. Although implementations can set up the HLR and AuC to be separate, it is more common to find the HLR and AuC integrated on the same platform.

Calls from another network, such as the PSTN, first arrive at a type of MSC known as a *Gateway MSC* (GMSC). The main purpose of the GMSC is to query the HLR to determine the location of the subscriber. The response from the HLR indicates to the MSC where the subscriber may be found. The call is then forwarded from the GMSC to the MSC serving the subscriber. A GMSC may be a full MSC/VLR such that it may have some BSCs connected to it. Alternatively, it may be a dedicated GMSC and its only function is to interface with the PSTN and query the HLR. The choice is dependent upon the amount and types of traffic in the network and the relative cost of a full MSC/VLR versus a pure GMSC.

In Figure 3-6, we also note the *Short Message Service Center* (SMSC). Strictly speaking, the correct term is *Short Message Service-Service Center* (SMS-SC), but that is a bit of a mouthful and is usually shortened to SMSC. The SMSC is a node that supports the storing and forwarding of short messages to and from mobile stations. Typically, these short messages are text messages up to 160 characters in length.

Logically, an SMSC has three components. First is the *Service Center* (SC) itself, which stores messages and interfaces with other systems such as e-mail or voice mail equipment. Second, there is the *SMS-Gateway MSC*

(SMS-GMSC) which is used for the delivery of short messages to a mobile subscriber. Much like a GMSC, the SMS-GMSC queries the HLR for the subscriber's location, and then forwards the short message to the appropriate visited MSC where it is relayed to the subscriber. Third is the SMS-Interworking MSC, which receives a short message from the MSC serving the subscriber. It forwards such messages to the SC, which then passes them on to the final destination. It is very common for the SC, SMS-GMSC, and SMS-IWMSC to be included within the same platform, though certain implementations enable a stand-alone SC. In such implementations, the SMS-GMSC function may be included within a GMSC and the SMS-IWMSC function may be included with an MSC/VLR.

In a GSM network, we may also find a node known as the *Equipment Identity Register* (EIR). As mentioned, it is not the handset that identifies a subscriber, rather it is the information on the SIM. Therefore, to some degree, the handset used by a particular subscriber is not relevant. On the other hand, it may be important for the network to verify that a particular handset (ME) or a model of ME is acceptable. For example, a network operator might want to restrict access from a handset that has not been fully type-approved. Also, a network operator might want to restrict access from a handset that is known to be stolen.

Stored in each handset is an *International Mobile Equipment Identity* number (IMEI, 15 digits) or the *International Mobile Equipment Identity and Software Version Number* (IMEISV, 16 digits). Both the IMEI and IMEISV have a structure that includes the *type approval code* (TAC) and the *final assembly code* (FAC). The TAC and FAC combine to indicate the make and model of the handset and the place of manufacture. The IMEI and IMEISV also include a specific serial number for the ME in question. The only difference between IMEI and IMEISV is the software version number.

Within the EIR are three lists—black, gray, and white. These lists contain values of TAC, TAC and FAC, or complete IMEI or IMEISV. If a given TAC, a TAC/FAC combination, or a complete IMEI appears in the black list, then calls from the ME are barred. If it appears in the gray list, then calls may or may not be barred at the discretion of the network operator. If it appears in the white list, then calls are allowed. Typically, a given TAC included in the white list has the model of handset that has been approved by the handset manufacturer. The EIR is an optional network element and some network operators have chosen not to deploy an EIR.

Finally, we find the *Interworking Function* (IWF). This is used for circuit-switched data and fax services and is basically a modem bank. Typical dial-up modems and fax machines are analog. For example, when one uses a computer with a 28.8 Kbps modem on a regular telephone line, the modem modulates the digital data from the computer to an analog format that appears like analog speech. The same cannot be done directly for a digital system such as GSM because all transmissions are digital and it is not possible to transmit data over the air in a manner that emulates analog voice. Furthermore, a remote dial-up modem, such as at an ISP, expects to be called by another modem. Therefore, a circuit-switched data call from an MS is looped through the IWF before being routed onwards by the IWF. Within the IWF, a modem is placed in the call path. The same applies for facsimile service, where a fax modem would be used rather than a data modem. GSM supports data and fax services up to 9.6 Kbps.

3.4.2 The GSM Air Interface

GSM is a TDMA system, with *Frequency Division Duplex* (FDD). It uses *Gaussian Minimum Shift Keying* (GMSK) as the modulation scheme. TDMA means that multiple users share a given RF channel on a time-sharing basis. FDD means that different frequencies are used in the downlink (from network to MS) and uplink (from MS to network) directions.

GSM has been deployed in numerous frequency bands—including the 900-MHz band, the 1800-MHz band, and the 1900-MHz band (in North America). Table 3-4 shows the frequency allocations for these three bands.

Of course, the amount of spectrum allocated in a given band in a given country is at the discretion of the appropriate regulatory authorities in that country. Moreover, even if the entire spectrum in a given band is made available in a given country, it is likely to be divided among several operators such that it is extremely rare for a single network operator to have access to a complete band.

Table 3-4

GSM Frequency Bands

	GSM 900	Extended GSM (E-GSM)	DCS 1800	PCS 1900
Uplink (MS to network)	890 MHz – 915 MHz	880 MHz – 915 MHz	1710 MHz – 1785 MHz	1850 MHz – 1910 MHz
Downlink (network to MS)	935 MHz – 960 MHz	925 MHz – 960 MHz	1805 MHz – 1880 MHz	1930 MHz – 1990 MHz

In GSM, a given band is divided into 200-kHz carriers or RF channels in both the uplink and downlink directions. In addition, a guard band of 200 kHz is located at each end of each frequency band. For example, in standard GSM 900, the first uplink RF channel is at 890.2 MHz and the last uplink RF channel is at 914.8MHz, allowing for a total of 124 carriers. Similarly, DCS 1800 has a maximum of 374 carriers and PCS 1900 has a maximum of 299 carriers.

As mentioned, in GSM, a given band is divided into a number of RF channels or carriers, each 200 kHz in both the uplink and downlink. Thus, if a handset is transmitting on a given 200-kHz carrier in the uplink, then it is receiving on a corresponding 200-kHz carrier in the downlink. Because the uplink and downlink are rigidly associated, when one talks about a carrier or RF channel, both the uplink and downlink are usually implied. A given cell can have multiple RF carriers—typically one to three in a normally loaded system, though as many as six carriers might exist in a heavily loaded cell in an area of very high traffic demand. Note that, when we talk about a cell in GSM terms, we mean a sector. Thus, a three-sector BTS implies three cells. This is a somewhat confusing distinction between GSM and some other technologies.

Each RF carrier is divided into eight timeslots, numbered 0 to 7, and these are transmitted in a frame structure. Each frame lasts approximately 4.62 ms, such that each time slot lasts approximately 576.9 μs. Depending on the number of RF carriers in a given cell, all eight timeslots on a given carrier might be used to carry user traffic. In other words, the RF carrier might be allocated to eight *traffic channels* (TCHs). There must be, however, at least one timeslot in a cell allocated for control channel purposes. Thus, if only one carrier is in a cell, then there is a maximum of seven TCHs, such that a maximum of seven simultaneous users can be accommodated.

3.4.3 Types of Air Interface Channels

The foregoing description of the RF interface suggests that only traffic channels and control channels exist. This is only partly correct. In fact, there are traffic channels, numerous types of control channels, and a number of other channels. To begin with, a number of broadcast channels are available:

- **Frequency Correction Channel (FCCH)** This is broadcast by the BTS and used for frequency correction of the MS.

- **Synchronization Channel (SCH)** This is broadcast by the BTS and is used by a mobile station for frame synchronization. It addition to frame synchronization information, it also contains the *Base Station Identity Code* (BSIC).
- **Broadcast Control Channel (BCCH)** This is used to broadcast general information regarding the BTS and the network in general. It is also used to indicate the configuration of the *Common Control Channels* (CCCH) described in the following section.

The CCCH is a bidirectional control channel used primarily for functions related to initial access by a mobile station. It has a number of components:

- **Paging Channel (PCH)** This is used for the paging of mobile stations.
- **Random Access Channel (RACH)** This is used only in the uplink direction. It is used by a mobile station to request the allocation of a *Stand alone Dedicated Control Channel* (SDCCH) described later.
- **Access Grant Channel (AGCH)** This is used in the downlink in response to an access request received on the RACH. It is used to allocate an MS to an SDCCH or directly to a *Traffic Channel* (TCH).
- **Notification Channel (NCH)** This is used with voice group call and voice broadcast services to notify mobile stations regarding such calls.

A number of dedicated control channels exist. These are channels that are used by one mobile station at a time, typically either during call establishment or while a call is in progress. The dedicated control channels are as follows:

- **Stand Alone Dedicated Control Channel (SDCCH)** This is a bidirectional channel used for communication with an MS when the MS is not using a TCH. The SDCCH is used, for example, for *Short Message Service* (SMS) when the MS is not in a call. It is also used for call establishment signaling prior to the allocation of a TCH for a call.
- **Slow Associated Control Channel (SACCH)** This is a unidirectional or bidirectional channel, used when the MS is using a TCH or SDCCH. For example, when an MS in engaged in a call on a TCH, power control messages from a BTS to an MS are sent on the SACCH. In the uplink, the MS sends measurement reports to the BTS on the SACCH. These reports indicate how well the MS can receive transmissions from other BTSs and the information is used in determination of if or when a handover should occur. The SACCH is also used for short message transfers when the MS is in on a TCH.

- **Fast Associated Control Channel (FACCH)** This is associated with a given TCH and thus is used when the mobile is involved in a call. It is typically used to transmit non-voice information to and from the MS. Such information would include, for example, handover instructions from the network, commands from the MS for generation of DTMF tones, supplementary service invocations, and so on.

3.4.4 Air Interface Channel Structure

Clearly, it does not make sense for these different types of channels to each be allocated one of the eight timeslots. Firstly, there would simply not be enough timeslots. Moreover, different data rates apply to the various types of channels. Instead, a sophisticated framing structure is used on the air interface to allocate the various channel types to the available timeslots. The structure includes frames, multiframes, superframes, and hyperframes.

As mentioned previously, a single frame lasts approximately 4.62 ms and contains eight timeslots. In standard GSM (as opposed to GPRS), two types of multiframes are used—a 26 multiframe (containing 26 frames and having a duration of 120 ms) and a 51 multiframe (containing 51 frames and having a duration of 235.4 ms). The 26 multiframe is used to carry TCHs and the associated SACCH and FACCH. The 51 multiframe is used to carry BCCH, CCCH (including PCH, RACH, and AGCH), and SDCCH (and its associated SACCH). A superframe lasts 6.12 seconds, corresponding to 51*26 multiframes or 26*51 multiframes. A hyperframe corresponds to 2,048 superframes (a total of 2,715,648 frames, lasting just under 3 hours, 28 minutes, and 54 seconds). When numbering frames over the air interface, each frame is a numbered modulo of its hyperframe. In other words, a frame can have a *frame number* (FN) from 0 to 2,715,467. The reason for the large hyperframe is to allow for a large value of FN, which is used as part of the encryption over the air interface.

Certain timeslots on a given RF carrier may be allocated to control channels, while the remaining timeslots are allocated for traffic channels. For example, timeslot 0 on the first carrier in a cell is used to carry the BCCH and CCCH. It may also carry four SDCCH channels. It is also common to find that timeslot 1 on the first RF carrier in a cell is used to carry eight SDCCH channels (with the associated SACCHs), with the remaining timeslots allocated as TCHs. Exactly how much SDCCH capacity is allocated is dependent upon the number of carriers and the amount of traffic in the cell. Figure 3-8 shows two typical arrangements.

As mentioned, the 26 multiframe is used for the TCH. The structure is depicted in Figure 3-9, where only one timeslot per frame is shown (only full-rate TCH is considered in the figure). A given timeslot carries user traffic (voice) for 24 out of 26 frames. One of the 26 frames is idle and one of the 26 frames carries the SACCH. The FACCH is transmitted by pre-empting half or all of the user traffic in a TCH.

This overall structure enables a TCH to have a gross bit rate of 22.8 Kbps. Of course, this rate is not allocated completely to user data (such as speech). Rather, a sophisticated coding and interleaving scheme is applied. This scheme adds a significant number of bits for error detection and correction, which reduces the bandwidth available for raw user data. In fact, for standard GSM *full rate* (FR) voice coding, the speech is carried at 13 Kbps and for *enhanced full rate* (EFR), the speech is carried at 12.2 Kbps. Although it may seem that a great deal of the gross 22.8 Kbps is consumed

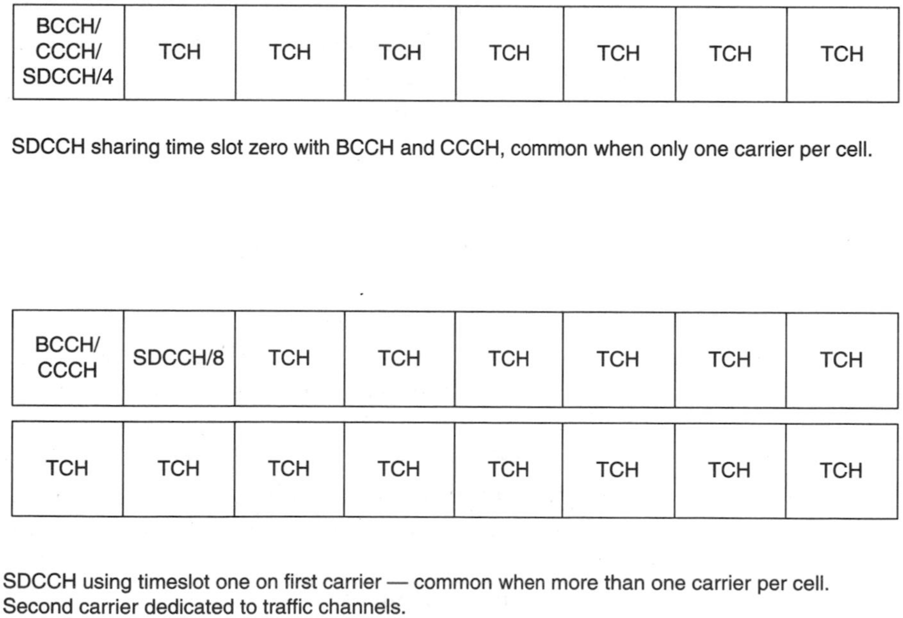

Figure 3-8 Example GSM Air Interface Timeslot Allocations.

BCCH/ CCCH/ SDCCH/4	TCH	TCH	TCH	TCH	TCH	TCH	TCH

SDCCH sharing time slot zero with BCCH and CCCH, common when only one carrier per cell.

BCCH/ CCCH	SDCCH/8	TCH	TCH	TCH	TCH	TCH	TCH
TCH	TCH	TCH	TCH	TCH	TCH	TCH	TCH

SDCCH using timeslot one on first carrier — common when more than one carrier per cell. Second carrier dedicated to traffic channels.

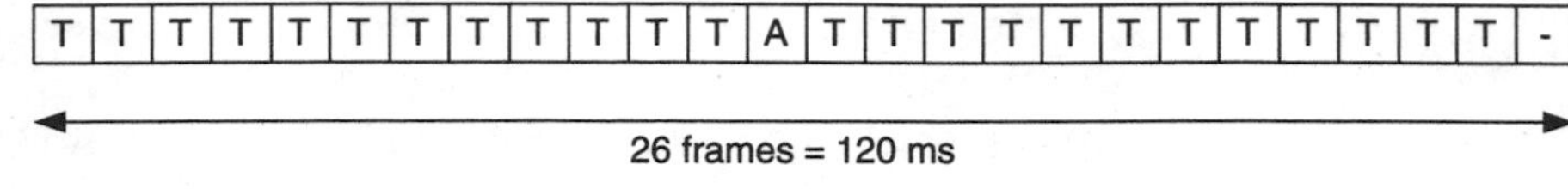

Figure 3-9 TCH/SACCH Framing Structure.

by coding overhead, it is worth remembering that an RF interface is unreliable at best, and error-correction overhead is necessary to overcome the limitations of the medium.

Since the control channels (with the exception of FACCH and SACCH) are carried on different timeslots from the TCHs, it is possible to have a different framing structure. In fact, a 51-multiframe structure is used for transmitting the control channels and this structure applies to any timeslot that is allocated to control channels.

3.4.5 GSM Traffic Scenarios

We will show a number of traffic examples for UMTS in later chapters. The following sections provide some straightforward examples of GSM traffic. This allows for an understanding of the differences between the technologies, the evolution from one to the other, and how compatibility can be achieved.

3.4.6 Location Update

When an MS is first turned on, it must first "camp on" a suitable cell. This largely involves scanning the air interface to select a cell with a suitably strong received signal strength and then decoding the information broadcast by the BTS on the BCCH. Generally, the MS will camp on the cell with the strongest signal strength, provided that cell belongs to the *home PLMN* (HPLMN) and provided that the cell is not barred. The MS then registers with the network, which involves a process known as location updating, as shown in Figure 3-10.

The sequence begins with a channel request issued by the MS on the RACH. This includes an establishment cause, such as location updating,

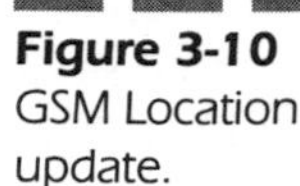

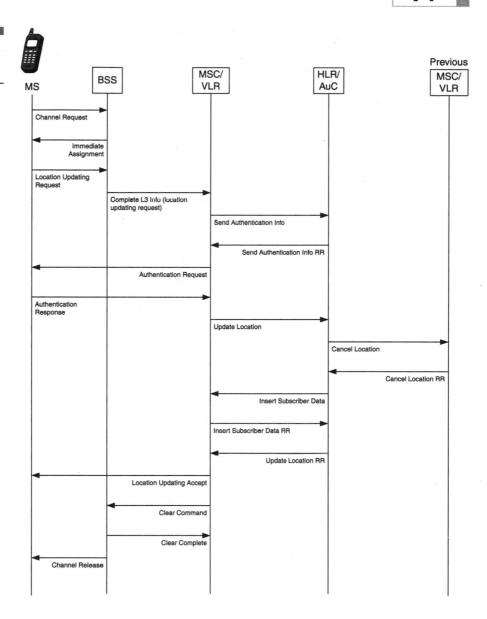

Figure 3-10 GSM Location update.

voice call establishment, and emergency call establishment. In the example of Figure 3-10, the cause is location updating.

The BSS allocates an SDCCH for the MS to use. It instructs the MS to move to the SDCCH by sending an Immediate Assignment message on the

AGCH. The MS then moves to the SDCCH and sends the Location Updating Request. This contains a set of information including the location area identity (as received by the MS on the BCCH) and the mobile identity. The mobile identity is usually either the *International Mobile Subscriber Identity* (IMSI) or the *Temporary Mobile Subscriber Identity* (TMSI). This is sent through the BSS to the MSC using a generic message known as Complete Layer 3 Info. This message is included as part of an SCCP Connection Request. Hence, it uses connection-oriented SCCP.

If the subscriber attempts to register with TMSI and the TMSI is unknown in the MSC/VLR, then the MSC/VLR may request the MS to send the IMSI (not shown in the figure). Equally, the MSC/VLR may request the MS to send the IMEI so that it can be checked (also not shown in the figure).

Upon receipt of the location updating request, the MSC/VLR may attempt to authenticate the subscriber. If the MSC/VLR does not already have authentication information for the subscriber, then it requests that information from the HLR, using the *Mobile Application Part* (MAP) operation Send Authentication Info. The HLR/AuC sends a MAP *return result* (RR) with up to five authentication vectors, known as triplets. Each triplet contains a *random number* (RAND) and a *signed response* (SRES).

The MSC sends an Authentication Request to the MS. This contains only the RAND. The MS performs the same calculations as were performed in the HLR/AuC and sends an Authentication Response containing an SRES parameter. The MSC/VLR checks to make sure that the SRES received from the MS matches that received from the HLR/AuC. If a match is made, then the MS is considered authenticated.

At this point, the MSC/VLR uses the MAP operation Update Location to inform the HLR of the subscriber's location. The message to the HLR includes the subscriber's IMSI and the SS7 *Global Title Address* (GTA) of the MSC and VLR. The HLR immediately sends a MAP Cancel Location message to the VLR (if any) where the subscriber had previously been registered. That VLR deletes any stored data related to the subscriber and issues a return result to the HLR.

The HLR uses the MAP operation Insert Subscriber Data to the VLR to inform the VLR about a range of data regarding the subscriber in question, including information regarding supplementary services. The VLR acknowledges receipt of the information. The HLR then issues a return result to the MAP Update Location.

Upon receipt of that return result, the MSC/VLR sends the DTAP message Location Updating Accept to the MS. It then clears the SCCP connection to the BSS. This causes the BSS to release the MS from the SDCCH by sending a Channel Release message to the MS.

A number of optional messages have been excluded in Figure 3-10. For a complete understanding of all the options, the reader is referred to GSM specification 04.08. A number of messages shown in Figure 3-10 (Channel Request, Immediate Assignment, Channel Release) are common to many traffic scenarios. For the sake of brevity, they are not shown in the following call examples.

3.4.7 Mobile-Originated Voice Call

Figure 3-11 shows a basic mobile-originated call to the PSTN. After the MS has been placed on an SDCCH by the BSS (not shown), the MS issues a CM Service Request to the MSC (CM = Connection Management). This includes information about the type of service that the MS wants to invoke (a mobile-originated call in this case, but it could also be another service such as SMS).

Upon receipt of the CM Service Request, the MSC may optionally invoke authentication of the mobile. Typically, an MSC is configured to authenticate a mobile whenever it performs an initial location update and every N transactions thereafter (every N calls). Next, the MSC initiates ciphering so that the voice and data sent over the air is encrypted. Since it is the BSS that performs the encryption and decryption, the MSC needs to pass the *Cypher key* (Kc) to the BSS. The BSS then instructs the MS to start ciphering. The MS, of course, generates the Kc independently, so that it is not passed over the air. Once the MS has started ciphering, it informs the BSS, which, in turn, informs the MSC.

Next, the MS sends a Setup message to the MSC. This includes further data about the call, including information such as the dialed number and the required bearer capability. Once the MSC has determined that it has received sufficient information to connect the call, it lets the MS know by sending a Call Proceeding message.

Next, using the Assignment Request message, the MSC requests the seizure of a circuit between the MSC and BSS. That circuit will be used to carry the voice to and from the MS. At this point, the BSS sends an Assignment Command message to the MS, instructing the MS to move from the SDCCH to a TCH. Further signaling between the MS and the network will now occur on the FACCH associated with the assigned TCH. The MS responds with an Assignment Complete message, indicating that it has

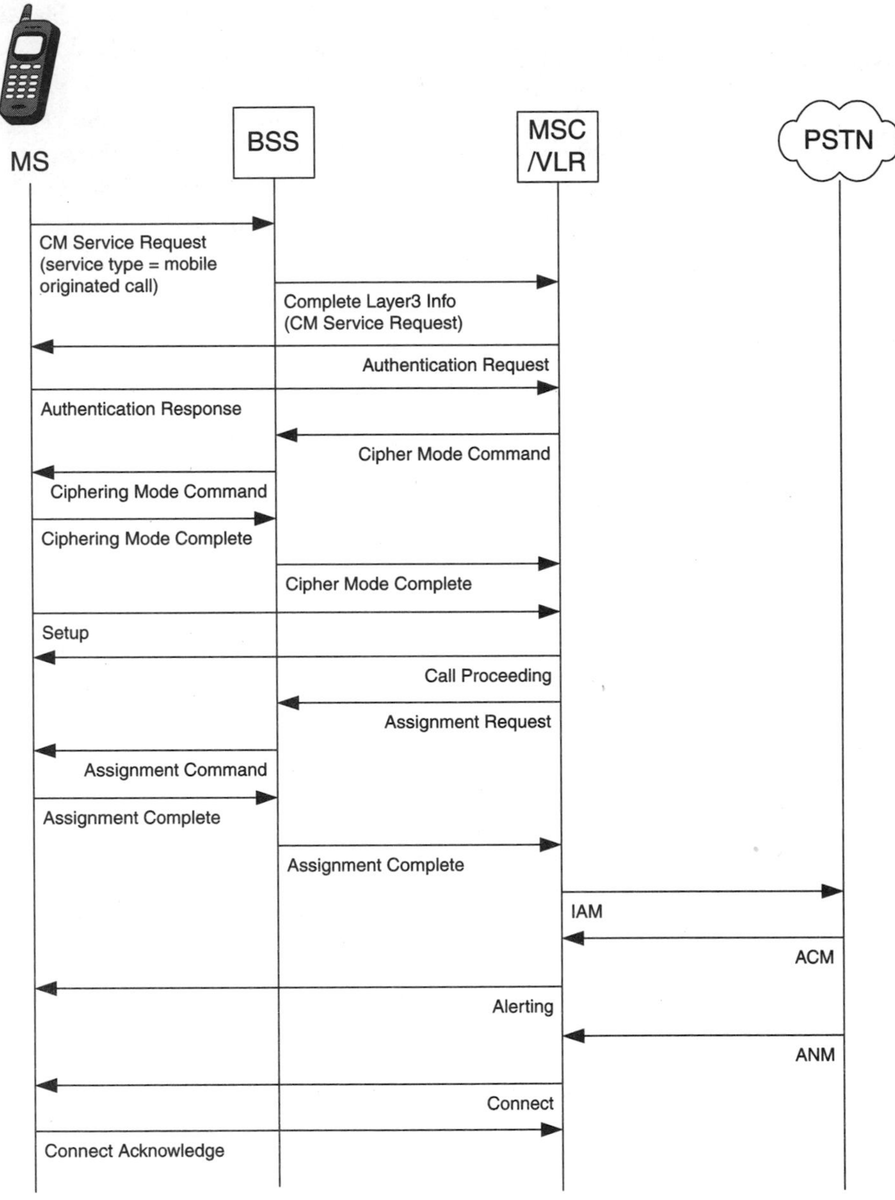

Figure 3-11 Mobile to land call flow diagram.

moved to the assigned TCH. Upon receipt of that message, the BSS sends an Assignment Complete message to the MSC, which indicates that a voice path is now available from the MS through to the MSC.

Upon receipt of the Assignment Complete message from the BSS, the MSC initiates the call setup towards the PSTN. This starts with issuance of an *Initial Address message* (IAM). A subsequent receipt of an *Address Complete Message* (ACM) from the destination end indicates that the destination phone is now ringing. The MSC informs the MS of that fact by sending an Alerting message. In addition, the ACM triggers a one-way path to be opened from the destination PSTN switch through to the MS and the ring-back tone heard at the MS is actually being generated at the destination PSTN switch.

Upon answer at the called phone, an *Answer Message* (ANM) is returned. This leads the MSC to open a two-way path to the MS and also causes the MSC to send a Connect message to the MS. Upon receipt of the Connect message, the MS responds with a Connect Acknowledge message. The two parties are now in conversation and, from a billing perspective, the clock is now ticking.

3.4.8 Mobile-Terminated Voice Call

Figure 3-12 shows a basic mobile-terminated call from the PSTN. It begins with the arrival of an IAM at the GMSC. The IAM contains the directory number of the called subscriber, known as the *Mobile Station ISDN Number* (MSISDN). The GMSC uses this information to determine the applicable HLR for the subscriber and invokes the MAP operation *Send Routing Information* (SRI) towards the HLR. The SRI contains the subscriber's MSISDN.

The HLR uses the MSISDN to retrieve the subscriber's IMSI from its database. Through a previous location update, the HLR knows the MSC/VLR that serves the subscriber, and it queries that MSC/VLR using the MAP operation *Provide Roaming Number* (PRN), which contains the subscriber's IMSI. From a pool, the MSC/VLR allocates a temporary number, known as a *Mobile Station Roaming Number* (MSRN) for the call and returns that number to the HLR. The HLR returns the MSRN to the GMSC.

The MSRN is a number that appears to the PSTN as a dialable number. Thus, it can be used to route a call through any intervening network between the GMSC and the visited MSC/VLR. In fact, that is exactly what the GMSC does. It routes the call to the MSC/VLR by sending an IAM, with the MSRN as the called party number. Upon receipt of the IAM, the MSC/VLR recognizes the MSRN and knows the IMSI for which the MSRN

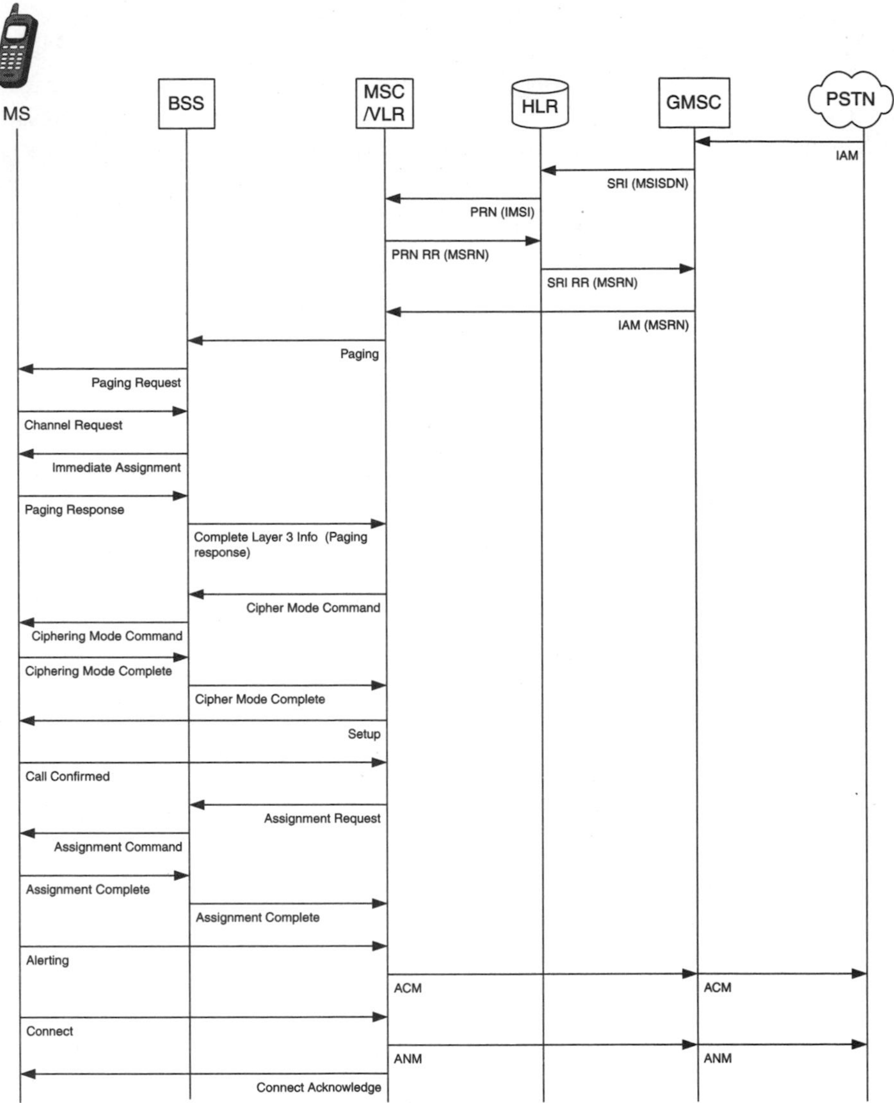

Figure 3-12 Land-to-mobile call flow diagram.

was allocated. At this point the MSRN can be returned to the pool for use with another call.

Next, the MSC requests the BSS to page the subscriber using the Paging Request message, which indicates the location area in which the subscriber should be paged. The BSS uses the PCH to page the MS.

Upon receipt of the page, the MS attempts to access the network using a Channel Request message on the RACH. The BSS responds with an Immediate Assignment message, instructing the MS to move to an SDCCH. The MS moves to the SDCCH and, once there, indicates to the network that it is responding to the page. The BSS passes the response to the MSC.

At this point, the MSC may optionally authenticate the MS (not shown). It will then proceed to initiate ciphering, which is done in the same manner as was described previously for a mobile-originated call. Once ciphering is started, the MSC sends a Setup message to the MS. This is similar to the Setup message that is sent from an MS for a mobile-originated call, including information such as the calling party number and the required bearer capability.

Upon receipt of the Setup message, the MS sends a Call Confirmed message to the MSC, indicating that it has the information it needs to establish the call. The Call Confirmed message acts as an instruction to the MSC to establish apath through to the MS. Therefore, the MSC begins the assignment procedure, which establishes a circuit between the MSC and the BSS, and a TCH between the BSS and the MS (rather than an SDCCH). Further signaling between the MS and the network will now use the FACCH associated with the TCH to which the MS has been assigned.

Once established on the TCH, the MS starts ringing to alert the user and informs the network by sending the Alerting message to the MSC. This triggers the MSC to open a one-way path back to the original caller, generate a ring-back tone, and send an ACM message back to the originating PSTN switch via the GMSC.

Once the called user answers, the MS sends a Connect message to the MSC. This triggers the MSC to send an ANM message back to the originating switch and to open a two-way path. Finally, it sends Connect Acknowledge to the MS and conversation begins.

3.4.9 Handover

A *handover* (also known as a handoff) is the process by which a call in progress is transferred from a radio channel in one cell to another radio channel, either in the same cell or in a different cell. A handover can occur within a cell, between cells of the same BTS, between cells of different BTSs connected to the same BSC, between cells of different BSCs, or between cells of different MSCs. Not only can a handover occur between TCHs, a handover is also possible from an SDCCH on one cell to an SDCCH on another cell. It is also possible from an SDCCH on one cell to a TCH on another cell. The most common, however, is a handover from TCH to TCH.

Depending on the source (the original cell) and the target (the destination cell) involved in the handover, the handover may be handled completely within a BSS or may require the involvement of an MSC. In the case where a handover occurs between cells of the same BSC, the BSC may execute the handover and simply inform the MSC after the handover has taken place. If, however, the handover occurs between BSCs, then the MSC must become involved, because no direct interface exists between BSCs.

A handover in GSM is known as a *mobile-assisted handover* (MAHO). This means that it is the network that decides if, when, and how a handover should take place. The MS, however, provides information to the network to enable the network to make the decision.

Recall that GSM is a TDMA system, with eight timeslots per frame in the case of full-rate speech. This means that the MS is transmitting for one-eighth of the time and receiving for one-eighth of the time. In fact, at the BTS, a given timeslot on the uplink is three timeslots later than the corresponding downlink timeslot, which means that the MS is not required to receive and transmit simultaneously. We note that this offset is specified at the BTS rather than at the MS because the distance of the MS from the BTS influences the exact instant at which the MS should transmit. For example, when an MS is close to the BTS it should transmit slightly later than if it were further from the BTS. This variation is known as time-alignment and is controlled by the BSS. In other words, the BSS periodically instructs the MS to change its time alignment as necessary.

Nonetheless, it is clear that for most of the time the MS is neither transmitting nor receiving. During this time, the MS has the opportunity to tune to other carrier frequencies and determine how well it can receive those signals. It can then relay that information to the network to allow the network to make a determination as to whether the MS would be better served by a different cell. Because of frequency reuse, it is possible that a number of nearby cells might be using the same BCCH frequency. Therefore, it is not sufficient for the MS to simply report signal strength for specific frequencies. Rather, the MS must be able to synchronize to the BCCH of neighboring cells and decode the information being transmitted. Exactly which frequencies the MS should check for are specified in system information messages transmitted by the BTS on the BCCH and the SACCH. The MS sends measurement reports to the BSS on the SACCH as often as possible. These reports include information on how well the MS can "hear" the serving cell as well as information about signal strength measurements on up to six neighboring cells. Specifically, for the serving cell, the MS reports the RXLEV (an indication of received signal strength) and the RXQUAL (an indication of the bit error rate on the received signal). For neighboring cells, the MS reports the BSIC, the BCCH frequency, and the RXLEV.

In addition to the measurements reported by the MS, the BTS itself makes measurements regarding the RXLEV and RXQUAL received from the MS. These measurements and those from the MS are reported to the BSC. Based on its internal algorithms, the BSC makes the decision as to whether a handover should occur, and if so, to what cell.

Figure 3-13 shows an inter-BSC handover. In this case, it is not sufficient for the BSC to handle the handover autonomously—it must involve the MSC. Therefore, once the serving BSC determines that a handover should take place, it immediately sends the message Handover Required to the MSC. This message contains information about the desired target cell (or the cells in the preferred order), plus information about the current channel that the MS is using. The MSC analyzes the information and identifies the target BSC associated with at least one of the target cells identified by the source BSC. It then sends a Handover Request message to the target BSC. This contains, among other items, information about the target cell, the type of channel required, and, in the case of a speech or data call, the circuit to be used between the MSC and the target BSC.

If the target BSC can accommodate the handover (if resources are available), then it allocates the necessary resources and responds to the MSC with the Handover Request Acknowledge message. This message contains a great deal of information regarding the cell and channel to which the MS is to be transferred, such as the cell identity, the exact channel to be used (including the type of channel), synchronization information, the power level to be used by the MS when accessing the new channel, and a handover reference. The MSC then sends the Handover Command message to the serving BSC. This message is used to relay the information received from the target BSC. On receipt of the Handover Command message from the MSC, the serving BSC passes the information to the MS in a Handover Command message over the air interface.

Upon receipt of the Handover Command message, the MS releases existing RF connections, tunes to the target channel, and attempts to access that channel. Upon access, it may send a Handover Access message to the target BSS. It will do so if it was commanded to do so in the Handover Command message. If the Handover Access message is received by the target BSS,

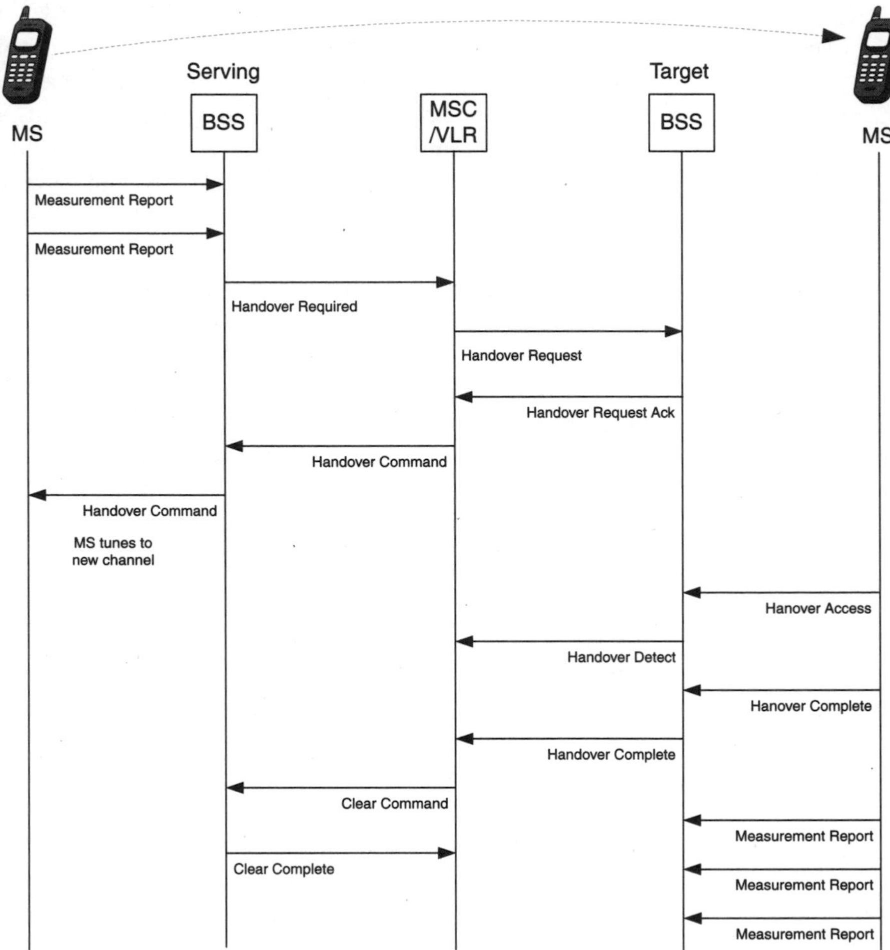

Figure 3-13
Inter-BSC Handover.

then it sends a Handover Detect message to the MSC. When the MS has established all lower layer connections on the target channel, it sends a Handover Complete message to the target BSC, which, in turn, sends a Handover Complete message to the MSC. At this point, the MS again starts taking measurements of neighboring cells. Meanwhile, the MSC instructs the old BSC to release all radio and terrestrial resources related to the MS.

3.4.10 Traffic Calculation Methods

Like any mobile communications technology, traffic calculation and system dimensioning for GSM begin with the estimation of how much traffic demand there will be and from where it will come. In other words, one must estimate the traffic demand in the coverage of each cell. This is rather an inexact science. One can certainly acquire demographic data such as population density, average household income, and so on. One can also acquire data related to vehicular traffic in order to estimate traffic demand for cells that cover roads. Based on these factors and others (such as how many competing operators exist), one makes an estimate of the peak traffic demand per cell. This estimate may well be incorrect. Fortunately, however, time is an ally. In a new network, traffic demand grows gradually, which provides the operator with sufficient time to monitor usage and more accurately predict traffic demand over time.

Because all GSM traffic is circuit-switched, network dimensioning is a relatively straightforward process once traffic demand per cell is specified. The process largely involves determining the amount of traffic to be carried in the busy hour and dimensioning the network according to Erlang tables.

The air interface, which represents the scarcest resource in the network, is dimensioned with the highest blocking probability. Typically, network designers dimension the air interface according to a two-percent blocking probability (Erlang B). For a one-TRX cell with seven TCHs (BCCH, CCCH, and SDCCH/4 are sharing timeslot 0), the cell can accommodate approximately 2.9 Erlangs. For a two-TRX cell with 14 TCHs (timeslot 0 on one carrier is used for BCCH and CCCH and timeslot 1 is used for SDCCH/8), the cell can accommodate approximately 8.2 Erlangs. For a three-TRX cell with 22 TCHs (one timeslot is allocated for SDCCH/8), the cell can accommodate approximately 14.9 Erlangs. It is important to note that the traffic-carrying capacity of each cell must be calculated independently.

Other interfaces in the network are usually dimensioned at much lower blocking probabilities. For example, the A interface would typically be designed for a 0.1-percent blocking probability. Similar blocking would apply to other network-internal interfaces such as the interface between the MSC and IWF. Typically, interfaces to other networks, such as the PSTN, are dimensioned at slightly higher blocking probabilities—such as 0.5 percent. Of course, the choice of blocking probability for any interface is a balance between cost and quality. The lower the blocking probability, the higher the quality and the higher the cost. The higher the lower blocking probability, the lower the quality and the lower the cost.

3.5 IS-136 System Description

IS-54 and IS-136 represent the most direct evolution from 1G systems. In fact, IS-54 and IS-136 were designed to allow significant compatibility with analog AMPS so that dual-mode handsets could be developed at a reasonable cost. Since IS-54, and then IS-136, initially began as islands of in a sea of AMPS coverage, it was important to have dual-mode phones so that subscribers could still obtain AMPS coverage when roaming outside of IS-54 or IS-136 coverage.

IS-54 represents the first step in moving from analog AMPS to digital technology and is often known as *Digital AMPS* (D-AMPS). IS-54 could be called a generation 2.5 technology because it is not completely digital. Only the voice channels are digital—the control channel is still analog. The introduction of the digital control channel came about with the introduction of IS-136. Nevertheless, IS-54 was an important step forward as it provided a number of significant advantages over AMPS, including increased system capacity and security through support for authentication. Support for authentication within analog AMPS had already been designed, but since it involved changes to the air interface, it required support within the handsets. Unfortunately, millions of handsets were already in the field and these did not support authentication. IS-54, however, required new handsets and these new phones incorporated authentication from the start.

3.5.1 The IS-54 Digital Voice Channel

IS-54 takes the existing 30-kHz AMPS voice channel and, applying *Time Division Multiplexing* (TDM), divides the 30-kHz channel into a number of time slots, as shown in Figure 3-14. Rather than having a full 30-kHz channel for a conversation, each user is assigned a number of time slots, each known as a *Digital Traffic Channel* (DTC). In IS-54, typically three users are supported on a given RF channel. Having three users per RF channel implies an obvious increase in capacity over analog AMPS, which supports just a single user on an RF channel.

3.5.1.1 Voice Channel Structure Associated with each DTC are two other channels—the *Fast Associated Control Channel* (FACCH) and the *Slow Associated Control Channel* (SACCH). The FACCH is a signaling channel used for the transmission of control and supervisory information between the mobile and the network. For example, if a mobile is to send

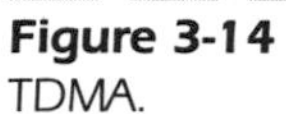

Figure 3-14
TDMA.

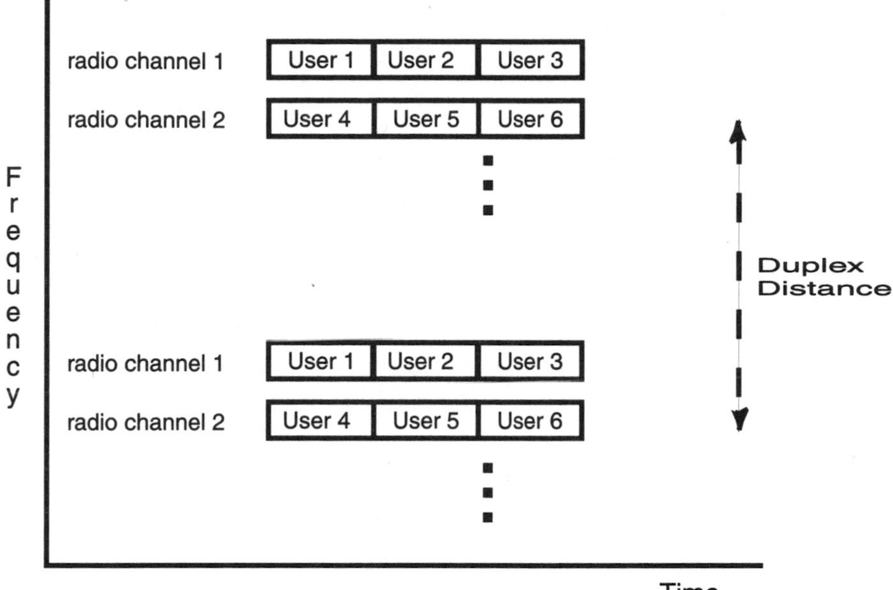

DTMF tones, then these are indicated on the FACCH. The SACCH is also used for the transmission of control and supervisory information between the mobile and the network. Most notably, the SACCH is used by the mobile to transmit measurement information to the network describing the mobile's experience of the RF conditions. This information is used by the network to determine when and how a handoff should occur.

Figure 3-15 shows the structure of the DTC. It is notable that the figure does not show the FACCH. This is because the DATA field, which is normally used to transmit voice, is also used to transmit FACCH information. In other words, if information is to be sent on the FACCH, then user data is briefly suspended while the FACCH information is being sent. Figure 3-15 also shows six time slots within the frame structure. In fact, IS-54 enables two types of mobiles: full-rate and half-rate. A full-rate mobile uses two of the timeslots in the frame (1 and 4, 2 and 5, or 3 and 6), while a half-rate mobile uses just a single time slot. A full-rate mobile transmits 260 bits of speech per time slot (520 bits per frame). Since there are 25 frames per second, this means that the gross bit rate for speech is 13 Kbps. In practice, only full-rate handsets are used.

Figure 3-15 Digital Traffic Channel (DTC).

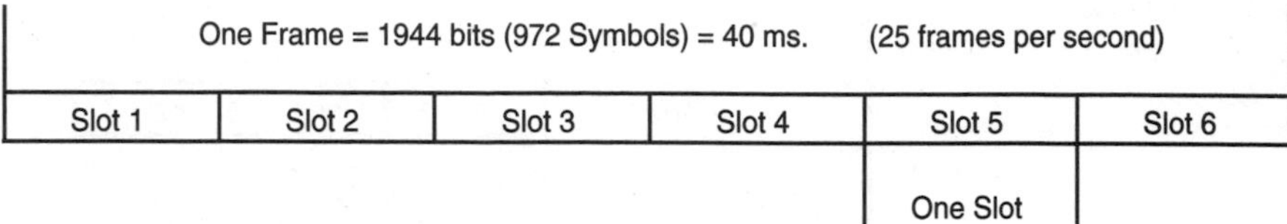

6	6	16	28	122	12	12	122
G	R	DATA	SYNC	DATA	SACCH	CDVCC	DATA

SLOT FORMAT MOBILE STATION TO NETWORK (uplink)

28	12	130	12	130	12
SYNC	SACCH	DATA	CDVCC	DATA	RSVD = 00...00

SLOT FORMAT NETWORK TO MOBILE STATION (downlink)

INTERPRETATION OF THE DATA FIELDS IS AS FOLLOWS:

G ñ Guard Time
R ñ Ramp Time
DATA ñ User Information or FACCH
RSVD ñ Reserved
SACCH ñ Slow Associated Control Channel
CDVCC ñ Coded Digital Verification Color Code
SYNC ñ Synchronization and Training

In addition to the user data and SACCH within the DTC, we see a number of other fields, as follows:

- **Guard Time** This field is three symbols (six bits) in duration. It is used as a buffer between adjacent time slots used by different mobiles and enables compensation for variations in distance between the mobile and the base station.
- **Ramp Time** This is a three-symbol duration allowing for a ramp up of the RF power.
- **Sync** This is a special synchronization pattern, which is unique for a given time slot. It is used for correct time alignment.
- **CDVCC** This is the Coded Digital Voice Color Code, which is analogous to the Supervisory Audio Tone used in analog AMPS. It is used to detect co-channel interference.

3.5.1.2 Offset Between Transmit and Receive IS-54 is a frequency duplex TDMA system. In other words, the mobile transmits on one frequency and receives on another frequency. In the uplink, the mobile transmits on a given pair of time slots, and on the downlink, it receives on the corresponding pair of time slots. If, for example, a given mobile transmits on time slots 1 and 4 on the uplink, then it receives on time slot 1 and 4 on the downlink. Time slots 1 and 4 on the downlink do not, however, correspond to the same instants in time as time slots 1 and 4 on the uplink. A time offset between the downlink and the uplink corresponds to one time slot plus 45 symbol periods (207 symbol periods total or 8.5185 ms), with the downlink lagging the uplink. Therefore, the mobile does not transmit and receive simultaneously. Rather, during a conversation, it receives a time slot on the downlink shortly after sending a time slot on the uplink. Figure 3-16 depicts this offset, showing the transmission and reception by a given mobile on time slots 1 and 4.

As can be seen from Figure 3-16, times will occur when the mobile is neither transmitting on a given time slot nor listening to the base station on the corresponding downlink time slot. So what does it do during these times? Rather than do nothing, the mobile tunes briefly to other base stations to measure the signal from those base stations. As described later in

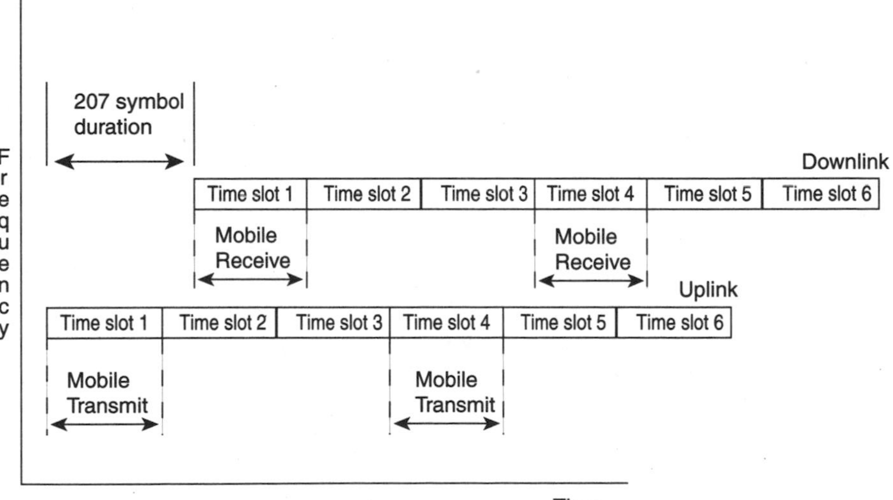

Figure 3-16 Transmit and receive offset.

this chapter, those measurements can be provided to the network to assist the network in determining when a handoff should take place.

3.5.1.3 Speech Coding Because the DTC is digital, it is necessary to convert the user speech from analog form to digital. In other words, the handset (and the network) must include a digital speech-coding scheme. In IS-54, the speech-coding technique uses *Vector Sum Excited Linear Prediction* (VSELP). This is a *linear predictive coding* (LPC) technique that operates on 20-ms speech samples at a time. For each 20-ms sample, the coding scheme itself generates 159 bits. Thus, the coder provides an effective bit rate of 7.95 Kpbs.

The RF interface, however, is an error-prone medium. Therefore, to ensure high speech quality, it is necessary to include mechanisms that mitigate against errors caused in RF propagation. Consequently, the 159 bits are subject to a channel-coding scheme designed to minimize the effects of errors. Of the 159 bits, 77 are considered class 1 bits (of greater significance to the speech perception) and 82 are considered class 2 bits. As shown in Figure 3-17, the 77 class 1 bits are passed through a convolutional coder, which results in 178 bits. These 178 bits are combined with the 82 class 2 bits to give a total of 260 bits, and the 260 bits are allocated across the time slots

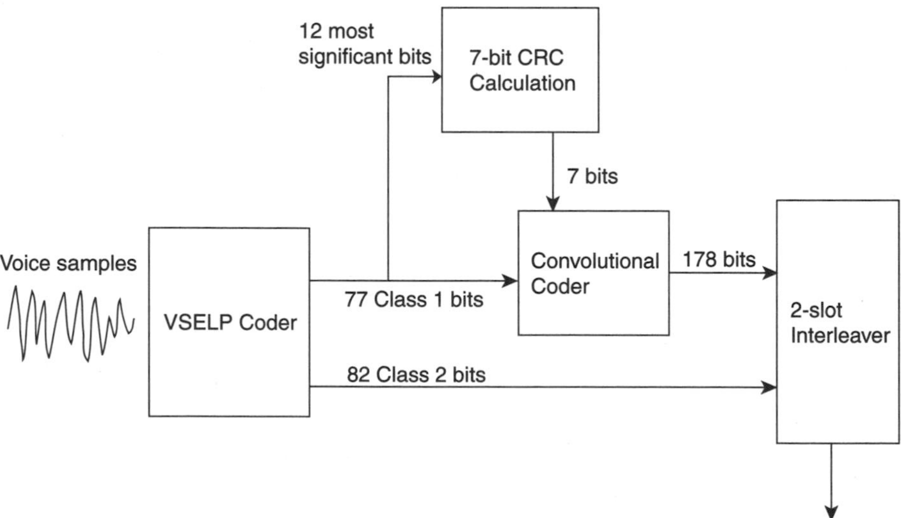

Figure 3-17
IS-S4 Speech Coding.

used by the subscriber. Thus, each 20 ms of speech gives rise to a transmission of 260 bits, resulting in a gross rate of 13 Kbps over the air interface.

3.5.1.4 Time Alignment Since three mobiles use a given RF channel on a time-sharing basis, it is necessary that they each time their transmissions exactly. Otherwise, their signals would overlap and cause interference at the base station receiver. Furthermore, a given cell may be many miles in diameter, and the time for transmission from one mobile to the base station may be different than the time taken by the transmission from another mobile. Therefore, if one mobile begins transmission immediately after another mobile stops transmission, it is possible that the two signals could collide at the base station.

For example, consider a situation where Mobile A is far away from the base station and Mobile B is close to the base station. It takes longer for Mobile B's transmission to reach the base station than that of Mobile A. Therefore, if Mobile A starts transmitting immediately after Mobile B stops transmitting, the transmission from Mobile B could still be arriving at the base station when Mobile A's transmission starts to arrive. Consequently, it is necessary not just to ensure that no two mobiles transmit at the same time, but it is also necessary to time transmissions such that no two transmissions arrive at the base station at the same time. The methodology for this timing is called time alignment, which involves advancing or retarding the transmission from a given mobile so that the transmission arrives at the base station at the correct time relative to transmissions from other mobiles using the same RF channel.

When a mobile first accesses the system, the network assigns it a traffic channel, including a Digital Voice Color Code (DVCC). At this point, however, the network has not provided any time alignment information. Given that the mobile could be close to the base station or far away, it needs the correct time alignment information before transmitting real user data, which means that the base station must determine roughly how far away the mobile happens to be and must send time alignment instructions. In order to help the base station determine what the time alignment instructions should be, the mobile sends a special sequence of 324-bit duration, called a shortened burst, as shown in Figure 3-18. The structure of the shortened burst is such that if the base station detects two or more sync words of the burst it can determine the mobile's distance from the base station. The base station then sends a Time Alignment Message instructing the mobile to adjust its transmission timing.

G1	R S D S D V S D W S D X S D Y S	G2

Figure 3-18 Shortened Burst Structure.

G1: 3 symbol length guard time.
R: 3 symbol length Ramp time.
S: 14 symbol length Sync Word; The mobile station uses its assigned sync word.
D: 6 symbol length CDVCC; The mobile station uses its assigned DVCC.
G2: 22 Symbol length guard time.

V = 0000
W = 00000000
X = 000000000000
Y = 0000000000000000

3.5.2 Control Channel

Even though the IS-54 control channel is analog, and even though IS-54 is designed to include a certain degree of compatibility with analog AMPS, the control channel contains a number of significant differences from the analog control channel. These changes were introduced to overcome known problems in AMPS and to provide control channel support for digital voice channels. For example, when assigning a mobile to a given traffic channel, the downlink control channel must specify the time slots to be used by the mobile. Obviously, such capability does not exist in the standard AMPS control channel.

Access to the TDMA system is either achieved through the primary control channel, utilized for analog communication, or the secondary dedicated control channel. During the initial acquisition phase, the mobile reads the overhead control message from the primary control channel and determines if the system is digital-capable. If the system is digital capable, a decision will be made whether to utilize the primary or secondary dedicated control channel. The secondary dedicated control channels are assigned as FCC channels 696 to 716 for the A band system and channels 717 through 737 for the B band system. The use of the secondary dedicated control channels enables a variety of enhanced features to be provided by the system operator to the subscribers.

IS-136 brings to the table the *Digital Control Channel* (DCCH) and it enables the delivery of adjunct features that in cellular were not really possible. The DCCH occupies two of the six time slots and therefore if a physical radio also has a DCCH assigned to it, only two subscribers can use the physical radio for communication purposes.

The DCCH's can be located anywhere in the allocated frequency band; however, certain combinations of channels are preferred to be used. The preference is based on the method that the subscriber unit scans the available spectrum looking for the DCCH.

The preferred channel sets are broken down into 16 relative probability blocks for each frequency band of operation, both cellular and PCS. The relative probability block #1 is the first group of channels the subscriber unit uses to find the DCCH for the system and cell. The subscriber unit will then scan through the entire frequency band, going through channel sets according to the relative probability blocks until it finds a DCCH. In the case of cellular, if no DCCH is found, it reverts to the control channel for a dual-mode phone and then acquires the system either through the control channel or is directed to a specific channel that has the DCCH.

3.5.3 MAHO

One of the unique features associated with TDMA is the capability for a *mobile assisted hand-off* (MAHO). The MAHO process enables the mobile to constantly report back to the cell site, indicating its present condition in the network. The cell site is also collecting data on the mobile through the reverse link measurements, but the forward link, base to mobile, is being evaluated by the mobile itself, therefore providing critical information about the status of the call.

For the MAHO process, the mobile measures the *received signal strength level* (RSSI) received from the cell site. The mobile also performs a *bit error rate* (BER) test and a *frame error rate* (FER) test as another performance metric.

The mobile also measures the signals from a maximum of six potential digital hand-off candidates, utilizing either a dedicated control channel or a beacon channel. The channels utilized by the mobile for the MAHO process are provided by the serving cell site for the call. The dedicated control channel is either the primary or secondary control channel and the measurements are performed on the forward link. The mobile can also utilize a beacon channel for the performance measurement. The beacon channel is either a TDMA voice channel or it is an analog channel, both of which are transmitting continuously with no dynamic power control on the forward link. The beacon channel is utilized when the setup or control channel for the cell site has an omni configuration and not a dedicated setup channel per sector.

3.5.4 Frequency Reuse

The modulation scheme utilized by the NADC TDMA system is a pi/4 DQPSK format. The C/I levels used for frequency management associated with IS-54 or IS-136 are the same for analog, 17 dB C/I. The C/I level desired is 17 dB and is the same for DCCH and the DTC. This is convenient because in all the cellular systems, the majority of the channels are analog and they too require a minimum of 17 dB C/I. The fundamental issue here is that the same D/R ratios can and are used when implementing the radio channel assignments for digital.

The additional parameters associated with IS-136/IS-54 involve SDCC, DCC, and DVCC. The DCC is the Digital Color Code, SDCC is the Supplementary Digital Color Code, and DVCC is the Digital Verification Color Code.

DCC and SDCC must be assigned to each sector, cell, or control channel of the system that utilizes IS-136/IS-54. The DCC is used by analog and dual-mode phones for accessing the system. The SDCC is used by dual-mode phones only and should be assigned to each control channel along with the DCC.

Parameter	Values
DCC	0, 1, 2, 3
SDCC	0–15

The DVCC is assigned to each DTC. A total of 255 different DVCC values exist, ranging from 1 to 255, leaving much room for variations in assignments.

3.6 IS-95 System Description

Code Division Multiple Access (CDMA), also known as IS-95 and J-STD-008, is a spread spectrum technology platform that enables multiple users to occupy the same radio channel, or frequency spectrum, at the same time. CDMA has and is being utilized for microwave point-to-point communication and satellite communication, as well as by the military. With CDMA, each of the subscribers, or users, utilize their own unique code to differentiate themselves from the other users. CDMA offers many unique features, including the capability to thwart interference and improved immunity to multipath affects due to its bandwidth. The IS-95 technology has been championed by many system operators in the United State and Asia.

IS-95 has two distinct versions, IS-95A and IS-95B, besides the J-STD-008. The J-STD-008 is compatible with both the IS-95A and B, with the exception of the frequency band of operation. However, the difference between IS-95A and IS-95B is that IS-95B enables ISDN-like data rates to exist. Although this would seem to be an interim step between 2G and 3G, for the purpose of this text the IS-95A and B are considered 2G only.

CDMA is based on the principal of *direct sequence* (DS) and is a wideband spread spectrum technology. The CDMA channel utilized is reused in every cell of the system and is differentiated by the *pseudorandom number* (PN) code that it utilizes. Depending on whether the system will be deployed in an existing AMPS or new PCS band system, the design concepts are fundamentally the same, with the exception of frequency band particulars that are directly applicable to the channel assignments in an existing cellular band. Beyond the nuances, the design principals for CDMA are the same for a cellular and PCS system.

The introduction of CDMA into an existing cellular network is not simple due to the issue of immediate capacity reduction, but with a long-term upside. Also, for PCS operators, a requirement specifies that they must relocate existing microwave links to clear the spectrum for their use. The degree of ease or difficulty for implementing CDMA into the PCS market will be directly impacted by the ability to clear microwave spectrum. The diagram shown in Figure 3-19 is a simplified version of the IS-95A/B architecture.

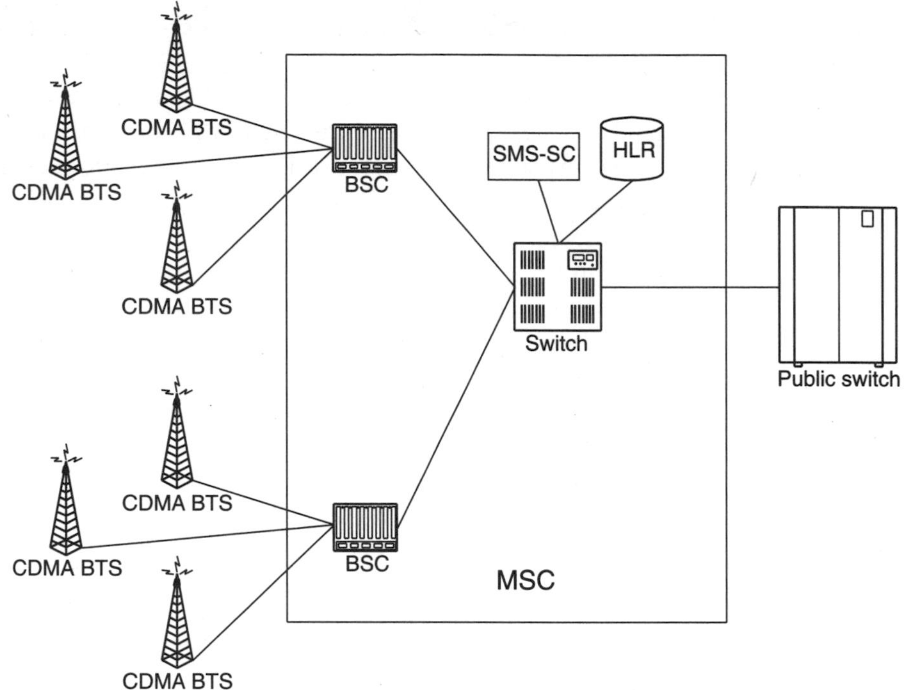

Figure 3-19 IS-95A/B simplified system architecture.

3.6.1 Standard CDMA Cell Site Configurations

Several general types of cell sites are currently usable at this time. The configuration is slightly different for both cellular and PCS due to colocation issues with the legacy systems. However, both cellular and PCS have the commonality of either being a omni or three-sector cell site; it is just the amount of antennas per sector that drive the difference.

It is important to note that the radio equipment for both cellular and PCS is fundamentally the same also. The difference between the two is that for PCS the frequency for transmitting and receiving is up-banded; that is, an additional mix is taking place. Typically, each cell or sector will require a separate transmit antenna per CMDA carrier per sector and two receive antennas. The reason for the separate transmit antennas per sector lies in the forward transmit power for the cell in that combing the channels either through use of a cavity or hybrid results in about a 3-dB loss.

The generic configurations that follow are meant for PCS and cellular CDMA-only cells and only a single sector, or omni site, is represented. The first configuration involves a PCS system deploying CDMA only in Figure 3-20.

Figure 3-20 depicts several situations that do occur for PCS operators. The first configuration is one that involves only a single carrier when three antennas can be installed on a per-sector or cell site basis. The second configuration is where, due to a multitude of reasons, only two antennas can be installed, thereby requiring the use of a duplexer. The third situation assumes that three antennas are used and shows how multiple carriers can be supported by three antennas.

Regarding cellular systems, initially the common use of the antennas at a cell site that had legacy 1G technology was promoted. However, after implementation, it was found that this might not have been the best choice. The reason for the error was the AMPS system and the CDMA system have different design requirements, and having the common antenna system restricts the flexibility of either system for optimization and expansion purposes.

Therefore, where possible, the use of a separate set of antennas for CDMA and AMPS systems is preferred. However, as the reader would surmise, the leasing, loading, roof space, and, of course, local ordinances may preclude this method of deployment.

Figure 3-21 illustrates a common situation when integrating 2G systems into a 1G environment. The first diagram shown represents the typical situation where only three antennas are available for use in a given sector, necessitating the use of duplexers. However, as discussed briefly, the sharing of antennas can lead to optimization problems because both systems

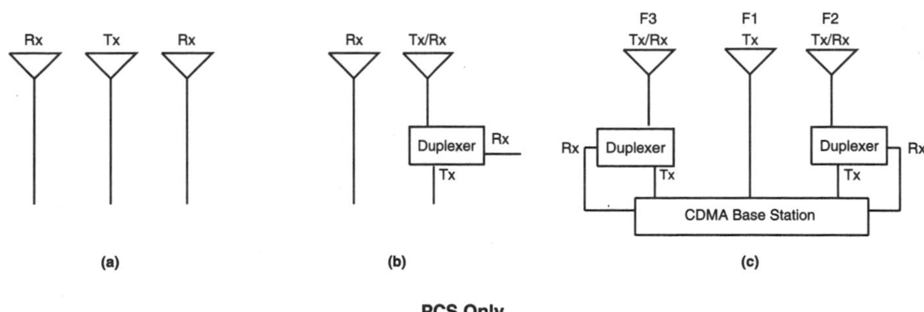

Figure 3-20 PCS system CDMA antenna configuration: (a) one carrier with three antennas, (b) single carrier with two antennas, (c) multiple carriers with three antennas

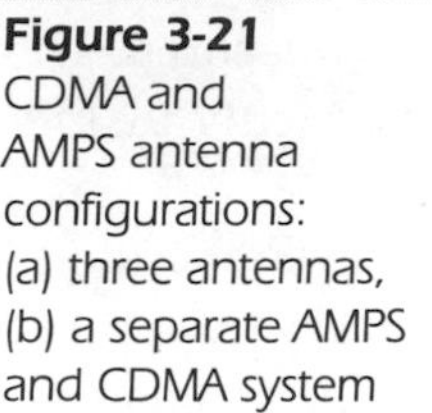

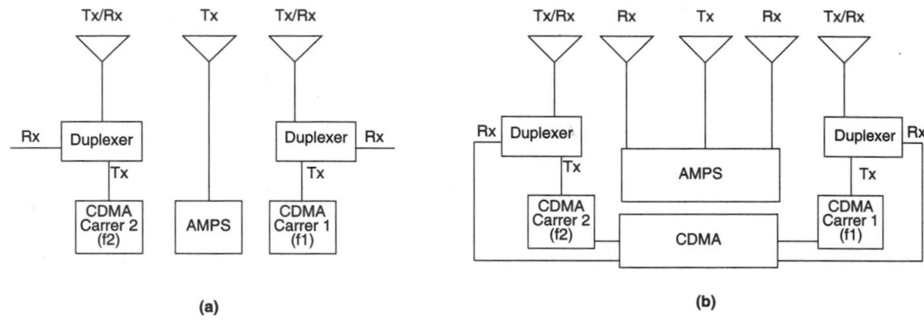

Figure 3-21 CDMA and AMPS antenna configurations: (a) three antennas, (b) a separate AMPS and CDMA system

have different design criteria. The second diagram shown in Figure 3-21 illustrates a configuration where the AMPS and CDMA systems share the same cell site location, but the systems utilize different antenna systems.

3.6.2 Pilot Channel Allocation

The location within the AMPS spectrum that the primary and secondary IS-95 pilot channels are supposed to operate at is shown in Figure 3-22, which is further clarified in Table 3-5, the CDMA channel designation channel table.

The CDMA channel assignment for cellular is defined as requiring the primary or secondary CDMA channel defined in the table to be utilized. The rational behind this issue lies in the initialization algorithm that is used for CDMA. Simply put, if the subscriber unit, dual mode, does not find a pilot channel on either the primary or secondary channel, then it reverts to an analog mode.

Figure 3-23 is a brief illustration of where a second CDMA carrier could be placed for, say, a B band operator. Specifically, the fact that a preferred channel is used enables the deployment of a second CDMA carrier that is more congenial for the operator. In this case, the second channel is planted next to the primary preferred channel and the guard band is now shifted up in frequency.

PCS, on the other hand, has a different set of preferred channels that are recommended. The initialization algorithm is simply when the subscriber powers up, it will search in its preferred block, A through F, for a pilot chan-

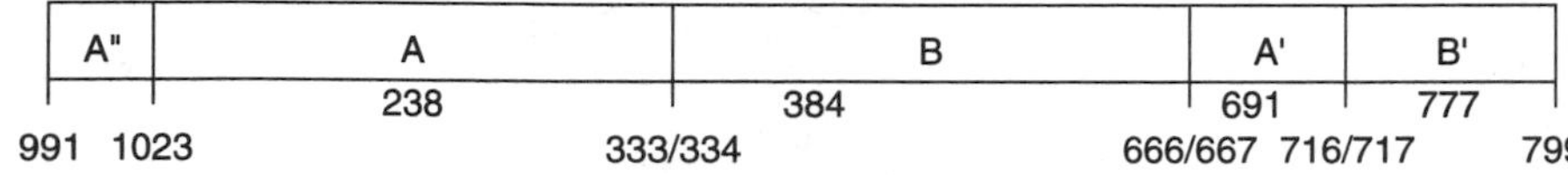

Figure 3-22 IS-95 pilot channel locations.

Table 3-5 CDMA Preferred Channels.

CDMA Channel Designation	A-Band	B-Band
Primary	238	384
Secondary	691	777

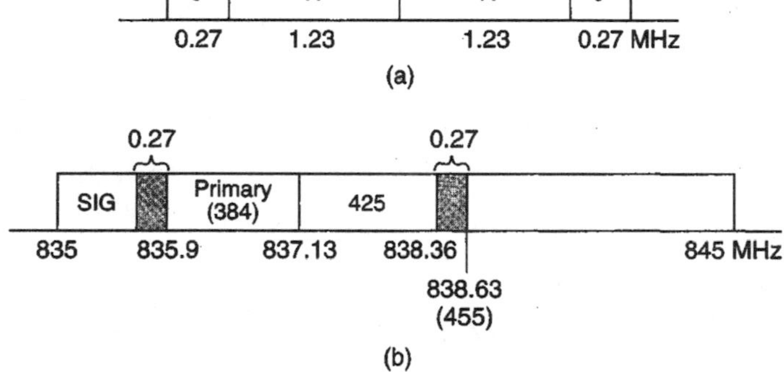

Figure 3-23 Multiple CDMA carriers.

nel using the preferred channel set located in Figure 3-24. The preferred channels are designated by the PCS operator that the subscriber has contracted mobile service from. The pilot channels can, like cellular, also exist in any of the valid ranges listed in the table.

Additionally, the comments listed as *conditionally valid* (cv), are based on the premise that the operator has control of the adjacent block of frequencies. The comments could also be based on the fact that both of the

Figure 3-24
PCS preferred pilot channel.

PCS block	CDMA channel no.	Valid CDMA assignment	Preferred set channel numbers
A (15 MHz)	0–24	NV	25, 50, 75, 100, 125, 150, 175, 200, 225, 250, 275
	25–275	V	
	276–299	CV	
D	300–324	CV	325, 350, 375
	325–375	V	
	376–399	CV	
B	400–424	CV	425, 450, 475, 500, 525, 550, 575, 600, 625, 650, 675
	425–675	V	
	676–699	CV	
E	700–724	CV	725, 750, 775
	725–775	V	
	776–799	CV	
F	800–824	CV	825, 850, 875
	825–875	V	
	876–899	CV	
C	900–924	CV	925, 950, 975, 1000, 1025, 1050, 1075, 1100, 1125, 1150, 1175
	925–1175	V	
	1176–1199	NV	

NV = not valid.
V = valid.
CV = conditionally valid.

adjacent blocks like C and F utilize CDMA technology, therefore eliminating the need for a guard band on each side of the allotted spectrum.

3.6.3 Forward CDMA Channel

The forward CDMA channel, shown in Figure 3-25, consists of the pilot channel, one sync channel, up to seven paging channels, and potentially 64 traffic channels. The cell site transmits the pilot and sync channels for the mobile to use when acquiring and synchronizing with the CDMA system. When this occurs, the mobile is in the mobile station initiation state. The paging channel also transmitted by the cell site is used by the subscriber unit to monitor and receive messages that might be sent to it during the mobile station idle state or system access state.

The pilot channel is continuously transmitted by the cell site. Each cell site utilizes a time offset for the pilot channel to uniquely identify the for-

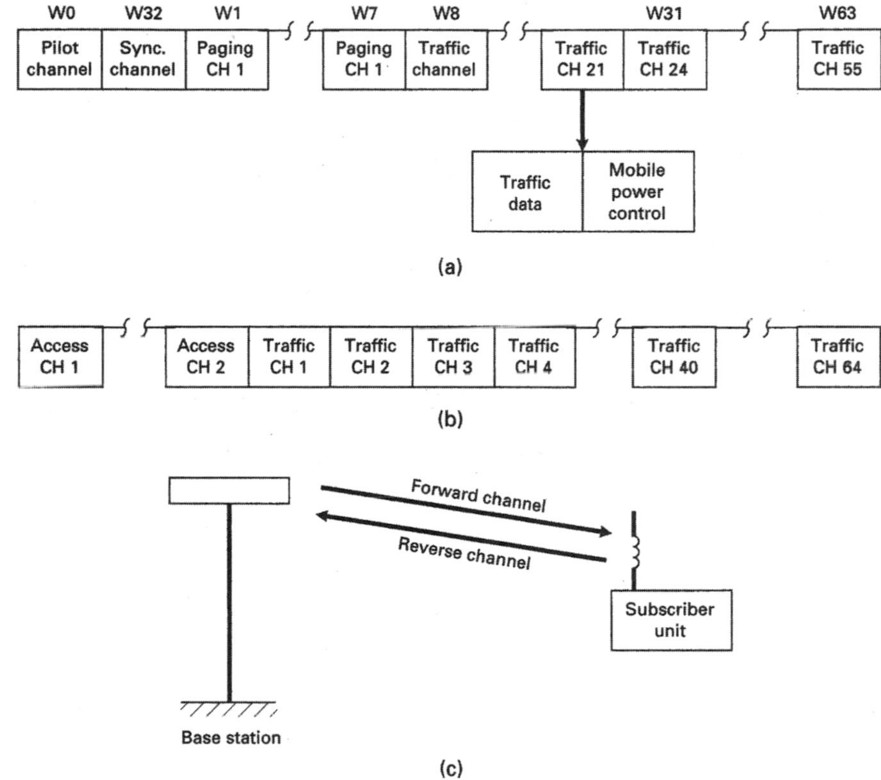

Figure 3-25 CDMA forward channel.

ward CDMA channel to the mobile unit. The cell site can utilize a possible 512 different time offset values. If multiple CDMA channels are assigned to a cell site, the cell will still utilize only one time offset value, which is utilized during the handoff process.

The sync channel is a forward channel that is used during the system acquisition phase. Once the mobile acquires the system, it will not normally reuse the synch channel until it powers on again. The sync channel provides the mobile with the timing and system configuration information. The sync channel utilizes the same spreading code, time offset, as the pilot channel for the same cell site. The sync channel frame is the same length as the pilot PN sequence. The information sent on the sync channel is the paging channel rate and the time of the base station's pilot PN sequence with respect to the system time.

The cell site utilizes the paging channel to send overhead information and subscriber-specific information. The cellsite will transmit at the minimum one paging channel for each supported CDMA channel that has a synch channel.

Once the mobile has obtained the paging information from the sync channel, the mobile will adjust its timing and begin monitoring the paging channel; each mobile, however, only monitors a single paging channel. The paging channel conveys four basic types of information. The first set of information conveyed by the paging channel is the overhead information. The overhead information conveys the system's configuration by sending the system and access parameter messages, the neighbor lists, and CDMA channel list messages.

Paging is another message type sent when a mobile unit is paged by the cell site for a land-to-mobile or mobile-to-mobile call. The channel assignment messages allow the base stations to assign a mobile to the traffic channel, alter the paging channel assignment, or redirect the mobile to utilize the analog FM system.

The forward traffic channel is used for the transmission of primary or signaling traffic to a specific subscriber unit during the duration of the call. The forward traffic channel also transmits the power control information on a subchannel continuously as part of the closed loop system. The forward traffic channel will also support the transmission of information at 9600, 4800, or 1200 bps, utilizing a variable rate that is selected on a frame-by-frame basis, but the modulation symbol rate remains constant.

3.6.4 Reverse CDMA Channel

The cell site contiguously monitors the reverse access channel to receive any message that the subscriber unit might send to the cell site during the system access state. The reverse CDMA channel consists of an access channel and the traffic channel. The access channel provides communication from the mobile to the cell site when the subscriber unit is not utilizing a traffic channel. One access channel is paired with a paging channel and each access channel has its own PN code. The mobile responds to the cell sites messages sent on the paging channel by utilizing the access channel.

The forward and reverse control channels utilize a similar control structure that can vary from 9600, 4800, 2400, or 1200 bps, which enables the cell or mobile to alter the channel rate dynamically to adjust for the speaker. When a pause occurs in the speech, the channel rate decreases so to reduce the amount of energy received by the CDMA system, increasing the overall system capacity.

Four basic types of control messages are used on the traffic channel. The four messages involve messages that will control the call itself, handoff messages, power control, security, and authentication.

CDMA power control is fundamentally different than that utilized for AMPS or IS-54. The primary difference is that the proper control of total power coming into the cell site, if limited properly, will increase the traffic-handling capability of that cell site. As more energy is received by the cell site, its traffic-handling capabilities will be reduced unless it is able to reduce the power coming into it.

The forward traffic power control is composed of two distinct parts. The first part is the cell site, which will estimate the forward links transmission loss, utilizing the mobile subscribers' received power during the access process. Based on he estimated forward link path loss, the cell site will adjust the initial digital gain for each of the traffic channels. The second part of the power control involves the cell site making periodic adjustments to the digital gain, which is done in concert with the subscriber unit.

The reverse traffic channel signals arriving at the cell site vary significantly and require a different algorithm to be used than that of the forward traffic power control. The reverse channel also has two distinct elements used for making power adjustments. The two elements are the open loop estimate of the transmit power, which is performed solely by the subscriber unit without any feedback from the cell site itself. The second element is the closed loop correction for these errors in the estimation of the transmit power. The power control subchannel is continuously transmitted on the forward traffic channel every 1.25 ms, instructing the mobile to either power up or power down, which affects the mean power output level. A total of 16 different power control positions are available.

Table 3-6 illustrates the CDMA subscriber power levels available by station class.

3.6.5 Call Processing

The call flows for 2G CDMA are shown next. It is important to note that 2G CDMA is primarily a voice and not a data-oriented system. However, data is available to be sent via circuit-switched methods, but the call processing flow is the same as voice since it still utilizes a traffic channel set up for voice transport. The first call-processing flow chart is for a mobile-to-land

Table 3-6
CDMA Subscriber Power Levels

Station Class	EIRP (max) dBm
I	3
II	0
III	−3
IV	−6
V	−9

call (origination), shown in Figure 3-26, while Figure 3-27 illustrates a land-to-mobile call (termination).

3.6.6 Handoffs

Several types of handoffs are available with CDMA. The types of handoffs involve soft, softer, and hard. The difference between the types is dependent upon what is trying to be accomplished.

Several user-adjustable parameters help the handoff process take place. The parameters that need to be determined involve the values to add or remove a pilot channel from the active list and the search window sizes. Several values determine when to add or remove a pilot from consideration. In addition, the size of the search window cannot be too small, nor can it be too large.

As mentioned previously, the handoff process for CDMA can take on several variants. Each of the handoff scenarios is a result of the particular system configuration and where the subscriber unit is in the network.

The hand-off process begins when a mobile detects a pilot signal that is significantly stronger than any of the forward traffic channels assigned to it. When the mobile detects the stronger pilot channel, the following sequence should take place. The subscriber unit sends a pilot strength measurement message to the base station, instructing it to initiate the handoff process. The cell site then sends a handoff direction message to the mobile unit, directing it to perform the handoff. Upon the execution of the handoff direction message, the mobile unit sends a handoff completion message on the new reverse traffic channel.

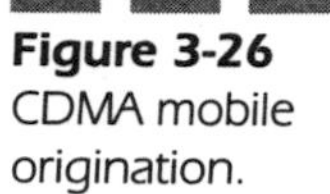

Figure 3-26 CDMA mobile origination.

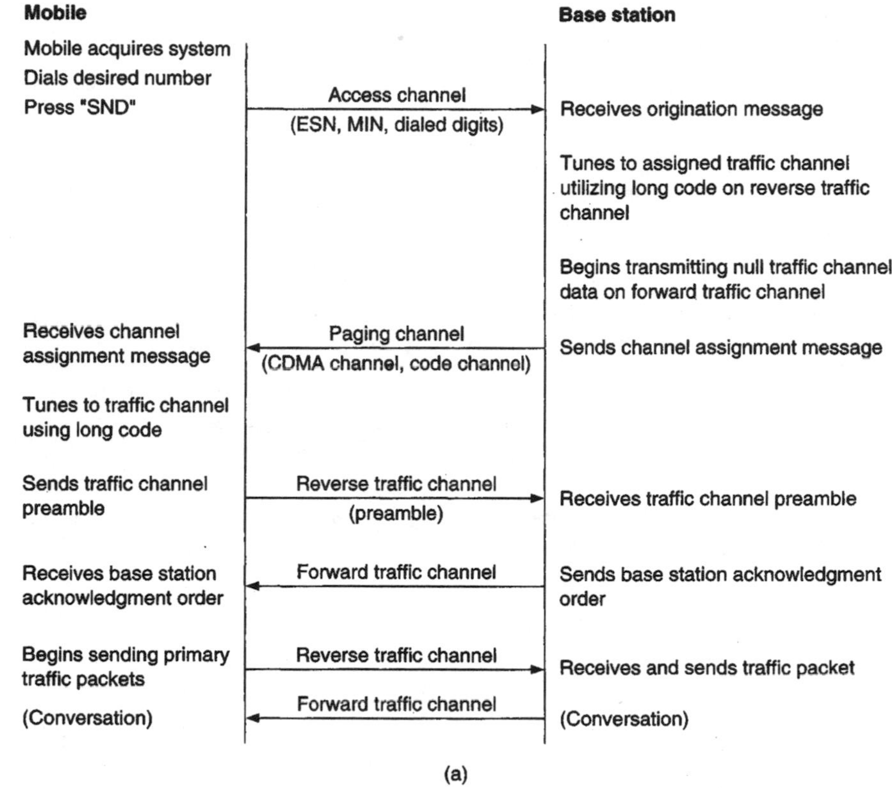

In CDMA, a soft handoff involves a inter-cell handoff and is a make-before-break connection. The connection between the subscriber unit and the cell site is maintained by several cell sites during the process. A soft handoff can only occur when the old and new cell sites are operating on the same CDMA frequency channel.

The advantage of the soft handoff is path diversity for the forward and reverse traffic channels. Diversity on the reverse traffic channel results in less power being required by the mobile unit, reducing the overall interference, which increases the traffic-handling capacity.

The CDMA softer handoff is an intracell handoff occurring between the sectors of a cell site and is a make-before-break type. The softer handoff occurs only at the serving cell site.

The hard handoff process is meant to enable a subscriber unit to handoff from a CDMA call to an analog call. The process is functionally a

Figure 3-27 CDMA mobile termination.

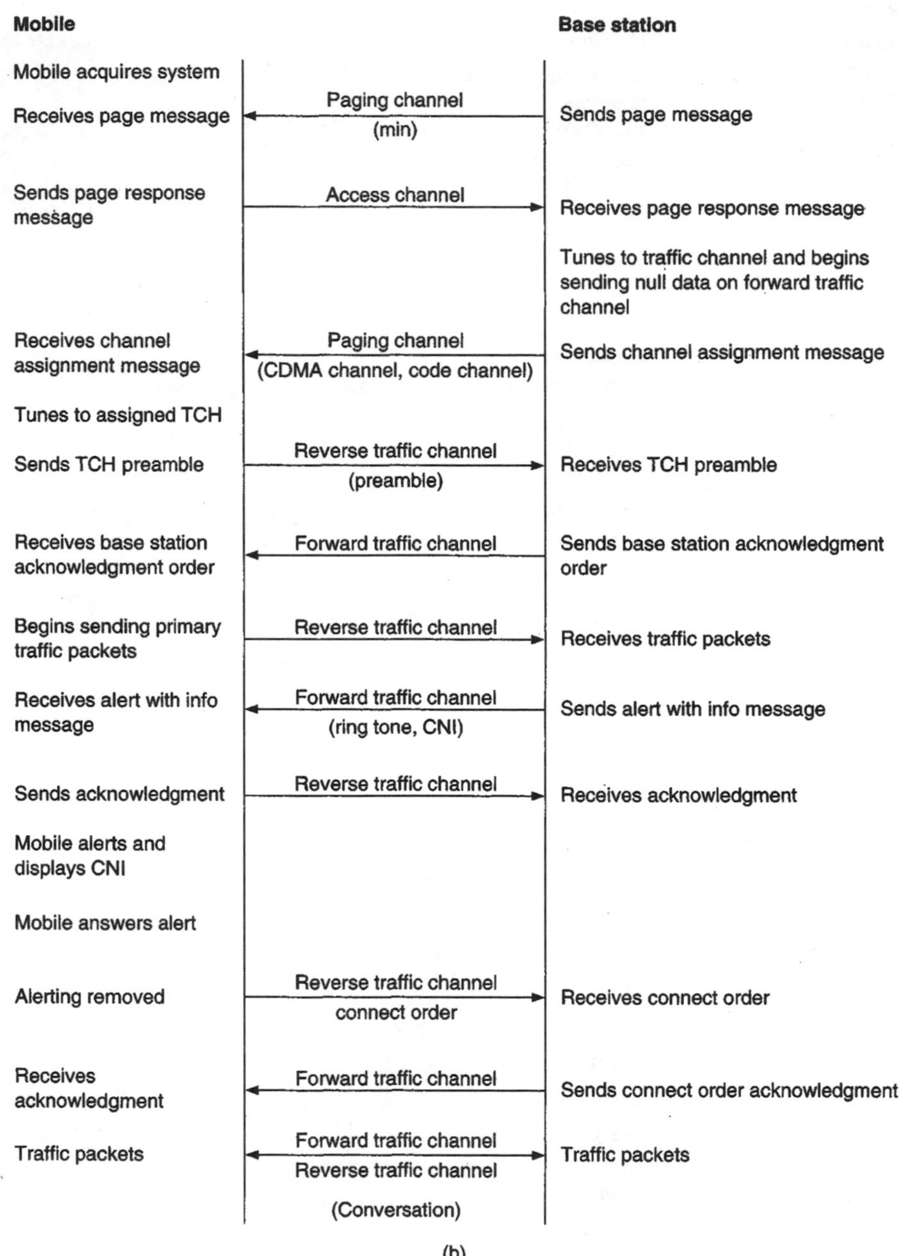
Mobile
Base station
Mobile acquires system
Receives page message
Paging channel
(min)
Sends page message
Sends page response message
Access channel
Receives page response message
Tunes to traffic channel and begins sending null data on forward traffic channel
Receives channel assignment message
Paging channel
(CDMA channel, code channel)
Sends channel assignment message
Tunes to assigned TCH
Sends TCH preamble
Reverse traffic channel
(preamble)
Receives TCH preamble
Receives base station acknowledgment order
Forward traffic channel
Sends base station acknowledgment order
Begins sending primary traffic packets
Reverse traffic channel
Receives traffic packets
Receives alert with info message
Forward traffic channel
(ring tone, CNI)
Sends alert with info message
Sends acknowledgment
Reverse traffic channel
Receives acknowledgment
Mobile alerts and displays CNI
Mobile answers alert
Alerting removed
Reverse traffic channel
connect order
Receives connect order
Receives acknowledgment
Forward traffic channel
Sends connect order acknowledgment
Traffic packets
Forward traffic channel
Reverse traffic channel
Traffic packets
(Conversation)

(b)

break-before-make and is implemented in areas where CDMA service is no longer available for the subscriber to utilize while on a current call. The continuity of the radio link is not maintained during the hard handoff.

A hard handoff can also occur between two distinct CDMA channels that are operating on different frequencies.

3.6.6.1 Search Window Several Search windows are used in CDMA. Each of the Search windows has its own role in the process and it is not uncommon to have different Search window sizes for each of the windows for a particular cell site. Additionally, the Search window for each site needs to be set based on actual system conditions; however, several system startup values are shown that can be used to get you in the ball park initially.

The Search windows needed to be determined for CDMA involve the Active, Neighbor, and Remaining windows. The Search window is defined as an amount of time, in terms of chips, that the CDMA subscriber's receiver will hunt for a pilot channel. A slight difference exists in how the receiver hunts for pilots depending on its type.

If the pilot is an Active Set, the receiver center for the Search window will track the pilot itself and adjust the center of the window to correspond to fading conditions. The other Search windows are set as defined sizes (see Table 3-7).

The size of the Search window is directly dependent upon the distance between the neighboring cell sites. How to determine what the correct Search window is for your situation can be extrapolated using the example shown in Figure 3-28 .

To determine the search window size, the following simple procedure is used:

1. Determine the distance between the sites A and B in chips.
2. Determine the maximum delay spread in chips.
3. Search window $\pm$ (cell spacing + maximum delay spread).

The Search window for the Neighbor and Remaining sets consists of parameters SRCH_WIN_N and SRCH_WIN_R, which represent the Search window sizes associated with the Neighbor Set and Remaining set pilots. The subscriber unit centers its Search window around the pilots' PN offset and compensates for time variants with its own time reference.

The SRCH_WIN_N should be set so that it encompasses the whole area in which a neighbor pilot can be added to the set. The largest the window should be set is 1.75 D + 3 chips, where D is the distance between the cells.

Table 3-7
Search Window Sizes

Search Window A,N,R	Window Size PN Chips
0	2
1	4
2	6
3	8
4	10
5	14
6	20
7	28
8	40
9	56
10	80
11	114
12	160
13	226
14	320
15	452

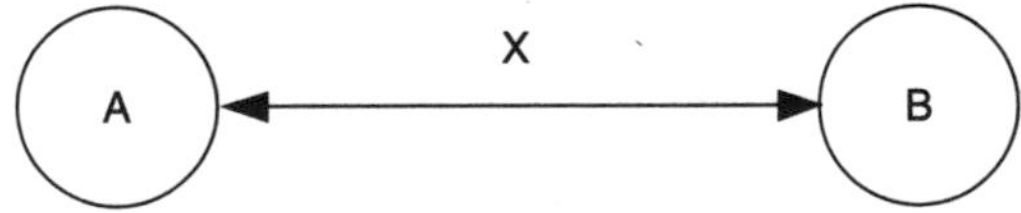

Figure 3-28
Search Window

X = 10 Chips

Therefore Search Window = +/− 10 chips

Search Window = 6 (20 chips)

SRCH_WIN_A is the value that is used by the subscriber unit to determine the Search window size for both the Active and Candidate sets. The difference between the Search window for the active and candidate sets and the neighbor and remaining sets is the Search window effectively floats with the active and candidate sets based on the first arriving pilot it demodulates.

3.6.6.2 Soft Handoffs Soft handoffs are an integral part of CDMA. The determination of which pilots will be used in the soft handoff process has a direct impact on the quality of the call and the capacity of the system. Therefore, setting the soft handoff parameters is a key element in the system design for CDMA.

The parameters associated with soft handoffs involve the determination of which pilots are in the Active, Candidate, Neighbor, and Remaining sets. The list of neighbor pilots is sent to the subscriber unit when it acquires the cell site or is assigned a traffic channel.

A brief description of each type of pilot set follows:

The *Active set* is the set of pilots associated with the forward traffic channels assigned to the subscriber unit. The Active set can contain more than one pilot because a total of three carriers, each with its own pilot, could be involved in a soft handoff process.

The *Candidate set* is made up of the pilots that the subscriber unit has reported are of a sufficient signal strength to be used. The subscriber unit also promotes the Neighbor set and Remaining set pilots that meet the criteria to the candidate set.

The *Neighbor set* is a list of the pilots that are not currently on the active or candidate pilot lists. The Neighbor set is identified by the base station via the Neighbor list and Neighbor list update messages.

The *Remaining set* consists of all possible pilots in the system that can possibly be used by the subscriber unit. However, the remaining set pilots that the subscriber unit looks for must be a multiple of the Pilot_Inc.

Figure 3-29 shows an example of a soft handoff region, which is an area between cells A and B. Naturally, as the subscriber unit travels farther away from cell A, cell B increases in signal strength for the pilot. When the pilot from cell B reaches a certain threshold, it is added to the active pilot list.

The process of how a pilot channel moves from a neighbor to a candidate, to active, and then back to neighbor is best depicted in Figure 3-30.

Here are the steps that a pilot channel takes:

1. Pilot exceeds T_ADD and the subscriber unit sends a *Pilot Strength Measurement Message* (PSMM) and a transfer pilot to the candidate set.
2. The base station sends an extended handoff direction message.

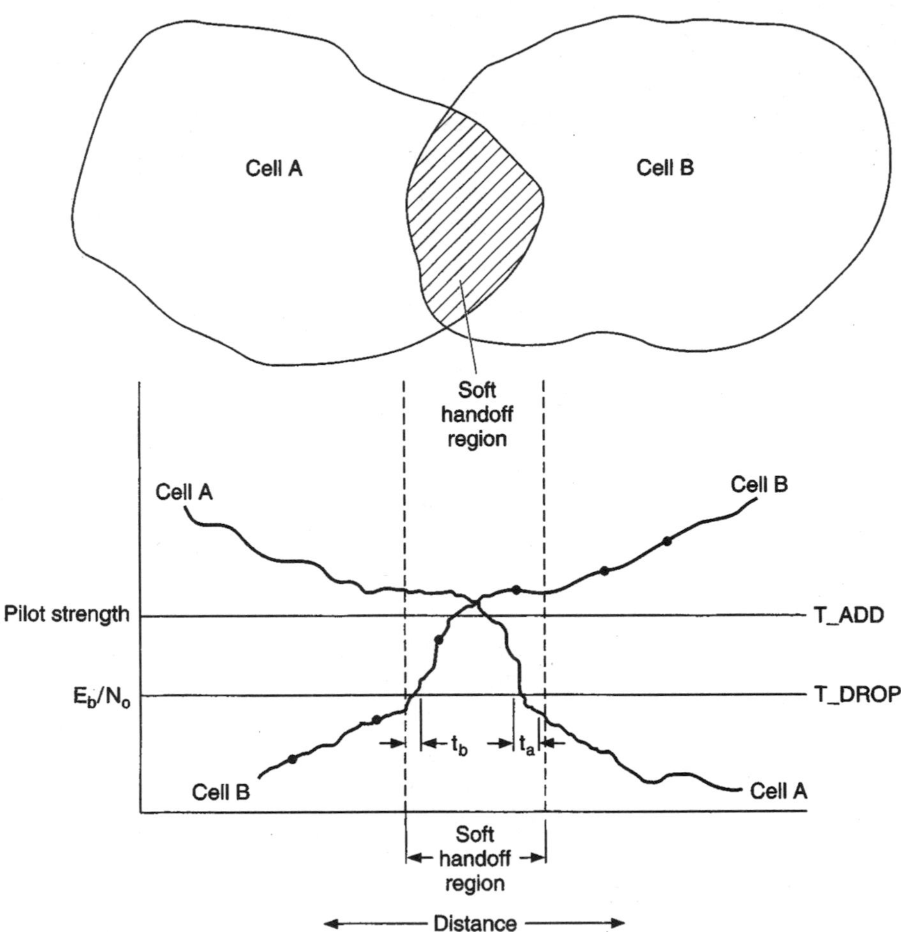

Figure 3-29
Soft handoff.

3. The subscriber unit transfers the pilot to active set and acknowledges this with a handoff completion message.
4. The pilot strength drops below T_DROP and the subscriber unit begins the handoff drop time.
5. The pilot strength goes above T_DROP prior to the handoff drop time expiring and the T_DROP sequences topping.
6. The pilot strength drops below T_DROP and the subscriber unit begins the handoff drop timer.

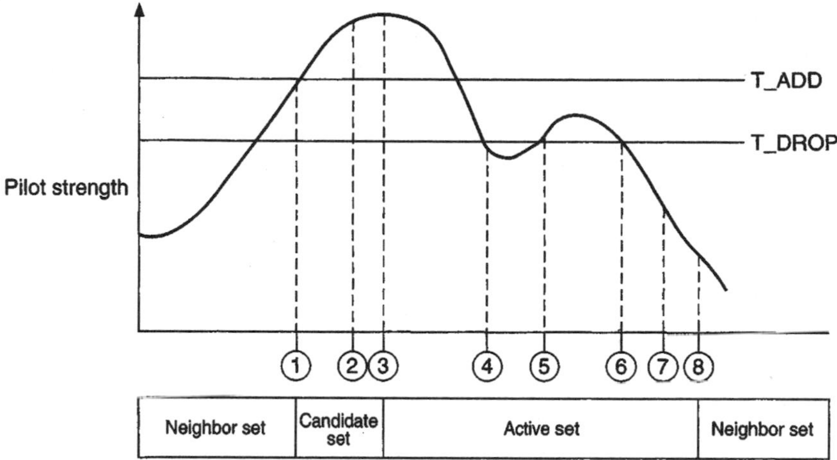

Figure 3-30 The pilot elevation and demotion process.

7. The handoff drop timer expires and the subscriber unit sends a PSMM.
8. The base station sends an extended handoff direction message.
9. The subscriber unit transfers the pilot from the Active set to the Neighbor set and acknowledges this with a handoff completion message.

To help augment the previous description, Figure 3-31 highlights how T_Comp is factored into the decision matrix for adding and removing pilots from the Neighbor, Candidate, and Active sets.

3.6.7 Pilot Channel PN Assignment

The pilot channel carries no data, but it is used by the subscriber unit to acquire the system and assist in the process of soft handoffs, synchronization, and channel estimation. A separate pilot channel is transmitted for each sector of the cell site. The pilot channel is uniquely identified by its PN offset or rather its PN short code that is used.

The PN sequence has some 32,768 chips that, when divided by 64, result in a total of 512 possible PN codes that are available for use. The fact that 512 potential PN short codes to pick from almost ensures that no problems will be associated with the assignment of these PN codes. However, some

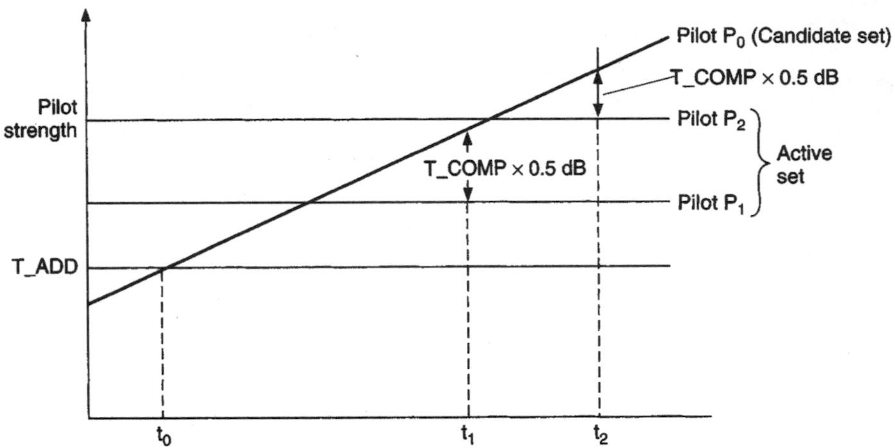

Figure 3-31 Active set.

simple rules must be followed in order to ensure that no problems are encountered with the selection of the PN codes for the cell and its surrounding cell sites.

$$(32768)/64 = 512 \, possible \, PN \, offsets$$

$$f_{chip} = 1.228 \times 10^6 \, Chips/s$$

$$Time = 1/f_{chip} = 0.8144 \, \mu s/chip$$

$$Distance = 244m/chip$$

Numerous perturbations exist for how to set the PN codes, but it is suggested that a reuse pattern be established for allocating the PN codes. The rational behind the establishment of a reuse pattern lies in the fact that it will facilitate the operation of the network for maintenance and growth. In addition, when adding a second carrier, the same PN code should be used for that sector.

Table 3-8 can be used for establishing the PN codes for any cell site in the network. The method that should be used is to determine whether you want to have a 4, 7 ,9, or 19 reuse pattern for the PN codes.

The suggested PN reuse pattern is an N = 19 pattern for a new PCS system, as shown in Figure 3-32. If you are overlaying the CDMA system on to

Table 3-8
PN Reuse Scheme

Sector	PN Code
Alpha	$3 \times P \times N - 2P$
Beta	$3 \times P \times N$
Gamma	$3 \times P \times N - P$
Omni	$3 \times P \times N$

Where N = reusing PN cell and P = PN code increment

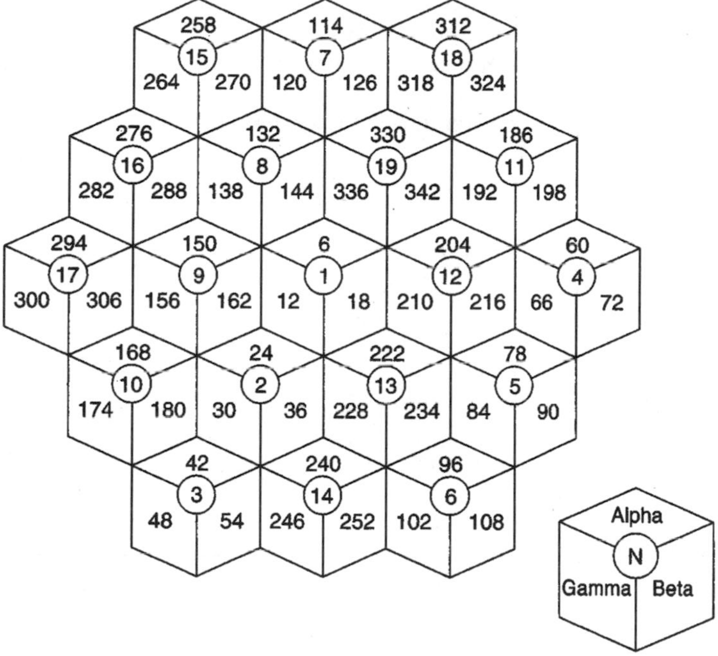

Figure 3-32
PN reuse pattern.

a cellular system, an N = 14 pattern should be used when the analog system utilizes an N = 7 voice channel reuse pattern.

Please note that not all the codes have been utilized in the N = 19 pattern. The remaining codes should be left in reserve for use when a PN code problem arises. In addition, a PN_INC value of 6 is also recommended for use.

The PN short code used by the pilot is an increment of 64 from the other PN codes and an offset value is defined. The Pilot_INC is the value that is used to determine the amount of chips or rather the phase shift that one pilot has versus another pilot.

The method that is used for calculating the PN offset is using the equations in the following example.

$$C/I = 10 \log_{10} \left(\frac{D(P, P_0)}{D(P, P_1)} \right)^{-3} \geq a$$

$$M \geq (R + S) \cdot (10^{a/(\alpha)10} - 1)$$

where M = offset
R = radius in chips
S = ½ Search window_A
a = C/I
α = attenuation factor, propagation exponent

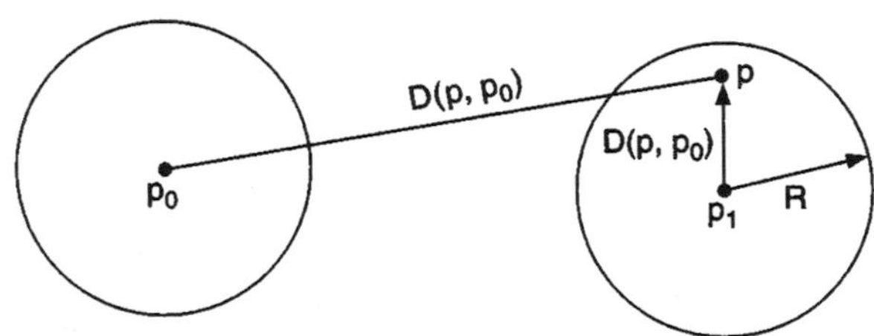

R, km	*R* (chips)	*S*	*C/I*	*m* (chips)	Pilot_INC	No. of offsets
25	103	14	24	622	10	50
20	82	12	24	499	8	64
15	61	12	24	390	6	85
12.5	51	10	24	325	5	102
10	41	10	24	271	4	128
7	29	10	24	207	4	128
5	21	10	24	165	3	170
3	12	10	24	117	2	256
2.5	10	10	24	106	2	256
2	8	10	24	96	2	256

Pilot_INC is valid from the range of 0 to 15. Pilot_INC is the PN sequence offset index and is a multiple of 64 chips. The subscriber unit uses the Pilot_INC to determine which are the valid pilots to be scanned. Included in the example is a simple table that can be used to determine the Pilot_INC as a function of the distance between reusing sites.

3.6.8 Link Budget

The Link Budget calculations directly influence the performance of the CDMA system since it is used to determine power settings and capacity limits for the network. Proper selection of the variables that comprise the link budget is a very obvious issue.

Two links are used: forward and reverse. The forward and reverse links utilize different coding and modulation formats. The first step in the link budget process is to determine the forward and the reverse links' maximum path losses. The forward links' maximum path loss is determined using Table 3-9a.

The data gathered shows that the maximum path loss sustainable is about −159.6 dB using the parameters selected. The reverse link calculations are shown in Table 3-9b.

The maximum path loss that is sustainable in the reverse direction is 139.dB, which shows that the base station is reverse link limited for the parameters inputted into the link budget.

3.6.9 Traffic Model

The capacity for a CDMA cell site is driven by several issues. The first and most obvious point for traffic modeling in a CDMA cell site involves how many channel cards the cell site is configured with. A total of 55 possible traffic channels are available for use at a CDMA cell site, but unless the channel cards are installed, the full potential is not realizable utilizing IS-95/J-STD-008 specifications.

Additionally, the other factor that fits into the traffic calculations for the site involves system noise. A simple relationship exists between system noise and the capacity of the cell site. Typically, the load of the cell site design is somewhere in the vicinity of 40 to 50 percent of the pole capacity, with a maximum of 75 percent.

Table 3-9 a

Forward Link Budget

Forward Link Budget	14.4 Kbps	Value (units)		Comment
Tx Power Distribution				
	Tx PA Power	39.0 dBm		8 Watts
	Pilot Channel Power	30.8 dBm	15.0%	% of Max Power per Channel
	Synch Channel Power	20.8 dBm	10.0%	% pilot power
	Paging Channel Power	26.2 dBm	35.1%	% pilot power
	Traffic Channel Power	**38.0 dBm**	78.2%	% of Max Power per channel
	Number Mobiles per Carrier	13		
	Soft/Softer Handoff Traffic	13		1.85 overhead factor
	Max # of Active Traffic Chs	26		
	Avg Traffic Channel Pwr	**23.8 dBm**		26 Total Traffic Channels
	Voice Activity Factor	0.479		voice = 0.479, data =1.0
	Peak Traffic Channel Pwr	27.0 dBm		Avg Traffic Ch Pwr/Voice Activity Factor
Base Station				
	Traffic Channel Tx Pwr	27.0 dBm		
	Duplexer Loss	0.5 dB		
	Jumper and Connector Loss	0.25 dB		
	Lightening Arrestor Loss	0.25 dB		
	Feedline Loss	1 dB		
	Jumper and Connector Loss	0.25 dB		
	Tower Top Amp Loss	0 dB		
	Antenna Gain	15 dBd		
	Net Base Station Tx Pwr	**39.8 dBm**		10 Watts ERP per Traffic Channel(voice)
	Total Base Station Tx Power	51.8 dBm		151 Watts ERP per carrier
Environmental	Fade Margin	5 dB		Log Normal
	Penetration Loss	10 dB		(street/vehicle/building)
	Cell Overlap	3 dB		
	External Losses	**−18 dB**		
Subscriber	Antenna gain	0 dBd		
	Cable Loss	2 dB		
	Rx Noise Figure	10 dB		

Table 3-9 a (cont.)

Forward Link Budget

	Reciever Noise Density	−174	dBm/Hz		
	Information Rate	60.90	dB	1230 Kbps	
	Rx Sensitivity	−101.1	dBm		
Subscriber Traffic Channel RSSI	Base Tx	39.8			
	Environmental Loss	−18			
	Max Path Loss	139.77			obtained from Uplink Path Analysis
	RSSI at Sub Antenna	−118.01			
Subscriber Total RSSI	Base Tx	51.8			
	Environmental Loss	−18			
	Max Path Loss	139.77			obtained from Uplink Path Analysis
	Total RSSI at Sub Antenna	−105.99			
Interference					
Internal Interference	Orthogonality Factor	−8	dB		0.16 same sector interference
	Total RSSI at Sub Antenna	−105.99			
	Other User Interference Level	−113.99			Orthoginal Factor × (RSSI total)
	Other Sector Interference	4	dB		
	Interference Density	−109.99			
Total Interference	Internal Interference	−109.99			
	external interference	−117	dBm		depends on local environment
	Total Interfernce on TCH	−107.95			external interference + Rx sensitivity +other user interference
RSSI	Mobile TCH RSSI	−118.01			
	Information Rate	41.58	dB	14.4	
	Traffic Channel Eb	−159.59			
	Total RSSI	−107.95			
	Information Rate	60.90	dB	1230	
	Traffic Channel No	−168.85			
Eb/No	Traffic Channel Eb	−159.59			
	Traffic Channel No	−168.85			
	Eb/No	9.26			

Table 3-9 b

Reverse Link Budget

Reverse Link Budget	14.4 Kbps	Value (units)		Comment
Subscriber Terminal	Tx Power	23 dBm		maximum power per traffic channel
	Cable Loss	2 dB		
	Antenna gain	0 dBd		
	Tx Power per Traffic Channel	21 dBm		
External Factors	Fade Margin	5 dB		Log Normal
	Penetration Loss	10 dB		(street/vehicle/building)
	External Losses	−15 dB		
Base Station	Rx Antenna Gain	15 dBd		(approx 17.25 dBi)
	Tower Top Amp Net Gain	0		
	Jumper and Connector Loss	0.25 dB		
	Feedline Loss	1 dB		
	Lightening Arrestor Loss	0.25		
	Jumper and Connector Loss	0.25		
	Duplexer Loss	0.5		
	Receive Configuration Loss	0		
	Handoff Gain	4 dB		
	Rx Diversity Gain	0 dB		
	Rx Noise Figure	5 dB		
	Receiver Interference Margin	3.4 dB	55% pole	
	Reciever Noise Density	−174 dBm/Hz		
	Information Rate	41.58 dB	14.4	
	Rx Sensitivity	−124.0 dBm		
	Eb/No	7 dB		
	Total Base Station	−140.77 dBm		
Eb/No	Eb/No	7.00 dB		
	Maximum Path Loss	139.77 dB		

The third major element in determining the capacity of a CDMA cell is the soft handoff factor. Since CDMA relies on soft handoffs as part of the fundamental design for the network, this must also be factored into the usable capacity at the site. The reason for factoring soft handoffs into capacity is that if 33 percent of the calls are in a soft handoff mode, then this will require more channel elements to be installed at the neighboring cell sites to keep the capacity at the desired levels.

With CDMA, the capacity of the site is dynamic because as the system noise floor is raised, the base station loading decreases. The specific capacity for any CDMA base station is typically achieved through computer simulation due to the dynamics of cell loading and interference levels, making a pure traffic calculation on a spreadsheet rather impractical. However, some rules of thumb should be followed for simple planning exercises that do not require a computer simulation to run.

As stated earlier, a total of 64 Walsh codes are available. Typically, the Walsh codes are allocated in the following manner:

Channel Type	Number of Walsh Codes
Pilot	1
Synch	1
Paging	1–7
Traffic channels	55

The pole capacity for CDMA is the theoretical maximum number of simultaneous users that can coexist on a single CDMA carrier. However, at the pole, the system will become unstable, and therefore operating at less than 100 percent of the pole capacity is the desired method of operation.

The effective traffic channels for a CDMA carrier are the number of CDMA traffic channels needed to handle the expected traffic load. However, since soft handoffs are an integral part of CDMA, they also need to be included in the calculation for capacity. In addition to each traffic channel that is assigned for the site, a corresponding piece of hardware is needed at the cell site also.

The actual traffic channels for a cell site are determined using the following equation:

$$\textit{Actual traffic channels} = (\textit{Effective traffic channels'} + \textit{soft handoff channels})$$

The maximum capacity for a CDMA cell site should be 75 percent of the pole, but typical loading in IS-95 systems has found that the pole point is really around 50 percent.

The physical limit for a CDMA system's capacity is dictated by the mutual interference driven by the forward channel. Therefore, the number of users that can be placed onto an CDMA system at any time is limited by mutual interference, which is directly related to power.

$$\begin{aligned} P\,(pole\ point) &= g/(a \times d \times [1 + B]) + 1 \\ a &= Voice\ activity\ factor \\ d &= Required\ Eb/No \\ g &= Processing\ gain \\ B &= Other\ cell/sector\ interference\ factor \end{aligned}$$

Looking at the pole point equation, it is obvious that it is unique for every site since it is dependant upon the local situation at that site. Additionally, due to the Eb/No factor, the cell can be allowed to degrade, allowing for the soft capacity factor, which of course impacts the pole point, leading to more dynamics and the need for computer simulation.

However, assuming the 50 percent pole point the following Erlangs of offered traffic, using Erlang B, can be derived for an individual CDMA carrier is shown in Figure 3-10.

The *Channel Elements* (CEs) are a pooled resource, and therefore equipping a full compliment of CEs for all sectors to be used simultaneously is not a practical approach. Instead it is typically recommended that only 95 percent of the CE's estimate be installed for the cell.

When more than one carrier is in a sector, the capacity can be estimated. In Table 3-10, it is assumed that the sector has two carriers; if more are in that sector, then it is a matter of multiplication to arrive at the new traffic levels since no trunking efficiency exists between CDMA carriers.

3.7 iDEN (Integrated Dispatch Enhanced Network)

The iDEN system is a unique wireless access platform because it involves integrating several mobile phone technologies together, which is based on a

Table 3-10
Channel Elements

Blocking Rate	Offered Traffic	CE Required per Sector	CE required per Cell (3 sector)
1%	7.35	14	40
2%	7.4	13	38
3%	7.48	12	35
5%	7.63	11	32
10%	8.06	10	29

Blocking Rate	# of Carriers	Offered Traffic-Erlangs	CE's Required per sector	CE Required per Cell (3 sector)
1%	2	14.7	28	80
2%	2	14.8	26	74
3%	2	14.96	24	69

modified GSM platform. The services that are integrated into iDEN involve a dispatch system, full-duplex telephone interconnections, data transport, and short messaging services.

The dispatch system involves a feature called group call, where multiple people can engage in a conference. The user list is preprogrammed and the conference call can be set up just like it is done in two-way or specialized mobile radio (SMR) with the exception that the connection can take place utilizing any of the frequencies that are available from the pool of channels where the subscriber is physically located.

The telephone interconnect and data transport are meant to offer conventional mobile communications. The short messaging service enables the iDEN phones to receive up to 140 characters for an alphanumeric message. An example of a typical iDEN system is shown in Figure 3-33.

The elements that comprise the iDEN system, as shown in Figure 3-33, are briefly listed here:

DAP—Dispatch Application Processor

EBTS—Enhanced Base Transceiver

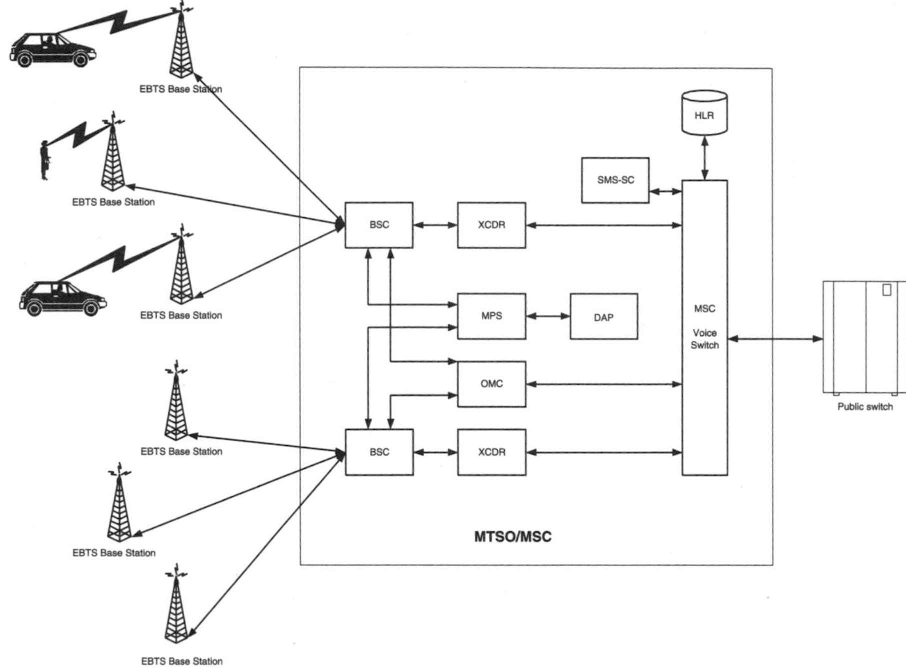

Figure 3-33 iDEN system architecture.

HLR—Home Location Register

MPS—Metro Packet Switch

MSC—Mobile Switching Center

OMC—Operations and Maintenance Center

SMS-SC—Short Message Service-Service Center

XCDR—Transcoder

In review of Figure 3-33, there are several differences with an iDEN system as compared to a typical mobile wireless system. iDEN is unique in wireless mobility because it combines both interconnect as well as dispatch services in the same wireless system. The two distinct systems, interconnect and dispatch, are effectively overlaid on top of each other but are integrated and share some common elements like the EBTS radio.

The BSC is responsible for traffic and control channel allocations in addition to handover data collection and controlling handovers between other BSCs.

The *Metro Packet Switch* (MPS) provides the connectivity for the dispatch calls. It also distributes the dispatch packets as well as the ISMI assignment.

The *Dispatch Application Processor* (DAP) is the processing entity responsible for overall coordination and control of the dispatch services. The DAP enables the following types of calls to take place:

- Talk group
- Private call
- Call alert

The radio access system used by an iDEN system is TDMA. The channel bandwidth is 25 kHz, which consists of four independent side bands, each being a 16QAM baseband signal. The center frequencies of these side bands are 4.5 kHz apart from each other, and they are spaced symmetrically about a suppressed RF carrier frequency, resulting in a 16-point data symbol constellation that carries four data bits per symbol. The location where iDEN is utilized in the spectrum is shown in Figure 3-34. The RF channel structure is shown in Figure 3-35.

iDEN was introduced using a 6:1 interleave for both dispatch and interconnect services. Later the system was upgraded, enabling a 3:1 interleave for interconnect-only service.

The wireless operator has the choice of offering 6:1 or 3:1 voice service in addition to dispatch. Capacity is affected by the selection of which interconnection method is used and the amount of dispatch traffic that is carried on a system. Looking at a simplistic example, the 3:1 voice call requires two TCHs, while a 6:1 or dispatch call requires only a single TCH. Of course, other issues related to signaling and call quality are factored into this.

iDEN utilizes several control channels similar in nature to GSM systems. The control channels used by iDEN systems are listed here for reference. In addition to the control channels, two other channels are used in iDEN; they are the TCH and PCH, also listed.

- **PCCH** The *primary control channel* is a multiple access channel used for the transmission of general system parameters. The outbound PCCH contains the *broadcast control channel* (BCCH) and the *common control channel* (CCCH), whereas the inbound PCCH is referred to as the *random access channel* (RCCH):
  - **Inbound** Service requests
  - **Outbound** Service grants
  - **BCCH**

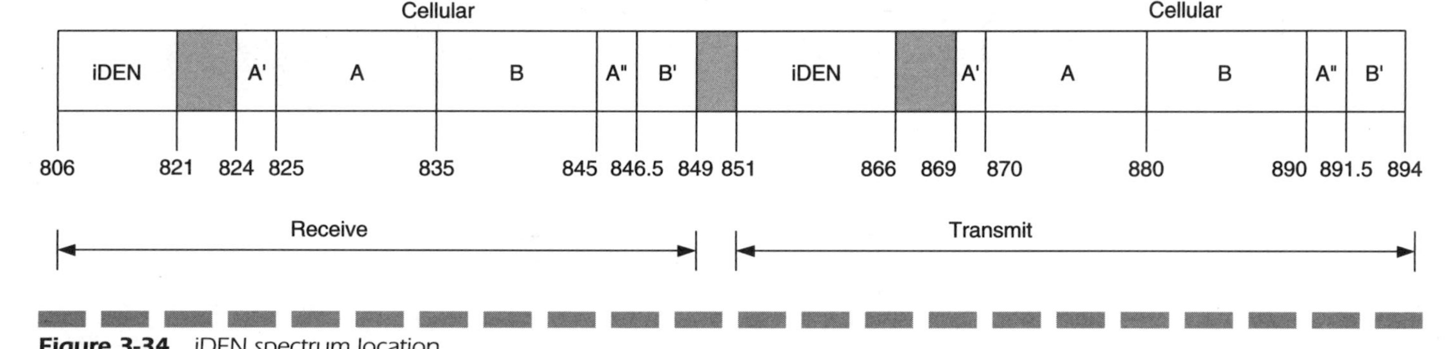

Figure 3-34 *iDEN spectrum location.*

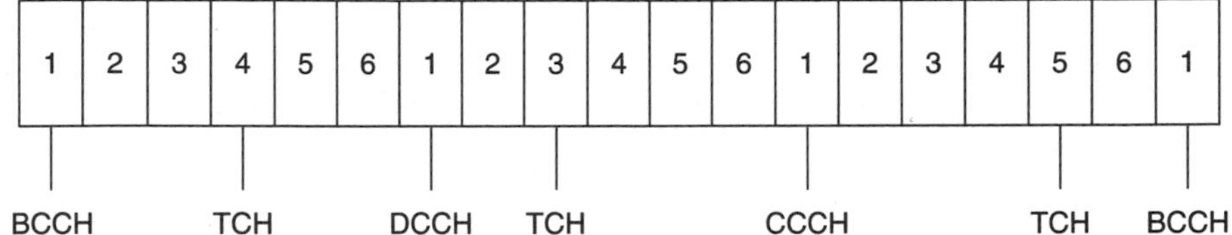

Figure 3-35 iDEN RF channel structure.

  - Neighbor cells
  - Control channels
  - Packet channels
  - Location areas
  - Common control channel
  - Paging subchannel
  - Service grants
- **TCCH** Temporary control channel
  - **Inbound** Dispatch reassignment requests
  - **Outbound** Handover target
- **DCCH** Dedicated control channel
  - **Inbound** Location updating
  - **Authentication**
  - **SMS**
  - **Registration**
  - **Outbound**
  - **Location updating**
  - **Authentication**
  - **SMS**
  - **Registration**
- **ACCH** Associated control channel
- **TCH** Traffic channel that provides circuit mode transmission for voice and data
  - **Inbound** Dispatch reassignment requests
  - **Outbound** Handover target
- **PCH** Packet channel provides for multi-access packet mode transmission

Many interesting issues are associated with the iDEN call processing for either dispatch or interconnection calls. From the time an iDEN mobile subscriber is initially powered up until it is powered down, a series of procedures are executed between the EBTS and the MOBILE to control the radio communications link. Before a call description flow chart is shown, a few terms or processes used in iDEN systems associated with the mobile need to be briefly covered.

- **Cell selection** At power up, the mobile scans a pre-programmed list of system frequencies called a bandmap looking for a PCCH. When the mobile hears a PCCH, outbound power and *Signal Quality Estimate* (SQE) measurements are taken and the frequency is added to a list. The mobile continues scanning channels until either 32 PCCHs are found or until the bandmap list is exhausted, which is market-specific. The PCCH list is sorted based on SQE and RSSI, and the subscriber then attempts to camp on the first cell on the list. If it fails, it will attempt to camp on the next cell, until it either succeeds in the camping or exhausts the list, requiring a new cell selection process to begin.
- **Cell reselection** Each serving cell will transmit its neighbor cell list to all the subscribers it serves and the MOBILE will take SQE measurements of the received power of the serving cell and of each neighbor cell. It will then sort the neighbor cell list according to received signal strength. When the mobile determines that the best neighbor cell is a better candidate for a serving cell than the current serving cell, a reselection occurs, making the formerly best neighbor cell the new serving cell.
- **Fast reconnect** During the duration of a dispatch call, the mobile continues to monitor the SQE and signal strength of the serving and neighbor cells. Under certain conditions, the mobile may decide to change its serving cell.

  When the mobile is on the traffic channel (during the talk phase of a call), the mobile initiates a reconnect if the serving cell's outbound SQE is less than desired or upon the failure or disconnect of the serving cell.
- **Power control** The mobile periodically adjusts its transmit power based on the power received at the FNE. The mobile periodically receives a power control constant and measures the serving cell's output power. The mobile then calculates the desired mobile transmit power by subtracting the serving cell output power from the power control constant and adjusts its transmit power accordingly.

- **Handoff** IDEN utilizes MAHO to assist in the handoff process. The handoff can either be mobile- or base station-initiated depending on the parameter settings. Handoffs are only possible with interconnection calls. However, for a dispatch, the location information supplied in the response also includes the neighbor list from cells that are on the beacon channel list. Therefore, if the DLA is set up incorrectly, it is possible that the subscriber will need to reacquire the system if it moves outside of the coverage area of the sites in the list.

  The mobile-assisted handover (MAHO) process is as follows:

1. The mobile monitors information on BCCH as to which cells to monitor for inclusion in MAHO list.
2. The mobile continues to monitor SQE, the *Receive Signal Strength Indicator* (RSSI) for the primary serving channel, and the channels in the MAHO list.
3. If the subscriber detects trouble in the primary service or a better neighbor cell, the mobiles sends a sample of its measurements.
4. The subscriber signals in the ACCH with an SQE measurement.
5. MSC/BSC/EBTS finds a new server to handover to and allocates a TCH for this process.
6. MSC/BSC/EBTS senses a handover command on ACCH with the initial power setting, channel, and TCH to tune to.
7. MS changes to an assigned channel.
8. MS uses the *random access procedure* (RAP) to get its timing information from the target EBTS.
9. The channel changes to TCH and conversation continues.

Lastly, SQE is used extensively in various cell site selection decisions and is based primarily on the outbound RSSI measurements of the serving cell as well as for neighboring cells which are potential handover candidates. SQE is very similar to C/(I + N) in the range of 15 to 23 dB. The Dispatch system involves the key components of the iDEN system (see Figure 3-36).

The Dispatch system basically has three primary service offering or functions:

- Private
- Talk Group
- Call Alert (twiddle)

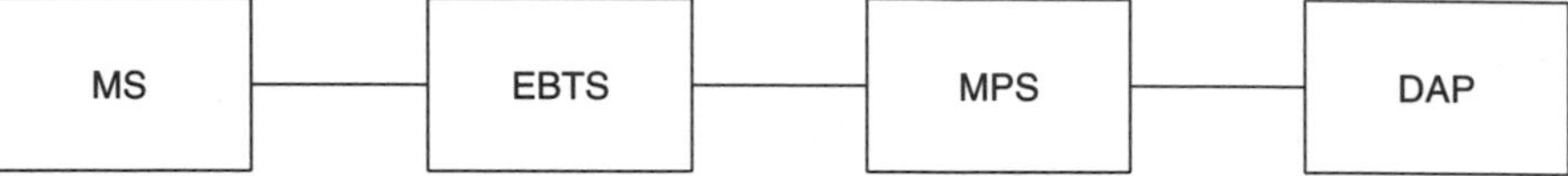

Figure 3-36
Dispatch only.

Whereas Private Dispatch is where the originating call uses PTT between one subscriber unit and another (classic two way), Call Alert is used to notify a subscriber that a voice communication is desired. However, talk groups involve a more extensive look.

Service areas (SAs) define talk groups, shown in Figure 3-39. The SA is used for dispatch group calls. When a dispatch call takes place, a single voice channel slot is used in any coverage area for a cell when one or more members of the call group are in that coverage area. Fleets are assigned to the same group and a mobile can be grouped into several talk groups in order to communicate between specific groups that comprise the entire fleet.

As briefly stated, a mobile can be grouped into several talk groups used to communicate with a group of mobiles in the fleet at the same time or all of the mobiles. For example, let's say there is a fleet for all of N.Y. City, but the subscriber only wants to talk with the Queens fleet. The mobile for the Queens fleet is assigned their own talk group, which is part of the overall fleet group. In doing so, a mobile can be part of numerous talk groups.

To help clarify, or further confuse the situation, a call-flow diagram for dispatch calls is shown in Figure 3-37.

Looking at the flow chart in Figure 3-37, the following text better explains some of the sequences:

1. *Push to Talk* (PTT) dispatches a call request.
2. The call request packet is routed to the DAP.
3. The DAP recognizes subscriber units' group affiliation and tracks the group members' current location area.
4. DAP sends a location request to each group member location area to obtain the various subscribers' cell/sector location information.
5. The subscriber units in the group responds with their current cell/sector location information.
6. The DAP instructs the originating EBTS with packet routing information for all group members.

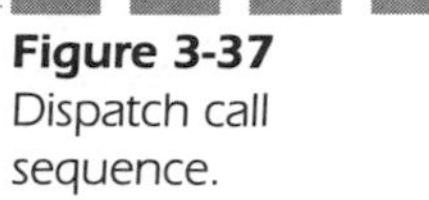

Figure 3-37 Dispatch call sequence.

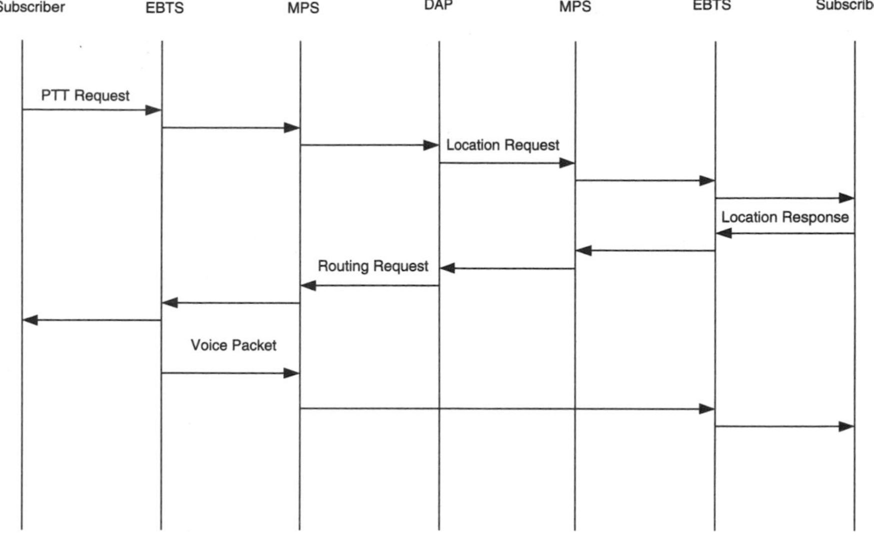

7. Call voice packets are received by the PD and then are replicated and distributed to the group's end node.

For interconnections, another portion of the iDEN system is utilized after the radio access. The general sequence of events for an interconnection call is the same, whether it is for a 3:1 or 6:1 call, with the exception of the amount of TCHs assigned.

Therefore, the interconnection sequence for a mobile-to-land call is listed here in brevity:

1. Call initiation
2. RAP on PCCH
3. DCCH assigned
4. Authentication
5. Call setup transaction
6. TCH assignment
7. Conversation
8. Call termination request via ACCH
9. Call is released

Figure 3-38 is a call flow diagram for a mobile land interconnection call sequence that should help bring the components together. It is interesting to note the differences between the interconnection call diagram and those for the dispatch sequence.

The interaction of sharing resources for radio access for both interconnections and dispatches involves the establishment of dispatch and interconnection location areas, referred to as DLA and ILA. The DLA and ILA are usually designed independently but have interactions that require joint considerations to be made for the selection of both the DLA and ILA bound-

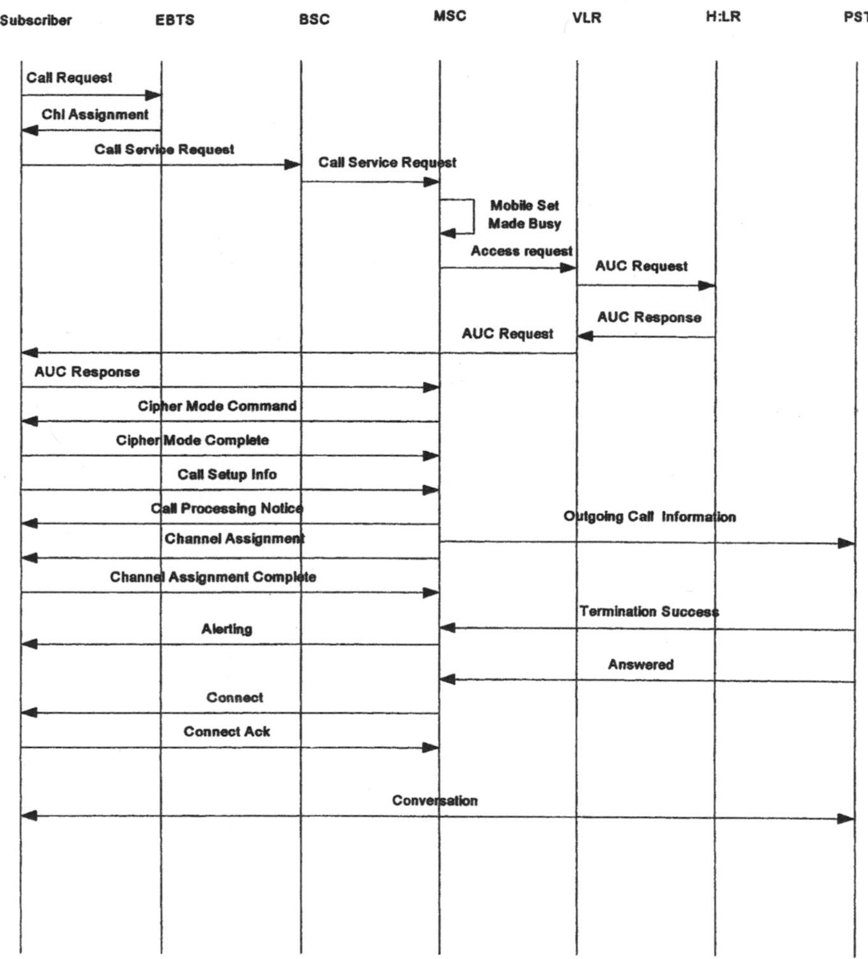

Figure 3-38 M-L interconnection call flow diagram.

aries. The DLA and ILA boundaries are in addition to BSC boundaries; however, the ILA or DLA needs to be inclusive of the EBTSs, which are connected to a the BSC.

An example of a DLA boundary is shown in Figure 3-39, which shows a total of four location areas associated with dispatch. Each location area is then folded into an SA. Keeping in mind the dispatch discussion regarding SAs, the design engineer must take care not only during the selection of location areas, but in what constitutes the service area. The location area is where the dispatch call is broadcast when the service area defines which location areas are possible for inclusion in the dispatch call.

Figure 3-40 is the corollary to the DLA boundaries and shows the *Interconnection Location Areas* (ILAs) for the same sample system. The ILA is used for call delivery and paging for the subscriber unit. The ILA boundaries should not be set up such that the subscriber units regularly transition from one ILA to another, increasing the amount of overhead signaling required to keep track of the mobile.

In looking at Figures 3-39 and 3-40, the differences between the ILA and DLA boundaries become evident. Next, Figure 3-41 shows the composite view of both ILA and DLA boundaries.

Figure 3-39 Dispatch location areas (DLAs).

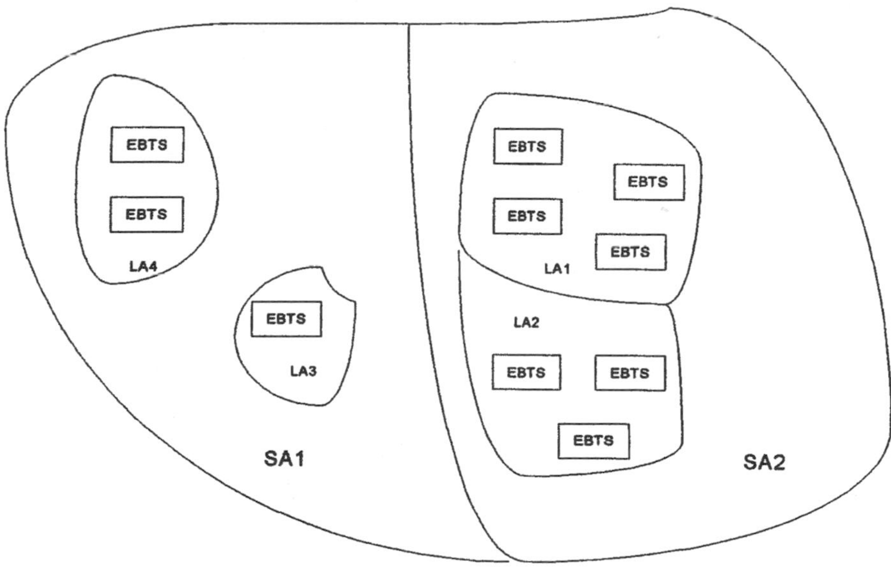

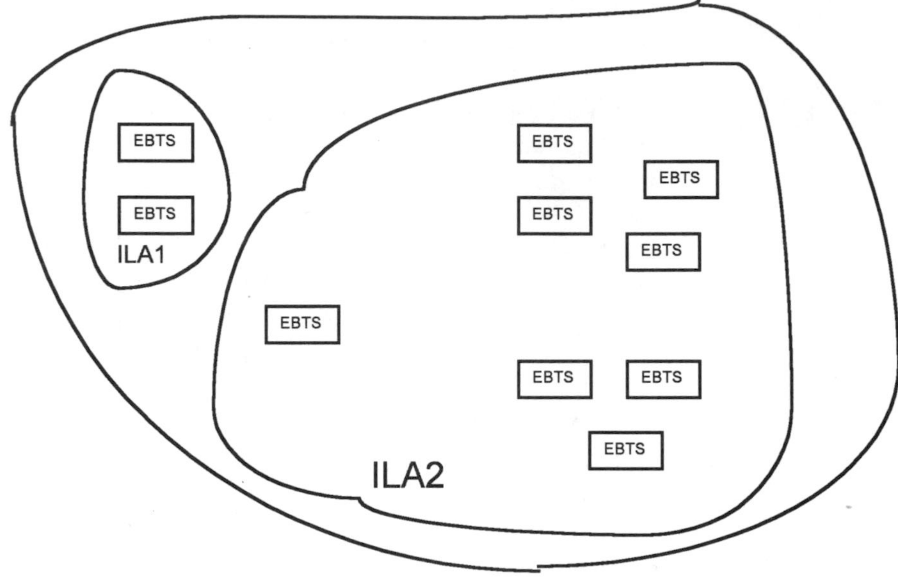

Figure 3-40
Interconnection Location Areas (ILAs).

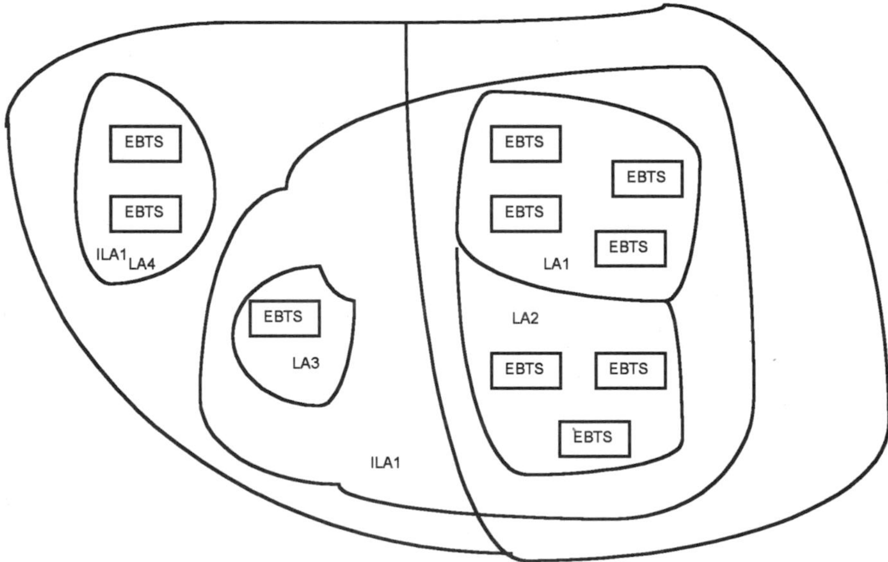

Figure 3-41
The ILA and DLA composite view.

3.8 CDPD

CDPD is a packetized data service utilizing its own air interface standard that is utilized by the cellular operators. CDPD is functionally a separate data communication service that physically shares the cell site and cellular spectrum.

CDPD has many applications but are most applicable for short, bursty-type data applications and not large file transfers. CDPD application of the short messages would consist of e-mail, telemetry applications, credit card validation, and global positioning, to mention a few potentials. CDPD is a pure data service designed for mobility; however, it cannot, nor was it ever designed to, supply data speeds needed for 3G services.

CDPD does not establish a direct connection between the host and server locations. Instead it relies on the OSI model for packet-switching data communications, and the model routes the packet data throughout the network. The CDPD network has various layers that comprise the system. Layer 1 is the physical layer, layer 2 is the data link itself, and layer 3 is the network portion of the architecture. CDPD utilizes an open architecture and has incorporated authentication and encryption technology into its airlink standard.

The CDPD system consists of several major components, and a block diagram of a CDPD system is shown in Figure 3-42.

The *Mobile End System* (MES) is a portable wireless computing device that moves around the CDPD network, communicating to the MDBS. The MES is typically a laptop computer or other personal data device that has a cellular modem.

The *Mobile Data Base Station* (MDBS) resides in the cell site itself and can utilize some of the same infrastructure that the cellular system does for transmitting and receiving packet data. The MDBS acts as the interface between the MES and the MDIS. One MDBS can control several physical radio channels, depending on the site's configuration and loading requirements. The MDBS communicates to the MDIS via a 56-Kbps data link. Often the data link between the MDBS and MDIS utilizes the same facilities as that for the cellular system, it but occupies a dedicated time slot.

The *Mobile Data Intermediate System* (MDIS) performs all the routing functions for CDPD. The MDIS performs the routing tasks utilizing the knowledge of where the MES is physically located within the network itself. Several MDISs can be networked together to expand a CDPD network.

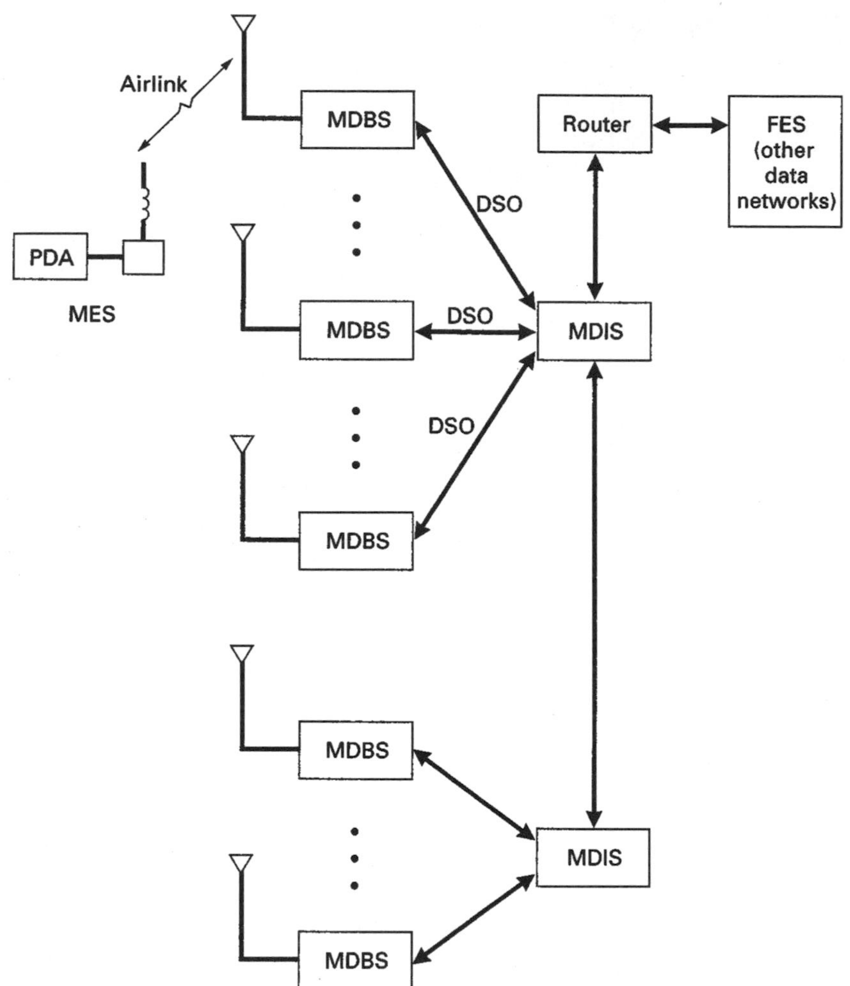

Figure 3-42
CDPD.

The MDIS also is connected to a router or gateway, which connects the MDIS to a *Fixed End System* (FES). The FES is a communication system that handles layer-4 transport functions and other higher layers.

The CDPD system utilizes a *Gaussian minimum-shift keying* (GMSK) method of modulation and is able to transfer packetized data at a rate of 19.2 Kbps over the 30-kHz-wide cellular channel. The frequency assignments for CDPD can take on two distinct forms. The first form of frequency assignment is a method of dedicating specific cellular radio channels to be utilized by the CDPD network for delivering the data service. The other method of frequency assignment for CDPD is to utilize channel hopping where the CDPD's *Mobile Data Base Station* (MDBS) utilizes unused channels for delivering its packets of data. Both methods of frequency assignments have advantages and disadvantages.

Utilizing a dedicated channel assignment for CDPD has the advantage of the CDPD system not interfering with the cellular system it is sharing the spectrum with. By enabling the CDPD system to operate on its own set of dedicated channels, no real interaction takes place between the packet data network and the cellular voice network. However, the dedicated channel method reduces the overall capacity of the network and, depending on the system loading conditions, this might not be a viable alternative.

If the method of channel hopping is utilized for CDPD, and this is part of the CDPD specification, the MDBS for that cell or sector will utilize idle channels for the transmission and reception of data packets. In the event the channel that is being used for packet data is assigned by the cellular system for a voice communication call, the CDPD MDBS detects the channel's assignment and instructs the *Mobile End System* (MES) to retune to another channel before it interferes with the cellular channel. The MDBS utilizes a scanning receiver or sniffer, which scans all the channels it is programmed to scan to determine which channels are idle or in use.

The disadvantage of the channel hopping method involves the potential interference problem to the cellular system. Coexisting on the same channels with the cellular system can create mobile-to-base-station interference. This kind of interference occurs because of the different handoff boundaries for CDPD and cellular for the same physical channel. The difference in handoff boundaries is due largely to the fact that CDPD utilizes a BER for handoff determination and the cellular system utilizes RSSI at either the cell site, analog, or MAHO for digital.

3.9 Summary

This chapter covered numerous radio access platforms that were built to improve the efficiency of mobility systems offering voice services. The advent of the Internet during the time that these services were beginning to be deployed has resulted in a desire to have a wireless mobility system capable of handling high-speed data traffic. However, as was the case with migrating from 1G to 2G, the path to 3G is not straightforward. It is hoped that the inclusion of the 2G systems will facilitate the introduction of 3G systems and the interim platforms that are currently being deployed, which are referred to as 2.5G.

References

AT&T. "*Engineering and Operations in the Bell System*," 2nd Ed., AT&T Bell Laboratories, Murry Hill, N.J., 1983.

Barron, Tim. "*Wireless Links for PCS and Cellular Networks*," Cellular Integration, Sept. 1995, pgs. 20–23.

DeRose. "*The Wireless Data Handbook*," Quantum Publishing Inc., Mendocino, CA, 1994.

Dixon. "*Spread Spectrum Systems*," 2nd Ed., John Wiley & Sons, New York, NY, 1984.

Harte, Hoenig and Kikta McLaughlin. "*CDMA IS-95 for Cellular and PCS*," McGraw-Hill, 1996.

Jakes, W.C. "*Microwave Mobile Communications*," IEEE Press, New York, NY, 1974.

Johnson, R.C. and Jasik, H. "*Antenna Engineering Handbook*," 2nd Ed., McGraw-Hill, New York, NY, 1984.

Kaufman, M., and A.H. Seidman. "*Handbook of Electronics Calculations*," 2nd Ed., McGraw-Hill, New York, NY, 1988.

Lee, W.C.Y. "*Mobile Cellular Telecommunications Systems*," 2nd Ed., McGraw-Hill, New York, NY, 1996.

Lynch, Dick. "*Developing a Cellular/PCS National Seamless Network*," Cellular Integration, Sept. 1995, pgs. 24–26.

MacDonald. "*The Cellular Concept*," Bell Systems Technical Journal, Vol. 58, No. 1, 1979.

Newton, Harry. "*Newton's Telcom Dictionary*," 14th Ed., Flatiron Publishing, 1998.

Pautet, Mouly. "*The GSM System for Mobile Communications*," Mouly Pautet, 1992.

Qualcomm. "*An Overview of the Application of Code Division Multiple Access (CDMA) to Digital Cellular Systems and Personal Cellular Networks*," Qualcomm, San Diego, CA, May 21, 1992.

Rappaport. "*Wireless Communications Principals and Practices*," IEEE, 1996.

"*Reference Data for Radio Engineers*," Sams, 6th Ed., 1983.

Smith, Clint. "*Practical Cellular and PCS Design*," McGraw-Hill, 1997.

Smith, Clint. "*Wireless Telecom FAQ*," McGraw-Hill, 2000.

Smith, Gervelis. "*Cellular System Design and Optimization*," McGraw-Hill, 1996.

Steele. "*Mobile Radio Communications*," IEEE, 1992.

GSM 01.02 Digital cellular telecommunications system (Phase 2+); General description of a GSM Public Land Mobile Network (PLMN)

GSM 02.09 Digital cellular telecommunications system (Phase 2+); Security aspects

GSM 02.17 Digital cellular telecommunications system (Phase 2+); Subscriber identity modules functional characteristics

GSM 03.01 Digital cellular telecommunications system (Phase 2+); Network functions

GSM 03.03 Digital cellular telecommunications system (Phase 2+); Numbering, addressing, and identification

GSM 03.18 Digital cellular telecommunications system (Phase 2+); Basic call handling; Technical realization

GSM 03.20 Digital cellular telecommunications system (Phase 2+); Security-related network functions

GSM 04.02 Digital cellular telecommunications system (Phase 2+); GSM Public Land Mobile Network (PLMN) access reference configuration

GSM 04.03 Digital cellular telecommunications system (Phase 2+); Mobile Station-Base Station System (MS-BSS) interface Channel structures and access capabilities

GSM 04.07 Digital cellular telecommunications system (Phase 2+); Mobile radio interface signalling layer 3 general aspects

GSM 04.08 Digital cellular telecommunications system (Phase 2+); Mobile radio interface layer 3 specification

GSM 05.02 Digital cellular telecommunications system (Phase 2+); Multiplexing and multiple access on the radio path

GSM 05.03 Digital cellular telecommunications system (Phase 2+); Channel coding

GSM 05.04 Digital cellular telecommunications system (Phase 2+); Modulation

GSM 05.05 Digital cellular telecommunications system (Phase 2+); Radio transmission and reception

GSM 05.08 Digital cellular telecommunications system (Phase 2+); Radio subsystem link control

GSM 09.02 Digital cellular telecommunications system (Phase 2+); Mobile Application Part (MAP) specification

CHAPTER 4

Third Generation (3G) Overview

4.1 Introduction

The rapid increase in the demand for data services, primarily IP, has been thrust upon the wireless industry. Over the years there has been much anticipation of the onslaught of data services, but the radio access platforms have been the inhibitor from making this a reality. Third generation (3G) is a term that has received and continues to receive much attention as the enabler for high-speed data for the wireless mobility market. 3G and all it is meant to be are defined in the ITU specification *International Mobile Telecommunications-2000* (IMT-2000). IMT-2000 is a radio and network access specification defining several methods or technology platforms that meet the overall goals of the specification. The IMT-2000 specification is meant to be a unifying specification, enabling mobile and some fixed high-speed data services to use one or several radio channels with fixed network platforms for delivering the services envisioned:

- Global standard
- Compatibility of service within IMT-2000 and other fixed networks
- High quality
- Worldwide common frequency band
- Small terminals for worldwide use
- Worldwide roaming capability
- Multimedia application services and terminals
- Improved spectrum efficiency
- Flexibility for evolution to the next generation of wireless systems
- High-speed packet data rates
    - 2 Mbps for fixed environment
    - 384 Mbps for pedestrian
    - 144 Kbps for vehicular traffic

Figure 4-1 shows the linkage between the various platforms that comprise the IMT-2000 specification group.

The definition of what exactly 3G encompasses is usually clouded in marketing terms, with the technical reader desiring a straightforward answer. The reason 3G is hard to pin down is primarily due to the fact that it involves radio access and network platforms that do not exist right now. The standard that everyone is striving for is IMT-2000 and it incorporates

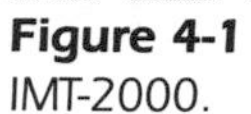

Figure 4-1
IMT-2000.

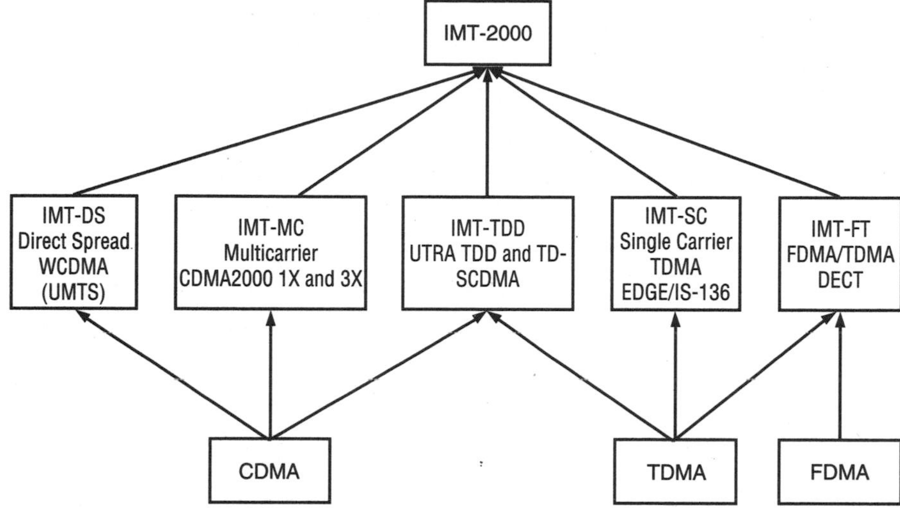

several competing radio access platforms, which will not achieve harmonization, if ever, until 4G or beyond. The radio access platforms that comprise the IMT-2000 specification are all different and it should be no wonder that it is difficult to obtain a simple answer when asked to describe what a 3G system will look like.

IMT2000/3G can be described as:

- Being used to reference a multitude of technologies covering many frequency bands, channel bandwidths, and, of course, modulation formats.
- No single 3G-infrastructure platform, technology, or application exists.
- 3G is applied to mobile and stationary wireless applications involving high-speed data. IMT-2000 mandates data speeds of 144 Kbps at driving speeds, 384 Kbps for outside stationary use or walking speeds, and 2 Mbps for indoors.

Coupled with the different platforms that comprise the IMT-2000 standard is the issue that the existing 1G/2G platforms need to transition into the 3G arena. The transition method that an operator must select and spend currency on is, of course, a difficult decision and will determine how successful the wireless operator will be in the future. The interim platform that bridges the 2G systems into a 3G environment is referred to as 2.5G.

Table 4-1 attempts to group some of the major technology platforms by Wireless Generation.

What follows is a brief visualization of the interaction between the major 1G, 2G, 2.5G, and 3G platforms. Obviously, if an operator chooses to implement more than one technology platform for marketing and strategic reasons, then the lines of transition become more complicated than those shown in Figure 4-2.

3G is a mobile radio and network access scheme that enables high-speed data to be utilized, allowing for true multimedia capabilities in a mobile wireless system. Presently, voice has been the primary wireless application

Table 4-1

3G

Wireless Generation	Systems	General Service	Comments
First (1G)	AMPS, TACS, NMT	Voice	Traditional Analog cellular deployment scheme
Second (2G)	GSM, TDMA, CDMA	Primarily voice with SMS	Digital Modulation Scheme implemented
			Deployment in 800, 900, 1800, and 1900 MHz bands Spectrum clearing required for 1900 MHz in U.S. Spectrum refarming required for existing 1G operators to implement 2G systems
Transition (2.5G)	CDMA,GPRS, EDGE	Primarily voice with packet data services being introduced	Overlay approach used except in new spectrum
			Packet Data enhancements to existing 2G operators
Third (3G)	CDMA2000/ WCDMA	Packet Data and Voice services Designed for high-speed multimedia data and voice True 3G platforms expected 2003–2005	Defined by IMT-2000 Europe (UMTS –WCDMA) America (UMTS / CDMA2000) Asia (UMTS / CDMA2000) Overlay Approach for existing operators of 2/2.5G networks

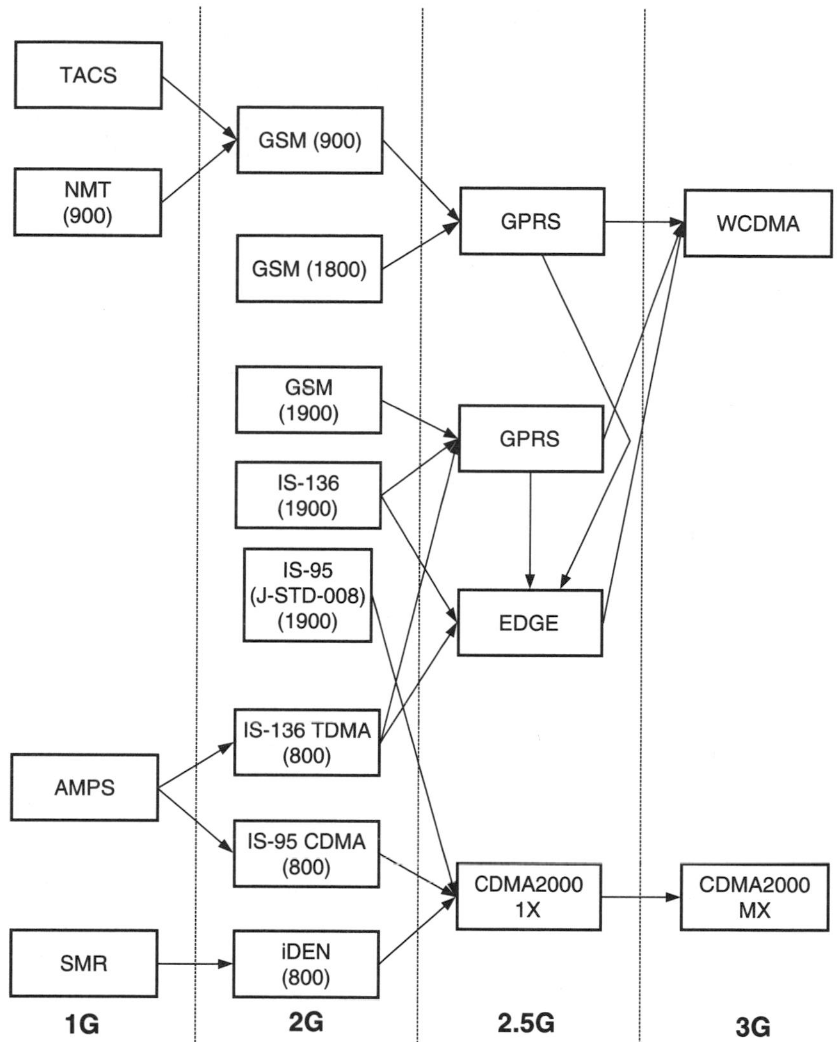

Figure 4-2
Migration path.

with the use of the short message service (SMS) being the largest packet data service.

Today's wireless cellular and *personal communications services* (PCS) systems have the same radio bandwidth allocated for both voice and data. Some of the 2.5G transition or migration plans call for the use of a dedicated spectrum just for data applications. The IMT-2000 specifies that data speeds of 144 Kbps for vehicular, 384K for pedestrian, and 2 Mbps for indoor applications are the desired goals and have been built into the specifications.

Table 4-2 is a brief grouping of the various major technology platforms and the data speeds that are associated with each.

In examining Table 4-2, it is apparent that several of the IMT-2000 platform standards are not included and that is on purpose. The platforms that are listed in both *Wideband Code Division Multiple Access* (WCDMA) and CDMA2000 are the two 3G platforms that will be discussed in some level of detail for the remainder of this textbook. The reason for the two-platform focus lies in the primary issue that a vast majority of wireless operators, both existing and new, are planning to utilize one of these two standards, which are part of the IMT-2000 specification.

Table 4-2

2.5G and 3G Comparison

2G Technology	**Data Capability**	**Spectrum Required**	**Comment**
GSM	9.6 Kbps or 14.4 Kbps	200 kHz	Circuit Switched data
IS-136	9.6 Kbps	30 kHz	Circuit Switched data
IDEN	9.6 Kbps	25 kHz	Circuit Switched data
CDMA (IS-95A/J-STD-008)	9.6 bps/14.4 Kbps 64bps (IS-95B)	1.25 MHz	Circuit Switched data
2.5G Technology	**Data Capability**	**Spectrum Required**	**Comment**
HSCSD	28.8/56 Kbps	200 kHz	Circuit/Packet Data
GPRS	128 Kbps	200 kHz	Circuit/Packet Data
Edge	384 Kbps	200 kHz	Circuit/Packet Data
CDMA2000-1XRTT	144 Kbps	1.25 MHz	Circuit/Packet Data
3G Technology	**Data Capability**	**Spectrum Required**	**Comment**
WCDMA	144 Kbps vechicular 384 Kbps outdoors		
	2 Mbps indoors	5 MHz	Packet Data
CDMA2000-3XRTT	144 Kbps vechicular 384 Kbps outdoors		
	2 Mbps indoors	5 MHz	Packet Data

4.2 Universal Mobile Telecommunications Service (UMTS)

When the International Telecommunications Union solicited solutions to meet the requirements laid down for IMT-2000, a number of technologies were proposed by various standards groups. These included both *Time Division Multiple Access* (TDMA) solutions and *Code Division Multiple Access* (CDMA) solutions. They also included both *Frequency Division Duplex* (FDD) and *Time Division Duplex* (TDD) solutions.

The *European Telecommunications Standards Institute* (ETSI) agreed on a WCDMA solution using FDD. In Japan, a WCDMA solution was also proposed, with both TDD and FDD options. In Korea, two different types of CDMA solution were proposed—one similar to the European and Japanese proposals, and one similar to a CDMA proposal being considered in North America (CDMA2000, which is an evolution of IS-95 CDMA).

It was clear that a number of groups were working on very similar technologies and it was fairly obvious that the most effective way forward was to pool resources. This led to the creation of two groups—the *Third Generation Partnership Project* (3GPP) and 3GPP2. 3GPP works on UMTS, which is based on WCDMA, and 3GPP2 works on CDMA2000. The following discussion provides a brief of UMTS.

4.2.1 Migration Path to UMTS and the Third Generation Partnership Project (3GPP)

The radio access for UMTS is known as *Universal Terrestrial Radio Access* (UTRA). This is a WCDMA-based radio solution, which includes both FDD and TDD modes. The *radio access network* (RAN) is known as UTRAN. It takes more than an air interface or an access network to make a complete system, however. The core network must also be considered. Because of the widespread deployment and success of *Global System for Mobile communications* (GSM), it is appropriate to base the UMTS core network upon an evolution of the GSM core network. In fact, as we shall see, the initial release of UMTS (3GPP Release 1999) makes use of the same core network architecture as defined for GSM/GPRS, albeit with some enhancements. Moreover, the core network is required to support both UMTS and GSM radio access networks (that is, both UTRAN and the GSM BSS).

The evolution of the GSM BSS has not stopped, however. As we shall see, enhancements such as the *Enhanced Data Rates for Global Evolution* (EDGE) have been made. With the requirements for the continued evolution of GSM and for the GSM to meet UMTS requirements, it makes sense for the continued maintenance and evolution of GSM specifications to be undertaken by 3GPP. Consequently, 3GPP, rather than ETSI, is now responsible for GSM specifications as well as UMTS-specific specifications.

For several years, the various enhancements to GSM have been developed according to yearly releases. Thus, for a given GSM specification, versions have been related to Release 1996, Release 1997, and Release 1998. Initially, 3GPP determined to continue with that approach. Therefore, the first release of specifications from 3GPP is known as 3GPP Release 1999. The release includes not only new specifications for the support of a UTRAN access, but also enhanced versions of existing GSM specifications (such as for the support of EDGE). The 3GPP Release 1999 specifications were completed in March of 2000. These, of course, will be subject to some revisions and corrections as errors and inconsistencies are discovered during test and deployment.

The next release of 3GPP specifications was originally termed 3GPP Release 2000. This included major changes to the core network. The changes were so significant, however, that they could not all be handled in a single step. Thus, Release 2000 was divided into two releases: Release 4 and Release 5. Going forward, the concept of yearly releases will no longer apply, and releases will be structured and timed according to defined functionality. The Release 4 specifications were frozen in the first half of 2001. This means that no new content is to be added and any changes to the specifications will occur only to correct errors or inconsistencies. For Release 5, it is expected that specifications will be frozen in December of 2001.

For the most part (although not exclusively), 3GPP Release 1999 focuses mainly on the access network (including a totally new air interface) and the changes needed to the core network to support that access network. Release 4 focuses more on changes to the architecture of the core network. Release 5 introduces a new call model, which means changes to user terminals, changes to the core network, and some changes to the access network (although the fundamentals of the air interface remain the same). Given that the air interface is new in Release 1999 and that it does not drastically change in later releases, it is best to begin our description of UMTS technology with the WCDMA air interface. The primary focus in this book will be on the FDD mode of operation, with less emphasis on TDD. First, however, a few words about the types of services that UMTS can offer.

4.3 UMTS Services

Of course, the most notable capability promised by UMTS is a high data rate—up to 2 Mbps. There is, however, more to a given service than just the data rate that the service demands. Depending on what the end user is trying to do, various considerations must be made, of which data rate is only one.

UMTS specifications define four service classes, where the services within a given class have a common set of characteristics. The service classes are as follows:

- **Conversational** This is characterized by low delay tolerance, low jitter (delay variation) and low error tolerance. The data rate requirement may be high or low, but is generally symmetrical. In other words, the data rate in one direction will be similar to that in the other direction. Voice, which is highly delay-sensitive, is a typical conversational application, one that does not require very high data rates. Video conferencing is also a conversational application. It has similar delay requirements to voice, but is less error-tolerant and generally requires a higher data rate.
- **Interactive** This consists of typically request/response-type transactions. Interactive traffic is characterized by low tolerance for errors, but with a larger tolerance for delays than conversational services. Jitter (delay variation) is not a major impediment to interactive services, provided that the overall delay does not become excessive. Interactive services may require low or high data rates depending on the service in question, but the data rate is generally significant only in one direction at a time.
- **Streaming** This concerns one-way services, using low- to high-bit rates. Streaming services have a low-error tolerance, but generally have a high tolerance for delay and jitter. That is because the receiving application usually buffers data so that it can be played to the user in a synchronized manner. Streaming audio and streaming video are typical streaming applications.
- **Background** This is characterized by little, if any, delay constraint. Examples include server-to-server e-mail delivery (as opposed to user retrieval of e-mail), SMS, and performance/measurement reporting. Background applications require error-free delivery.

4.3.1 UMTS Speech Service

Although UMTS will be used for a variety of data services, speech may well remain the most widely used service. Speech has certain requirements in terms of data rate, delay, jitter, and error-free delivery, all of which are derived from human perceptions and expectations. Moreover, speech quality in UMTS needs to be comparable to that in fixed telephony networks and certainly no worse than that experienced in 2G wireless networks.

UMTS uses the *Adaptive Multirate* (AMR) speech coder. This is actually several coders in one and provides coding rates of 12.2 Kbps, 10.2 Kbps, 7.95 Kbps, 7.40 Kbps, 6.70 Kbps, 5.90 Kbps, 5.15 Kbps, and 4.75 Kbps. The 12.2-Kbps rate is the same coding scheme as used in the GSM Enhanced Full-Rate coding scheme. The 7.4-Kbps rate is the same coding scheme as used in IS-136 TDMA networks. The reuse of existing coders means that the voice-coding scheme of UMTS should at least offer the same levels of quality as experienced in existing 2G networks.

The AMR coder allows for the speech bit rate to change dynamically during a call. As we shall describe later, the higher the bit rate of any service, the smaller the effective footprint of a cell. Thus, a user at the edge of a cell could change from a high speech-coding rate to a lower speech-coding rate to effectively extend the coverage for speech service. Each AMR speech frame is 20 ms in duration and it is possible to change the speech-coding rate from one speech frame to the next. Thus, the coding rate could change as often as every 20 ms, although that is unlikely to ever happen in reality.

The AMR coder also supports *voice activity detection* (VAD) and *discontinuous transmission* (DTX), with comfort noise generation. The net effect is that little or nothing is sent over the air interface when nothing is being said. Given that typical speech involves one person speaking, followed by the other, it is possible to reduce the amount of transmission over the air interface by as much as 50 percent. Of course, VAD and DTX are supported by most modern wireless technologies.

Many of the services supported by UMTS are packet-switched data services. Speech, on the other hand, at least in 3GPP Release 1999 and 3GPP Release 4, is a circuit-switched service. This means that a user in a speech call has access to dedicated resources throughout the call. In effect, a dedicated pipe is used between the two parties in a speech conversation. This is similar to the way speech is handled in a GSM/GPRS network, where a speech call uses a dedicated timeslot on the air interface and uses a dedicated transport and switching in the core network. Although the concept of timeslots does not map well to WCDMA radio access, the assignment of dedicated resources still applies.

4.4 The UMTS Air Interface

The UMTS air interface is a *Direct-Sequence CDMA* (DS-CDMA) system. Given that this is a radical departure from the TDMA techniques of GSM, GPRS, and EDGE, it is worth briefly describing the concepts involved.

4.4.1 WCDMA Basics

DS-CDMA means that user data is spread over a much wider bandwidth through multiplication by a sequence of pseudo-random bits called chips. Figure 4-3 provides a conceptual depiction of this spreading. One can see that the user data, at a relatively low rate compared to the rate of the spreading code, is spread over a signal that has a higher bit rate. We can also see that the signal that is transmitted has pseudo-random characteristics. When transmitted over a radio interface, the spread signal looks like noise.

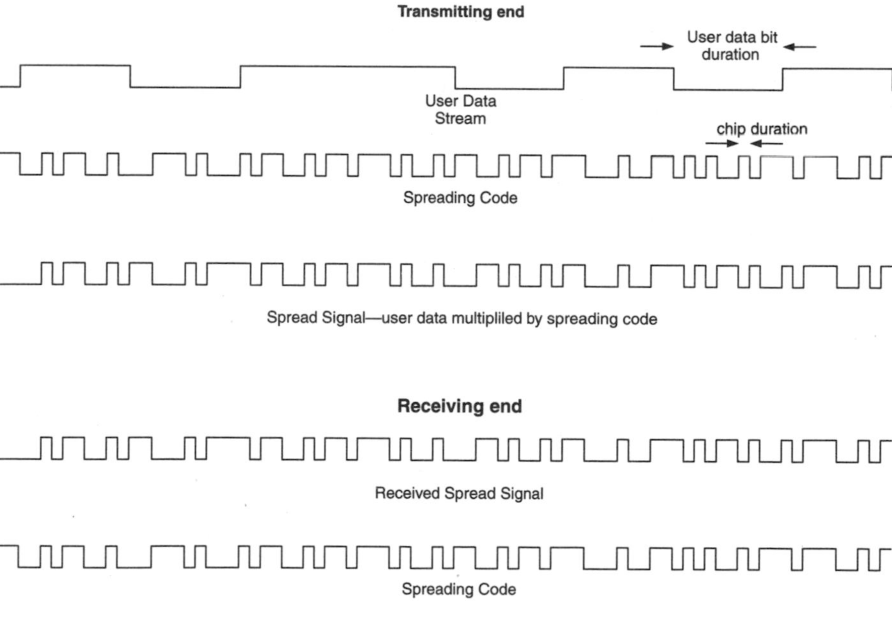

Figure 4-3
CDMA basic concept.

If multiple users transmit simultaneously on the same frequency, then the stream of data from each user needs to be spread according to a different pseudo-random sequence. In other words, each user data stream needs to be spread according to a different spreading code. At the receiving end, the stream of data from a given user is recovered by despreading the set of received signals with the appropriate spreading code. Of course, what is being despread is the complete set of signals received from all users that are transmitting.

Imagine, for example, two users (A and B) that are transmitting on the same frequency, but with two different spreading codes. If, at the receiving end, the received signal is despread with the spreading code applicable to user A, then the original data stream from user A is recovered. The data stream that is recovered does have some noise created by the fact that the received signal also contains user data from user B. The noise, however, is small.

Similarly, if the received signal is despread according the spreading code used by user B, then the original data stream from user B is recovered, with a little noise generated by the presence of user A's data within the spread signal. Provided that the rate of the spreading signal (the chip rate) is far larger than the user data rate, then the noise (that is, the interference) generated by the presence of other users will be sufficiently small to not inhibit the recovery of the data steam from a given user. Of course, as the number of simultaneous users increases, so does the interference and it eventually becomes impossible to recover a specific user's data with any confidence.

In other words, for a given bit of recovered user data, the signal-to-noise ratio must be sufficiently high. In CDMA, we refer to E_b/N_o, where E_b is the power density per bit of recovered user data and N_o is the noise power density. Provided that E_b/N_o is sufficiently large, then the user data can be recovered.

The ratio of the chip rate to the user data symbol rate is known as the spreading factor. The capability to recover a given user's signal is directly influenced by the spreading factor. The higher the spreading factor, the greater the capability to recover a given user's signal. In terms of transmission and reception, a higher spreading factor has an equivalent effect as transmitting at a higher power. Thus, the magnitude of the spreading factor can be considered a type of gain and is known as the processing gain. In dB, the processing gain is given by $10 \times 10\text{Log}_{10}$ (spreading rate/user rate). In some cases, this can be quite a large number and can help to overcome the effect of interference generated by the presence of other users.

If, for example, the processing gain for a given CDMA service were 20 dB and if an E_b/N_o value of 5 dB were needed, then for a given user, the signal-to-interference ratio can be as low as -15 dB and the user's signal can still be recovered. This is because the despreading benefits from the processing gain of 20 dB. Note that, for a given chip rate, the processing gain for low-bit-rate user applications is greater than for high-bit-rate applications, which often means that lower-bit-rate applications can tolerate more interference than high-bit-rate applications.

The WCDMA air interface of UMTS (hereafter simply WCDMA) has a nominal bandwidth of 5 MHz. While 5 MHz is the nominal carrier spacing, it is possible to have a carrier spacing of 4.4 MHz to 5 MHz in steps of 200 kHz. This enables spacing that might be needed to avoid interference, particularly if the next 5-MHz block is allocated to another carrier.

The chip rate in WCDMA is 3.84×10^6 chips/second (3.84 Mcps). In theory, for a speech service at 12.2 Kbps (and, for now, assuming no extra bandwidth for error correction), the spreading factor would be $3.84 \times 10^6/12.2 \times 10^3 = 314.75$. This would equate to a processing gain of 25 dB. In reality, however, WCDMA does include extra coding for error correction. Consequently, a spreading factor as high as 314.75 is not supported, at least not in the uplink. The supported uplink spreading factors are 4, 8, 16, 32, 64, 128, and 256. The highest spreading factor (256) is used mostly by the various control channels. Some control channels can also use lower spreading factors, while user services generally use lower spreading factors.

Table 4-3 provides a summary of the spreading factors and the corresponding data rates on the uplink.

Table 4-3

Uplink Spreading Factors and Data Rates

Spreading Factor	Gross Data Rate (Kbps)	User data rate (Kbps) (assuming half-rate coding for error correction)
256	15	7.5
128	30	15
64	60	30
32	120	60
16	240	120
8	480	240
4	960	480

At first glance, it appears that the lowest spreading factor (4) provides a gross rate of only 960 Kbps and a usable rate of only 480 Kbps. This does not meet the requirements of IMT-2000, which states that a user should be able to achieve speeds of 2 Mbps. In order to meet that requirement, UMTS supports the capability for a given user to transmit up to six simultaneous data channels. Thus, if a user wants to transmit user data at a user rate greater than 480 Kbps, then multiple channels are used, each with a spreading factor of four. With six parallel channels, each at a spreading factor of four, a single user can obtain speeds of over 2 Mbps.

In the downlink, the same spreading factors are available, with a spreading factor of 512 also possible. One difference between the uplink and downlink, however, is the number of bits per symbol. As will be described in Chapter 6, "Universal Mobile Telecommunications Service (UMTS)," the uplink effectively uses one bit per user symbol, while the downlink effectively uses two bits per user symbol. Consequently, for a given spreading factor, the user bit rate in the downlink is greater than the corresponding bit rate in the uplink. The user rate in the downlink is not quite twice that in the uplink, however, due to differences in the way that control channels and traffic channels are multiplexed on the air interface. The details of uplink and downlink transmissions are provided in Chapter 6. Table 4-4 provides a summary of the spreading factors and the corresponding data rates on the downlink.

Table 4-4

Downlink Spreading Factors and Data Rates

Spreading Factor	Gross air interface bit rate (Kbps)	User data rate (Kbps) (including coding for error correction)	Approximate net user data rate (Kbps) (assuming half rate coding)
512	15	3–6	1–3
256	30	12–24	6–12
128	60	42–512	21–25
64	120	90	45
32	240	210	105
16	480	432	216
8	960	912	456
4	1920	1,872	936

As is the case for the uplink, WCDMA supports multiple simultaneous user data channels in the downlink, so that a single user can achieve rates of over 2 Mbps. It should be noted, however, that Table 4-4 does not tell the whole story of possible data rates on the downlink. WCDMA supports a concept known as compressed mode, whereby gaps exist in downlink transmission so that the terminal can take measurements on other frequencies. When compressed mode is used, a reduction will take place in the data rate compared to that shown in Table 4-4.

An important capability of WCDMA is that user data rates do not need to be fixed. In WCDMA, channels are transmitted with a 10-ms frame structure. It is possible to change the spreading factor on a frame-by-frame basis. Thus, within one frame, the user data rate is fixed, but the user data rate can change from frame to frame. This capability means that WCDMA can offer bandwidth on demand. Note that rate changes every 10 ms do not apply to AMR speech as each speech packet is 20 ms in duration, so that the speech rate can change every 20 ms if needed, but not every 10 ms.

4.4.2 Spectrum Allocation

With the WCDMA FDD option, the paired 5-MHz carriers in the uplink and downlink are as follows: uplink—1920 MHz to 1980 MHz; downlink—2110 MHz to 2170 MHz. Thus, for the FDD mode of operation, a separation of 190 MHz exists between uplink and downlink. Although 5 MHz is the nominal carrier spacing, it is possible to have a carrier spacing of 4.4 MHz to 5 MHz in steps of 200 kHz. This enables spacing that might be needed to avoid interference, particularly if the next 5-MHz block is allocated to another carrier.

For the TDD option, a number of frequencies have been defined, including 1900 MHz to 1920 MHz, and 2010 MHz to 2025 MHz. Of course, with TDD, a given carrier is used in both the uplink and the downlink so that no separation exists.

Of course, there is no reason why WCDMA could not be deployed at other frequencies. In fact, the use of other frequencies may well be necessary in some countries. You may have noticed that the frequency bands defined previously overlap significantly with frequencies used for PCS in North America. Therefore, in North America, it will be necessary to move some existing users from the PCS band and/or acquire a new spectrum in some other band. The movement of existing PCS users is likely only to happen when a given carrier that wants to implement UMTS already has an existing PCS system and uses some of the spectrum for UMTS. The net result for such an operator will, of course, be limited spectrum for both PCS and UMTS.

4.5 Overview of the 3GPP Release 1999 Network Architecture

Figure 4-4 shows the network architecture for 3GPP Release 1999, the first set of specifications for UMTS. Working our way from the top left, we see that a user device is termed the *User Equipment* (UE). Strictly speaking, the UE contains the *Mobile Equipment* (ME) and the *UMTS Subscriber Identity Module* (USIM). The USIM is a chip that contains some subscription-related information, plus security keys. It is similar to the SIM in GSM.

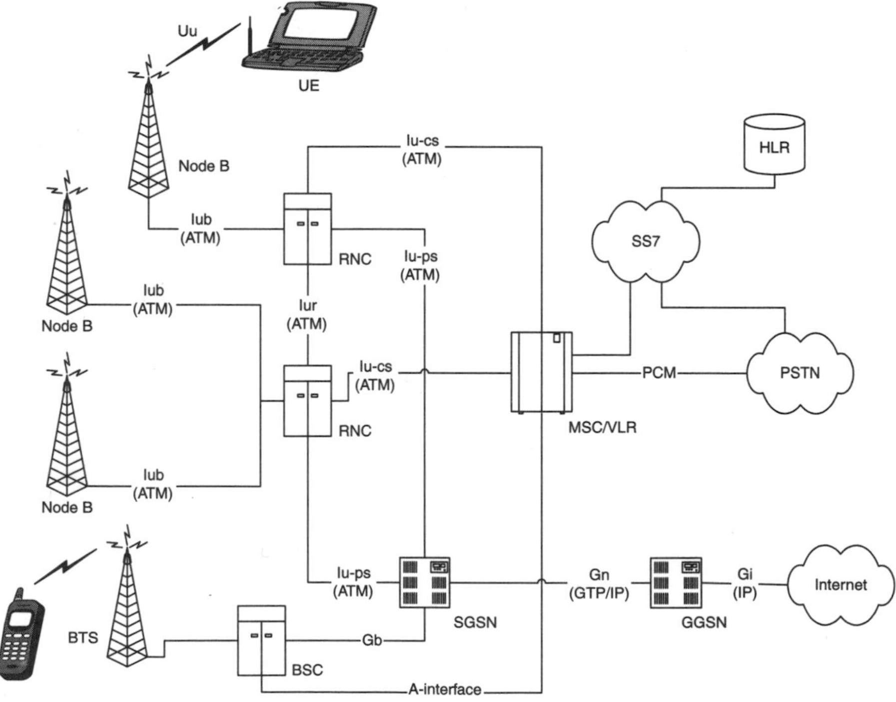

Figure 4-4 3GPP Release 1999 Network Architecture.

The interface between the UE and the network is termed the Uu interface. This is the WCDMA air interface previously described. Strictly speaking, the WCDMA interface, at least at the physical layer, is between the UE and the BTS. In 3GPP specifications, the base station is known as Node B. This was originally a temporary name that somehow stuck.

A Node B is connected to a single *Radio Network Controller* (RNC). The RNC controls the radio resources of the Node Bs that are connected to it. The RNC is analogous to a BSC in GSM. Combined, an RNC and the Node Bs that are connected to it are known as a *Radio Network Subsystem* (RNS). The interface between a Node B and an RNC is the Iub interface. Unlike the equivalent Abis interface in GSM, the Iub interface is fully standardized and open. It is possible to connect a Node B to an RNC of a different vendor.

Unlike in GSM, where BSCs are not connected to each other, in the UMTS RAN (officially, the *UMTS Terrestrial Radio Access Network*, or UTRAN), an interface exists between the RNCs. This interface is termed Iur. The primary purpose of this interface is to support inter-RNC mobility and soft handover between Node Bs connected to different RNCs. The Iur signaling in support of soft handoff is described in more detail later in Chapter 6.

The UTRAN is connected to the core network via the Iu interface. The Iu interface, however, has two different components. The connection from UTRAN to the circuit-switched part of the core network is via the Iu-CS interface, which connects an RNC to a single *Mobile Switching Center* (MSC)/*Visitor Location Register* (VLR). The connection from UTRAN to the packet-switched part of the core network is termed Iu-PS. This connection is from an RNC to an SGSN.

It can be seen from Figure 4-4 that all of the interfaces in the UTRAN of 3GPP Release 1999 are based on *Asynchronous Transfer Mode* (ATM). ATM was chosen because of its capability to support a range of different service types (such as a variable bit rate for packet-based services and a constant bit rate for circuit-switched services).

One can see from Figure 4-4 that the core network uses the same basic architecture as that of GSM/GPRS. This was purposely done so that the new radio access technology could be supported by an established, robust core network technology. It should be possible for an existing core network to be upgraded to support UTRAN, so that a given MSC, for example, could connect to both a UTRAN RNC and a GSM BSC.

In fact, UMTS specifications include support for a hard handover from UMTS to GSM and vice-versa. This is an important requirement, since the widespread rollout of UMTS coverage will take time to complete, and if holes exist in UMTS coverage, it is desirable that a UMTS subscriber should receive service from the more ubiquitous GSM coverage. If UTRAN and the GSM BSS are supported by different MSCs, then an inter-system handover could be achieved through an inter-MSC handover. Given that many of the functions of the MSC/VLR are similar for UMTS and GSM, however, it makes sense for a given MSC to be able to support both types of access simultaneously. Similar logic suggests that a given SGSN should be able to simultaneously support an Iu-PS connection to an RNC and a Gb interface to a GPRS BSC.

In most vendor implementations, many of the network elements are being upgraded to simultaneously support GSM/GPRS and UMTS. Such network elements include the MSC/VLR, the *Home Location Register* (HLR), the SGSN, and the GGSN. For some vendors, the base stations deployed for GSM/GPRS have been designed so that they can be upgraded to support both GSM and UMTS simultaneously. This is a major consideration for those network operators that want to deploy a UMTS network in parallel with an existing GSM network. For some vendors, the BSC is being upgraded to act as both a GSM BSC and a UMTS RNC. This configuration is rare, however. The different interfaces and functions (such as a soft handover) required of a UMTS RNC mean that its technology is quite different from that of a GSM BSC. Consequently, it is normal to find separate UMTS RNCs and GSM BSCs.

4.6 Overview of the 3GPP Release 4 Network Architecture

Figure 4-5 shows the basic network architecture for 3GPP Release 4. The main difference between the Release 1999 architecture and the Release 4 architecture is that the core network becomes a distributed network. Rather than having traditional circuit-switched MSCs, as has been the case in previous network architectures, a distributed switch architecture is introduced.

Basically, the MSC is divided into an MSC server and a *media gateway* (MGW). The MSC server contains all of the mobility management and call control logic that would be contained in a standard MSC. It does not, however, contain a switching matrix. The switching matrix is contained within

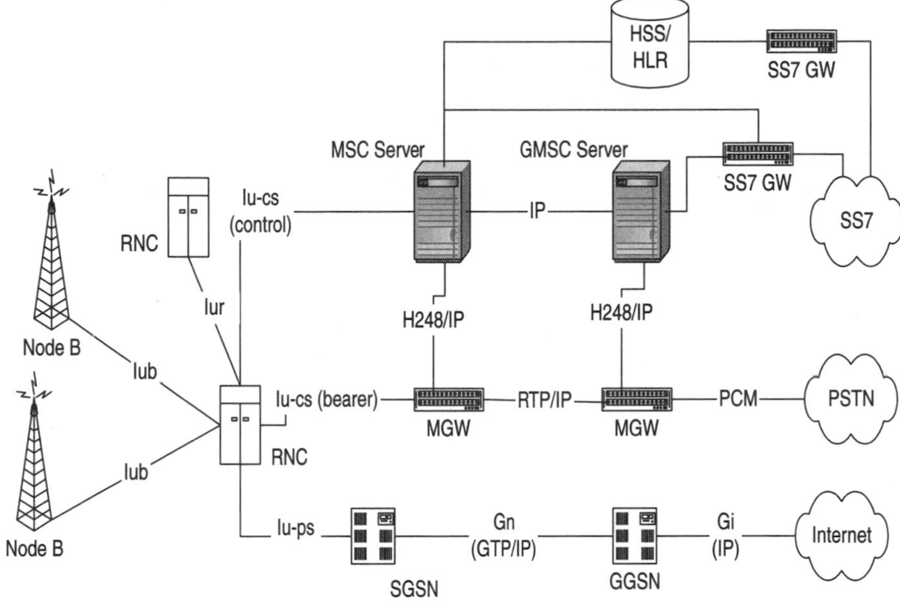

Figure 4-5
3GPP Release 4 Distributed Network Architecture.

the MGW, which is controlled by the MSC server and can be placed remotely from the MSC.

Control signaling for circuit-switched calls is between the RNC and the MSC server. The media path for circuit-switched calls is between the RNC and the MG. Typically, an MG will take calls from the RNC and routes those calls towards their destinations over a packet backbone. In many cases, that packet backbone will use the *Real-Time Transport Protocol* (RTP) over the *Internet Protocol* (IP). As can be seen from Figure 4-5, packet data traffic from the RNC is passed to the SGSN and from the SGSN to the GGSN over an IP backbone. Given that data and voice can both use IP transport within the core network, a single backbone can be constructed to support both types of service. This can mean significant capital and operating expenses compared to the construction and operation of separate packet and circuit-switched backbone networks.

At the remote end, where a call needs to be handed off to another network, such as the PSTN, another *media gateway* (MGW) is controlled by a *Gateway MSC server* (GMSC server). This MGW will convert the packetized voice to standard PCM for delivery to the PSTN. It is only at this point that transcoding needs to take place. Assuming, for example, that speech over the air interface is carried at 12.2 Kbps, then the voice does not need to be converted up to 64 Kbps until it reaches the MGW that interfaces with the PSTN. This packetized transport can mean significant bandwidth savings on the backbone network, particularly if the two MGWs are some significant distance apart.

The control protocol between the MSC server or GMSC server and the MGW is the ITU H.248 protocol. This protocol was developed jointly by the ITU and the *Internet Engineering Task Force* (IETF). It also goes by the name media gateway control (MEGACO). The call control protocol between the MSC server and the GMSC server can be any suitable call control protocol. The 3GPP standards suggest but do not mandate the *Bearer Independent Call Control* (BICC) protocol, which is based on the ITU-T recommendation, Q.1902.

In many cases, an MSC server will also support the functions of a GMSC server. Moreover, one MGW may have the capability to interface both with the RAN and with the PSTN. In that case, calls to or from the PSTN can be handed off locally. This can represent another major saving.

Consider, for example, a scenario where an RNC is located in one city (City A) and is controlled by an MSC in another city (City B). Let's assume that a subscriber in City A makes a local phone call. Without a distributed architecture, the call needs to travel from City A to City B (where the MSC is), only to be connected back to a local PSTN number in City A. With a distributed architecture, the call can be controlled by an MSC server in City B, but the actual media path can remain within City A, thereby reducing transmission requirements and reducing network operations costs.

One will notice that, in Figure 4-5, the HLR may also be known as a *Home Subscriber Server* (HSS). The HSS and HLR are functionally equivalent, with the exception that interfaces to an HSS will use packet-based transports such as IP, whereas an HLR is likely to use standard *Signaling System 7* (SS7)-based interfaces. Although not shown, a logical interface exists between the SGSN and HLR/HSS and between the GSN and HLR/HSS.

Many of the protocols used within the core network are packet-based, using either IP or ATM. The network must, however, interface with traditional networks—through the use of media gateways. Moreover, the network must also interface with standard SS7 networks. This interface is achieved through the use of an *SS7 gateway* (SS7 GW). This is a gateway that on one side supports the transport of a SS7 message over a standard SS7 transport. On the other side, it transports SS7 application messages over a packet network such as IP. Entities such as the MSC server, the GMSC server, and HSS communicate with the SS7 gateway using a set of transport protocols specially designed for carrying SS7 messages in an IP network. This suite of protocols is known as Sigtran.

Many of the protocols mentioned in this brief discussion (RTP, H.248, and Sigtran) are described in greater detail in Chapter 8.

4.7 Overview of the 3GPP Release 5 All-IP Network Architecture

The next step in the UMTS evolution is the introduction of an all-IP multimedia network architecture (see Figure 4-6). This step in the evolution represents a change in the overall call model. Specifically, both voice and data are largely handled in the same manner all the way from the user terminal to the ultimate destination. This architecture can be considered the ultimate convergence of voice and data.

As we can see from Figure 4-6, voice and data no longer need separate interfaces; just a single Iu interface can carry all the media. Within the core

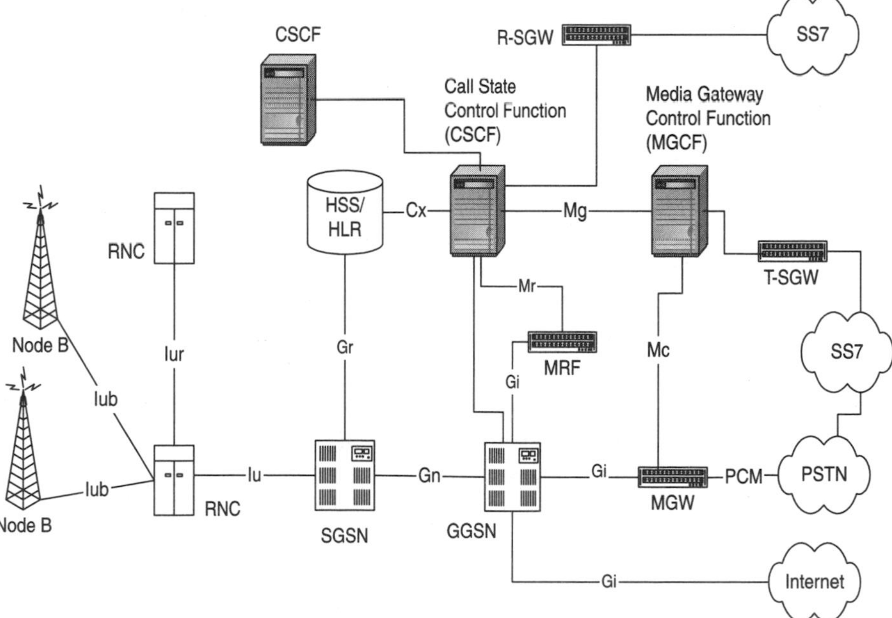

Figure 4-6 3GPP IP Multimedia Network Architecture.

network, that interface terminates at the SGSN—there is no separate media gateway.

We also find a number of new network elements, notably the *Call State Control Function* (CSCF), the *Multimedia Resource Function* (MRF), the *Media Gateway Control Function* (MGCF), the *Transport Signaling Gateway* (T-SGW), and the *Roaming Signaling Gateway* (R-SGW).

An important aspect of the all-IP architecture is the fact that the user equipment is greatly enhanced. Significant logic is placed within the UE. In fact, the UE supports the *Session Initiation Protocol* (SIP), which is described in Chapter 8, "Voice over IP Technology." The UE effectively becomes a SIP user agent. As such, the UE has far greater control of services than previously.

The CSCF manages the establishment, maintenance, and release of multimedia sessions to and from user devices. This includes functions such as translation and routing. The CSCF acts like a proxy server/registrar, as defined in the SIP architecture described in Chapter 8.

The SGSN and GGSN are enhanced versions of the same nodes used in GPRS and UMTS Release 1999 and Release 4. The difference is that these nodes, in addition to data services, now support services that have traditionally been circuit-switched—such as voice. Consequently, appropriate *Quality of Service* (QOS) capabilities need to be supported either within the SGSN and GGSN or, at a minimum, in the routers immediately connected to them.

The *Multimedia Resource Function* (MRF) is a conference bridging function used to support features such as multi-party calling and meet-me conference service.

The *Transport Signaling Gateway* (T-SGW) is an SS7 gateway that provides SS7 interworking with standard external networks such as the PSTN. The T-SGW will support Sigtran protocols. The *Roaming Signaling Gateway* (R-SGW) is a node that provides signaling interworking with legacy mobile networks that use standard SS7. In many cases, the T-SGW and R-SGW will exist within the same platform.

The *media gateway* (MGW) performs interworking with external networks at the media path level. The MGW in the 3GPP Release 5 network architecture is the same as the equivalent function within the 3GPP Release 4 architecture. The MGW is controlled by a *Media Gateway Control Function* (MGCF). The control protocol between these entities is ITU-T H.248. The MGCF also communicates with the CSCF. The protocol of choice for that interface is SIP.

It should be noted that the Release 5 all-IP architecture is an enhancement to an existing Release 1999 or Release 4 network. It is effectively the addition of a new domain in the core network—the IP-Multimedia (IM) domain. This new domain, which enables both voice and data to be carried over IP all the way from the handset, uses the services of the PS domain for transport purposes. That is, it uses the SGSN, GGSN, Gn, Gi, etc.—nodes and interfaces that belong to the PS domain.

4.8 Overview CDMA2000

CDMA2000 is a wireless platform that is part of the IMT-2000 specification and is an extension of the CDMAOne wireless platforms using the IS-95A/B and J-STD-008 standards. CDMA2000, being a IMT-2000 standard, is geared toward the transport and treatment of 3G wireless services supporting multimedia applications for fixed as well as mobile situations.

In the existing 2G platforms that are operational today for both cellular and PCS, the same radio bandwidth is allocated for voice and data. The data services are, of course, really circuit-switched services, without the capability to overbook the data service and thus increase the capacity of a wireless system through the appropriate use of data services.

4.8.1 Migration Path

The migration path that a wireless operator must take to realize CDMA2000 as envisioned for 3G is usually thought of as a staged approach for implementation. The concept behind the phased approach is to enable wireless operators using IS-95 platforms to migrate toward 3G without having to either forklift their existing platforms or acquire a new spectrum. CDMA2000 also is backward-compatible with existing 2G CDMA systems, thereby speeding time to market.

From an operator's point of view, the migration from 2G to 3G and the realization of 3G must include

- It needs to be cost effective based on the capital infrastructure already in place
- Increased capacity and throughput both in voice and data services that utilize existing spectrum allocations
- Standard systems that enable backward as well as forward compatibility with other network and data platforms
- The flexibility to meet the ever-changing market conditions

CDMA2000 phase1 is an interim step between IS-95B and full realization of the IMT-2000 MC specification. CDMA200 can be deployed in an existing IS-95 channel or system and will exhibit the numerous enhancements, some of which are included here:

- 1X and 3X 1.25-MHz channel support
- 144-Kbps packet data rates
- 2X increase in voice capacity
- 2X increase in standby time
- Improved handoff

It is envisioned that IS-95, CDMA200 1xRTT, and CDMA2000-3xRTT can and will coexist in the same market and possibly at the same cell site. Obviously, one can take numerous approaches in the course of implementing any technology platform, and CDMA2000 is by no means unique to this situation. However, several common migration paths are being pursued for implementing CDMA2000. The migration path, of course, is dependent upon whether the operator is currently utilizing IS-95A/B or J-STD-008 and upgrading to CDMA2000, or in the process of either installing a new system or segmenting the existing spectrum to facilitate the introduction of CDMA2000 into the network (see Table 4-5).

Table 4-5
CDMA Path

Standard	Salient Issues
IS-95A	9600 bps or 14.4 Kbps
IS-95B	Primarily voice, data on forward link, improved handoff and data speeds of 64/56 Kbps
CDMA2000 phase1	SR1 (1.2288 Mcps), Voice and data (packet data via separate channel) 128 Walsh codes 2X voice capacity over IS-95 144 Kbps using 1xRTT with SR1
CDMA2000 phase2	SR3 (3.6864 Mcps) Packet Data oriented Higher data rate 144 Kbps – mobility 384 Kbps – pedestrian 2 Mbps – fixed 256 walsh codes

The following are three, 3, possible migration paths an operator may persue.

CDMAOne (IS-95A)—CDMA2000 (phase1)—CDMA2000 (phase2)

CDMAOne (IS-95A)—CDMAOne (IS-95B)—CDMA2000 (phase1)—CDMA2000 (phase2)

CDMA2000 (phase1)—CDMA2000 (phase2)

To complicate matters a little more for migration issues, several interim steps within the CDMA2000 implementation process bear mentioning relative to the single carrier (1x) aspects. The expected migration path or, rather, the options for possible deployment of a CDMA2000-1x system are shown in Figure 4-7.

4.8.2 System Architecture

The system architecture that will comprise a CDMA2000 network is a logical extension of an existing CDMAone network with the fundamental difference being the introduction of packet data services. The implementation of a CDMA2000 system is meant to involve upgrades to the BTS and BSC for the purpose of handling the packet data services. Additionally, the use of packet data services also necessitates the introduction of a packet server complex that may exist already to support services like CDPD.

However, it is recommended that the existing packet data network that exists should not by default be considered for inclusion into the CDMA2000 network architecture. The system architecture for a CDMA2000 network, due to packet data services, can be either centralized or distributed. The decision as to whether the system utilizes a distributed or centralized system is dependant upon the design requirements as well as operational issues. Figure 4-8 is an example of a standalone CDMA2000 system that has the inclusion of a PDSN for handling packet data services.

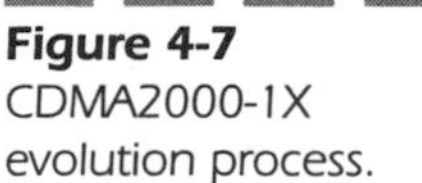

Figure 4-7
CDMA2000-1X evolution process.

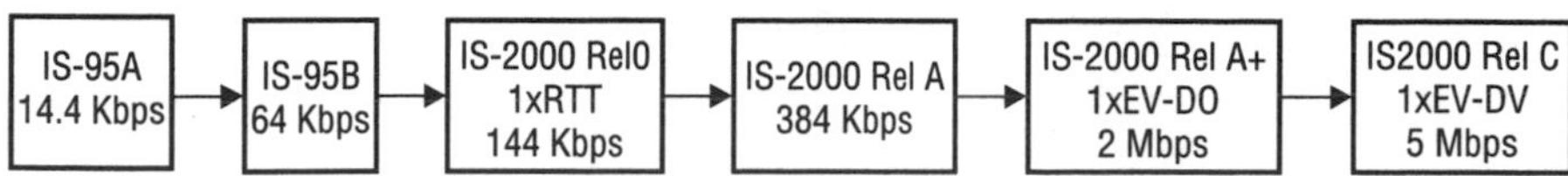

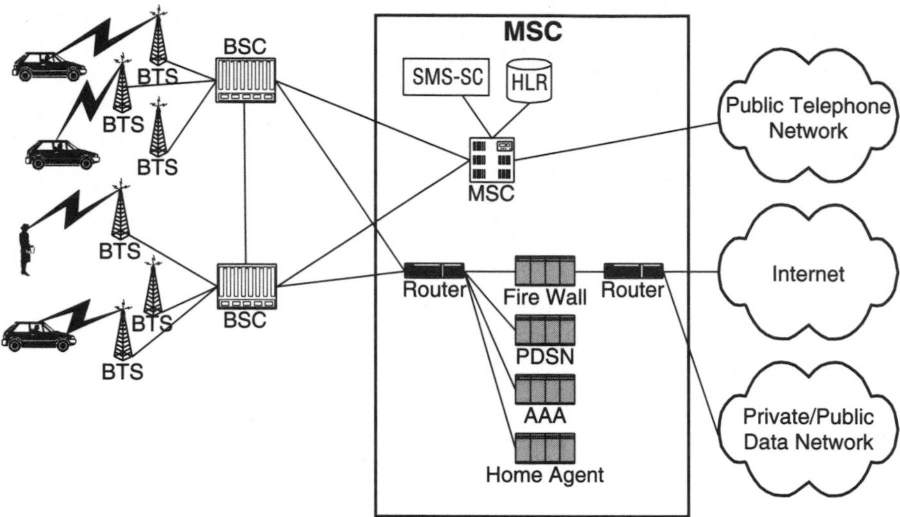

Figure 4-8 Generic CDMA2000 system architecture.

4.8.3 Spectrum

The spectrum requirements for a CDMA2000 system have their roots in IS-95, but some differences exist. A comparison for spectrum requirements between IS-95, CDMA2000-1x, and CDMA2000-3x carriers is shown in Figure 4-9.

The channel depicted in Figure 4-9 indicates that for whatever version of CDMA2000-1x the operator decides to deploy, it can be overlaid onto the existing IS-95 channel, through a 1:1 or N:1 upgrade. The CDMA2000-1x introduction of a reverse pilot channel as well as Walsh codes are covered in more detail in Chapter 7, "CDMA2000."

When the decision is made to migrate to a CDMA2000-3x system, the operator can make two effective choices. Either the 3x system will be allocated its own specific spectrum or the existing 1x channels will be part of the 3X platform offering. Depending on the plan, the CDMA2000-1x channel locations will need to be thought through in advance and this issue is covered in Chapter 7.

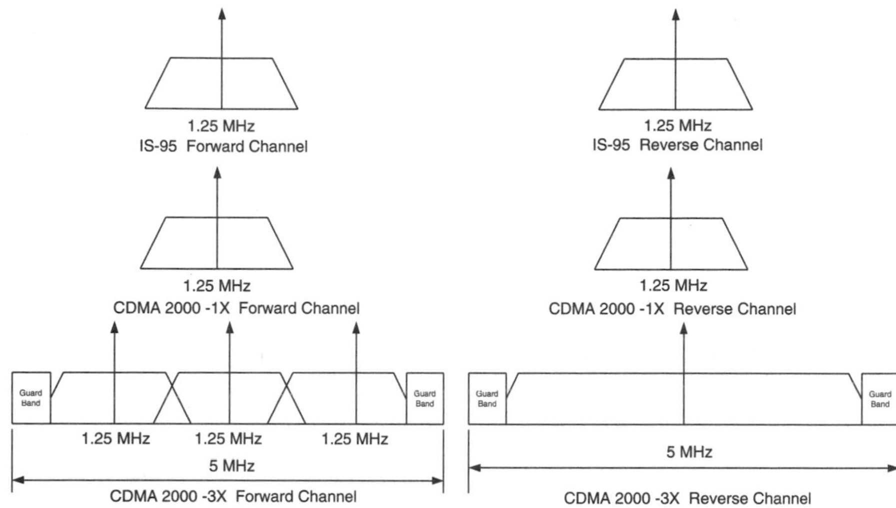

Figure 4-9
1X and 3X carriers.

4.9 Commonality Between WCDMA/CDMA2000/CDM

Both WCDMA and CDMA2000 share several commonalties that are part of the IMT2000 platform specification. Both systems utilize CDMA technology and both require, in their final version, a total of 5 MHz of spectrum. Both systems will be able to interoperate with each other and it is possible for a wireless operator to deploy both a CDMA2000 network as well as a WCDMA system, barring, of course, the capital cost issues.

Both systems have a migration path from existing 2G platforms to that of 3G. However, the path both systems take is different and is driven by the imbedded infrastructure the existing operator has already deployed. Since the end game is to offer high-speed packet data services to the end user, the real issue between both of these standards within the IMT2000 specification is the methodology for how they realize the desired speed.

WCDMA utilizes a wide band channel, while CDMA2000 utilizes both a wideband and several narrow band channels in the process of achieving the required throughput levels. Additionally, both WCDMA and CDMA2000 are designed to operate in multiple frequency bands. Both systems can operate in the same frequency bands provided the spectrum is available.

Therefore, the commonalties between WCDMA and CDMA2000 can be summed up in the following brief bullet points that were introduced at the beginning of this chapter:

- Global standard
- Compatibility of service within IMT-2000 and other fixed networks
- High quality
- Worldwide common frequency band
- Small terminals for worldwide use
- Worldwide roaming capability
- Multimedia application services and terminals
- Improved spectrum efficiency
- Flexibility for evolution to the next generation of wireless systems
- High-speed packet data rates
  - 2 Mbps for fixed environment
  - 384 Mbps for pedestrian
  - 144 Kbps for vehicular traffic

References

Barron, Tim. "*Wireless Links for PCS and Cellular Networks*," Cellular Integration, Sept. 1995, pgs. 20–23.

Bates, Gregory. "*Voice and Data Communications Handbook*," Signature Ed., McGraw-Hill, 1998.

Brewster. "*Telecommunications Technology*," John Wiley & Sons, New York, NY, 1986.

Brodsky, Ira. "*3G Business Model*," Wireless Review, June 15, 1999, pg. 42.

Daniels, Guy. "*A Brief History of 3G*," Mobile Communications International, Issue 65, Oct. 99, pg. 106.

DeRose. "*The Wireless Data Handbook*," Quantum Publishing, Inc., Mendocino, CA, 1994.

Dixon. "*Spread Spectrum Systems*," 2nd Ed., John Wiley & Sons, New York, NY, 1984.

Gull, Dennis. "*Spread-Spectrum Fool's Gold?*" Wireless Review, Jan. 1, 1999, pg. 37.

Harte, Hoenig, and Kikta McLaughlin. "*CDMA IS-95 for Cellular and PCS*," McGraw-Hill, 1996.

Harter, Betsy. "*Putting the C in TDMA?*" Wireless Review, Jan., 2001, pgs. 29–34.

Held, Gil. "*Voice & Data Interworking*," 2nd Ed., 2000, McGraw-Hill.

Hoffman, Wayne. "*A new Breed of RF Components*," Glenayre.

Homa, Harri and Toskala, Antti. "*WCDMA for UMTS*," John Wiley & Sons, 2000.

LaForge, Perry M. "*cdmaOne Evolution to Third Generation: Rapid, Cost-effective Introduction of Advanced Services*," CDMA World, June, 1999.

Louis, P.J. "*M-Commerce Crash Course*," McGraw-Hill, 2001.

McClelland, Stephen. "*Europe's Wireless Futures*," Microwave Journal, Sept. 1999, pgs. 78–107.

McDysan, Spohn. "*ATM Theory and Applications Signature Edition*," McGraw-Hill, 1999.

Molisch, Andreas F. "*Wideband Wireless Digital Communications*," Prentice Hall, New Jersey, 2001.

Mouly, Pautet. "*The GSM System for Mobiel Communications*," Mouly Pautet, 1992.

Muratore, Flavio. "*UMTS Mobile Communications for the Future*," John Wiley & Sons, Sussex, England, 2000.

Newton, Harry. "*Newton's Telcom Dictionary*," 14th Ed., Flatiron Publishing, 1998.

Oba, Junichi. "*W-CDMA Systems Provide Multimedia Opportunities*," Wireless System Design, July 1998, pg. 20.

Ramjee, Prasad, Werner Mohr, and Walter Konhauser. "*Third Generation Mobile Communication Systems*," Artech House, 2000.

Rusch, Roger. "*The Market and Proposed Systems for Satellite Communication*," Applied Microwave & Wireless, Fall 1995, pgs. 10–34.

Salter, Avril. "*W-CDMA Trial&Error*," Wireless Review, Nov. 1, 1999, pg. 58.

Shank, Keith. "*A Time to Converge*," Wireless Review, Aug. 1, 1999, pg. 26.

Smith, Clint. "*LMDS*," McGraw-Hill, 2000.

Smith, Clint. "*Wireless Telecom FAQ*," McGraw-Hill, 2000.

Webb, William. "*CDMA for WLL*," Mobile Communications International, Jan. 1999, pg. 61.

Webb, William. "*Introduction to Wireless Local Loop, Second Editions: Broadband and Narrowband Systems*," Artech House, Boston, 2000.

Wesley, Clarence. "*Wireless Gone Astray*," Telecommunications, Nov. 1999, pg. 41.

Willenegger, Serge. "*cdma2000 Physical Layer: An Overview*," Qualcomm 5775, San Diego, CA.

William, C.Y. Lee. "*Lee's Essentials of Wireless Communications*," McGraw-Hill, 2001.

3GPP TS 23.002	Network Architecture (Release 1999)
3GPP TS 23.002	Network Architecture (Release 4)
3GPP TS 23.002	Network Architecture (Release 5)
3GPP TS 25.101	UE Radio Transmission and Reception (FDD)
3GPP TS 25.104	UTRA (BS) FDD; Radio Transmission and Reception
3GPP TS 25.211	Physical channels and mapping of transport channels onto physical channels (FDD)
3GPP TS 25.212	Multiplexing and channel coding (FDD)
3GPP TS 25.213	Spreading and modulation (FDD)
3GPP TS 25.214	Physical layer procedures (FDD)
3GPP TS 25.301	Radio Interface Protocol Architecture
3GPP TS 25.302	Services provided by the physical layer
3GPP TS 25.401	UTRAN overall description
3GPP TS 26.090	AMR speech codec; Transcoding functions
IETF RFC 2543	Session Initiation Protocol (SIP)
IETF RFC 768	User Datagram Protocol (STD 6)
IETF RFC 791	Internet Protocol (STD 5)
IETF RFC 793	Transmission Control Protocol (STD 7)
ITU-T H.248	Media Gateway Control Protocol

CHAPTER 5

The Evolution Generation (2.5G)

5.1 What Is 2.5G?

As the question implies, just what is *2.5 Generation* (2.5G)? Well, 2.5G, or the next generation transitional technology, is the method or methodology from which existing cellular and *Personal Communications Service* (PCS) operators are migrating to the next generation wireless technology referenced in the *International Mobile Telecommunications-2000* (IMT-2000) specification. 2.5G enables the wireless operators whether they utilize in cellular, PCS, or *Universal Mobile Telecommunications System* (UMTS) spectrum to deploy digital packet services prior to the availability of 3G platforms. The specific technology and implementation path that each operator must make or has made follows a similar decision path. The decision path that is followed is driven largely based on the existing infrastructure that has been previously deployed, the spectrum that is available and will be available, the growth rate, and of course the expected services being offered.

Obviously, the decision on which platform to utilize involves guesswork and decisions based on a fundamental belief that particular technology platforms will enable services that are yet to be developed. The 2.5G platforms are meant to provide the bridge between the existing 2G systems that have already been deployed and those envisioned for 3G.

Several platforms are leading the 2.5G effort; they are as follows:

- *General Packet Radio Service* (GPRS)/*High Speed Circuit Switched Data* (HSCSD)
- *Enhanced Data Rates for Global Evolution* (EDGE)
- *Code Division Multiple Access* (CDMA2000) (phase 1)

The 2.5G platform chosen for the operating system needs to involve the following fundamental issues independent on the technology platform:

- The underlying technology platform in existence
- The overlay approach (only for existing wireless operators)
- The introduction of packet data services
- The new user devices required
- New modifications to existing infrastructure

This chapter will attempt to cover the vast array of 2.5G issues that an operator needs to factor in to the decision process. Obviously, not all the issues that need to be addressed by a wireless operator can or will be covered in this chapter. Because the practical design issues for a 3G system are interrelated with 2.5G systems, the design examples are included in Chapters 12, "UMTS System Design," for UMTS, and Chapter 13, "CDMA2000 System Design," for CDMA2000. However, having a fundamental understanding of the major platforms being deployed will help proper technological and business decisions to be made that can exploit the advantages of each of the infrastructure platforms.

Some of the key concepts that need to be kept in mind when establishing a wireless technology transition plan from 2G to 3G is the methodology associated with the realization of the transition itself. The key concepts associated with a 2.5G transition are as follows:

- Existing wireless and fixed network access platforms.
- Transition platforms required.
- Overlay implementation.
- No one specific standard chosen for transition.
- New user devices required.
- 2.5G is primarily a data-play only.
- Additional base station and support infrastructure required.
- 2.5G is an application enabler only and can support a host of applications offered of which few, if any, are defined.

5.2 Enhancements over 2G

The introduction of 2.5G has many enhancements over the present 2G systems that are in place. The specific advantages of each 2.5G system are directly related to the market and services that the wireless operator current serves and wants to serve in the near future. The enhancements lie primarily in the use and delivery of packet data services with speeds exceeding the existing 14.4K barrier with 2G systems.

Table 5-1 illustrates the relative advantages that each of the 2.5G platforms has over its fundamental underlying technology platform.

Table 5-1 again is not meant to be all-inclusive but rather is a guide to illustrate what the new technology platform offers. The reference used for the 2G to 2.5 platform is not a prerequisite. For example, the deployment of GPRS can be enabled with an underlay system using IS-136 or even CDMA, provided the spectrum is available and the required infrastructure is deployed properly.

Table 5-1
2G and 2.5G

2G Technology	2.5G Technology	Enhancements	Migration-to-3G Platform
GSM	GPRS	• High speed packet data services (144.4K) • Uses existing radio spectrum	WCDMA
IS-136	EDGE	• High speed packet data services (144.4K) • Uses existing radio spectrum	WCDMA
CDMA	CDMA2000 (phase1)	• High speed packet data services (144.4K) • Uses existing radio spectrum • 1XRTT used	CDMA2000 – MC multi carrier

5.3 Technology Platforms

The migration path that an operator must or should make from an exiting 2G to a 3G wireless platform needs to be chosen with extreme care to ensure the best allocation of the company's resources, including capital, spectrum, and manpower. In order to determine the best utilization of resources, the choice of which 3G platform to use needs to be decided upon. The decision as to which platform to choose is often, and correctly, based on the existing system that is in place. However, which 3G platform to use does not necessarily need to be dictated based on the existing 2G platform. The 2.5 platform will in all cases require some change to the existing infrastructure. The commonality for the 3G systems decided upon is the packet data network that the operator will need to deploy, and this will need to be done regardless of which platform is chosen. Therefore, the migration strategy is really directly related to the radio frequency network that is in place or will be in place.

Several access platforms are referred to as 2.5G. The objective for a 2.5G platform is to bridge an existing network that is using 1G or 2G radio access platforms to that of 3G. The obvious question is, "why not transition to 3G right from 1G or 2G?" The brutal reality is that 3G systems are not currently deployed or even really available at this writing; however, a vast array of 2.5G platforms are available that can deliver many of the required data rates envisioned for 3G services.

The general platforms that will be briefly discussed are as follows:

- EDGE/GPRS
- *High-Speed-Circuit Switched Data* (HSCSD)
- CDMA2000

In order to make an informed decision as to which interim platform to utilize, fundamental knowledge of the interim platforms needs to be understood. Therefore, what follows is an overview of several major technology platforms that are referenced as 2.5G.

5.4 General Packet Radio Service (GPRS)

As discussed in Chapter 3, "Second Generation (2G)," the *Global System for Mobile communications* (GSM) provides voice and data services that are circuit-switched. For data services, the GSM network effectively emulates a modem between the user device and the destination data network. Unfortunately, however, this is not necessarily an efficient mechanism for the support of data traffic. Moreover, standard GSM supports user data rates of up to 9.6 Kbps. In these days of the Internet, such a speed is considered very slow. Consequently, the need exists for a solution that provides more efficient packet-based data services at higher data rates. One solution is the *General Packet Radio Service* (GPRS). Although GPRS does not offer the high-bandwidth services envisioned for 3G, it is an important step in that direction.

In this chapter, we spend some time describing the operation of GPRS. As we shall see in later chapters, UMTS Release 1999 reuses a great deal of GPRS functionality. Therefore, a solid understanding of GPRS will greatly help in understanding UMTS.

5.4.1 GPRS Services

GPRS is designed to provide packet data services at higher speeds than those available with standard GSM circuit-switched data services. In theory, GPRS could provide speeds of up to 171 Kbps over the air interface, though such speeds are never achieved in real networks (because, among other considerations, there would be no room for error correction on the *radio frequency* [RF] interface). In fact, the practical maximum is actually a little over 100 Kbps, with speeds of about 40 Kbps or 53 Kbps more realistic. Nonetheless, once can see that such speeds are far greater than the 9.6-Kbps maximum provided by standard GSM.

The greater speeds provided by GPRS are achieved over the same basic air interface (that is, the same 200-kHz channel, divided into eight timeslots). With GPRS, however, the *mobile station* (MS) can have access to more than one timeslot. Moreover, the channel coding for GPRS is somewhat different than that of GSM. In fact, GPRS defines a number of different channel coding schemes. The most commonly used coding scheme for packet data transfer is *Coding Scheme 2* (CS-2), which enables a given timeslot to carry data at a rate of 13.4 Kbps. If a single user has access to multiple timeslots, then speeds such as 40.2 Kbps or 53.6 Kbps become available to that user. Table 5-2 lists the various coding schemes available and the associated data rates for a single timeslot.

The air interface rates in Table 5-2 give the user rates over the RF interface. As we shall see, however, the transmission of data in GPRS involves a number of layers above the air interface, with each layer adding a certain amount of overhead. Moreover, the amount of overhead generated by each layer depends on a number of factors, most notably the size of the application packets to be transmitted. For a given amount of data to be transmit-

Table 5-2

GPRS Coding Schemes and Data Rates

Coding Scheme	Air Inteface Data Rate (Kbps)	Approximate Usable Data Rate (Kbps)
CS-1	9.05	6.8
CS-2	13.4	10.4
CS-3	15.6	11.7
CS-4	21.4	16.0

ted, smaller application packet sizes cause a greater net overhead than larger packet sizes. The result is that the rate for usable data is approximately 20 to 30 percent less than the air interface rate.

As mentioned, the most commonly used coding scheme for user data is CS-2. This scheme provides reasonably robust error correction over the air interface. Although CS-3 and CS-4 provide higher throughput, they are more susceptible to errors on the air interface. In fact, CS-4 provides no error correction at all on the air interface. Consequently, CS-3 and particularly CS-4 generate a great deal more retransmission over the air interface. With such retransmission, the net throughput may well be no better than that of CS-2.

Of course, the biggest advantage of GPRS is not simply the fact that it allows higher speeds. If that were the only advantage, then it would not be any more beneficial than *High-Speed Circuit-Switched Data* (HSCSD), described later in this chapter. Perhaps the greatest advantage of GPRS is the fact that it is a packet-switching technology. This means that a given user consumes RF resources only when sending or receiving data. If a user is not sending data at a given instant, then the timeslots on the air interface can be used by another user.

Consider, for example, a user that is browsing the Web. Data is transferred only when a new page is being requested or sent. Nothing is being transferred while the subscriber contemplates the content of a page. During this time, some other user can have access to the air interface resources, with no adverse impact to our Web-browsing friend. Clearly, this is a very efficient use of scarce RF resources.

The fact that GPRS enables multiple users to share air interface resources is a big advantage. This means, however, that whenever a user wants to transfer data, then the MS must request access to those resources and the network must allocate the resources before the transfer can take place. Although this appears to be the antithesis of an "always-connected" service, the functionality of GPRS is such that this request-allocation procedure is well hidden from the user and the service appears to be "always-on."

Imagine, for example, a user that downloads a Web page and then waits for some time before downloading another page. In order to download the new page, the MS requests the resources, is granted the resources by the network, and then sends the Web page request to the network, which forwards the request to the external data network (such as the Internet). This happens quite quickly, however, so that the delay is not great. Quite soon, the new page appears on the user's device and at no point does the user have to dial-up to the ISP.

5.4.2 GPRS User Devices

GPRS is effectively a packet-switching data service overlaid on the GSM infrastructure, which is primarily designed for voice. Furthermore, although certainly a demand exists for data services, voice is still the big revenue generator—at least for now. Therefore, it is reasonable to assume that users will require both voice and data services, and that operators will want to offer such services either separately or in combination. Consequently, GPRS users can be grouped into three classes:

- **Class A** Supports the simultaneous use of voice and data services. Thus, a Class-A user can hold a voice conversation and transfer GPRS data at the same time.
- **Class B** Supports simultaneous GPRS attach and GSM attach, but not the simultaneous use of both services. A Class-B user can be "registered" on GSM and GPRS at the same time, but cannot hold a voice conversation and transfer data simultaneously. If a Class-B user has an active GPRS data session and wants to establish a voice call, then the data session is not cleared down. Rather it is placed on hold until such time as the voice call is finished.
- **Class C** Can attach to either GSM or GPRS, but cannot attach to both simultaneously. Thus, at a given instant, a Class-C device is either a GSM device or a GPRS device. If attached to one service, then the device is considered detached from the other.

In addition to the three classes described, other aspects of the MS are important. Most notable is the multi-slot capability of the device, which directly affects the supported data rate. For example, one device might support three timeslots, whereas another might only support two. Note also that GPRS is asymmetric—it is possible for a single MS to have different numbers of timeslots in the downlink and uplink. Normal usage patterns (such as Web browsing) generally require more data transfer in the downlink direction. Consequently, it is common for a user device to have different multi-slot capabilities between the uplink and downlink. For example, many of today's handsets support just a single timeslot in the uplink direction, while supporting three or four timeslots in the downlink direction.

5.4.3 The GPRS Air Interface

The GPRS air interface is built upon the same foundations as the GSM air interface—the same 200-kHz RF carrier and the same eight timeslots per carrier. This allows GSM and GPRS to share the same RF resources. In fact, if one considers a given RF carrier, then at a given instant, some of the timeslots may be carrying GSM traffic, while some are carrying GPRS data. Moreover, GPRS enables the dynamic allocation of resources, such that a given timeslot may be used for standard voice traffic and subsequently for GPRS data traffic, depending on the relative traffic demands. Therefore, no special RF design or frequency planning is required by GPRS above that required for GSM. Of course, GPRS demand may require the addition of additional carriers in a cell. In such a situation, additional frequency planning effort may be required, but this is no different from the frequency planning that is required with the addition of an RF carrier to support additional GSM voice traffic.

Although GPRS uses the same basic structure as GSM, the introduction of GPRS means the introduction of a number of new logical channel types and new channel coding schemes to be applied to those logical channels. When a given timeslot is used to carry GPRS-related data traffic or control signaling, then it is known as a *Packet Data Channel* (PDCH). As shown in Figure 5-1, such channels use a 52-multiframe structure as opposed to a 26-multiframe structure for GSM channels. In other words, for a given timeslot (that is, PDCH), the information that is being carried at a given instant is dependent upon the position of the frame within an overall 52-frame structure. Of the 52 frames in a multiframe, 12 radio blocks carry user data and signaling, 2 idle frames are used, and 2 *Packet Timing Control Channels*

Figure 5-1 GPRS Air Interface Frame Structure.

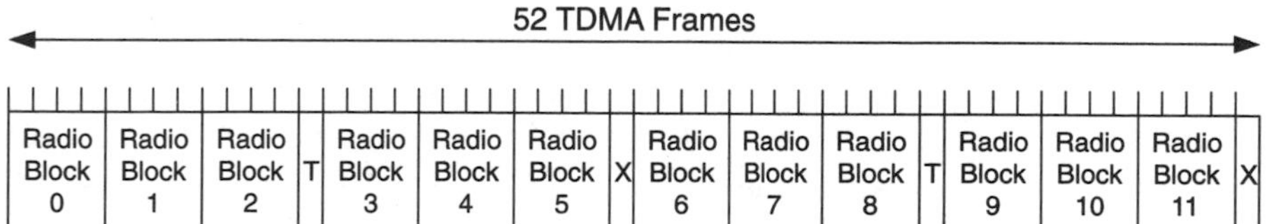

X = Idle Frame
T = Frame Used for PTCCH

(PTCCHs) as described in the following section. Each radio block occupies four TDMA frames, such that 12 radio blocks are used in a multiframe. In other words, a radio block is equivalent to four consecutive instances of a given timeslot. The idle frames in the multiframe can be used by the MS for signal measurements.

5.4.4 GPRS Control Channels

Similar to GSM, GPRS requires a number of control channels. To begin with, the *Packet Common Control Channel* (PCCCH), like the CCCH in GSM, is comprised of a number of logical channels. The logical channels of the PCCCH include

- **Packet Random Access Channel (PRACH)** Applicable only in the uplink, this is used by an MS to initiate a transfer of packet signaling or data.
- **Packet Paging Channel (PPCH)** Applicable only in the downlink, this is used by the network to page an MS prior to a downlink packet transfer.
- **Packet Access Grant Channel (PAGCH)** Applicable only in the downlink, this is used by the network to assign resources to the MS prior to packet transfer.
- **Packet Notification Channel (PNCH)** This is used for *Point-to-Multipoint Multicast* (PTM-M) notifications to a group of MSs.

The PCCCH must be allocated to a different RF resource (that is, a different timeslot) from the CCCH. The PCCCH, however, is optional. If it is omitted, then the necessary GPRS-related functions are supported on the CCCH.

Similar to the BCCH in GSM, GPRS includes a *Packet Broadcast Control Channel* (PBCCH). This is used to broadcast GPRS-specific system information. Note, however, the PBCCH is optional. If the PBCCH is omitted, then the BCCH can be used to carry the necessary GPRS-related system information. If the PBCCH is provisioned in a cell, then it is carried on the same timeslot as the PCCCH in the same way that a CCCH and BCCH can be carried on the same timeslot in GSM.

In the case where a given timeslot is used to carry control channels (PBCCH or PCCCH), then radio block 0 is used to carry the PBCCH, with up to three additional radio blocks allocated for PBCCH. The remaining radio blocks are allocated to the various PCCCH logical channels such as PPCH or PAGCH.

Similar to GSM, GPRS supports some *Dedicated Control Channels* (DCCHs). In GPRS, these DCCHs are the *Packet Associated Control Channel* (PACCH) and the *Packet Timing Control Channel* (PTCCH). The PTCCH is used for the control of the timing advance for MSs. The PACCH is a bidirectional channel used to pass signaling and other information between the MS and the network during packet transfer. It is associated with a given *Packet Data Traffic Channel* (PDTCH) described in the following section. The PACCH is not permanently assigned to any given resource. Rather, when information needs to be sent on the PACCH, part of the user packet data is pre-empted, in much the same manner as is done for the FACCH in GSM.

5.4.4.1 Packet Data Traffic Channels (PDTCHs) The PDTCH is the channel that is used for the transfer of actual user data over the air interface. All PDTCHs are unidirectional—either uplink or downlink. This corresponds to the asymmetric capabilities of GPRS. One PDTCH occupies a timeslot and a given MS with multislot capabilities may use multiple PDTCHs at a given instant. Furthermore, a given MS may use a different number of PDTCHs in the downlink versus the uplink. In fact, an MS could be assigned a number of PDTCHs in one direction and zero PDTCHs in the other.

If an MS is assigned a PDTCH in the uplink, it must still listen to the corresponding timeslot in the downlink, even if that timeslot has not been assigned to the MS as a downlink PDTCH. Specifically, it must listen for any PACCH transmissions in the downlink. The reason is the bidirectional nature of the PACCH, which in the downlink is used to carry signaling from the network to the MS, such as acknowledgements.

5.4.5 GPRS Network Architecture

GPRS is effectively a packet data network overlaid on the GSM network. It provides packet data channels on the air interface as well as a packet data switching and transport network that is largely separate from the standard GSM switching and transport network.

5.4.5.1 GPRS Network Nodes Figure 5-2 shows the GPRS network architecture. One can see a number of new network elements and interfaces. In particular, we find the *Packet Control Unit* (PCU), the *Serving*

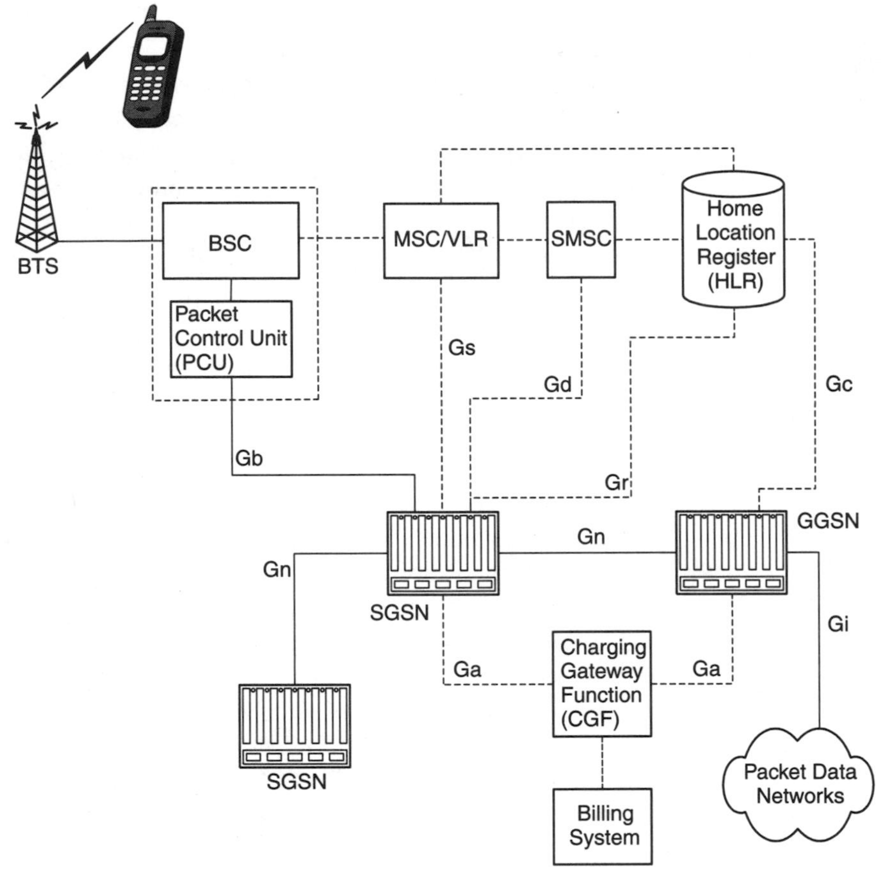

Figure 5-2 GPRS Network Architecture.

GPRS Support Node (SGSN), the *Gateway GPRS Support Node* (GGSN), and the *Charging Gateway Function* (CGF).

The PCU is a logical network element that is responsible for a number of GPRS-related functions such as the air interface access control, packet scheduling on the air interface, and packet assembly and re-assembly. Strictly speaking, the PCU can be placed at the BTS, at the BSC, or at the

SGSN. Logically, the PCU is considered a part of the BSC and in real implementations, one finds the PCU physically integrated with the BSC.

The SGSN is analogous to the *Mobile Switching Center* (MSC)/*Visitor Location Register* (VLR) in the circuit-switched domain. Just as the MSC/VLR performs a range of functions in the circuit-switched domain, the SGSN performs the equivalent functions in the packet-switched domain. These include mobility management, security, and access control functions.

The service area of an SGSN is divided into *routing areas* (RAs), which are analogous to location areas in the circuit-switched domain. When a GPRS MS moves from one RA to another, it performs a routing area update, which is similar to a location update in the circuit-switched domain. One difference, however, is that an MS may perform a routing area update during an ongoing data session, which in GPRS terms is known as a *Packet Data Protocol* (PDP) context. In contrast, for an MS involved in a circuit-switched call, a change of location area does not cause a location update until after the call is finished.

A given SGSN may serve multiple BSCs, whereas a given BSC interfaces with only one SGSN. The interface between the SGSN and the BSC (in fact, the PCU within the BSC) is the Gb interface. This is a Frame Relay-based interface, which uses the *BSS GPRS protocol* (BSSGP). See Figure 5-3. The Gb interface is used to pass signaling and control information as well user data traffic to or from the SGSN.

The SGSN also interfaces to a *Home Location Register* (HLR) via the Gr interface. This is an SS7-based interface and it uses MAP, which has been enhanced for support of GPRS. The Gr interface is the GPRS equivalent of the D interface between a VLR and HLR. The Gr interface is used by the SGSN to provide location updates to the HLR for GPRS subscribers and to retrieve GPRS-related subscription information for any GPRS subscriber that is located in the service area of the SGSN.

An SGSN may optionally interface with an MSC via the Gs interface. This is a SS7-based interface that uses the *Signaling Connection Control Part* (SCCP). Above SCCP is a protocol known as BSSAP+, which is a modified version of the *Base Station Subsystem Application Part* (BSSAP), as used between an MSC and a BSC in standard GSM. The purpose of the Gs interface is to enable coordination between an MSC/VLR and a SGSN for those subscribers that support both circuit-switched services controlled by the MSC/VLR (such as voice) and packet data services controlled by the SGSN. For example, if a given subscriber supports both voice and data services, and is attached to an SGSN, then it is possible for an MSC to page the subscriber for a voice call via the SGSN by using the Gs interface.

The SGSN interfaces with a *Short Message Service Center* (SMSC) via the Gd interface. This enables GPRS subscribers to send and receive short messages over the GPRS network (including the GPRS air interface). The Gd interface is an SS7-based interface using MAP.

A GGSN is the point of interface with external packet data networks (such as the Internet). Thus, the user data enters and leaves the Public *Land Mobile Network* (PLMN) via a GGSN. A given SGSN may interface with one or more GGSNs and the interface between an SGSN and GGSN is known as the Gn interface. This is an IP-based interface used to carry signaling and user data. The Gn interface uses the *GPRS Tunneling Protocol* (GTP), which tunnels user data through the IP backbone network between the SGSN and GGSN.

A GGSN may optionally use the Gc interface to an HLR. This interface uses MAP over SS7. This interface would be used when a GGSN needs to determine the SGSN currently serving a subscriber, similar to the manner in which a *Gateway MSC* (GMSC) queries an HLR for routing information for a mobile-terminated voice call. One difference between the scenarios, however, is the fact that a given data session is usually established by the MS rather than by an external network. If the MS establishes the session, then the GGSN knows which SGSN is serving the MS because the path from the MS to the GGSN passes via the serving SGSN. Therefore, in such situations, the GGSN does not need to query the HLR. A GGSN will query the HLR when the session is initiated by an external data network. This is an optional capability and a given network operator may choose not to support that capability. In many networks, the capability is not implemented as it requires that the MS has a fixed packet protocol address (an IP address). Given that address space is often limited (at least for IP version 4), a fixed address for each MS is not often possible.

An SGSN may interface with other SGSNs on the network. This inter-SGSN interface is also termed the Gn interface and also uses GTP. The primary function of this interface is to enable the tunneling of packets from an old SGSN to a new SGSN when a routing area update takes place during an ongoing PDP context. Note that such forwarding of packets from one SGSN to another occurs only briefly—just as long as it takes for the new SGSN and the GGSN to establish the PDP context directly between them, at which point the old SGSN is removed from the path. This is different from, for example, an inter-MSC handover for a circuit-switched call, where the first MSC remains as an anchor until the call is finished.

5.4.5.2 Transmission Plane Not only does the SGSN interface with a BSC for packet transfer to and from a given MS, direct logical interfaces are also used between an MS and an SGSN—for signaling (signaling plane) and for packet data transfer (transmission plane), even though the interfaces pass physically through the BSS. The overall interface structure for the transmission plane is shown in Figure 5-3.

At the MS, we first have the RF interface, above which are the *Radio Link Control* (RLC) and *Medium Access Control* (MAC) functions. Above these, we find the *Logical Link Control* (LLC), which provides a logical link and framing structure for communication between the MS and the SGSN. Any data between the MS and SGSN is sent in *Logical Link Protocol Data Units* (LL-PDUs). The LLC supports the management of this transfer, including mechanisms for the detection and recovery from lost or corrupted LL-PDUs, ciphering, and flow control. It is worth noting that ciphering in GPRS is somewhat more extensive than in standard GSM. In standard GSM, only the radio link between the MS and BTS is ciphered. In GPRS, ciphering is applied between the MS and the SGSN, such that information is encrypted across the radio interface, the Abis interface, and the Gb interface.

Above the LLC, we find the *SubNetwork Dependent Convergence Protocol* (SNDCP), which resides between the LLC and the network layer (such as IP or X.25). The purpose of SNDCP is to enable support for multiple network protocols without having to change the lower layers such as LLC. Not only does SNDCP provide a buffer between the higher and lower layers, it enables several packet streams to be multiplexed onto a single logical link

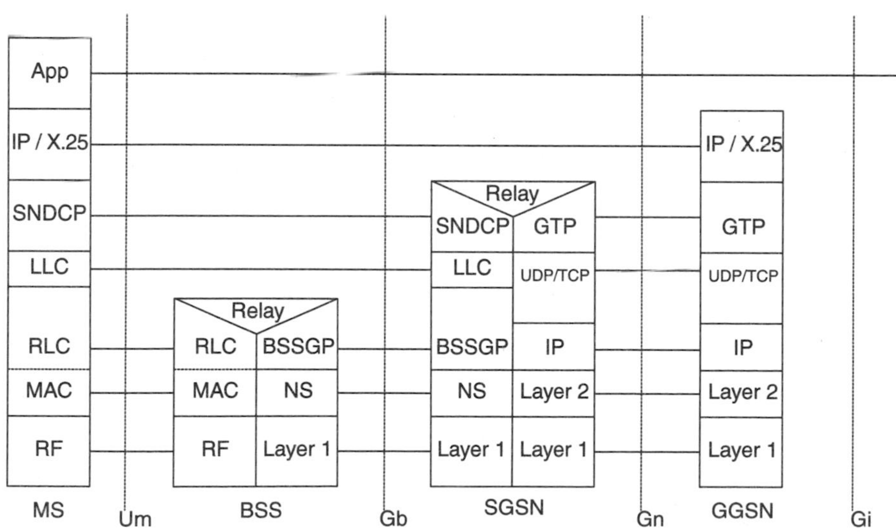

Figure 5-3 GPRS transmission plane.

between the MS and SGSN. It also optionally performs compression (such as TCP/IP header compression and/or V.42bis data compression). Such compression, in particular V.42bis, can make a noticeable difference to the throughput.

At the BSS, a relay function relays LL-PDUs from the Gb interface to the air interface (the Um interface). Similarly, at the SGSN, a relay function relays PDP PDUs between the Gb interface and the Gn interface.

When one looks at Figure 5-3 initially, one finds that the IP layer appears to be repeated. In fact, it can be. Recall that GTP is a tunneling protocol. As far as the applications at either end are concerned, only one IP connection exists—the one directly below the application layer, as shown in Figure 5-3. GTP effectively places this connection and its associated packets in a wrapper for transmission through the IP network between GGSN and SGSN. Thus, the IP network nodes (routers) between SGSN and GGSN consider the GTP packets to be the application, and those routers do not examine the contents of the GTP layer. At the SGSN, the wrapper is removed and the packet is passed to the MS using SNDCP, LLC, and lower layers. For packets from the MS to the external network (such as the Internet), the GGSN removes the wrapper and forwards the IP packets.

5.4.5.3 Signaling Plane Figure 5-4 shows the signaling plane from MS to SGSN. At the lower layers, it is identical to the transmission plane. However, at the higher layers, we find the *GPRS Mobility Management and Session Management* (GMM/SM) protocol instead of the SNDCP. This is the protocol that is used for routing area updates, security functions (such as authentication), session (that is, the PDP context) establishment, modification, and deactivation.

5.4.6 GPRS Traffic Scenarios

The following sections provide some straightforward examples of GPRS traffic. This allows for an understanding of the differences between GSM and GPRS, and later, the differences between GPRS and UMTS. Prior to describing these, we need to familiarize ourselves with some terms.

Temporary Block Flow (TBF) is the physical connection between the MS and the network for the duration of the data transmission. The TBF can be considered to use a number of radio blocks over the air interface.

Temporary Flow Identity (TFI) is an identifier assigned to a given TBF and is used for distinguishing one TBF from another. A TFI is used in con-

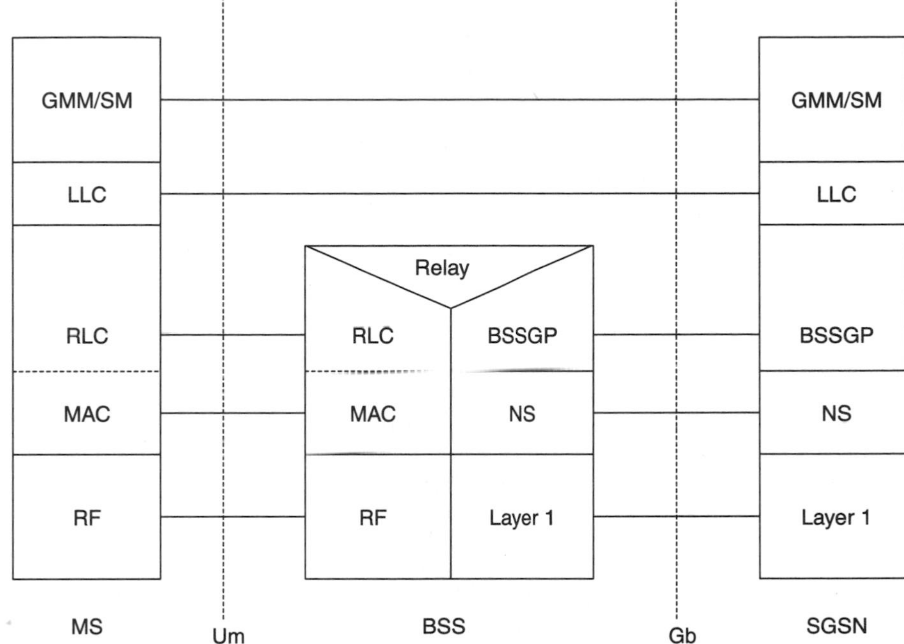

Figure 5-4 GPRS Signaling Plane MS-SGSN.

trol messages (such as acknowledgements) related to a given TBF, so that the entity receiving the control message can correlate the message with the appropriate TBF.

Temporary Logical Link Identity (TLLI) is an identifier that uniquely identifies an MS within a routing area. The TLLI is sent in all packet transfers over the air interface. The TLLI is derived from the *Packet Temporary Mobile Station Identity* (P-TMSI) assigned by an SGSN, provided that the MS has been assigned a P-TMSI. In case the MS has never been assigned a P-TMSI, then the MS may generate a random TLLI.

An *Uplink State Flag* (USF) is an indicator used by the network to specify when a given MS is entitled to use a given uplink resource. In GPRS, resources are shared in both the downlink and the uplink. The downlink is under the control of the network, which can schedule transmissions for a given user on a given downlink PDTCH as appropriate. On the uplink, however, a mechanism is necessary to ensure that only a given MS transmits on a given uplink resource at a given time. This can be done in two ways, through fixed allocation or dynamic allocation.

With fixed allocation, the network allocates some number of uplink timeslots to a user, some number of radio blocks that the MS may transmit, and specifies the TDMA frame when the user may begin transmission. Thus, the MS is provided with exclusive access to the timeslot for a particular period of time. With dynamic allocation, the network does not allocate a specific time upfront for the user to transmit. Rather, it allocates the user a particular value of USF for each timeslot that the user may access. Then on the downlink, the network transmits a USF value on each radio block. This value indicates which MS has access to the next radio block on the corresponding timeslot in the uplink. Thus, by examining the value of USF received on the downlink, the MS can schedule its uplink transmissions. The USF is a three-bit field and thus has eight possible values. Thus, with dynamic allocation, up to eight MSs can share a given uplink timeslot.

5.4.6.1 GPRS Attach GPRS functionality in an MS can be activated either when the MS itself is powered on, or perhaps when the browser is activated. Whatever the reason for the initiation of GPRS functionality within the MS, the MS must attach to the GPRS network, so that the GPRS network (and specifically the serving SGSN) knows that the MS is available for packet traffic. In the terms used in GPRS specifications, the MS moves from an idle state (not attached to the GPRS network) to a ready state (attached to the GPRS network and in a position to initiate a PDP context). When in the ready state, the MS can send and receive packets. Also, a standby state is available, which the MS enters after a time-out in the ready state. If, for example, the MS attaches to the GPRS network but does not initiate a session, then it will remain attached to the network, but move to a standby state after a time-out.

Figure 5-5 shows the simple case of a Class-C MS performing a GPRS attach. In this figure, we have included a great deal of the air interface signaling. Many of the air interface messages shown in the figure are applicable to any access to or from the MS, whether or not that access is just for signaling of the transfer of user packets. For the sake of brevity, they will not be repeated in every subsequent scenario we describe.

A GPRS Attach is somewhat similar in functionality to a location update in GSM. The process begins with a packet channel request from the MS. In the request, the MS indicates the purpose of the request, such as a page response, a *Mobility Management* (MM) procedure, or two-phase access, which would be used in the case of transferring user data. In the scenario of Figure 5-5, a MM procedure is indicated. The network responds with a packet uplink assignment, which allocates a specific timeslot or timeslots to

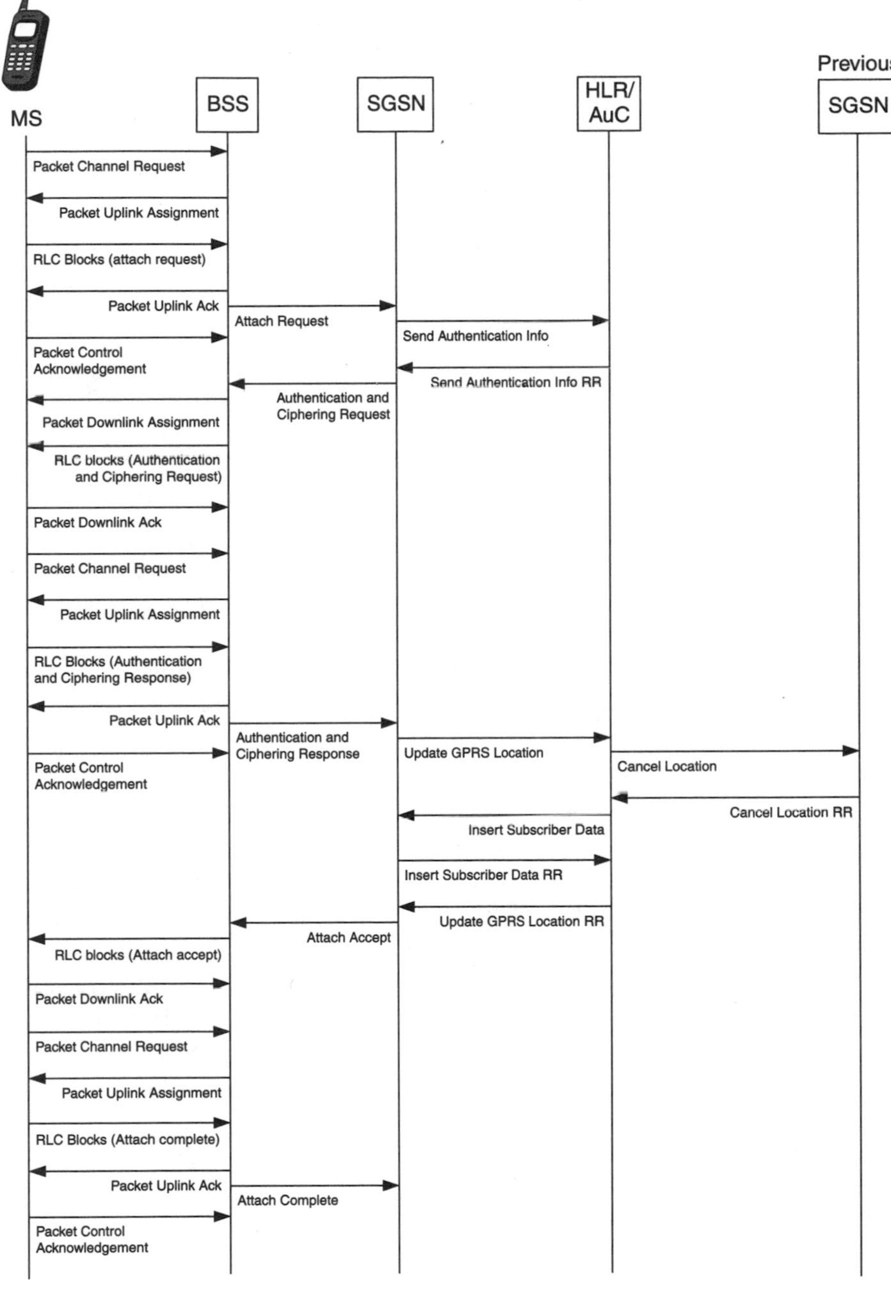

Figure 5-5
GPRS Attach.

the MS for the message that the MS wants to send. The network includes a TFI to be used by the mobile, a USF value for the mobile on the timeslot(s) assigned (in the case of dynamic allocation), and an indication of the number of RLC blocks granted to the MS for the TBF in question.

The MS proceeds to send the attach request in one or more radio blocks to the network on the assigned resources. The MS can send no more than the number of blocks that have been allocated by the network. In the case of MM messages, the assigned resources will typically be sufficient for the MS to send the necessary data. If not, as might be the case when the MS wants to send user packet data, then the MS can request additional resources through a Packet Resource Request message.

Upon receipt of the attach request at the BSS, the BSS uses the PACCH to acknowledge the receipt. In case the MS has sent all of the information it wants to send, which would be the case in our example, then this is indicated in the transmission from the MS to the network. In this case, the acknowledgement from the network is a final acknowledgement, which is indicated in the acknowledgement message itself. This causes the MS to send a Packet Control Acknowledgement message back to the network and release the assigned resources.

Meanwhile, the BSS forwards the attach request to an SGSN. The SGSN may choose to invoke security procedures, in which case it fetches triplets from the HLR. Note, however, that a slight difference can be seen in GPRS regarding authentication and ciphering. Specifically, ciphering in GPRS takes place between the MS and the SGSN such that the whole link from MS to SGSN is encrypted. In standard GSM, only the air interface is encrypted. The authentication and ciphering is initiated by the issuance from the SGSN of the authentication and ciphering request to the MS via the BSS.

The BSS first sends a Packet Downlink Assignment message to the MS. This message can be sent either on the PCCCH or the PACCH. Which one is chosen depends upon whether the MS currently has an uplink PDTCH. If it does, then the PACCH is used. The Packet Downlink Assignment instructs the MS to use a given resource in the downlink—including the timeslot(s) to be used and a downlink TFI value. The BSS subsequently forwards the Authentication and Ciphering request as received from the SGSN.

Upon receipt of the request, the MS acknowledges the downlink message and then requests uplink resources so that it can respond. Thus, it sends another Packet Channel Request, much like the one it sent initially. Once again, the network assigns resources to the MS, which the MS uses to send its authentication and ciphering response to the network. That response is forwarded from the BSS to the SGSN. The BSS also sends an acknowledgment to the MS, and the MS confirms receipt of the acknowledgement, just as it did for the acknowledgement associated with the initial attach request.

Once the MS is authenticated by the SGSN, the SGSN performs a GPRS Update Location towards the HLR. This is similar to a GSM location update, including the download of subscriber information from the HLR to the SGSN. Once the Update Location is accepted by the HLR, the SGSN sends the message Attach Accept to the MS. As for other messages, the BSS first assigns resources so that the MS can receive the message. Similarly, once the MS receives the message, it requests resources in the uplink so that it can respond with an Attach Complete message. The BSS acknowledges receipt of the RLC data containing the Attach Complete and forwards the message to the SGSN. The MS confirms receipt of the acknowledgement.

Note that throughout the procedure just described, the MS requests access to resources for each message that it sends towards the network. This is typical of the manner in which GPRS manages resources and is one of the main reasons why GPRS enables multiple users to share limited resources. Of course, in our example, only signaling is occurring, which consumes very little RF capacity (very few radio blocks). In the case of a packet data transfer, many more data blocks would be transmitted for a given TBF. Not every block needs to be acknowledged, however. In fact, GPRS enables both acknowledged and unacknowledged operations. In the case of an acknowledged operation, acknowledgements are sent only periodically, with each acknowledgement indicating all the correctly received RLC blocks up to an indicated block sequence number.

5.4.6.2 Combined GPRS/GSM Attach Figure 5-6 depicts a simple GPRS attach scenario that would apply to a Class-C MS. In the case of a Class-A or Class-B MS, the MS may want to simultaneously attach to the GSM network and the GPRS network. In this case, the MS can attach to the MSC/VLR during the GPRS attach procedure. This assumes, of course, that the network supports a combined attach (which it broadcasts in System Information messages) and that the network includes the Gs interface.

If, for example, a Class-B MS is powered up and needs to attach to both the GSM and GPRS services, then the sequence would be as depicted in Figure 5-6. For the sake of brevity, we have omitted some air interface signaling, which would be the same in the example of Figure 5-6, as already shown in Figure 5-5.

Figure 5-6 Combined GPRS/GSM Attach.

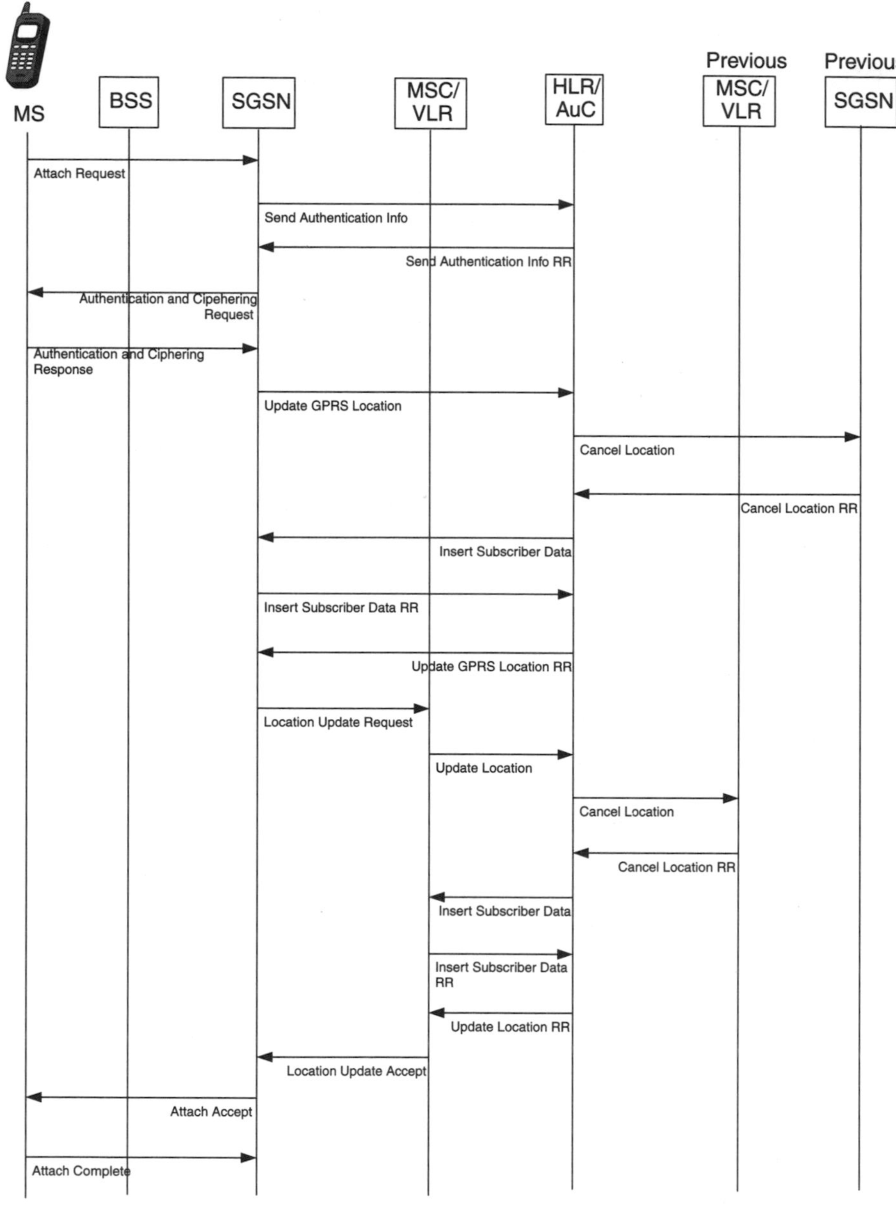

In this case, the MS instigates an attach to the SGSN, but it also indicates that it wants to perform a GSM attach. In this case, the new SGSN, in addition to performing the procedures required of a GPRS attach, also interacts with the VLR to initiate a GSM attach. Specifically, we note the use of the BSSAP+messages Location Update Request and Location Update Accept between the SGSN and the VLR. The Location Update Request message from the SGSN is similar to the equivalent message that would be received from an MS that performs a normal GSM location update. Therefore, the MSC/VLR performs similar mobility management functions (see also Figure 3-10 in Chapter 3) such as performing a MAP Update Location to the HLR. One difference in this scenario, however, is the fact that the MSC/VLR does not attempt to authenticate the MS itself, as the authentication has already been performed by the SGSN.

Note that in Figure 5-5 and Figure 5-6 certain optional functions have not been shown. These functions include an *International Mobile Equipment Identity* (IMEI) check and the allocation of a new *Packet Temporary Mobile Subscriber Identity* (P-TMSI).

5.4.6.3 Establishing a PDP Context The transfer of packet data is through the establishment of a *Packet Data Protocol* (PDP) context, which is effectively a data session. Normally, such a context is initiated by the MS, as would happen, for example, when a browser on the MS is activated and the subscriber's home page is retrieved from the Internet. When an MS or the network initiates a PDP context, the MS moves from the standby state to the ready state. The initiation of a PDP context is illustrated in Figure 5-7.

An MS-initiated PDP context begins with a request from the MS to activate a PDP context. This request includes a number of important information elements, including a requested *Network Service Access Point Identifier* (NSAPI), a requested LLC *Service Access Point Identifier* (SAPI), a requested *Quality of Service* (QoS), a requested PDP address, and a requested *Access Point Name* (APN).

The NSAPI indicates the specific service within the MS that wants to use GPRS services. For example, one service might be based on IP; another might be based on X.25.

The LLC SAPI indicates the requested service at the LLC layer. Recall that LLC is used both during data transfer and during signaling. Consequently, at the LLC layer, it is necessary to identify the type of service being requested, such as GPRS mobility management signaling, a user data transfer, or a *short message service* (SMS), which can be supported over GPRS as well as over GSM.

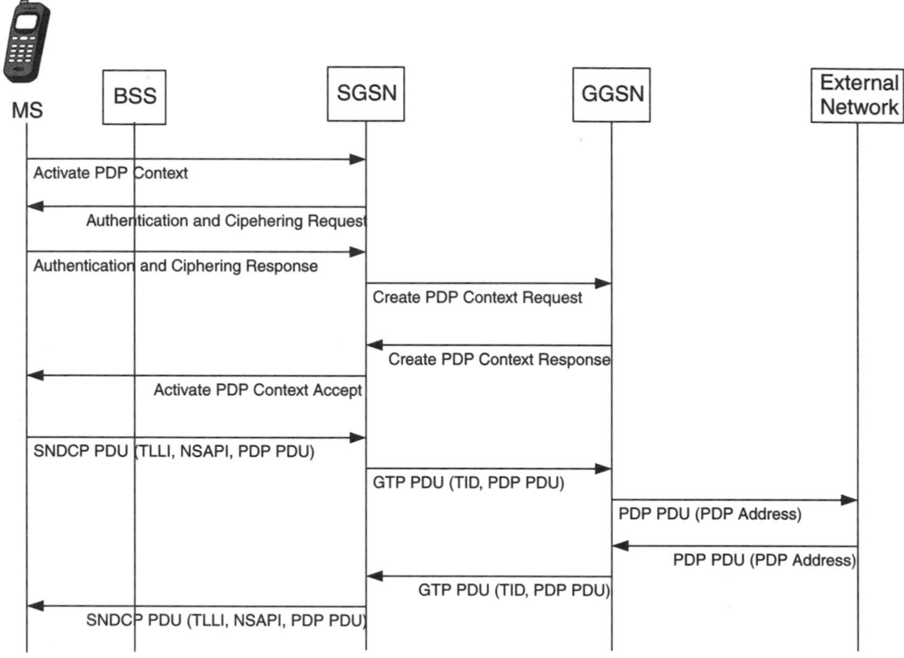

Figure 5-7 PDP Context Activation.

The requested QoS indicates the desires of the MS regarding how the session should be handled. Components of QoS include reliability (including the maximum acceptable probabilities of packet loss, packet corruption, and out-of-sequence delivery), delay, mean throughput, peak throughput, and precedence (which is used to determine the priority of the MS's packets in case of network congestion where packets may need to be discarded).

The requested PDP address will typically be either an IP address or will be empty. The network will interpret an empty address as a request that the network should assign an address. In such a case, the *Dynamic Host Configuration Protocol* (DHCP) should be supported in the network. The address is assigned by the GGSN, which must either support DHCP capabilities itself or must interface with a DHCP server.

The access point name indicates the GGSN to be used and, at the GGSN, it may indicate the external network to which the MS should be connected. The APN contains two parts—the APN network identifier and the APN operator identifier. The APN network identifier appears like a typical Internet URL according to *Domain Name Service* (DNS) conventions—a number of strings separated by dots, such as host.company.com. The APN operator identifier is optional. When present, it has the format operator.operator-group.gprs. Each operator has a default APN operator identifier, which has the form MNC.MCC.GPRS. The *Mobile Country Code* (MCC) and the *Mobile Network Code* (MNC) are part of the *International Mobile Subscriber Identity* (IMSI) that identifies the subscriber and is available at the SGSN. The default APN operator identifier is used to route packets from a roaming subscriber to a GGSN in the home network in case the APN from the subscriber does not include an APN operator identifier.

Based on the APN received from the subscriber, the SGSN determines the GGSN that should be used. The SGSN normally does this by sending a query to a DNS server (not shown in Figure 5-7). The query contains the APN, and the DNS server responds with an IP address for the appropriate GGSN.

Next, the SGSN creates a *tunnel ID* (TID) for the requested PDP context. The TID combines the subscriber IMSI with the NSAPI received from the MS and uniquely identifies a given PDP context between the SGSN and the GGSN. The SGSN sends a Create PDP Context Request message to the GGSN. This contains a number of information elements, including the TID, the PDP address, the SGSN address, and the QoS profile. Note that the QoS profile sent from the SGSN to the GGSN may not match that received from the MS. The SGSN may choose to override the QoS parameters received from the MS based upon the QoS subscribed (as received from the HLR) or based upon the resources available at the SGSN. If the PDP address is empty, then the GGSN is required to assign a dynamic address.

The GGSN returns the message, Create PDP Context Response to the SGSN. Provided that the GGSN can assign a dynamic address and provided that it can support connection to the external network as specified by the APN, then the response is a positive one. In that case, the response includes, among other items, GGSN addresses for user traffic and for signaling, an end user address (as received from DHCP), the TID, a QoS profile, a charging ID, and a charging gateway address.

Upon receipt of the Create PDP Context Response message from the GGSN, the SGSN sends Activate PDP Context Accept to the MS. This contains the PDP address for the MS (in the case that a dynamic address has been assigned by the network), the negotiated QOS, and the radio priority (which indicates the priority the MS shall indicate to lower layers, and which is associated with the negotiated QOS). Note that the network shall attempt to provide the MS with the requested QoS, or at least come close. If the QoS returned by the SGSN is not acceptable to the MS, then the MS can deactivate the PDP context.

Once the MS has received the PDP Context Accept message from the SGSN, then everything necessary is in place to route packets from the MS through the SGSN to the GGSN and on to the destination network. The MS sends the user packets as SNDCP PDUs. Each such PDU contains the TLLI for the subscriber and the NSAPI indicates the service being used by the subscriber, plus the user data itself. The TLLI and NSAPI enable the SGSN to identify the appropriate GTP tunnel towards the correct GGSN. The SGSN encapsulates the user data within a GTP PDU, including a TID, and forwards the user data to the GGSN. At the GGSN, the GTP tunnel "wrapper" is removed and the user data is passed to the remote data network (such as the Internet).

Packets from the external network back to the MS first arrive at the GGSN. These packets include a PDP address for the MS (such as an IP address), which enables the GGSN to identify the appropriate GTP tunnel to the SGSN. The GGSN encapsulates the received PDU in a GTP PDU, which it forwards to the SGSN. The SGSN uses the TID to identify the subscriber and service in question (that is, the TLLI and NSAPI). It then forwards an SNDCP PDU to the MS via the BSS.

Note again, that access to and from the MS over the air interface requires the request and allocation of resources for use by the MS. In other words, the PDTCH(s) that the MS may be using are not dedicated solely to the MS either during the PDP context establishment or during a packet transfer to or from the external packet network.

5.4.7 Inter-SGSN Routing Area Update

In GPRS, each PDU to or from the MS is passed individually and no permanent resource is established between the SGSN and MS. Thus, if a subscriber moves from the service area of one SGSN to that of another, it is not necessary for the first SGSN to act as an anchor or relay of packets for the duration of the PDP context. This is fortunate as the PDP context could last for a long time. Thus, no direct equivalent of a handover, as it is known in circuit-switching technology, takes place, where the first MSC acts as an anchor until a call is finished. Nonetheless, as an MS moves from one SGSN to another during an active PDP context, special functions need to be invoked so that packets are not lost as a result of the transition.

The process is illustrated in Figure 5-8, where an MS moves from the service area of one SGSN to that of another during an active PDP context. The MS notices, from the PBCCH (or BCCH), that it is in a new routing area. Consequently, it sends a routing area update to the new SGSN. Among the information elements in the message are the TLLI, the existing P-TMSI, and the old *Routing Area Identity* (RAI). Based on the old RAI, the new SGSN derives the address of the old SGSN and sends an SGSN Context Request message to the old SGSN. This is a GTP message, passed over an IP network between the two SGSNs.

The old SGSN validates the P-TMSI and responds with an SGSN Context Response message, with information regarding any PDP context and MM context currently active for the subscriber, plus the subscriber's IMSI. PDP context information includes GTP sequence numbers for the next PDUs to be sent to the MS or tunneled to the GGSN, the APN, the GGSN

Figure 5-8 Inter-SGSN Routing Area Update during an active context.

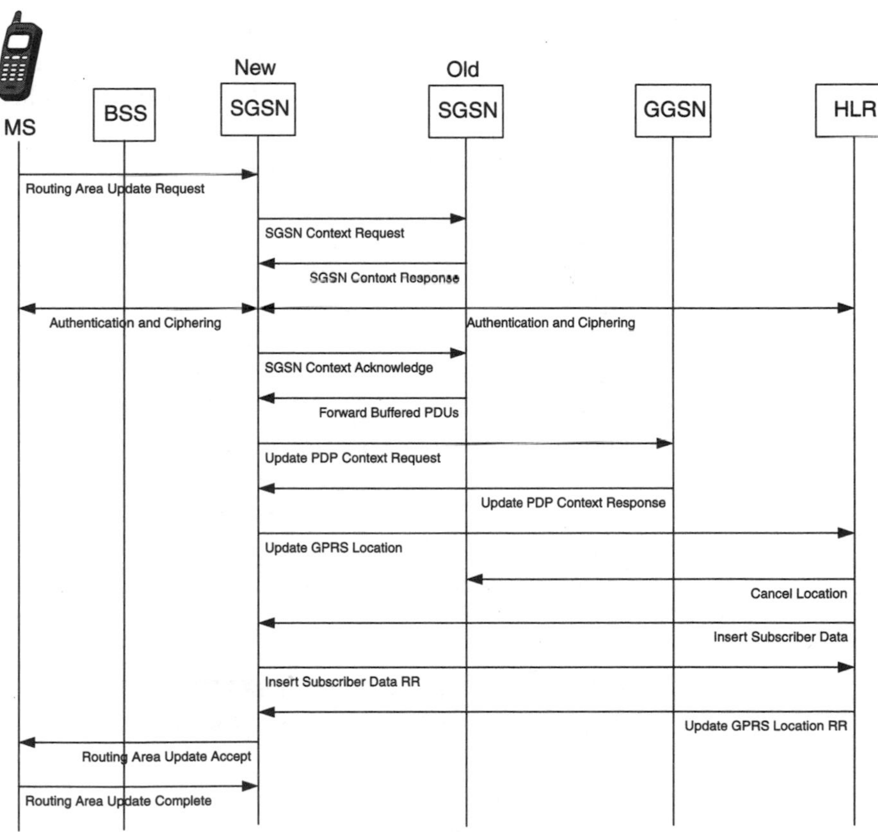

address for control plane signaling, and QOS information. The old SGSN stops the transmission of PDUs to the MS, stores the address of the new SGSN, and starts a timer.

The MM context sent from the old SGSN to the new one may include unused triplets, which the new SGSN will use to authenticate the subscriber. If the old SGSN has not sent such triplets, then the new SGSN can fetch triplets from the HLR in order to perform authentication and ciphering.

The new SGSN responds to the old one with the GTP message, SGSN Context Acknowledge. This indicates to the old SGSN that the new one is ready to take over the PDP context. Consequently, the old SGSN forwards any packets that may have been buffered at the old SGSN so that the new SGSN can forward them. The old SGSN continues to forward to the new SGSN any additional PDUs that are received from the GGSN.

The new SGSN sends an Update PDP Context request to the GGSN to inform the GGSN of the new serving SGSN for the PDP context. The GGSN responds with the Update PDP Context Response message. Any subsequent PDUs from the GGSN to the MS are now sent via the new SGSN.

The new SGSN then invokes an Update GRPS Location operation towards the HLR. This causes the HLR to send a MAP Cancel Location to the old SGSN. Upon receipt of the Cancel Location, the old SGSN stops the timer and deletes any information regarding the subscriber and the PDP context.

Once the MAP Update Location procedure is complete, the new SGSN accepts the Routing Area update from the MS, which the MS acknowledges with a Routing Area Complete message. The new SGSN proceeds to send and receive PDUs to and from the MS.

In a combined GSM/GPRS network, it is common for location area boundaries and routing area boundaries to coincide. In such a case, an inter-SGSN routing area update might also coincide with the need to perform a location update towards a new MSC/VLR. In that case, the SGSN can communicate with the MSC over the Gs interface and can trigger a location update at the MSC in much the same manner as shown in Figure 5-6 for a combined GSM/GPRS Attach.

5.4.8 Traffic Calculation and Network Dimensioning for GPRS

Dimensioning a GPRS network involves the dimensioning of a number of network elements (such as SGSNs and GGSNs) and a number of network interfaces (air interface, Gb, and Gn). Of each of the resources (nodes and interfaces) available in a GPRS network, the most limited is the air interface. Moreover, the air interface resources must be shared with GSM. One does not want to inhibit GSM voice traffic for the sake of GPRS data traffic or vice versa, so some planning is required.

5.4.8.1 Air Interface Dimensioning The most straightforward way to determine the required GPRS air interface capacity is to estimate the amount of data traffic (in terms of bits/second) that a given cell will be required to handle in the busy hour. This can be done by estimating the number of GPRS users in the cell and estimating the usage requirements of those users (which will be linked to handset capabilities and the commercial agreements between users and the network operator). From this demand estimate, we can estimate an average GPRS throughput requirement in the busy hour. In order to allow for usage spikes within the busy hour, it is appropriate to add an overhead of 20 to 30 percent. From this, we can then determine the number of channels that are needed to support that load. For example, using CS-2, a single timeslot can carry about 10 Kbps of user data.

This approach, however, does not account for the fact that a given cell will most likely be used to support both GPRS data traffic and GSM voice traffic. When a cell's resources are shared between GPRS and GSM, it is quite inefficient to independently determine GPRS and GSM resource requirements (based on some blocking criteria) and simply add the two together. To do so would result in over-dimensioning of the cell. The reason for this is because voice traffic follows an Erlang distribution, which requires that there be more channels in a cell than are used, on average, by the voice traffic.

If, for example, we have a cell with three RF carriers and a total of 22 TCHs (one TCH for BCCH and one TCH for SDCCH/8), then at 2-percent blocking, the 22 TCHs can carry approximately 15 Erlangs. In other words, at any given instant, we can expect 15 of the 22 TCHs to be occupied with voice traffic, leaving seven channels available. This is not to say that voice traffic will never use more than 15 TCHs during the busy hour—just that there will be an average of seven TCHs available during this time. These seven TCHs can be used for GPRS traffic. At CS-2, this corresponds to a gross data rate on the air interface of over 90 Kbps for GPRS traffic and a usable rate of about 70 Kbps. Thus, we can accommodate an average of 70 Kbps of GPRS traffic in the cell during the busy hour without increasing the number of RF channels. Whether this will be sufficient to accommodate the needs of the GPRS users (including any buffer for usage spikes) is dependent upon what those needs happen to be. If it is insufficient, then more RF capacity will need to be added. This RF capacity can be dedicated for GPRS or can be shared between GSM and GPRS. For that matter, any cell that supports both GSM and GPRS can be configured so that all resources are shared or that certain resources are reserved for one service or the other with any remaining resources shared.

This approach whereby inefficiently used GSM capacity is used by GPRS does not necessarily tell the whole story of RF dimensioning. First, the approach assumes that the GSM network is correctly dimensioned for voice to begin with, which may not be the case in heavily loaded cells. Secondly, what we have described implies an assumption that may not be true in reality—the assumption that the GPRS busy hour and the GSM busy hour coincide. If they do not coincide, then the approach described above will err on the conservative side.

As of this writing, relatively few GPRS networks have been deployed (when compared with the number of GSM networks) and relatively few GPRS subscribers exist. Therefore, there is not a great deal of real-world experience to draw upon. This is unfortunate as it means that real-world rules of thumb have not yet been developed. On the other hand, it is fortunate that we have not had to deal with a sudden explosion in the number of GPRS subscribers. As the number of subscribers grows, we will be able to monitor traffic patterns to see which types of transactions subscribers require, the typical file sizes, burstiness, and so on. Such monitoring will enable trending so that RF dimensioning decisions can be made in advance of subscriber demands.

5.4.8.2 GPRS Network Node Dimensioning Among the nodes that need to be dimensioned for GPRS traffic are the BSC, the SGSN, and the GGSN. Generally, the capacity of a BSC is limited by the number of cells, the number of BTS sites (or interfaces to BTS sites), the number of transceivers (regardless of whether those transceivers are used for voice or data), and the number of simultaneous PDCHs. In addition, one needs to dimension the Gb interface, which is related to the number of Gb ports (T1 or E1) supported by the BSC. In most implementations, one finds that the number of supported PDCHs is sufficiently large so that other limitations, such as the maximum number of transceivers, will be reached first, particularly in a combined GSM/GPRS network.

An SGSN has a number of capacity limitations—the number of attached subscribers, the number of cells, the number of routing areas, the number of

Gb ports, and the total throughput capacity. Typically, one finds that the key capacity limitations are the number of attached subscribers and the total throughput, as these limits are likely to be met before any of the others.

For the GGSN, the key limitations are the number of simultaneous PDP contexts and the total throughput. These key dimensioning factors are listed in Table 5-3.

5.5 Enhanced Data Rates for Global Evolution (EDGE)

EDGE once stood for the term 'Enhanced Data Rates for GSM Evolution.' Not long after the technology was proposed, however, it was also suggested that it be used as part of the evolution of IS-136 TDMA networks. In fact, for a while, the accepted evolution path for IS-136 networks was IS-136 to EDGE to something called UWC-136, a wideband TDMA technology. More recently, however, some of the world's largest IS-136 network operators have abandoned that migration path and have opted to move towards UMTS. In fact, those same operators are in the process of complementing and/or replacing their existing networks with GSM/GPRS as a stepping

Table 5-3
GPRS Node Dimensioning Factors

Network Node	GPRS-Specific Dimensioning Factors
BSC	• Number of PDCHs • Number of Gb ports
SGSN	• Number of attached subscribers • Total throughput • Number of Gb ports • Number of cells • Number of routing areas
GGSN	• Number of simultaneous PDP contexts • Total throughput

stone towards UMTS. Consequently, UWC-136 is unlikely to be widely deployed. Moreover, the deployment of EDGE with IS-136 will certainly not happen on the scale once envisaged, if at all. Thus, although the G in EDGE still officially means Global, it may well be that it will only ever be associated with GSM. In this chapter, we focus only on the use of EDGE in a GSM environment.

The basic goal with EDGE is to enhance the data throughput capabilities of a GSM/GPRS network. In other words, the objective is to squeeze more bits per second out of the same 200-kHz carrier and eight-timeslot TDMA. This is done primarily by changing the air interface modulation scheme from *Gaussian Minimum Shift Keying* (GMSK), as used in GSM, to *8 Phase Shift Keying* (8-PSK). The result is that EDGE can theoretically support speeds of up to 384 Kbps. Thus, it is clearly more advanced than GPRS, but still does not meet the requirements for a true 3G system (which should support speeds of up to 2 Mbps). Consequently, one might call EDGE a 2.75G technology.

Whether EDGE will see widespread deployment is a matter of some debate—a debate that revolves around timing, user demand for high-speed data services, the availability of EDGE-capable terminals, and cost. From a timing perspective, the development of EDGE and UMTS technologies are occurring in the same timeframe. In fact, the specification of EDGE standards is done within the *Third Generation Partnership Project* (3GPP) as part of a set of specifications known as 3GPP Release 1999—the same set of specifications that includes UMTS. From a user-demand perspective, it is still unclear as to exactly what the killer applications will be for wireless data and whether the speeds afforded by UMTS will really be required, or whether EDGE speeds will be sufficient.

Then there is the issue of cost. To deploy a UMTS network, one first requires the acquisition of UMTS spectrum. In some countries, this spectrum has been auctioned to the highest bidder, with billions of dollars committed by network operators simply for the right to use a certain amount of UMTS spectrum. Once the spectrum is acquired, one then has to build a completely new radio access network—something which can cost billions of dollars more. To deploy EDGE instead does not require a new spectrum (at least not in the bands set aside for UMTS) and does not require as drastic changes to the network. Consequently, EDGE can be deployed at far less cost than UMTS. It remains to be seen whether EDGE will be widely deployed as a psuedo-3G system, as a stepping stone towards UMTS, or whether operators will decide to leapfrog EDGE and move directly from GPRS to UMTS.

5.5.1 The EDGE Network Architecture

The network architecture for EDGE is basically the same as that for GPRS —largely the same network elements, the same interfaces, the same protocols, and the same procedures. We use the term "largely" because some minor differences exist in the network, but these are insignificant when compared to the enhancements to the air interface, which is where we shall focus.

5.5.1.1 Modulation As mentioned, EDGE uses the same 200-kHz channels and eight-timeslot structure as used for GSM and GPRS. With EDGE, however, 8-PSK modulation is introduced in addition to the 0.3 Gaussian Minimum Shift Keying (GMSK) used in GSM.

0.3 GMSK means that the modulator has a bandpass filter with a 3dB bandwidth of 81.25 kHz. In GSM, the symbol rate is a 270.833 ksymbols/second, with each symbol representing one bit, leading to 270.833 Kbps. The value of 81.25 is 0.3 times 270.833, which is why it is called 0.3 GMSK. The 270.833 Kbps is carried on a 200-kHz carrier, so that GSM provides a bandwidth efficiency of 270.833/200, which equals approximately 1.35 bits/s/Hz.

The objective with EDGE is to offer higher bandwidth efficiency, so that we can squeeze more user data from the same 200-kHz channel. This higher bandwidth efficiency is achieved through 8-PSK. In general, PSK involves a phase change of the carrier signal according to the incoming bit stream. The simplest form of PSK involves a 180° phase change at every transition from 0 to 1, or vice versa, in the incoming bit stream. With 8-PSK, we treat the incoming bit stream in groups of three bits at a time and allow phase changes of 45°, 90°, 135°, 180°, 225°, 270°, or 315°. The specific phase change of the signal represents the change from one set of three bits to the next, as shown in Figure 5-9. With EDGE, the symbol rate is still 270.833 ksymbols/second, as it is in GSM. Each symbol, however, is three bits, such that we have a bit rate of 812.5 Kbps.

Of course, we do not get this great increase in bandwidth efficiency for free. In addition to any extra cost associated with producing devices that can support 8-PSK modulation, we must also contend with the fact that 8-PSK is more sensitive to noise than GMSK. Noise in a signal can make it more difficult for a receiver to determine the exact phase change when the signal changes from one state to another. Because of the fact that the states in 8-PSK are quite close together, the amount of noise required for errors to occur can be relatively small—certainly smaller than the amount of noise

Figure 5-9
8-PSK Relative Phase Positions.

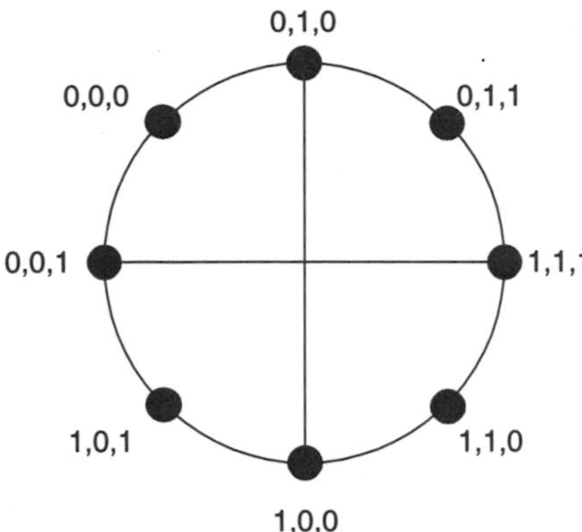

that GMSK can handle. The direct result of this is that if a BTS supports both GMSK and 8-PSK modulation and has the same output power for both, then the cell footprint is smaller for 8-PSK than for GMSK. Recognizing this limitation, however, the specifications for EDGE are such that both the coding scheme and modulation scheme can be changed in response to RF conditions. Thus, as a user moves towards the edge of a cell, the effect of lower signal to noise will mean that the network can reduce the user's throughput, either by changing the modulation scheme to GMSK or by changing the coding scheme to include greater error detection. All that the user will notice is somewhat slower throughput.

5.5.1.2 Air Interface Coding Schemes and Channel Types With the advent of EDGE, we find a number of new channel coding schemes in addition to the coding schemes that exist for GSM voice and GPRS. For packet data services in an EDGE network, we refer to *Enhanced GPRS* (EGPRS) and the new coding schemes for EGPRS are termed *Modulation and Coding Scheme-1 to Modulation and Coding Scheme-9* (MCS-1 to MCS-9). The reason why they are not just called coding schemes is the fact that for MCS-1 to MCS-4, GMSK modulation is used, whereas 8-PSK modulation is used for MCS-5 to MCS-9.

Table 5-4 shows the modulation scheme and data rate applicable to each MCS.

It should be noted that MCS-4 offers no error protection for the user data, nor does MCS-9. Given that MCS-4 offers no error protection and uses GMSK, one would expect that it would provide the same data rate as CS-4, as used in standard GPRS. The difference is due to the fact that in EGPRS, the RLC/MAC header is coded differently from the rest of the PDU and contains additional bits for error correction. The objective is to ensure that at least the header can be decoded. The same does not apply for CS-4.

The channel types applicable to EGPRS are the same as those applicable to GPRS—we have a number of PDCHs that carry PCCCH, PBCCH, PDTCHs, and so on. In fact, these channels are shared among GPRS and EGPRS users. Thus, both GPRS users and EGPRS users can be multiplexed on a given PDTCH. Of course, during those radio blocks when the PDTCH is used by an EGPRS user, the modulation may be either GMSK or 8-PSK, whereas it must be GMSK when used by a GPRS user.

Similar to the manner in which the network controls the coding scheme to be used by a GPRS user, the network also controls the MCS to be used by an EGPRS user in both the uplink and downlink. This is done through the addition of new information elements in the Packet Uplink Assignment and Packet Downlink Assignment messages.

One important aspect of GPRS and EGPRS users sharing the same PDTCH on the uplink is the use of the USF. Recall that the USF is used with dynamic allocation, is sent on the downlink, and is used to indicate which MS has access to the next RLC/MAC block on the uplink. If a given PDTCH is being used for both GPRS and EGPRS MSs, then it is important

Table 5-4

Modulation and Coding Schemes for EGPRS

Scheme	Modulation	RLC Blocks per Radio Block (20 ms)	Input Data payload (bits)	Data Rate (Kbps)
MCS-1	GMSK	1	176	8.8
MCS-2	GMSK	1	224	11.2
MCS-3	GMSK	1	296	14.8
MCS-4	GMSK	1	352	17.6
MCS-5	8-PSK	1	448	22.4
MCS-6	8-PSK	1	592	29.6
MCS-7	8-PSK	2	2 × 448	44.8
MCS-8	8-PSK	2	2 × 544	54.4
MCS-9	8-PSK	2	2 × 592	59.2

that both types of MS be able to decode the USF, so that they may appropriately schedule uplink transmissions. Consequently, when a PDTCH is used for both GPRS and EGPRS, GMSK modulation must be used for any radio block that assigns uplink resources to a GPRS MS. All other radio blocks may use 8-PSK modulation. Note that this forced use of GMSK for radio blocks destined for an 8-PSK MS only applies with dynamic allocation.

5.6 High-Speed Circuit Switched Data (HSCSD)

Prior to the arrival of GPRS or EDGE, the need for higher speeds of data service was well recognized. At the time, GSM supported only data services of up to 9.6 Kbps—the maximum that could be provided on a single timeslot. In order to support higher data rates, the most obvious approach was a solution whereby a given MS could use more than one timeslot, which is basically what HSCSD offers.

Like GPRS, HSCSD enables the asymmetric allocation of resources on the air interface. Unlike HSCSD, however, those resources cannot be shared among multiple users. After all, the connection is circuit-switched. Consequently, HSCSD is a rather inefficient use of valuable RF bandwidth, particularly if the data session is bursty in nature. HSCSD provides for the modification of allocated resources during a call, which can be useful if the network needs to reclaim some of the resources that are being consumed by HSCSD. This flexibility, however, does not approach the efficient use enabled by GPRS.

The initial versions of HSCSD allowed for multiple timeslots, each offering up to 9.6 Kbps of user data. Thus, four timeslots, for example, could offer up to 38.4 Kbps. Subsequently, a change in the channel coding scheme was proposed to allow a single timeslot to carry 14.4 Kbps of user data. One of the main reasons for this change was to enable the mobile fax service to support a fax transmission at 14.4 Kbps over just a single timeslot. Concatenation of four such timeslots could therefore offer speeds up to 57.6 Kbps.

With the advent of the 8-PSK modulation that EDGE can provide, it is possible for HSCSD to achieve high throughput levels with fewer timeslots. For example, depending on the coding scheme chosen, a given timeslot can support 28.8 Kbps, 32.0 Kbps, or 43.2 Kbps, as well as 14.4 Kbps, and it is still possible to concatenate timeslots. An upper limit of 64 kpbs is imposed, however, not because of air interface limitations, but because of limitations within the network. Specifically, the A interface (between the MSC and BSC) is not designed for a given call to occupy more than one 64 Kbps channel.

Although HSCSD has seen some deployment in GSM networks, it cannot be considered widely used. That situation is unlikely to change. With the arrival of 8-PSK modulation, HSCSD will become somewhat more efficient. Nonetheless, the packet-switching technologies of GPRS and EGPRS are still vastly more efficient. Given a choice between HSCSD and the efficiencies of GPRS and EGPRS, it makes sense for a network operator to opt for a packet-switched solution and not HSCSD.

5.7 CDMA2000 (1XRTT)

Phase one of CDMA2000, for the purposes of this discussion, is a 2.5G technology platform because it offers some but not all of the IMTS-2000 requirements that are envisioned for CDMA2000, like full mobility. For the purist at heart, IS-95B has been widely publicized as being 2.5G, whereas CDMA2000 is a 3G platform. However, IS-95B has seen limited implementation and the industry has moved to deploy a 1xRTT platform instead. The CDMA2000 platform is predominantly a non-European platform and is meant to transition IS-95A/B systems from a voice system to a high-speed packet data network capitalizing on the existing radio base station and spectrum allocations that the operators have.

CDMA2000 1xRTT or phase 1, as will be 3xRTT, is fully backward-compatible with the IS-95 infrastructure and subscriber units. It also supports all of the IS-95 existing services such as voice, circuit-switched data, SMS, over the air provisioning, and activation. CDMA2000 1xRTT supports handoffs with IS-95, which uses the same carrier as well as different carriers.

From an operator's point of view, the migration from 2G to 3G via a 2.5G strategy, regardless of the access technology platform chosen, needs to address the following major issues:

- Capacity
- Coverage
- Clarity
- Cost
- Compatibility

This chapter discusses many of the implementation issues associated with the introduction of CDMA2000-1xRTT whether it is for *data and voice* (DV), or just *data only* (DO). The design specifics associated with a1xRTT system will be covered in Chapter 7, "CDMA2000." Numerous purtebations can and will exist with the deployment of a CDMA2000-1x that are, of course, market-specific in nature. Therefore, what follows should provide the necessary guidance from which to begin making design decisions.

5.7.1 Deployment Issues

As implied previously, several deployment issues are associated with the introduction of CDMA2000-1x into a wireless system. Some of the obvious issues relate to the current spectrum usage that the operator has license control of. The spectrum usage considerations take on a different meaning depending on whether the system is new, that is, not a current infrastructure, or if it may or may not have the available spectrum from which to deploy the CDMA2000-1x channels.

Of course, an operator also needs to factor other issues into the decision process when deploying CDMA2000. Capacity is one of those topics and is driven by both the current capacity and utilization of the existing radio spectrum for the system. For example, if the system has the operator contemplating deploying CDMA2000-1x with IS-95A/B deployed also, then the decision of whether to convert existing IS-95 channels to that of IS-2000 needs to be made as well as, of course, which carriers are involved. Another issue that could come about is when an existing operator chooses to change or augment his or her existing wireless offering by introducing CDMA2000-1x into a GSM or IS-136 environment.

The coverage of the system with regards to 1x needs to be addressed and well thought out. The decision based on whether to deploy the 1xRTT channels in a 1:1 or N:1 scenario needs to be decided upon from the onset of the design process. Some of the decisions could be centered around a high-speed packet data offering for the core or heavy commuter locations like an airport or large industry park. Ultimately, the cost of the deployment will drive the final decision metric of when, how, and why.

One of the first issues that comes about in the determination of how to deploy a CDMA2000-1x system, besides estimating demand, is how channel assignment process are determined. The channel assignment method for 1xRTT can take on a slight variation when the decision is to deploy only 1x, as opposed to 1x and 3x. The variant is due primarily to the guard band issues that are different for 1x and 3x. The recommended channel assignment scheme for both cellular and PCS frequency bands is shown in Tables 5-5 and 5-6 .

The channel chart listed here requires a guard band, and the guard band for a single CDMA2000-1x, which is the same as IS-95, is shown in Figure 5-10a. Figure 5-10b shows the requirement when implementing a second channel and the overall impact to the spectrum or rather the existing channel plan that may exist in a wireless system.

However, one important issue needs to be reaffirmed and that is, for a cellular system, F1 needs to be deployed first in the system for any geographic area because the subscriber units hunt for the preferred channel set in the cellular band for CDMA systems. The preferred CDMA carriers are shown in Figure 5-11.

When looking at Figure 5-11, the secondary channels, 691 and 777, while initially assigned and defined for IS-95 systems, are not recommended to be

Table 5-5

Cellular CDMA2000-1x Carrier Assignment Scheme

IS-95 System Sector	Carrier	PN Offset	CDMA2000-1x Carrier	PN Offset
Alpha	1	6	1	6
	2	6	2	6
Beta	1	18	1	18
	–	–	2	18
Gamma	1	12	1	12
	–	–		

Table 5-6

PCS CDMA2000-1x Carrier Assignment Scheme

Source	Target	Destination Traffic Channel Type
IS-95	IS-95	IS-95
IS-95	IS-2000	IS-95
IS-2000	IS-95	IS-95
IS-2000	IS-2000	IS-2000

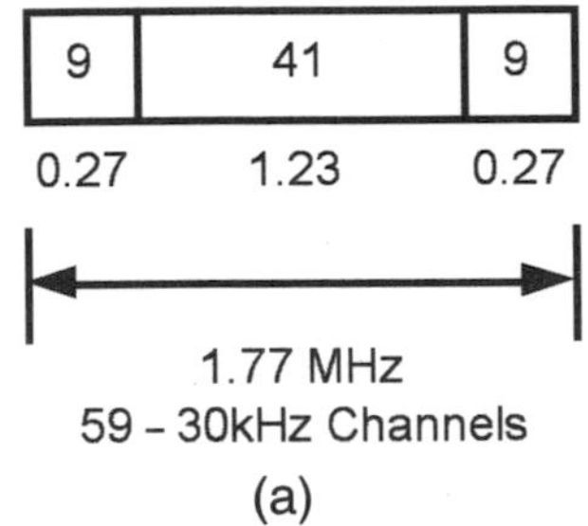

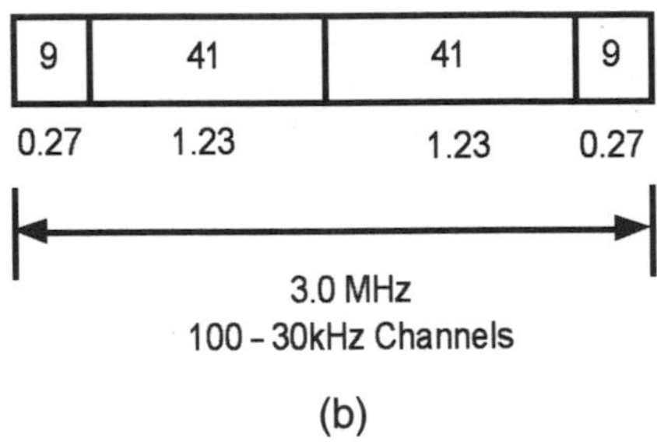

Figure 5-10 CDMA2000-1x guard band: (a) single CDMA2000-1x (b) two CDMA2000-1x channels

Figure 5-11 Preferred CDMA carriers for cellular systems.

used due to out-of-band emissions that create a rise in the noise floor, degrading the CDMA system performance for those particular channels.

Figure 5-11 illustrates the channel deployment scheme in a cellular or PCS system; however, CDMA2000 will be also implemented in the SMR band. Therefore, Figure 5-12 is an indication of the spectrum requirement for implementing CDMA2000-1x into the *specialized mobile radio* (SMR) band, which has a 25-kHz channel bandwidth. It is important to note that the spectrum requirement requires the control of contiguious channels within in the defined service area as well as within the guard zone itself.

5.7.2 System Architecture

The architecture that will be used for a CDMA2000 deployment is effectively the same as that used for an existing IS-95 system with the exception of the *Packet Data Serving Node* (PDSN) network, which is introduced with CDMA2000 systems. Additionally, it is important to note that 1xRTT in CDMA2000-1x utilizes a *spreading rate of 1* (SR1), which is directly compatible with IS-95 because that system utilizes the same spreading rate. However, CDMA2000-1xRTT now incorporates packet data sessions, and

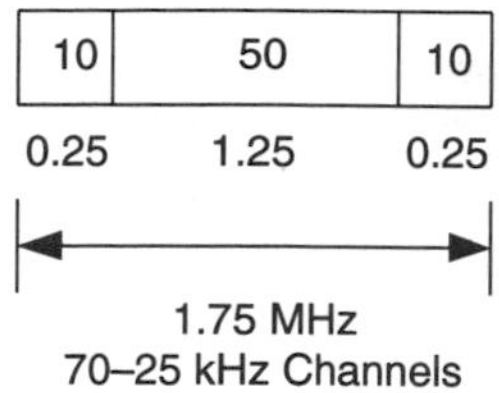

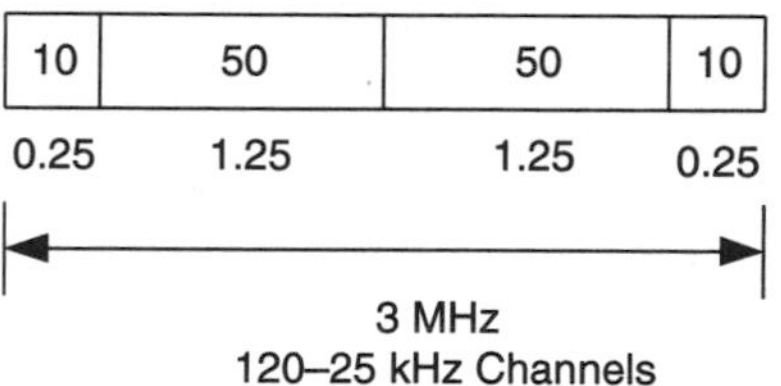

Figure 5-12 CDMA2000-1x spectrum requirements for SMR.

this change, as discussed previously, is implemented through the use of new vocoders as well as channel element cards.

Because the majority of CDMA2000 systems have IS-95 deployed, the choice or, rather, design is determined by the decision to use and how to deploy DO or DV channels, their coverage areas, and of course the number of carriers that will exhibit these new features.

A wireless operator can choose three basic scenarios, assuming two or fewer IS-95 CDMA carriers are deployed in the network. Obviously, if more IS-95 carriers are deployed, the concept discussed next can easily be expanded upon, but two carriers were chosen to ensure the clarity of the concept.

There are six general scenarios for deploying CDMA2000-1x into a wireless system, whether it is an existing system that has IS-95 deployed or not, and the are identified below:

- CDMA2000-1x into existing IS-95 (F1 replacement)
- CDMA2000-1x into existing IS-95 (F2/F3 or greater)
- CDMA2000-1xEV-DO into existing IS-95 (F2/F3 or greater)
- CDMA2000-1xEV-DV into existing IS-95 (F1 replacement and possibly F2)
- CDMA2000-1x and 1xEV-DO intonew system
- CDMA2000-1xEV-DO and 1xEV-DV into new system

Some recommended channel deployment methods for cellular and PCS systems are shown in Figures 5-13 through 5-16. Figure 5-13 is the recommended channel deployment scheme for both a cellular A and B band operator. The methodology used for deciding upon the channel deployment enables legacy systems to still exist and remain in the AMPS band, besides the EAMPS band. The reason for the AMPS band is to enable existing 1G- and 2G-capable phones to still have the capability to access the network for ROAMing as well as emergency services like 911. The EAMPS portion of the band can be used for analog, *Cellular Digital Packet Data* (CDPD), and IS-136 services.

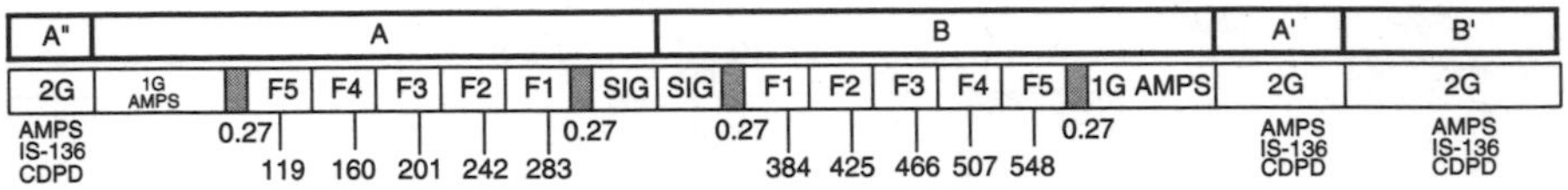

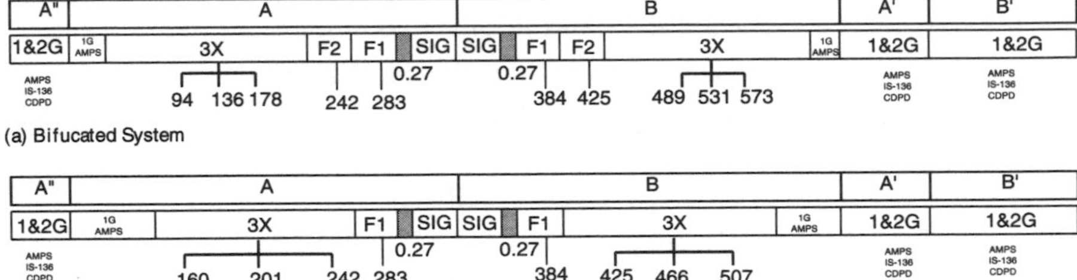

Figure 5-13 Cellular CDMA2000-1x channel deployment scheme.

Figure 5-14 Cellular CDMA2000 channel deployment scheme: (a) bifucated (b) overlay.

A" A B A' B'
1&2G 1G AMPS 3X F2 F1 SIG SIG F1 F2 3X 1G AMPS 1&2G 1&2G
AMPS IS-136 CDPD 94 136 178 242 283 0.27 0.27 384 425 489 531 573 AMPS IS-136 CDPD AMPS IS-136 CDPD
(a) Bifucated System

A" A B A' B'
1&2G 1G AMPS 3X F1 SIG SIG F1 3X 1G AMPS 1&2G 1&2G
AMPS IS-136 CDPD 160 201 242 283 0.27 0.27 384 425 466 507 AMPS IS-136 CDPD AMPS IS-136 CDPD

Figure 5-14 is similar to that shown for a CDMA2000-1x deployment scheme with the exception that it involves the allocation of channels so that a 3X system can be deployed in the network at a future date. The obvious impact is on the channel bandwidth requirements due to the none contiguous channel deployment scheme if a carrier designated for eventual 3X service is initially implemented for 1x capacity relief.

Figure 5-14 also represents two different methods for deploying the future 3X into a network. Figure 5-14a is an example of allowing for guard bands between the 1X and 3X platforms, whereas Figure 5-14b follows the current recommendation for channel allocations.

However, the recommended method for deployment is to deploy the 1x systems in a contiguous fashion and then at a future date, when 3x is available, migrate the channels to the new designations to support 3x.

Table 5-7 may be of some help in determining the assignment of a 1x channel. However, the ultimate decision is based on the marketing plan, the services offered, the current-capacity requirements, and the expected take rates of both voice and data.

The next set of figures is meant to illustrate the recommended PCS channel deployment schemes that can be implemented. Because the channel assignment scheme is operator-dependant, that is, the channel set is pre-programmed, the specific channel numbers associated with F1, F2, and F3 are not defined as in cellular systems.

Table 5-7
CDMA2000-1X Assignment

Existing IS-95 Carriers	CDMA2000-1X 1x	CDMA2000-1X DO	Comments
0	1 (F1)	—	New
0	1 (F1)	1 (F2)	New
1	1 (F1)	—	Overlay
2	2 (F1 and F2)	—	Overlay
2	2 (F1 and F2)	1 (F3)	Overlay and Expansion

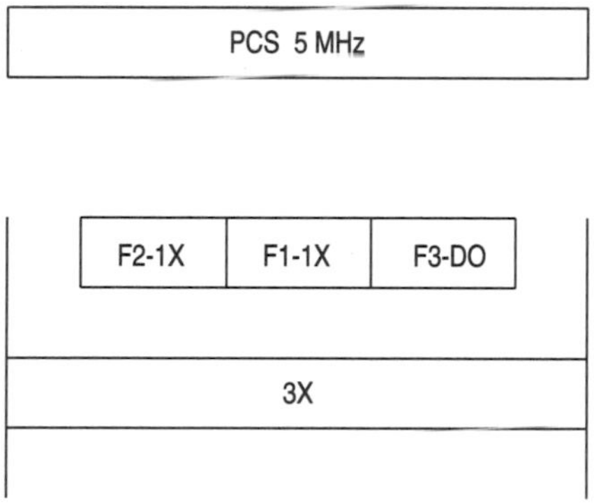

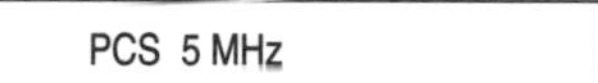

Figure 5-15
PCS 5-MHz channel deployment scheme.

Figure 5-15 is an example of a PCS system that is allocated 5 MHz of duplexed spectrum. The deployment scheme shown in Figure 5-15 is the preferred method that happens to be the same whether 1X is the only deployment contemplated or if a 3X deployment is planned for some date in the future.

The next channel deployment scheme is shown in Figure 5-16 and represents a methodology that bridges 2G, 2.5G, and of course 3G deployment schemes. Without performing any traffic studies, it is recommended that the initial deployment of CDMA2000-1X into any market involves a 1x channel and not the DO, unless there is a infrastructure vendor restriction. The rational behind this methodology of deployment lies in the uncertainty of the take rate and data throughput requirements for wireless packets. By implementing a 1x channel, the voice network is still served, while data is available for delivery. Once the packet data's take rate and usage patterns are better understood, then the possibility of a more robust deployment of DO channels can be envisioned.

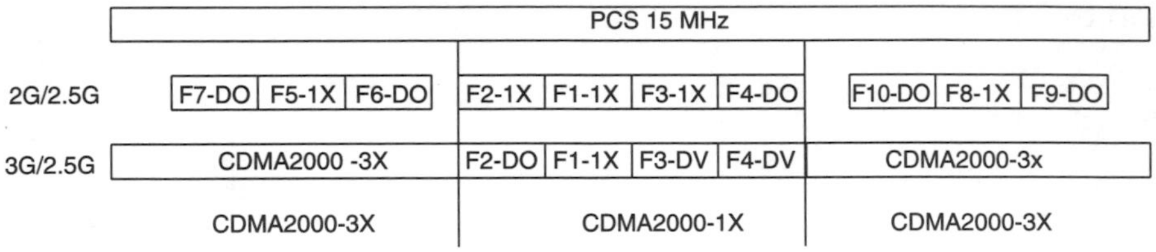

Figure 5-16 CDMA2000 PCS 15-MHz channel deployment scheme.

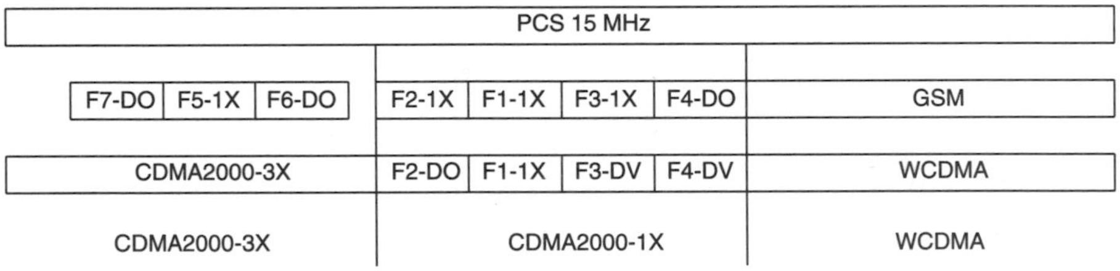

Figure 5-17 CDMA2000 PCS 15-MHz dual system channel deployment scheme.

Figure 5-17 shows an alternative method for implementing 3G services into the PCS band. More specifically, the channel deployment scheme enables the operation of both IMT2000-MC and DS systems side by side. Obviously, the primary channels for CDMA2000-1x are in the middle of the band and assume the obvious that no microwave clearance issues are left to be addressed at this time in the system's life cycle.

5.7.3 Frequency Planning

The frequency planning for CDMA2000-1x is the same as that done with IS-95. What is important is that the PN offset that is used for the existing sector, if IS-95 is deployed, be used for CDMA2000-1x. The PN offset value should also remain the same for all subsequent carriers that are deployed in the same sector. The PN offset reuse scheme utilized is the same that is referred to in Figure 3-32 and does not need to be repeated here.

To summarize the frequency planning scheme for CDMA2000, Table 5-8 helps illustrate the relative ease of inserting a CDMA2000-1x channel, spectrally speaking, into the existing system.

Table 5-8
PN Offsets

IS-95 System			CDMA2000-1x	
Sector	Carrier	PN Offset	Carrier	PN Offset
Alpha	1	6	1	6
	2	6	2	6
Beta	1	18	1	18
	—	—	2	18
Gamma	1	12	1	12
	—	—		

5.7.4 Handoff

CDMA2000, whether implemented as 1x only or a 1x/3x environment, needs to have the capability for handoffs and hangovers to exist in the system either between similar systems or legacy systems. Numerous decisions need to be made by the design engineer besides channel assignment schemes. The issue of how a handoff or hangover takes place within the system has a profound impact on the system performance of the network and obviously the subscriber's view of their service. Table 5-9 best illustrates the interaction between different traffic channel types in SR1 only.

The same types of handoffs occur with CDMA2000 as they do with IS-95 systems with the exception of packet data situations. The types of handoffs involved with CDMA2000 are as follows:

- Soft
- Softer
- Hard

In addition to the handoff situations with a mixed network listed in Table 5-9, CDMA2000 systems can also interact with AMPS analog channels like IS-95 systems by performing a hard handoff. Obviously, if a packet

Table 5-9
Handoff Compatibility Table

Source	Target	Destination Traffic Channel Type
IS-95	IS-95	IS-95
IS-95	IS-2000	IS-95
IS-2000	IS-95	IS-95
IS-2000	IS-2000	IS-2000

session is in place, the packet session will be lost and will be lost anytime the source channel is downgraded or if the mobile transitions out of the PSDN's effective coverage area.

Many scenarios are possible with CDMA2000-1x and legacy systems that are directly dependant upon the CDMA2000 system deployment and the logical BSC and PSDN boundaries that are established. To further add to the mix of possibilities, the BTS can also force a subscriber unit to a lower RC when

- The resource request is not a handoff
- The resource request is not available
- Alternative resources are available

5.7.5 Traffic Calculation Methods

An operator can pursue several methods for estimating the amount of voice and packet traffic with regards to implementing CDMA2000 1xRTT. A more robust discussion with some examples is included in Chapter 13, "CDMA2000 System Design," but the following concepts are approached. However, the key point to address is that without specific applications that the system is trying to address, the estimation of traffic is rather dubious since it bases several assumptions upon each other.

Traffic calculations are done in two methods: the forecast and discovery approaches. The *forecast approach* involves a detailed analysis of existing voice traffic and, by working with the marketing and subscriber sales force, a take rate and an estimated bandwidth for each subscriber can be achieved. This then is distributed across the regions or appropriate BTSs to arrive at the appropriate forecasted traffic volume, which then can be equated into channel elements, 1x/DO deployment schemes, and so on.

The other approach is the discovery approach where, in the core of the network, a 1x channel replaces either the F1 or F2 channels that are already deployed. Here you determine the number of mobiles that will be 1xRTT-capable and then multiply that number by 70 Kbps. You can assume that each mobile will be operational during the busy hour initially and you can weigh the traffic volume over the BTSs involved with upgrading to CDMA2000.

5.7.6 Deployment

The deployment of CDMA2000 into a new network is different than integrating it into an existing network. To be more specific, the traffic volumes and usage patterns are undefined in an initial system, leading to a homogenous traffic distribution from the onset since the focus is more coverage-oriented. For the existing system, the focus is on the capacity release and the introduction of new services, and the subscriber patterns are already known. However, the patterns have not been developed for the new services, therefore leading to much speculation as to where the usage will come from and how the subscribers will utilize the new service.

Regardless of the situation, the deployment of the system requires a decision as to where CDMA2000 will be introduced since, for all practical purposes, the possibility of a complete 1:1 deployment of CDMA2000 will not take place because of capital and implementation issues, which always arise during any capital build program or expansion.

Figure 5-18 is an example of a hypothetical system that has IS-95 fully deployed in the sample system with either one or two carriers per site. The choice of omni cells has been done for illustrative purposes, and in order to facilitate the discussion, in real life the sites would most likely be sectored, making the diagram very cluttered.

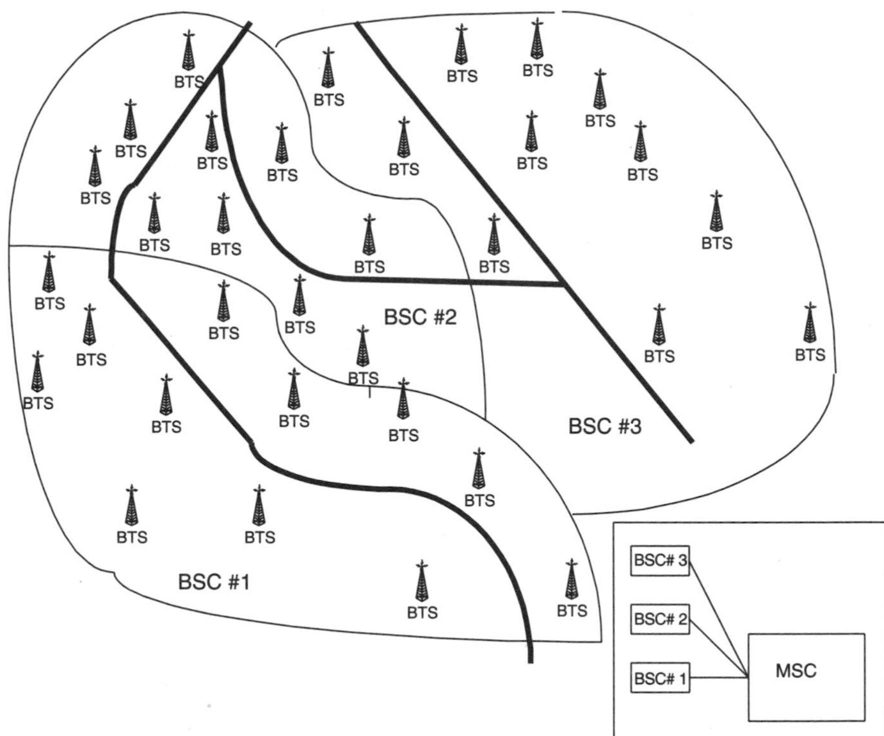

Figure 5-18 Sample IS-95 system layout.

The layout depicted in Figure 5-18 involves a total of three BSCs that are all connected to the same MSC. Several class 1 roads are shown in the figure, indicated by darkened lines that pass between the BTSs. BSC boundaries are also shown and it is assumed for the discussion that they are optimally located.

Figure 5-19 shows in a more visual method the BTSs that are associated with the BSCs and the amount of carriers each has in operation. A review of the diagram clearly indicates that the system has soft, softer, and hard handoffs that can take place within the network with IS-95.

Therefore, Figure 5-20 is an example of how CDMA2000-1x could be deployed into an existing IS-95 network. A quick comparison between Figures 5-19 and 5-20 illustrates several key issues. The first major item is that no new carriers are added in the expansion, and the carriers are upgraded from IS-95 to being CDMA2000-1x-capable. The second issue is

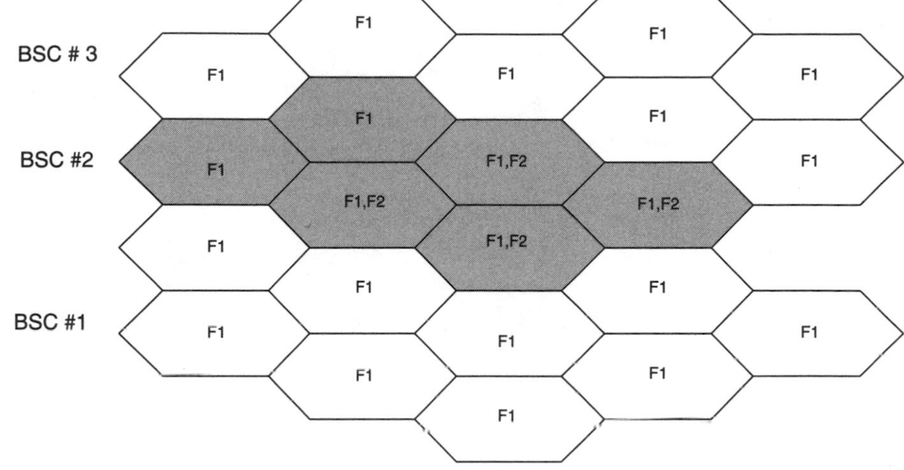

Figure 5-19 IS-95 carrier deployment for a sample system.

the CDMA2000 is only added at sites that have a second CDMA carrier or are adjacent to a site having a second CDMA carrier. The BSC boundaries remain the same, but in reality they would be altered to minimize the potential for BSC-BSC handoffs either for voice or packet data sessions. The concept illustrated is that not all the sites within the system need to be immediately upgraded to CDMA2000 from the start.

It does not take long to envision many different issues in the example shown, leading to a strong need to coordinate the CDMA2000 deployment within an existing system with the sales and marketing departments in order to best manage the capital resources of the network.

5.8 WAP

The *Wireless Application Protocol* (WAP) is one of the many protocols being implemented into the wireless arena for the purpose of increasing mobility by enabling mobile users to surf the internet. WAP is being implemented by numerous mobile equipment vendors since it is meant to provide a universal open standard for wireless phones, that is, cellular/GSM, and PCS for the purpose of delivering Internet content and other value-added services. Besides various mobile phones, WAP is also designed for PDAs to also utilize this protocol.

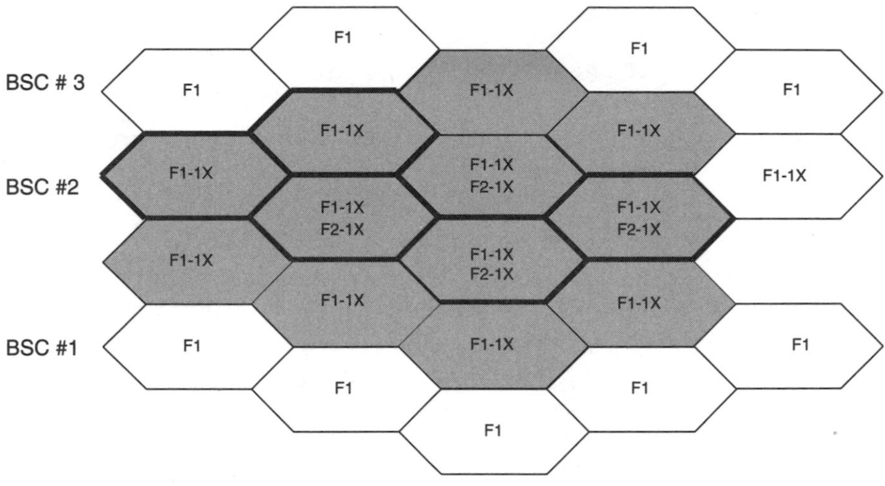

Figure 5-20 CDMA2000-1x carrier deployment scheme within a IS-95 system.

WAP enables mobile users to surf the Internet in a limited fashion; that is, they can send and receive e-mails and surf the net in a text format only-without graphics, which 2.5G systems will enable with the requisite handset. For WAP to be utilized by a mobile subscriber, the wireless operator, be it cellular or PCS, needs to implement WAP in his or her system as well as ensure that the subscriber units, that is, the phones, are WAP-capable.

WAP is meant to be utilized by the following cellular/PCS system types:

- GSM-900, GSM-1800, GSM-1900
- CDMA IS-95
- TDMA IS-136
- 3G systems

It is important to note that although WAP enables the user to send and receive text, it does not require additional spectrum and is a service enhancement that can and does coexist with the 2G technology platforms. WAP is not really a 2.5G platform for delivering high-speed wireless data due fundamentally to the fact that it uses 2G radio platforms to deliver it service and does not have the bandwidth0. However, WAP will increase the mobility of many subscribers and enable a host of data applications to be delivered for enhanced services to subscribers.

5.9 Migration Path from 2G to 2.5G to 3G

The specific migration path from any of the 2G platforms that an operator has deployed in a network to the 3G system involves the establishment of a migration path. The migration path involves numerous issues and technical challenges that will fundamentally define the character and services of the wireless system.

The end goal for the operator to be able to properly implement a 3G solution that follows the IMT-2000 specification involves the obvious and painful decision as to which IMT-2000 specification to utilize. For instance, the IMT-2000 specification that defines the 3G wireless mobility system has several platforms from which the existing wireless operator must make a decision as to which to utilize. In a situation when the overseeing regulatory agency dictates the IMT-2000 platform to utilize, the decision is academic. However, the difficulty begins when the decision is left to the operator to make. The difficulty lies in the amount of capital infrastructure that needs to be deployed for any of these systems in order to take them from a concept into a physical reality.

A decision from, say, a IS-95B CDMA may be to migrate to a WCDMA system, but the path from IS-95B to a WCDMA platform does not involve the commonality of the radio base station equipment, as it would in a CDMA2000 platform. Alternatively, if a GSM operator chose a CDMA2000 platform, a separate network, as in the previous example, would need to be deployed in order to provide the radio transport system needed. However, the operators using IS-136 need to make a fundamental decision as to which IMTS-2000 platform to utilize, WCDMA or CDMA2000. Either case requires the deployment of new radio base stations in order to realize the transition.

Lastly, and very important, to the overall discussion of migration path decisions is the spectrum that is available to the operator itself. The spectrum includes not only the bandwidth, but also the fundamental frequency of operation. The radio spectrum in the United States is not the same as that used in Europe or Asia. Therefore, in the decision and migration strategy from a 2G to a 3G platform, the operator needs to factor in the interoperability considerations usually available with existing tri-band mobile phones.

But no matter which 3G technology is chosen, the operator is left with two fundamental choices. The first is to continue utilizing the existing technology platforms, wait until the availability of a 3G platform, and transition directly from 2G to 3G. The other choice is to choose an interim platform that hopefully will be compatible with the 3G platform chosen and allow for enhanced data services to be deployed in advance of 3G, thus trying to capture the market share.

References

AT&T. "*Engineering and Operations in the Bell System*," 2nd Ed., AT&T Bell Laboratories, Murry Hill, N.J., 1983.

Barron, Tim. "*Wireless Links for PCS and Cellular Networks*," Cellular Integration, Sept. 1995, pgs. 20–23.

Brewster. "*Telecommunications Technology*," John Wiley & Sons, New York, NY, 1986.

Channing, Ian. "*Full Speed GPRS from Motorola*," Mobile Communications International, Issue 65, Oct. 1999, pg. 6.

Code of Federal Regulations. CFR 47 Parts 1, 17, 22, 24, and 90.

Collins, Daniel. "*Carrier Grade Voice Over IP*," McGraw-Hill, 2001.

DeRose. "*The Wireless Data Handbook*," Quantum Publishing, Inc., Mendocino, CA, 1994.

Dixon. "*Spread Spectrum Systems*," 2nd Ed., John Wiley & Sons, New York, NY, 1984.

Harte, Hoenig, Kikta McLaughlin. "*CDMA IS-95 for Cellular and PCS*," McGraw-Hill, 1996.

Held, Gil. "*Voice & Data Interworking*," 2nd Ed., 2000, McGraw-Hill.

Homa, Harri, and Antti Toskala. "*WCDMA for UMTS*," John Wiley & Sons, 2000.

McClelland, Stephen. "*Europe's Wireless Futures*," Microwave Journal, Sept. 1999, pgs. 78–107.

Molisch, Andreas F. "*Wideband Wireless Digital Communications*," Prentice Hall, New Jersey, 2001.

Pautet, Mouly. "*The GSM System for Mobile Communications*," Mouly Pautet, 1992.

Prasad, Ramjee, Werner Mohr, and Walter Konhauser. "*Third Generation Mobile Communication Systems*," Artech House, 2000.

Qualcomm. "*An Overview of the Application of Code Division Multiple Access (CDMA) to Digital Cellular Systems and Personal Cellular Networks*," Qualcomm, San Diego, CA, May 21, 1992.

Salter, Avril. "*W-CDMA Trial & Error*," Wireless Review, Nov. 1, 1999, pg. 58.

Schwartz, Bennett, Stein. "*Communication Systems and Technologies*," IEEE, New York, NY, 1996.

Shank, Keith. "*A Time to Converge*," Wireless Review, Aug. 1, 1999, pg. 26.

Smith, Clint. "*Practical Cellular and PCS Design*," McGraw-Hill, 1997.

Smith, Clint. "*Wireless Telecom FAQ*," McGraw-Hill, 2000.

Smith, Gervelis. "*Cellular System Design and Optimization*," McGraw-Hill, 1996.

Webb, William. "*Introduction to Wireless Local Loop, Second Editions: Broadband and Narrowband Systems*," Artech House, Boston, MA, 2000.

Wesley, Clarence. "*Wireless Gone Astray*," Telecommunications, Nov. 1999, pg. 41.

Willenegger, Serge. "*cdma2000 Physical Layer: An Overview*," Qualcomm 5775, San Diego, CA.

William, C.Y. Lee. "*Mobile Cellular Telecommunications Systems*," 2nd Ed., McGraw-Hill, New York, NY, 1996.

3GPP TS 05.01 Physical layer on the radio path, General description

3GPP TS 05.04 Modulation

3GPP TS 05.05 Radio transmission and reception (Release 1999)

3GPP TS 05.08 Radio subsystem link control

GSM 02.34 High-Speed Circuit-Switched Data (HSCSD)—Stage 1

GSM 02.60	General Packet Radio Service (GPRS); Service description; Stage 1
GSM 03.03	Numbering, addressing, and identification
GSM 03.07	Restoration procedures
GSM 03.20	Security-related network functions
GSM 03.22	Functions related to Mobile Station (MS) in idle mode and group receive mode
GSM 03.34	High-Speed Circuit-Switched Data (HSCSD)—Stage 2
GSM 03.64	Overall description of the General Packet Radio Service (GPRS) Radio interface; Stage 2
GSM 04.60	Mobile Station (MS)—Base Station System (BSS) interface; Radio Link Control/Medium Access Control (RLC/MAC) protocol
GSM 04.64	Mobile Station—Serving GPRS Support Node (MS—SGSN) Logical Link Control (LLC) layer specification
GSM 04.65	Mobile Station (MS)—Serving GPRS Support Node (SGSN); Subnetwork Dependent Convergence Protocol (SNDCP)
GSM 05.08	Radio subsystem link control
GSM 07.60	Mobile Station (MS) supporting GPRS
GSM 08.14	Base Station System (BSS)—Serving GPRS Support Node (SGSN) interface; Gb interface layer 1
GSM 08.16	Base Station System (BSS)—Serving GPRS Support Node (SGSN) interface; Network Service
GSM 08.18	Base Station System (BSS)—Serving GPRS Support Node (SGSN); BSS GPRS Protocol (BSSGP)
GSM 09.02	Mobile Application Part (MAP) specification
GSM 09.16	Serving GPRS Support Node (SGSN)—Visitors Location Register (VLR); Gs interface network service specification
GSM 09.18	Serving GPRS Support Node (SGSN)—Visitors Location Register (VLR); Gs interface layer 3 specification
GSM 09.60	GPRS Tunnelling Protocol (GTP) across the Gn and Gp Interface

GSM 12.15	GPRS charging
IETF RFC 768	User Datagram Protocol (STD 6).
IETF RFC 791	Internet Protocol (STD 5).
IETF RFC 793	Transmission Control Protocol (STD 7).

CHAPTER 6

Universal Mobile Telecommunications Service (UMTS)

6.1 Introduction

Universal Mobile Telecommunications Service (UMTS) represents an evolution of *Global System for Mobile communications* (GSM) to support *third-generation* (3G) capabilities. In this chapter, we examine the details of the UMTS, including the air interface and network architecture. The initial focus is on the air interface and *radio access network* (RAN), as these represent the greatest change from the technologies of GSM, the *General Packet Radio Service* (GPRS), and the *Enhanced Data Rates for Global Evolution* (EDGE). Subsequently, we delve into the specifics of the core network and examine the planned evolution of the UMTS over the next few years. First, however, some general information on UMTS technology.

6.2 UMTS Basics

As described briefly in Chapter 4, "Third Generation (3G) Overview," UMTS includes two of the air interface proposals submitted to the *International Telecommunications Union* (ITU) as proposed solutions to meet the requirements laid down for *International Mobile Telephony 2000* (IMT-2000). These both use *Direct Sequence Wideband CDMA* (DS-WCDMA). One solution uses *Frequency Division Duplex* (FDD) and the other uses *Time Division Duplex* (TDD). The FDD solution is likely to see the greatest deployment—particularly in Europe and the Americas. The TDD solution is likely to see deployment primarily in Asia. In this chapter, we focus mainly on the FDD option.

In the FDD option, paired 5-MHz carriers are used in the uplink and downlink as follows: uplink—1920 MHz to 1980 MHz; downlink—2110 MHz to 2170 MHz. Thus, for the FDD mode of operation, a separation of 190 MHz is used between the uplink and downlink. Although 5 MHz is the nominal carrier spacing, it is possible to have a carrier spacing of 4.4 MHz to 5 MHz in steps of 200 kHz. This enables spacing that might be needed to avoid interference, particularly if the next 5-MHz block is allocated to another carrier.

For the TDD option, a number of frequencies have been defined, including 1900 MHz to 1920 MHz and 2010 MHz to 2025 MHz. Of course, with TDD, a given carrier is used in both the uplink and the downlink so that no separation exists.

In any CDMA system, user data is spread to a far greater bandwidth than the user rate by the application of a spreading code, which is a higher-bandwidth, pseudo-random sequence of bits, known as chips. The transmission from each user is spread by a different spreading code, and all users transmit at the same frequency at the same time. At the receiving end, the signal from one user is separated from that of other users by despreading the set of received signals with the spreading code applicable to the user in question. The result of the despreading operation is the retrieval of the user data in question, plus some noise generated as a result of the transmissions from other users.

The ratio of the spreading rate (the number of chips per second) to the user data rate (the number of user data symbols per second) is known as the spreading factor. The greater the spreading factor, the greater the ability to extract a given user's signal from that of all others. In other words, for a given user data rate, the higher the chip rate, the more users can be supported. Alternatively, for a set number of users, the higher the chip rate, the higher the data rates that can be supported for each user. Thus, the spreading rate is of major significance. Of course, one gets nothing for nothing—the higher the chip rate, the greater the occupied spectrum. The chip rate in WCDMA is 3.84×10^6 chips/second (3.84 Mcps), which leads to a carrier bandwidth of between 4.4 MHz and 5 MHz.

From a network architecture perspective, UMTS borrows heavily from the established network architecture of GSM. In fact, many of the network elements used in GSM are reused (with some enhancements) in UMTS. This commonality means that a given *Mobile Switching Center* (MSC), *Home Location Register* (HLR), *Serving GPRS Support Node* (SGSN), or *Gateway GPRS Support Node* (GGSN) can be upgraded to support UMTS and GSM simultaneously.

The radio access, however, is significantly different from that of GSM, GPRS, and EDGE. In UMTS, the RAN is known as the *UMTS Terrestrial Radio Access Network* (UTRAN). The components that make up the UTRAN are significantly different from the corresponding elements in the GSM architecture. Therefore, the reuse of existing GSM base stations and *Base Station Controllers* (BSCs) is limited.

For some vendors, GSM base stations were planned in advance to be upgradable to support WCDMA as well as GSM. Thus, for some vendors, it is possible to remove some number of GSM transceivers from a base station and replace them with some number of UMTS transceivers. For other vendors, a completely new base station is needed. A similar situation applies to BSCs. For most vendors, the technology of a UMTS *Radio Network Controller* (RNC) is so different from that of a GSM BSC that the BSC cannot be upgraded to act simultaneously as a GSM BSC and a UMTS RNC. Cases will occur, however, where a BSC can be upgraded to simultaneously support both GSM and UMTS, but that situation is less common.

6.3 The WCDMA Air Interface

As mentioned, WCDMA uses a chip rate of 3.84 Mcps. As also mentioned, CDMA technology in general uses a spreading code to separate one user's transmissions from those of another. In reality, however, there will be multiple simultaneous data streams from multiple users and multiple simultaneous data streams from a singe base station. Therefore, not only is it necessary to separate the transmissions of one user or base station from those of another, it is also necessary to separate the various transmissions that a single user might generate. In other words, if a single user (user A) is transmitting both user data and control information, the base station must first separate the set of transmissions from user A from the transmissions of all other users. It must then separate the control information from the user data.

In order to support this requirement, WCDMA takes a two-step approach to the transmission from a single user, as shown in Figure 6-1. First, each individual data stream is spread to the chip rate by the application of a spreading code, also known as a channelization code, and which operates at the chip rate of 3.84 Mcps. Then the combined set of spread signals is scrambled by the application of a scrambling code, which also operates at the chip rate. The channelization spreads the individual data streams and hence increases the required bandwidth. Since the scrambling code also operates at the chip rate, however, it does not further increase the required bandwidth. At the receiving end, the combined signal is first descrambled by application of the appropriate scrambling code. The individual user data streams are then recovered through the application of the appropriate channelization codes. Clearly, it is important that different users use different scrambling codes. Multiple users, however, can use the same channelization codes, provided, however, that no two transmissions from the same user use the exact same channelization code.

6.3.1 Uplink Spreading, Scrambling, and Modulation

A physical channel is what carries the actual user data or control information over the air interface. A physical channel can be considered a combination of frequency, scrambling code, and channelization code, and in the uplink, as we shall describe later, the relative phase is also significant. For

Figure 6-1 Spreading and Scrambling, Basic Concept.

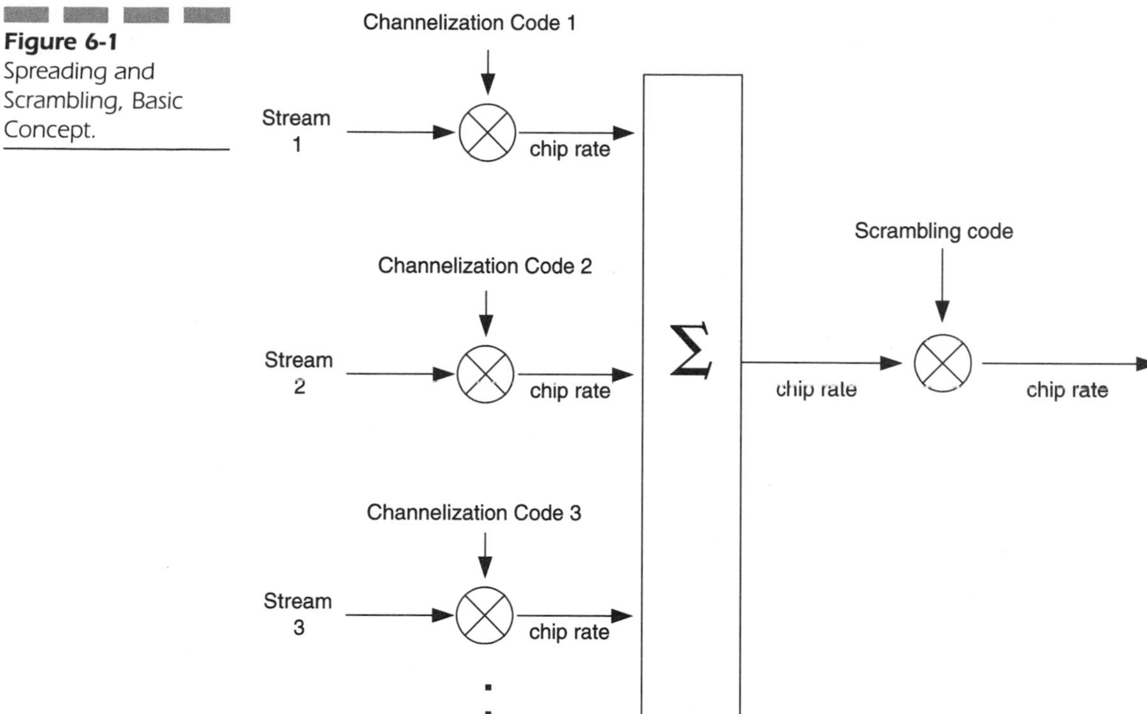

example, if a given user is transmitting user data and control information, then the user data stream will be carried on one physical channel and the control information will be on a different physical channel.

A number of different physical channels are used in the uplink, with a given type of channel selected according to what the *user equipment* (UE) is attempting to do—such as simply request access to the network, send just a single burst of data, or send a stream of data. These channels are described in further detail later in this chapter. For now, let us focus on the situation where a user is transmitting a stream of data, which would happen in a voice conversation. In such a situation, the terminal will normally use at least two physical channels—a *Dedicated Physical Data Channel* (DPDCH) and a *Dedicated Physical Control Channel* (DPCCH). The DPDCH carries the user data and the DPCCH carries control information. Depending on the amount of data to be sent, a single user can use just a single DPDCH, which will support up to 480 Kbps of user data or as many as six DPDCHs, which will support up to 2.3 Mbps of user data.

A DPDCH can have a variable spreading factor. This simply means that the user bit rate does not have to be fixed to a specific value. The spreading factor for a DPDCH can be 4, 8, 16, 32, 64, 128, or 256. These correspond to DPDCH bit rates of 15 Kbps ($3.84 \times 10^6/256 = 15 \times 10^3$) and up to 960 Kbps ($3.84 \times 10^6/4 = 960 \times 10^3$). Of course, these are not the actual user data rates, because a significant amount of coding overhead is included in the DPDCH to support forward error correction. In general, the user data rate is approximately half (or less) of the DPDCH rate. Thus, for example, a DPDCH operating at a spreading rate of 4 will carry data at a rate of 960 Kbps. Of this, however, only about 480 Kbps will correspond to usable data. The rest is consumed by additional coding required for error correction. If a single user wants to transmit user data at a rate greater than 480 Kbps, then multiple DPDCHs can be used (up to a maximum of 6).

Figure 6-2 shows how multiple DPDCHs are handled. Also shown is the DPCCH, which is also sent whenever one or more DPDCHs are sent. The channelization codes ($C_{d,1}$ to $C_{d,6}$) represent the channelization codes applied to each of the six DPDCHs. The channelization code applied to the DPCCH is represented as C_c. Each of the DPDCHs is spread to the chip rate by a channelization code. DPDCHs 1, 3, and 5 are channelized and weighted by a gain factor b_d. These DPDCHs are on the so-called I (in-phase) branch. DPDCHs 2, 4, and 6 plus the DPCCH are on the so-called Q (quadrature) branch. These are also channelized. These spread DPDCHs are also weighted by the gain factor b_d, whereas the spread DPCCH is weighted by the gain factor b_c. The two gain factors are specified as 4-bit words that represent steps from 0 to 1. Thus, 0000 = off, 0001 = 1/15, 0010 = 2/15, and 1111 = 15/15 = 1. At any given instant, one of the two gain factors has the value of 1 (binary 1111 = 15/15 = 1).

Mathematically, the spread signals on the Q branch are treated as a stream of imaginary bits. These are summed with the stream of real bits on the I branch to provide a stream of complex-valued chips at the chip rate. This stream of complex-valued chips is then subjected to a complex-valued scrambling code, which is aligned with the beginning of a radio frame.

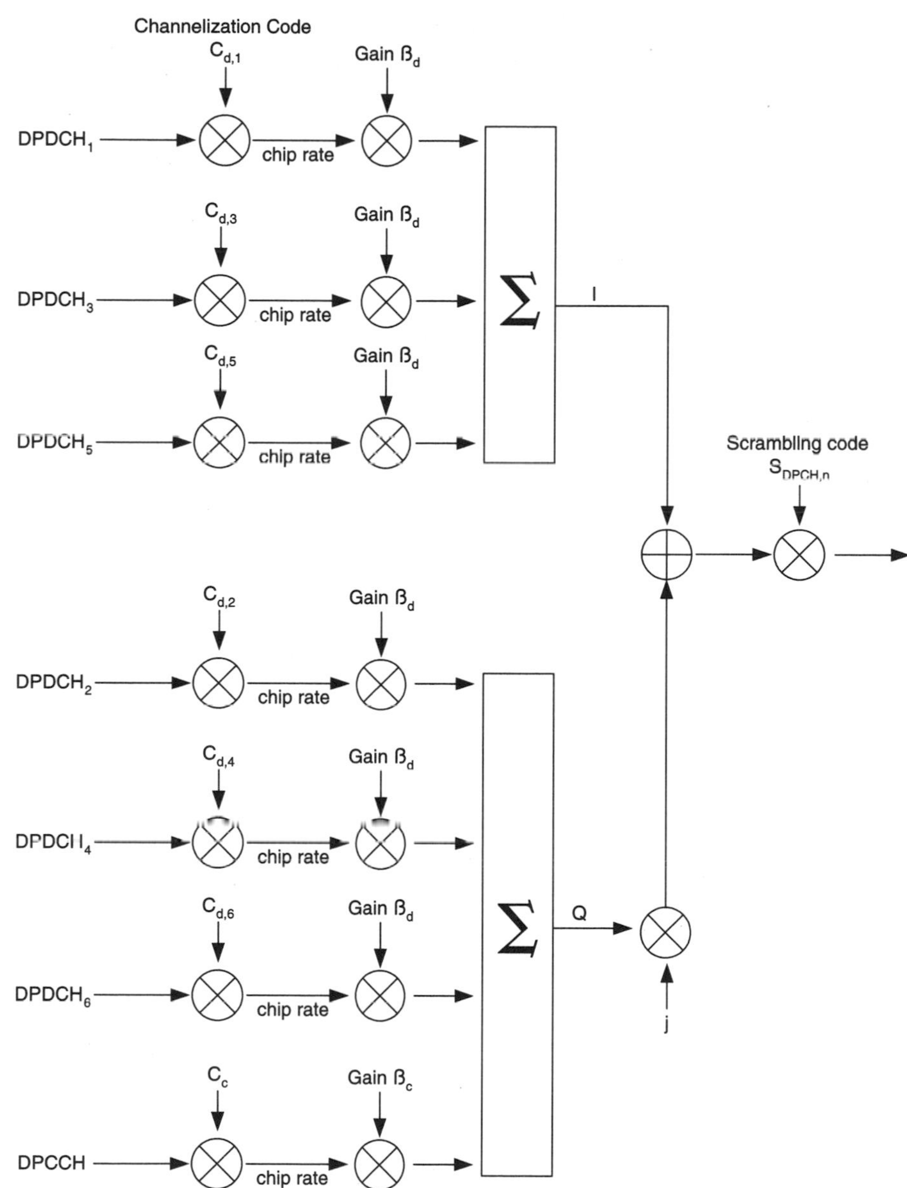

Figure 6-2 Uplink Channelization and Scrambling.

6.3.1.1 Channelization Codes As mentioned, the channelization codes are used to separate multiple streams of data from a given user, whereas the scrambling codes are used to separate transmissions from different users.

The channelization codes are known as *Orthogonal Variable Spreading Factor* (OVSF) codes. They are taken from the code tree shown in Figure 6-3. The generation of channelization codes is given by the following equations:

$$C_{ch,1,0} = (1)$$

$$[C_{ch,2,0}] = [C_{ch,1,0} \quad C_{ch,1,0}] = (1, 1)$$

$$[C_{ch,2,1}] = [C_{ch,1,0} \quad -C_{ch,1,0}] = (1, -1)$$

$$[C_{ch,2}{}^{(n+1)}{}_{,0}] = [C_{ch,2}{}^{(n)}{}_{,0} \quad C_{ch,2}{}^{(n)}{}_{,0}]$$

$$[C_{ch,2}{}^{(n+1)}{}_{,1}] = [C_{ch,2}{}^{(n)}{}_{,0} \quad -C_{ch,2}{}^{(n)}{}_{,0}]$$

$$[C_{ch,2}{}^{(n+1)}{}_{,2}] = [C_{ch,2}{}^{(n)}{}_{,1} \quad C_{ch,2}{}^{(n)}{}_{,1}]$$

$$[C_{ch,2}{}^{(n+1)}{}_{,3}] = [C_{ch,2}{}^{(n)}{}_{,1} \quad -C_{ch,2}{}^{(n)}{}_{,1}], \text{ etc.}$$

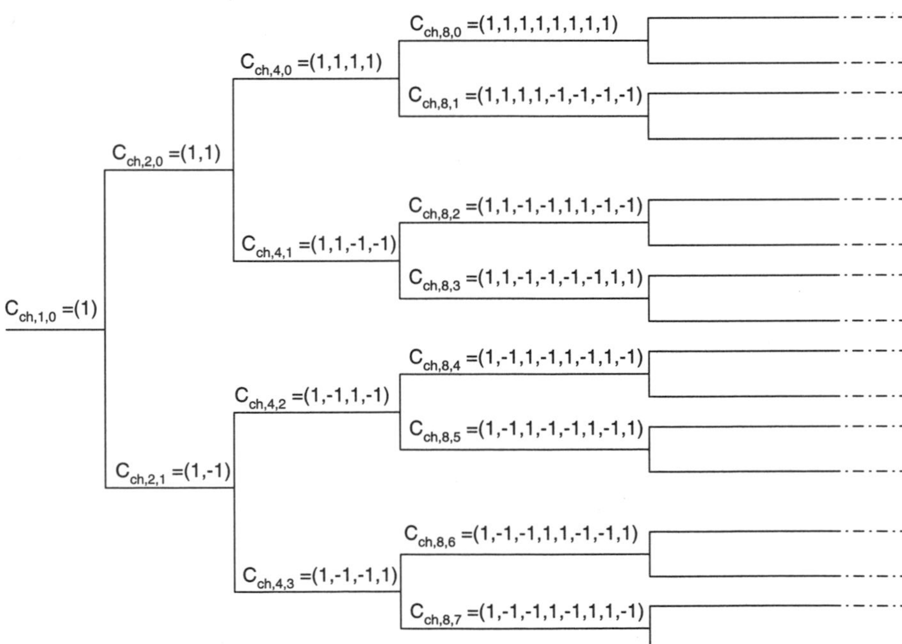

Figure 6-3 Channelization Code Tree.

In general, a given physical channel uses a channelization code that is related to the spreading factor being used for the channel. When only one DPDCH is to be transmitted, then the channelization code is $C_{ch,SF,k}$, where SF is the spreading factor and K = SF/4. Therefore, if the spreading factor is 128 (as determined by the user data rate plus coding overhead), then the code to be used shall be $C_{ch,128,32}$. The spreading factor for the DPCCH is always 256 and the channelization code is $C_{ch,256,0}$.

When more than one DPDCH is to be transmitted (greater than 960 Kbps of combined user data and coding overhead), then each DPDCH shall have a spreading factor of 4 and the channelization code for each DPDCH shall be $C_{ch,4,k}$. K = 1 for $DPDCH_1$ and $DPDCH_2$, K=2 for $DPDCH_3$ and $DPDCH_4$, and K = 3 for $DPDCH_5$ and $DPDCH_6$. For example, $DPDCH_3$ and $DPDCH_4$ would both use the channelization code $C_{ch,4,2} = (1, -1, 1, -1)$. Given that channelization codes are used to separate different transmissions from a single user, the fact that certain codes can be simultaneously used on two channels is, at first glance, troubling. The fact, however, that those two channels will always be on separate I and Q branches means that they can still be separated.

Some important restrictions apply to the use of channelization codes. That is because, in the case where more than one channel is being transmitted, the chosen channelization codes must be orthogonal. For example, consider the channelization code $C_{ch,4,0}$. This code is simply the sequence 1,1,1,1 repeated over and over, with each sequence of four bits repeated 960,000 times per second. Consider the channelization code $C_{ch,8,0}$. This is simply the sequence 1,1,1,1,1,1,1,1 repeated over and over, with each sequence of eight bits repeated 480,000 times per second. Clearly, if one data stream from a given user is spread with the code $C_{ch,4,0}$, and a second data stream from the same user is spread with the code $C_{ch,8,0}$, the net effect is that they are spread in the same way and cannot be distinguished at the receiver. Consequently, channelization codes must be selected in a manner that ensures that each channel is spread differently.

6.3.1.2 Scrambling Codes Once the different channels have been spread with appropriate channelization codes, they are combined, as shown in Figure 6-2, and then scrambled by a particular scrambling code. Two types of scrambling codes exist—long and short scrambling codes, with 2^{24} possibilities for each type. The choice of a particular code is determined by the type of physical channel in question (as we shall see, other physical channels are available besides the DPDCH and DPCCH) and by the higher layer that requires the use of a channel in the first place. Depending on higher-layer requirements, a DPDCH or DPCCH can use either a short scrambling code or a long scrambling code.

Clearly, the channelization codes are far from random, and they do not need to have pseudo-random properties. Scrambling codes, however, must appear to be random and thus must have pseudo-random characteristics. The easiest way to generate a pseudo-random sequence is through the use of a linear feedback shift register, such as that shown in Figure 6-4. This is basically a set of flip-flops that are clocked at a particular rate and the output of the last flip-flop is copied back into one or more of the other flip-flops, possibly after an addition. In Figure 6-4, each of the gain values (g_n) is simply a 1 or a 0. Depending on the values of each g_n (that is, exactly where the output is fed back), a different output pattern can be achieved. This output pattern can be described by a polynomial, known as a generator polynomial.

It is possible to produce maximum-length sequences, known as m-sequences. This means that if a register has m elements, then it can produce a sequence of length $2^m - 1$. For example, if a shift register has 10 elements, then it can produce a sequence of length $2^{10} - 1$ (1,023). This is a pattern that repeats after every 1023 bits. An m-sequence has a number of properties, including the property that, over the period of the sequence, there will be exactly 2^{m-1} ones and $2^{m-1} - 1$ zeros.

The long scrambling codes used in WCDMA are known as Gold codes and are constructed from the modulo 2 addition of portions of two binary m-sequences. The portions used are segments of length 38,400. This is due to the fact, as shall be explained later in this chapter, that the frame length in WCDMA is 10 ms, which corresponds to 38,400 chips. Because the long scrambling codes are generated from m-sequences, they have pseudo-random characteristics. The short scrambling codes also have pseudo-random characteristics. These, however, are much shorter, at a length of length 256 chips. Long scrambling codes are used in the case where the base station uses a rake receiver. Short scrambling codes can be used when the base station uses advanced multi-user detection techniques such as a *Parallel Interference Cancellation* (PIC) receiver.

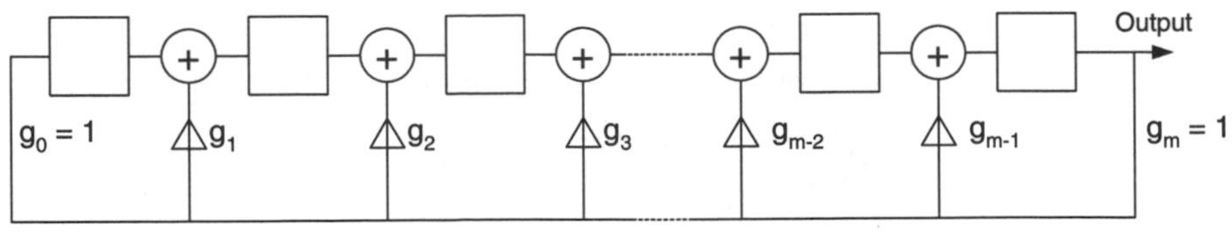

Figure 6-4 Linear Feedback Shift Register.

6.3.1.3 Uplink Modulation WCDMA uses *Quadrature Phase Shift Keying* (QPSK) modulation in the uplink. This technique is depicted in Figure 6-5. The stream of spread and scrambled signals, such as the output shown in Figure 6-2, forms the complex-valued input stream of chips. The real and imaginary parts are separated, with the real part of a given complex chip forming the in-phase (I) branch and the imaginary part forming the quadrature phase (Q) branch in the modulator.

6.3.2 Downlink Spreading, Scrambling, and Modulation

As is the case for the uplink, a number of channels are used in the downlink. In fact, more channels are defined for the downlink than for the uplink. That is because the downlink includes pilot channels, synchronization channels, channels used for the broadcast of system information, channels used for the paging of subscribers, and so on.

6.3.2.1 Downlink Spreading With the exception of the *synchronization channels* (SCHs), the downlink channels are spread to the chip rate and scrambled, as shown in Figure 6-6. Each channel to be spread is split into two streams—the I branch and the Q branch. The even symbols are mapped to the I branch and the odd symbols are mapped to the Q branch. The I branch is treated as a stream of real-valued bits, whereas the Q branch

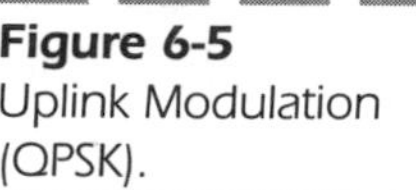

Figure 6-5
Uplink Modulation (QPSK).

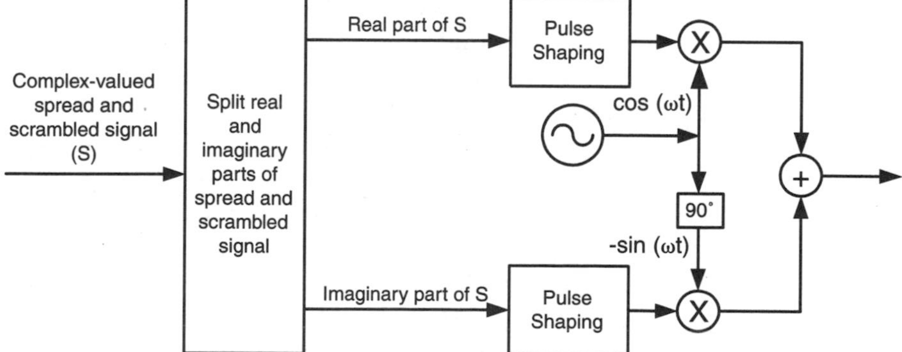

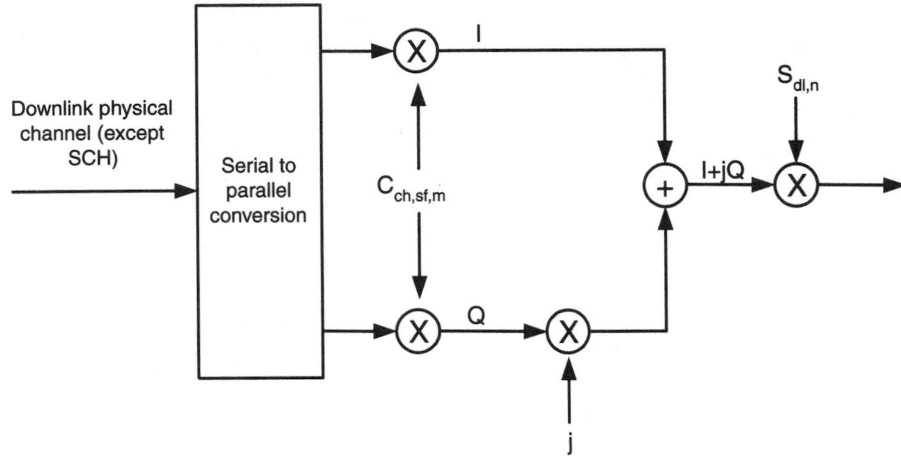

Figure 6-6 Downlink Scrambling.

is treated as a stream of imaginary bits. Each of the two streams is spread by the same channelization code. The spreading code/channelization code to be used is taken from the same code tree as used in the uplink—that is, OVSF codes that are chosen to maintain the orthogonality between different channels transmitted from the same base station. The spreading rate for a given channel depends on the channel in question.

The I and Q streams are then combined such that each I and Q pair of chips is treated as a single complex value, such that the result of combining them is a stream of complex-valued chips. This stream of chips is then subjected to a complex downlink scrambling code, identified as $S_{dl,n}$ in Figure 6-6.

One important difference occurs between spreading in the uplink and downlink. In the uplink, the data for a given physical channel (such as a single DPDCH) is directed either to the I branch or the Q branch, as shown in Figure 6-2. Thus, on the uplink, no serial-to-parallel conversion takes place. Therefore, for a spreading factor of, say, 8, the data rate of the physical channel is simply 3,840,000/8 = 480 Kbps.

On the downlink, however, each channel (with the exception of the synchronization channel) is subjected to a serial-to-parallel conversion, as shown in Figure 6-6. For a given spreading factor, the serial-to-parallel conversion effectively doubles the data rate of the physical channel. In other words, half of the channel's data is carried on the I branch with half on the Q branch, and both of these are spread with the same spreading factor.

If, for example, we have a spreading factor of 8, then the data rate on the I channel is 480 Kbps and the data rate on the Q channel is also 480 Kbps. The net data rate is 960 Kbps—twice that achieved on the uplink for the same value of spreading factor. In reality, however, the data rate on the downlink is not quite twice that on the uplink. This is due to the fact, as is explained later, that control information is time-multiplexed with a user data on the downlink. This reduces the net throughput for a given downlink data channel. Nonetheless, for a given spreading factor on the downlink, the effective throughput is significantly greater than the corresponding throughput on the uplink for the same spreading factor.

6.3.2.2 Downlink Scrambling The downlink scrambling codes are used to separate the transmissions of one cell from those of another. The downlink scrambling codes are Gold codes similar to the long scrambling codes used in the uplink. As is the case for the long codes used on the uplink, the codes used on the downlink are limited to a 10-ms duration. There is a total of $2^{18} - 1$ (262,143) downlink scrambling codes. Not all of these are used, however. If all possible codes were to be useable, then one could find a situation where a terminal would have to check a received signal against all 262,143 codes. This could occur, for example, during cell selection. Clearly, checking against so many scrambling codes is impractical.

Therefore, the available downlink scrambling codes are separated into 512 groups. Each group contains one primary scrambling code and 15 secondary scrambling codes. Thus, 512 primary scrambling codes exist and 7,680 secondary scrambling codes exist, for a total of 8,192 downlink scrambling codes. Table 6-1 shows the allocation of secondary downlink scrambling codes to primary downlink scrambling codes.

A cell is allocated one, and only one, primary scrambling code, which, of course, has 15 secondary scrambling codes associated with it. A given base station will use its primary scrambling code for the transmission of channels that need to be heard by all terminals in the cell. Thus, paging messages need to be scrambled by the cell's primary scrambling code. For that matter, all transmissions from the base station can simply use the cell's primary scrambling code. After all, it is the scrambling code that identifies the cell, while the various channelization codes are used to separate the various transmissions (physical channels) within the cell.

A cell can, however, choose to use a secondary scrambling code for channels that are directed to a specific user and do not need to be decoded by other users. In general, it is a good idea for all transmissions from a cell to use the cell's primary scrambling code, as this helps to minimize interference.

Table 6-1

Allocation of Secondary Scrambling Codes to Primary Scrambling Codes

Primary Scrambling Code Number	Secondary Scrambling Codes Numbers
0	1–15
16	17–31
32	33–47
48	49–63
⋮	⋮
8,160	8,161–8,175
8,176	8,177–8,191

As described, 512 primary scrambling codes are available. These are divided into 64 groups, each consisting of 8 scrambling codes, as shown in Table 6-2.

As mentioned, downlink spreading and scrambling are applied to all downlink physical channels transmitted on a cell, with the exception of the *synchronization channel* (SCH). This channel is added to the downlink stream, as shown in Figure 6-7. In fact, as explained later in this chapter, the SCH contains two subchannels—the primary SCH and secondary SCH. The reason why these are transmitted without scrambling is the fact that they are the first channels decoded by a terminal. If they were scrambled, then the terminal would first have to know the scrambling code of the base station just to synchronize.

6.3.2.3 Downlink Modulation As is the case for the uplink, the downlink uses QPSK modulation. The process in the downlink is the same as that shown in Figure 6-5 for the uplink. Each complex-valued chip is split into its constituent real and imaginary parts. The real part is sent on the I branch of the modulator and the imaginary branch is sent on the Q branch of the modulator.

Table 6-2
Primary Scrambling Code Groups

Primary Scrambling Group Number	Primary Scrambling Code Numbers
0	0; 16; 32; 48; 64; 80; 96; 112
1	128; 144; 160; 176; 192; 208; 228; 240
2	256; 272; 288; 304; 320; 336; 352; 368
⋮	⋮
62	7,936; 7,952; 7,968; 7984; 8,000; 8,016; 8,032; 8,048
63	8,064; 8,080; 8,096; 8,112; 8,128; 8,144; 8,160; 8,176

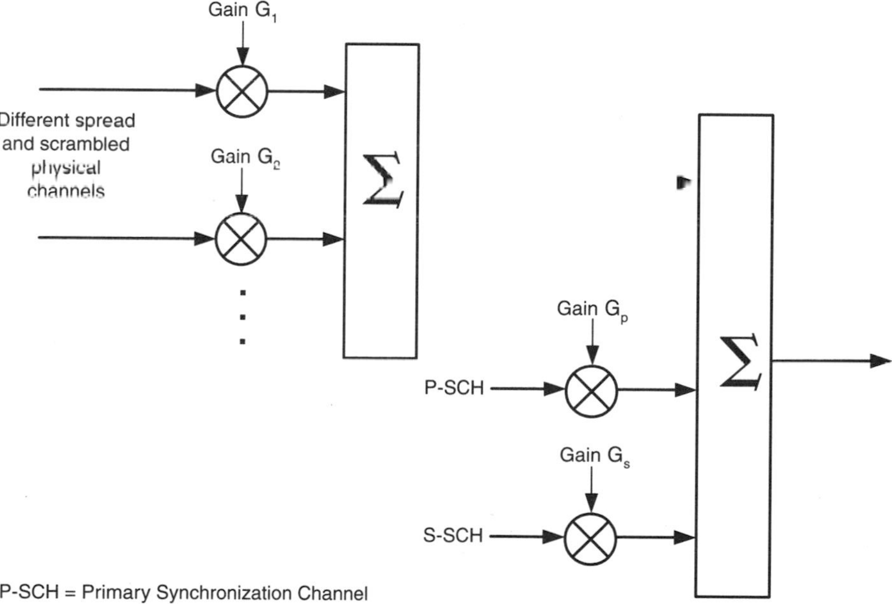

Figure 6-7 Downlink Multiplexing of Synchronization Channel.

6.3.3 WCDMA Air Interface Protocol Architecture

We have already mentioned some of the types of physical channels defined in WCDMA. In fact, many different channel types exist, and the various types of channels are defined in a logical hierarchy.

Figure 6-8 shows the overall logical structure of the WCDMA air interface. At the lowest level, we have the physical layer. The functions of the physical layer include RF processing, spreading, scrambling and modulation, coding and decoding for support of forward error correction, power control, timing advance, and soft handover execution. Physical channels, such as those already mentioned, exist at the physical layer and are used for

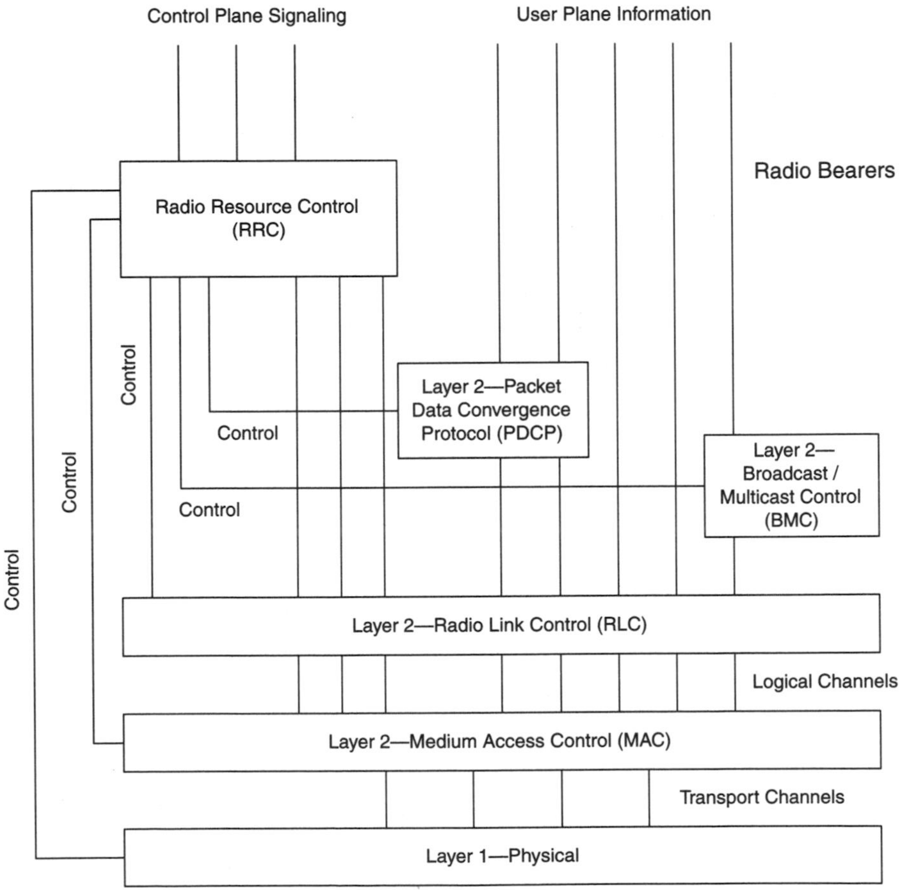

Figure 6-8 WCDMA Air Interface Protocol Structure.

transmission across the RF interface. A given physical channel is defined by a combination of frequency, scrambling code, channelization code, and, in the uplink, phase. Some physical channels exist solely for the correct operation of the physical layer. Other physical channels are used to carry information provided to or from higher layers.

Higher layers that want to transmit information across the RF interface pass information to the physical layer through the *Medium Access Control* (MAC) layer using a number of logical channels. MAC maps these logical transport channels to channels.

The physical layer maps transport channels to physical channels.

Above the MAC layer, we find the *Radio Link Control* (RLC) layer. Among the services provided by RLC are the following:

- **RLC connection establishment and release** A given upper layer may request the use of a certain radio bearer. For each radio bearer, an RLC connection is established between the MS and the network.
- **Error detection** RLC includes a sequence number check function that enables the detection of errors in received *protocol data units* (PDUs).
- **Ensuring error-free delivery through acknowledgements (if the upper layer protocol has requested an acknowledged service)** RLC can request that the peer entity retransmit in the event that a PDU is received incorrectly, lost, or received out of sequence. Note that this type of error correction is different to the error correction that is achieved through coding schemes on the air interface.
- **In-sequence delivery** This ensures that PDUs are passed to the upper layer in the correct order.
- **Unique delivery** This ensures that a given PDU is passed to an upper layer only once, even if erroneously received twice at RLC.
- **Quality of service (QoS) management** Upper layers can request a certain QoS. It is RLC that ensures that the QoS is controlled.

RLC supports both acknowledged and transparent services. With transparent service, any errors in received PDUs will cause the PDU to be discarded, in which case it is up to the upper layer to recover from the loss according to its own capabilities. With acknowledged service, RLC recovers from errors in received data by requesting a retransmission by the peer entity (the UE or the network).

One of the protocols above the RLC layer is the *Packet Data Convergence Protocol* (PDCP). The main objective of PDCP is to enable the lower layers

(RLC, MAC, and the physical layer) to be common regardless of the type or structure of the user data. For example, packet data transfer from a UE could use either IPv4 or IPv6. One does not want the RLC and lower layers to be different depending on which of those two protocols a subscriber uses. Moreover, if new protocols are introduced, one would want them to be supported by the same radio interface. PDCP meets these objectives by maintaining a standard interface to RLC regardless of the type of user data. PDCP is similar to the *Subnetwork Dependent Convergence Protocol* (SNDCP) of GPRS.

In Figure 6-8, we also find *Broadcast/Multicast Control* (BMC). This is a function that handles the broadcast of user messages across the cell. In other words, BMC supports the cell broadcast function, similar to cell broadcast, as defined in GSM. This enables users in a cell to receive broadcast messages, such as traffic warnings and weather information. In GSM, cell broadcast has also been used as a means of informing users of the geographical zone that they are in as part of zone-based tariffing.

One of the most important components depicted in Figure 6-8 is the *Radio Resource Control* (RRC). RRC can be considered the overall manager of the air interface and, as such, is responsible for the management of radio resources, including the determination of which radio resources shall be allocated to a given user. As can be seen, all control signaling to or from users passes through RRC. This is necessary so that requests from a user or from the network can be analyzed and radio resources can be allocated as appropriate. Also, a control interface exists between RRC and each of the other layers. Among the functions performed or controlled by RRC are

- The broadcast of system information.
- The establishment of initial signaling connections between the UE and the network. When the user and network want to communicate, an RRC connection is first established. It is this RRC connection that is used for the transfer of signaling information between the UE and the network for the purpose of allocation and management of the radio resources to be used.
- The allocation of radio bearers to a UE. A given UE may be allocated multiple radio bearers for the transfer of user data.
- Measurement reporting. RRC determines what needs to be measured, when it should be measured, and how it should be reported.
- Mobility management. It is the RRC that determines when, for example, a call should be handed over. RRC also executes cell reselection and location area or routing area updates.

- Quality of Service (QoS) control. The allocation of radio resources has a direct consequence for the QoS perceived by the user. Since RRC controls the allocation of radio resources, it has a direct influence on QoS. The resources allocated by RRC must be aligned with the QoS offered to the subscriber.

6.3.4 WCDMA Channel Types

At the physical layer, the UE and the network communicate via a number of physical channels. Many of these physical channels are used to carry information that is passed to the physical layer from higher layers. Specifically, information is passed to the physical layer from the MAC layer. The interface between the physical layer and the MAC layer is comprised of a number of transport channels, which are mapped to physical channels. Moreover, information from the RLC layer to the MAC layer is passed in the form of logical channels. These logical channels are mapped to transport channels. The following sections consider these various channels in a little more detail. We start with the transport channels.

6.3.4.1 Transport Channels In general, two types of transport channels exist. These are *common transport channels* and *dedicated transport channels*. Common transport channels may be applicable either to all users in a cell or to one or more specific users. In the case when a common transport channel is used to transmit information to all users, then no specific addressing information is required. When a specific user needs to be addressed by a common transport channel, then the user identification is included in-band (within the message being sent). For example, the *Broadcast Channel* (BCH), which is a common transport channel, is used to transmit system information to all users in a cell and is not specific to any given user. On the other hand, the Paging Channel, which is also a common transport channel and which is used to page a specific mobile, contains the identification of the user within the message being transmitted.

As we describe the various types of channels supported, we will make references to frames and slots. Basically, the various channels use a 10-ms frame structure, which corresponds to 38,400 chips. Each frame is divided into 15 slots, each with a length of 2,560 chips, as shown in Figure 6-9. The content of each frame, and for that matter the content of each slot, is dependent upon the type of channel in question.

Figure 6-9
WCDMA Channel Framing Structure.

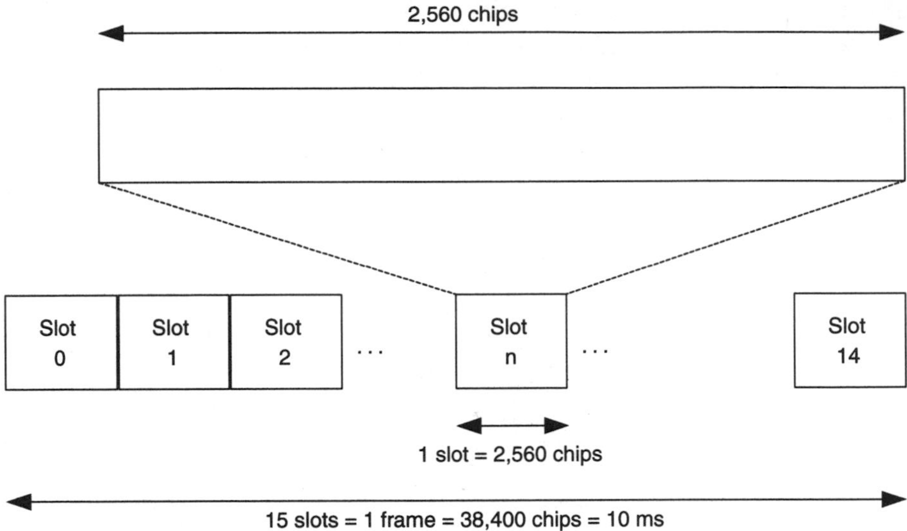

The following common transport channels are defined:

- The *Random Access Channel* (RACH) is used in the uplink when a user wants to gain access to the network. It may also be used when a user wants to send a small amount of data to the network. The amount of data sent on the RACH is small—it lasts either 10 or 20 ms. This is in accordance with the fact that the RACH is used primarily for signaling related to initial system access. It must be possible for the RACH to be heard at the base station from any user in the cell coverage area—even from at the edge, at least when the RACH is used for initial access to the network. Because, as we shall see in Chapter 12, "UMTS System Design," the effective coverage area of a cell decreases with the increasing bandwidth, it is necessary for the data rate on the RACH to be quite low. The RACH is available to all users in the cell. Consequently, the possibility of collision arises when multiple users attempt to access the RACH. UTRAN includes procedures at the physical layer for collision detection.
- The *Broadcast Channel* (BCH) is used in the downlink to transmit system information over the entire coverage area of a cell. For this reason, it is sent with a relatively high power level. Moreover, the data

rate on the BCH is quite low compared to some other channels. When sent over the air interface, the BCH information is sent at 30 Kbps, including coding overhead.

- The *Paging Channel* (PCH) is used in the downlink to page a given UE when the network wants to initiate communication with a user. A page for a given UE may be sent on a single cell or multiple cells, depending on the location area/routing area configuration of the network. In a given cell, the PCH must be heard over the whole cell area.
- The *Forward Access Channel* (FACH) is used to send downlink control information to one or more users in a cell. If, for example, a user attempts to access the network on the RACH, then the response to the access request will be sent on the FACH. The FACH can also be used to send small amounts of packet data to a mobile. It is possible to have more than one FACH in a cell. At least one FACH, however, must have a sufficiently low data rate that all users in the cell can hear it.
- The uplink *Common Packet Channel* (CPCH) is similar to the RACH but can last for several frames. Thus, it enables a greater amount of data to be sent than is allowed by the RACH. It can be used, for example, when the terminal wants to send a single burst of data that cannot be accommodated on the RACH. The CPCH is available to all users in the cell. Consequently, the possibility of collision occurs when a user attempts to access the CPCH. The UTRAN includes procedures at the physical layer for minimizing the likelihood of collision and detecting collision when it does occur.
- The *Downlink Shared Channel* (DSCH) is used to carry dedicated user data or control signaling to one or more users in a cell. It is similar to the FACH, but does not have to be transmitted over the entire cell area. Moreover, it supports higher data rates than the FACH and the data rate on the DSCH can change on a frame-by-frame basis. The DSCH is always associated with one or more downlink dedicated channels described later.

Only a single dedicated transport channel type exists, known as the *Dedicated Channel* (DCH). This is a channel that carries user data and is specific to a single user. Although other channels can carry small amounts of bursty user data, they are not designed for large amounts of data or for extended data sessions. The DCH is used for those types of sessions.

For example, in a voice conversation, the coded voice uses the DCH. The DCH exists in the uplink and the downlink and is mapped to the physical channels DPDCH and DPCCH, previously described. In the uplink, at the physical layer, the combination of frequency, the scrambling code, the channelization code, and the phase is used to indicate a particular DPDCH or DPCCH. In the downlink, the DCH is mapped to a *Dedicated Physical Channel* (DPCH), which is identified in the downlink by a particular channelization code. The downlink DPDCH and DPCCH are time multiplexed onto the downlink DPCH. The data rate on a DCH can vary on a frame-by-frame basis.

6.3.4.2 Physical Channels As mentioned, information from upper layers is passed to the physical layer through a number of transport channels. These transport channels are mapped to a number of physical channels on the air interface. In general, a physical channel is identified by a specific frequency, scrambling code, channelization code, duration, and, in the uplink, phase. In addition to those physical channels that are mapped to or from transport channels, a number of physical channels exist only for the correct operation of the physical layer. Such channels are not visible to higher layers. The following are the physical channels:

- The *Synchronization Channel* (SCH) is transmitted by the base station and is used by a UE during the cell search procedure. In order for a UE to retrieve broadcast information sent from the base station, it must first be properly synchronized with the base station. That synchronization is the primary purpose of the SCH. The SCH contains two subchannels—the primary SCH and the secondary SCH, as shown in Figure 6-7. The primary SCH contains a specific 256-chip codeword, known as the *primary synchronization code* (PSC), which is identical in every cell. This specific codeword is created from a set of 16-bit chip sequences as follows:

  Let a = (1, 1, 1, 1, 1, 1, −1, −1, 1, −1, 1, −1, 1, −1, −1, 1). Then the primary SCH contains a sequence of (1 + j) × (a, a, a, −a, −a, a, −a, −a, a, a, a, −a, a, −a, a, a).

  The secondary SCH is comprised of 16 codewords, each with a length of 256 chips. These 16 codewords are arranged into 64 different sequences of length 15. In other words, a sequence is a set of 15 codewords in a particular order and there are 64 such sequences. The 64 available sequences are mapped to the 64 downlink primary scrambling code groups. Thus, when a terminal receives a particular secondary SCH sequence, it can identify the primary scrambling code group of the cell in question. Since only eight primary scrambling codes are in each primary scrambling code group, the UE then has relatively few primary scrambling codes to check before being able to decode transmissions from the base station. The SCH is transmitted in conjunction with the *Primary Common Control Physical Channel* (Primary CCPCH) described later.

- The *Common Pilot Channel* (CPICH) is a channel always transmitted by the base station and is scrambled with the cell-specific primary scrambling code. It uses a fixed spreading factor of 256, which equates to 30 Kbps on the air interface.

  An important function of the CPICH is in measurements by the terminal for handover or cell reselection, as the measurements made by the terminal are based on reception of the CPICH. Consequently, manipulation of the transmitted power on the CPICH can be used to steer terminals towards a given cell or away from a given cell.

  For example, if the CPICH power transmitted on given cell is reduced, the effect is to make the CPICH reception from neighboring cells appear stronger, which may trigger a handover to a neighboring cell. This can be useful for load-balancing in the RF network. It is possible to have more than one CPICH in a given cell. The primary CPICH is transmitted over the entire cell area. The secondary CPICH can be transmitted over the whole cell area or can be restricted by transmission on narrow-beam antennas to specific areas of the cell, such as areas of high traffic. The channelization code for the Primary CPICH is fixed to $C_{ch,256,0}$. An arbitrary channelization code of SF =256 is used for the S-CPICH.

- The *Primary Common Control Physical Channel* (Primary CCPCH) is used on the downlink to carry the BCH transport channel. It operates at a spreading factor of 256, equivalent to 30 Kbps on the air interface. In fact, the actual rate is reduced to 27 Kbps on the air interface because of the fact that the Primary CCPCH is time-multiplexed with the SCH, as shown in Figure 6-10. For many of the channels on the air interface, all of the chips in a slot are allocated to a particular physical channel. The Primary CCPCH is an exception in that it shares every slot with the SCH. The first 256 chips of each slot are used by the SCH. The remaining 2,304 chips are used by the Primary CCPCH to carry the BCH transport channel. The 2,304 chips allocated to the Primary CCPCH correspond to 18 bits of primary CCPCH data. Moreover, the 18 bits include half-rate convolutional coding (to support forward error correction) so that the actual data rate is approximately 13.5 Kbps.

Figure 6-10 Multiplexing SCH and CCPCH.

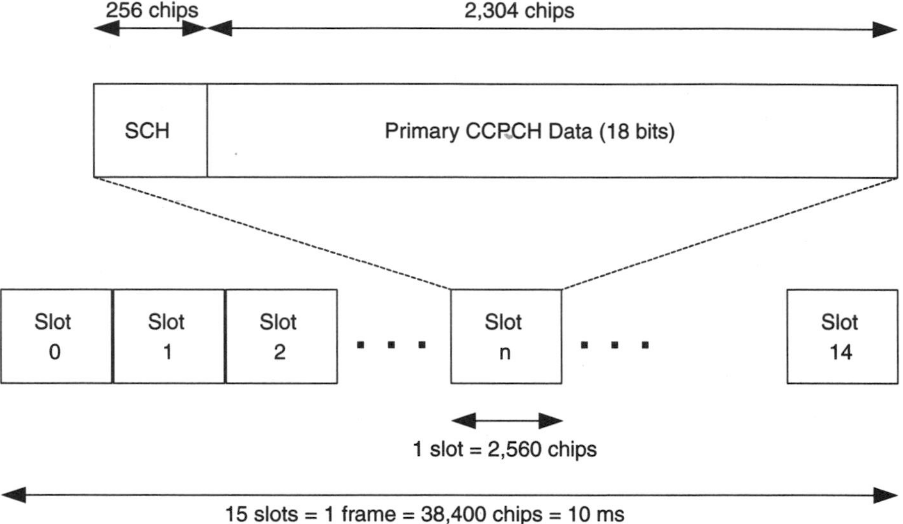

- The *Secondary Common Control Physical Channel* (Secondary CCPCH) is used on the downlink to carry two common transport channels—the FACH and the PCH. The FACH and the PCH can share a single secondary CCPCH or each can have a secondary CCPCH of its own. The secondary CCPCH carrying the PCH must be transmitted over the whole cell area, which applies regardless of whether the physical channel carries just the PCH or both PCH and FACH. If a secondary CCPCH is used just for the FACH, then it does not necessarily have to reach the whole cell coverage area.
- The *Physical Random Access Channel* (PRACH) is used in the uplink to carry the RACH transport channel. The uplink the PRACH has 15 access slots, each with a duration of 5,120 chips. These access slots are arranged in different combinations, known as RACH subchannels, for which certain scrambling codes and signatures are available. A given UE may be allowed to use one or more RACH subchannels according to the class of UE. The signatures and scrambling codes available for a particular RACH subchannel are broadcast on the BCH transport channel.

The process for accessing the uplink begins with the transmission from the terminal of a specific preamble sent on a specific access slot. This preamble is 4,096 chips long and comprises 256 repetitions of a 16-chip signature. The preamble is scrambled by one of 8,192 long scrambling codes. These 8,192 scrambling codes are grouped into 512 groups of 16 codes. A correspondence exists between a specific group of preamble scrambling codes and the primary downlink scrambling code used in the cell.

Once the base station detects the preamble, it uses the *Acquisition Indicator Channel* (AICH) to indicate to the UE that the preamble has been detected and that the UE either is or is not allowed uplink access. The AICH is also structured in slots, each of which is 5,120 chips long. Each slot indicates a number of PRACH signatures and an indication for each as to whether the UE is allowed access to the uplink. The UE checks the AICH to see whether it has been granted access (as determined by checking for the signature it has just used). Assuming that the UE has been granted access, then it transmits the actual RACH message (10 ms or 20 ms duration) on subsequent access slots.

- The *Physical Common Packet Channel* (PCPCH) is used in the uplink to carry the uplink CPCH transport channel. Given that the CPCH is somewhat similar to the RACH, the process for using the PCPCH is similar to that for using the PRACH. A preamble is first sent using a specific signature. The terminal then waits for a response from the base station on the *Access Preamble-Acquisition Indicator Channel* (AP-AICH), similar to what is done on the AICH for an access attempt on the PRACH.

  When the response is received on the AP-AICH, however, the terminal does not yet proceed to transmit the desired data. The reason is that the CPCH can support longer durations of data than the RACH. Thus, if there is a collision, a greater amount of data is lost. Therefore, the terminal next sends a specific *collision detection* (CD) signature and waits for this to be echoed back from the base station on the *Collision Detection / Channel Assignment-Indication Channel* (CD/CA-ICH). At this point, the terminal can send the CPCH data on the PCPCH. The duration of the data transfer can last several 10-ms frames. The spreading factor can take any value from 4 to 256.

  As an option, the base station can support the CPCH Status Indication Channel, which is used to indicate the current state of affairs for any CPCH defined in the cell. By monitoring this channel, the terminal can determine, in advance, if resources are available to support the terminal's use of a CPCH. This avoids access attempts from the mobile that are doomed to fail.

- The *Physical Downlink Shared Channel* (PDSCH) is used in the downlink to carry the DSCH transport channel. Because the DSCH transport channel can be shared among several users, the PDSCH has a structure that enables it to be shared among users. A PDSCH has a root channelization code and there may be multiple PDSCHs with channelization codes at or below the root channelization code. These various PDSCHs may be allocated to different UEs on a radio-frame-by-radio-frame basis. Within one radio frame, UTRAN may allocate different PDSCHs under the same PDSCH root channelization code to different UEs. Within the same radio frame, multiple parallel PDSCHs with the same spreading factor may be allocated to a single user. PDSCHs allocated to the same user on different radio frames may have different spreading factors.
- The Indicator Channels include the AICH, AP-AICH, and CD/CA-ICH already mentioned. In addition, there is the *Paging Indicators Channel* (PICH). The purpose of the PICH is to let a given terminal know when it might expect a paging message on the PCH (carried on the secondary PCPCH). When a user device registers with the network, it is assigned to a paging group. These paging groups are indicated through the use of paging indicators carried on the PICH. When a terminal is to be paged on the PCH, a paging indicator corresponding to the paging group in question is carried on the PICH. If a terminal decodes the PICH and finds that its paging group is indicated, then at least one terminal in its paging group is being paged, which means that the terminal must decode the PCH (carried on the secondary PCPCH) to determine if it is being paged. If a given terminal's paging group is not indicated on the PICH, then the terminal need not decode the PCH.
- The DCH transport channel is mapped to the two physical channels—DPDCH and DPCCH, as previously mentioned. The DPDCH carries the actual user data and can have a variable spreading factor, whereas the DPCCH carries control information.

The mapping between the transport channels and the physical channel is shown in Figure 6-11.

6.3.4.3 Logical Channels As shown in Figure 6-8, information is passed from the MAC layer to the physical layer in the form of transport channels. That information, however, can begin higher in the protocol stack, in which

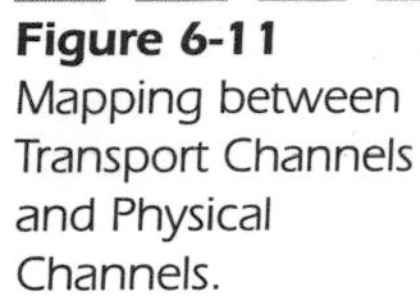

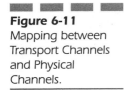

Figure 6-11 Mapping between Transport Channels and Physical Channels.

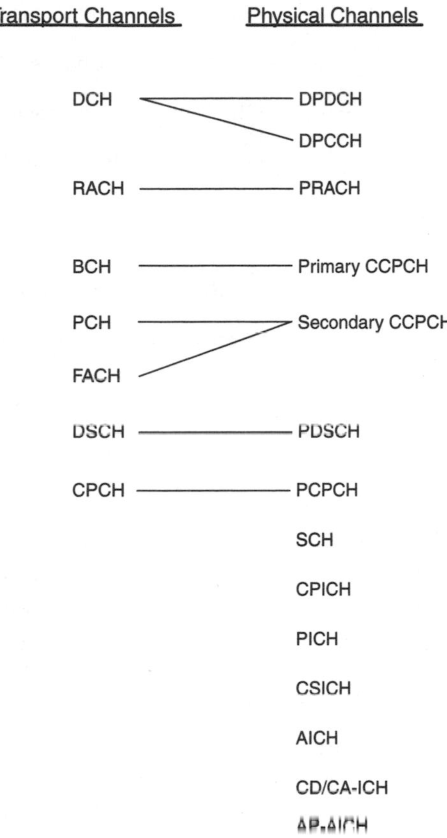

case it is passed from the RLC layer to the MAC layer in the form of logical channels. The logical channels are mapped to transport channels, which in turn are mapped to physical channels.

As mentioned, RLC interfaces with MAC through a number of logical channels. MAC maps those logical channels to the transport channels previously described. Logical channels relate to the information being transmitted, while transport channels relate largely to the manner in which the information is transmitted. Basically, two groups of logical channels exist—control channels and traffic channels. These are shown in Figure 6-12.

The *Broadcast Control Channel* (BCCH) is used for the downlink transmission of system information. The *Paging Control Channel* (PCCH) is used

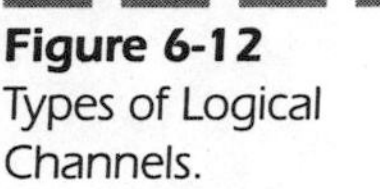

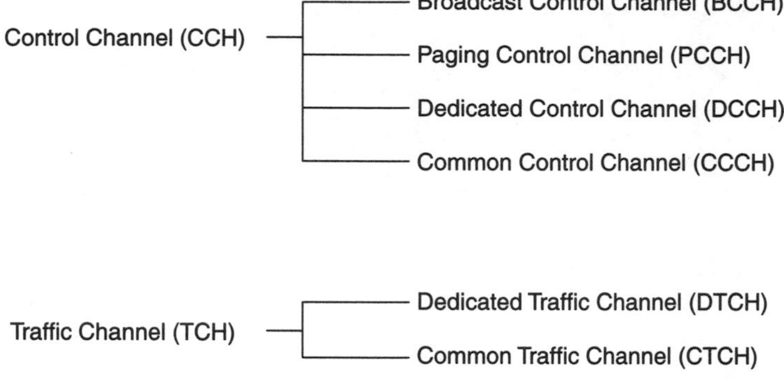

Figure 6-12 Types of Logical Channels.

for the paging of an MS across one or more cells. The *Common Control Channel* (CCCH) is used in the uplink by terminals that want to access the network but do not already have any connection with the network. The CCCH can be used in the downlink to respond to such access attempts. The *Dedicated Control Channel* (DCCH) is a bidirectional point-to-point control channel between the MS and the network for sending control information. WCDMA also defines the Shared Channel Control Channel, but that channel is used only in TDD mode.

Two types of logical traffic channels are available. The *Dedicated Traffic Channel* (DTCH) is a point-to-point channel, dedicated to one UE, for the transfer of user data. DTCHs apply to the uplink and the downlink. The *Common Traffic Channel* (CTCH) is point-to-multipoint unidirectional channel for the transfer of user data to all UEs or just to a single UE. The CTCH exists in the downlink only.

Numerous options are available for mapping between logical channels and transport channels. This mapping depends on a range of criteria such as the types of information to be sent, whether it is to be sent to multiple UEs (in the downlink), and whether the UE has already an established connection with the network. The possible mapping between logical channels and transport channels for the FDD mode of operation is shown in Figure 6-13.

6.3.5 Power Control in WCDMA

In any CDMA system, power control is of critical importance. Because all users share the same frequency at the same time, it is important that one

Figure 6-13 Mapping between Logical Channels and Transport Channels.

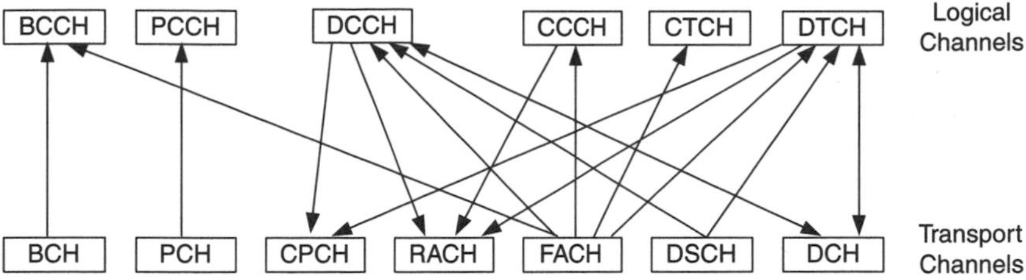

Mapping between logical channels and transport channels, as seen from the UE perspective

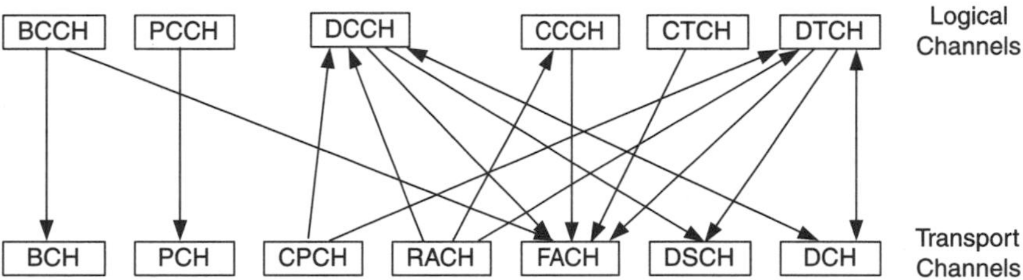

Mapping between logical channels and transport channels, as seen from the UTRAN perspective

user not transmit at such a high power that other users are drowned out. If, for example, a user near the base station were to transmit at the same power level as a user at the cell edge, then at the base station, the signal from the nearby user would be so great that it would completely overpower the signal from the far-away user. The result is that the signal from the far-away user would be impossible to recover. This is known as the near-far problem.

To avoid this problem, mechanisms are required whereby the UE can be instructed to adjust its transmit power up or down so that all transmissions from all users in the cell arrive at the base station with the same power level. Not only is power control required to combat the near-far problem, it is also required to combat the effects of Raleigh fading, where the received signal can suddenly drop by many decibels as a result of multi-path propagation, which results in multiple copies of a signal arriving at the receiver out of phase. Thus, power control is deployed both in the uplink and downlink.

In general, power control in WDCMA uses two main techniques—open-loop power control and closed-loop power control. With open-loop power control, the terminal estimates the required transmission power based upon the signal power received from the base station and information broadcast from the base station regarding the transmit power from the base station. Specifically, the base station broadcasts the transmit power used on the CPICH, and the terminal uses this information in conjunction with the received power level to determine the power that should be used on the uplink. In general, however fading in the uplink and fading in the downlink are unrelated. Consequently, open-loop power control provides only a very rough estimate of the ideal power that the terminal should use. For this reason, open-loop power control is used only when the UE is making initial access on the PRACH or PCPCH. In all other situations, closed-loop power control is used.

Closed-loop power control means that the receiving entity (the base station or UE) measures the received *Signal-to-Interference Ratio* (SIR) and compares it with a target SIR value. The base station or UE then instructs the far end to increase the transmitted power if the SIR is too low or decrease the power if the SIR is too high. Closed-loop power control is also known as fast power control since it operates at a rate of 1,500 Hz. In other words, power control commands and changes happen at a rate of 1,500 times per second. This rate is sufficiently fast to overcome path loss changes and Rayleigh fading effects for all situations except where the UE is travelling at high speed.

Closed-loop power control commands are sent on physical control channels that are associated with physical data channels. Recall, for example, that in the uplink, the DPDCH has an associated DPCCH. Among other pieces of information, the DPCCH carries transmit power control commands back to the base station. A power control command is sent in every slot. Because 15 slots are available for each 10 ms, we have a rate of 1,500 power control commands per second. Each power control command can instruct the sender to leave the transmitted power unchanged or to increase or decrease the transmitted power in steps of 1dB, 2dB, or 3dB. Similarly, for the downlink DPDCH, an associated DPCCH sends power control instructions to the UE, along with other functions.

There is also another form of power control known as outer-loop power control, with the primary objective being maintaining the service quality at the optimum level. In general, the objective of power control is to maintain the SIR at the receiver at the optimum level—not too high and not too low. The target SIR value, however, is a function of the required quality for the service to be supported. If we measure service quality in terms of *Frame Error Rate* (FER) on the air interface (as determined by a *cyclic redundancy check* (CRC), then the SIR can be considered a function of FER.

The acceptable FER can vary from service to service. Speech service using the *Adaptive Multirate* (AMR) coder at 12.2 Kbps, for example, could support a FER of one percent without noticeable service degradation. A non-real time data service could support much higher FER rates before retransmission, allowing retransmission to correct errors. The impact to such a service is greater delay and a lower overall throughput, but such impact can be perfectly acceptable for a non-real time service.

A real-time data service, however, may have a far more stringent FER requirement, perhaps 1×10^{-3} or better. Consequently, depending on the service requirements, the FER may need to vary, which means that the required SIR may need to vary. This variation in the required SIR is known as outer loop power control. It uses closed-loop power control to instruct the sender to vary the transmit power. With outer-loop power control, however, the reason for the change is due to a new SIR requirement.

6.3.6 User Data Transfer

WCDMA is designed to offer great flexibility in the transmission of user data across the air interface. For example, data rates can change on a frame by frame basis (every 10 ms). Moreover, it is possible to mix and match different types of service. For example, a subscriber may be sending and receiving packet data while also involved in a voice call. When sending information over the air interface, physical control channels are used in combination with physical data channels. Although the physical data channels carry the user information, the physical control channels carry information to support the correct interpretation of the data carried on the corresponding DPDCH frame, plus power control commands, and feedback indicators.

6.3.6.1 Uplink DPDCH and DPCCH Figure 6-14 shows the structure of the uplink DPCCH as used with the uplink DPDCH. The DPCCH is transmitted in parallel with the DPDCH and the information in a given DPCCH frame relates to the corresponding DPDCH frame.

The DPCCH always uses a spreading factor of 256. Thus, each slot (2,560 chips) corresponds to 10 bits of DPCCH information. These 10 bits are divided into pilot bits, *Transport Format Combination Indicator* (TFCI) bits,

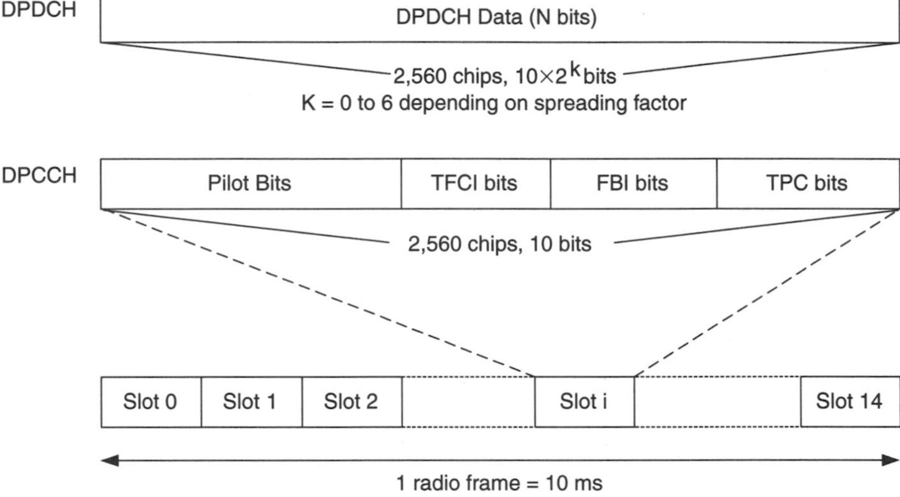

Figure 6-14 Uplink DPDCH and DPCCH Frame and Slot Structure.

Feedback Indicator (FBI) bits, and *Transmit Power Control* (TPC) bits.

The pilot information bits are used for channel estimation purposes and include specific bit patterns for frame synchronization. The TFCI bits indicate the bit rate and channel coding for the DPDCH. A single DPDCH can carry multiple DCH transport channels.

If, for example, a user were invoking multiple simultaneous services, the associated DCH transport channels could be multiplexed together on a single DPDCH. In that case, the DPDCH is said to carry a *Coded Composite Transport Channel* (CCTrCH). The TFCI is used to indicate the format of each of the transport channels within the CCTrCH. The FBI bits are used in conjunction with transmit diversity at the base station. WCDMA supports downlink transmit diversity, whereby two antennas can be used for downlink transmission. When transmit diversity is used, it is possible for the power and/or phase on one transmit antenna to differ from that on the other. The FBI bits are used in the uplink to instruct the base station to change the power or phase difference associated with transmit diversity. Finally, the TPC bits are used to command the base station to change the transmit power when necessary.

The number of bits in each of the uplink DPCCH fields depends upon the slot format for the DPCCH. A number of slot formats are possible, as shown in Table 6-3.

Table 6-3

Downlink DPCCH Slot Formats

Slot Format	Channel Bit Rate (Kbps)	Channel Symbol Rate (Ksps)	SF	Bits/ Frame	Bits/ Slot	Pilot Bits	TPC Bits	TFCI Bits	FBI Bits	Transmitted Slots per Radio Frame
0	15	15	256	150	10	6	2	2	0	15
0A	15	15	256	150	10	5	2	3	0	10–14
0B	15	15	256	150	10	4	2	4	0	8–9
1	15	15	256	150	10	8	2	0	0	8–15
2	15	15	256	150	10	5	2	2	1	15
2A	15	15	256	150	10	4	2	3	1	10–14
2B	15	15	256	150	10	3	2	4	1	8–9
3	15	15	256	150	10	7	2	0	1	8–15
4	15	15	256	150	10	6	2	0	2	8–15
5	15	15	256	150	10	5	1	2	2	15
5A	15	15	256	150	10	4	1	3	2	10–14
5B	15	15	256	150	10	3	1	4	2	8–9

As can be seen from Table 6-3, in some slot formats, the full 15 slots are not used in every radio frame. The reason for less than 15 slots per frame is because of the use of compressed mode. In compressed mode, gaps exist in both the uplink and downlink transmissions. These gaps are included to enable the UE to take measurements on other frequencies. By taking measurements on other frequencies and reporting those measurements, the UE enables the network to enable an inter-frequency handover either to another UMTS frequency or perhaps an inter-system handover to a GSM system.

Also, a number of different slot formats exist for the DPDCH, but these simply reflect the different spreading factors that can be applied to the DPDCH data. For example, a *spreading factor* (SF) of 256 for the uplink DPDCH means 10 bits per slot, whereas a spreading factor of 4 means 640 bits per slot.

6.3.6.2 Downlink DPDCH and DPCCH Figure 6-15 shows the structure of the downlink DPDCH and DPCCH. The most notable characteristic is that the DPCCH is time-multiplexed with the DPDCH rather than being transmitted separately. In each slot on the downlink, two fields contain DPDCH user data, while three other fields maintain information on the pilot bits, the TFCI, and the TPC. As is the case for the uplink, a number of slot formats can be applied to the downlink DPDCH/DPCCH. Table 6-4 shows the various combinations.

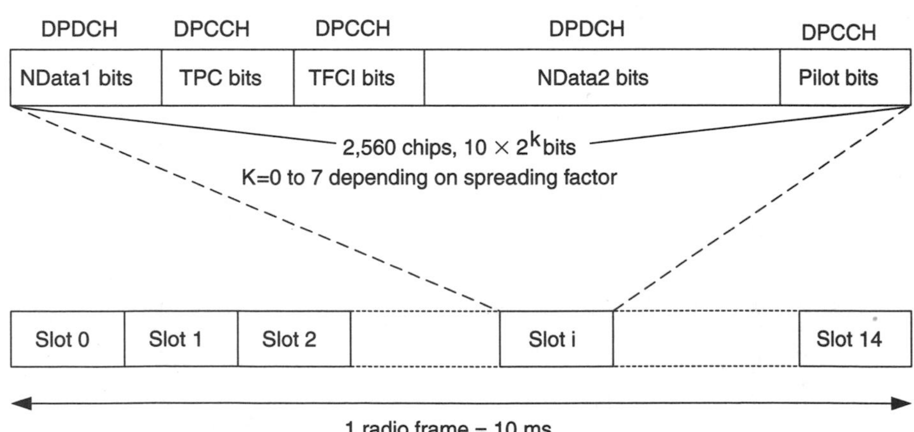

Figure 6-15 Downlink DPDCH and DPCCH Frame and Slot Structure.

Table 6-4

Downlink DPDCH/DPCCH Slot Formats

Slot Format	Channel Bit Rate (Kbps)	Channel Symbol Rate (Ksps)	SF	Bits/ Slot	DPDCH Bits/Slot		DPCCH Bits/Slot			Transmitted Slots per Radio Frame
					NDATA1	NDATA2	TPC	TFCI	Pilot	
0	15	7.5	512	10	0	4	2	0	4	15
0A	15	7.5	512	10	0	4	2	0	4	8–14
0B	30	15	256	20	0	8	4	0	8	8–14
1	15	7.5	512	10	0	2	2	2	4	15
1B	30	15	256	20	0	4	4	4	8	8–14
2	30	15	256	20	2	14	2	0	2	15
2A	30	15	256	20	2	14	2	0	2	8–14
2B	60	30	128	40	4	28	4	0	4	8–14
3	30	15	256	20	2	12	2	2	2	15
3A	30	15	256	20	2	10	2	4	2	8–14
3B	60	30	128	40	4	24	4	4	4	8–14
4	30	15	256	20	2	12	2	0	4	15
4A	30	15	256	20	2	12	2	0	4	8–14
4B	60	30	128	40	4	24	4	0	8	8–14
5	30	15	256	20	2	10	2	2	4	15
5A	30	15	256	20	2	8	2	4	4	8–14
5B	60	30	128	40	4	20	4	4	8	8–14
6	30	15	256	20	2	8	2	0	8	15
6A	30	15	256	20	2	8	2	0	8	8–14
6B	60	30	128	40	4	16	4	0	16	8–14
7	30	15	256	20	2	6	2	2	8	15
7A	30	15	256	20	2	4	2	4	8	8–14
7B	60	30	128	40	4	12	4	4	16	8–14
8	60	30	128	40	6	28	2	0	4	15
8A	60	30	128	40	6	28	2	0	4	8–14
8B	120	60	64	80	12	56	4	0	8	8–14
9	60	30	128	40	6	26	2	2	4	15

Table 6-4 (cont.)

Downlink DPDCH/DPCCH Slot Formats

Slot Format	Channel Bit Rate (Kbps)	Channel Symbol Rate (Ksps)	SF	Bits/ Slot	DPDCH Bits/Slot NDATA1	DPDCH Bits/Slot NDATA2	DPCCH Bits/Slot TPC	DPCCH Bits/Slot TFCI	DPCCH Bits/Slot Pilot	Transmitted Slots per Radio Frame
9A	60	30	128	40	6	24	2	4	4	8–14
9B	120	60	64	80	12	52	4	4	8	8–14
10	60	30	128	40	6	24	2	0	8	15
10A	60	30	128	40	6	24	2	0	8	8–14
10B	120	60	64	80	12	48	4	0	16	8–14
11	60	30	128	40	6	22	2	2	8	15
11A	60	30	128	40	6	20	2	4	8	8–14
11B	120	60	64	80	12	44	4	4	16	8–14
12	120	60	64	80	12	48	4	8*	8	15
12A	120	60	64	80	12	40	4	16*	8	8–14
12B	240	120	32	160	24	96	8	16*	16	8–14
13	240	120	32	160	28	112	4	8*	8	15
13A	240	120	32	160	28	104	4	16*	8	8–14
13B	480	240	16	320	56	224	8	16*	16	8–14
14	480	240	16	320	56	232	8	8*	16	15
14A	480	240	16	320	56	224	8	16*	16	8–14
14B	960	480	8	640	112	464	16	16*	32	8–14
15	960	480	8	640	120	488	8	8*	16	15
15A	960	480	8	640	120	480	8	16*	16	8–14
15B	1920	960	4	1280	240	976	16	16*	32	8–14
16	1920	960	4	1280	248	1000	8	8*	16	15
16A	1920	960	4	1280	248	992	8	16*	16	8–14

*TFCI on the downlink is optional. If TFCI bits are not used, then discontinuous transmission is used in the TFCI field.

As can be seen from Table 6-4, the actual DPDCH user throughput is dependent on the slot format used. Moreover, one can clearly see the effect of compressed mode, where less than 15 slots are used in the downlink.

6.4 The UTRAN Architecture

In most mobile communications networks, the network architecture can be split into two main parts—the access network and the core network. The access network is specific to the access technology being used, whereas the core network is shielded from the vagaries of the access technology and should ideally be able to handle multiple different access networks. This split applies quite well to UMTS, where the access network is known as the *UMTS Terrestrial Radio Access Network* (UTRAN). It is supported by a core network that is based upon the core network used for GSM. In fact, the GSM core network can be upgraded to simultaneously support both UTRAN and a GSM radio access network.

The UTRAN architecture is shown in Figure 6-16 as it applies to the first release of UMTS specification—3GPP Release 1999. The UTRAN comprises two types of nodes—the *Radio Network Controller* (RNC) and the Node B, which is the base station. The RNC is analogous to the GSM *Base Station Controller* (BSC). The RNC is responsible for the control of the radio resources within the network. It interfaces with one or more base stations, known as Node Bs. The interface between the RNC and the Node B is the Iub interface. Unlike the equivalent Abis interface in GSM, the Iub interface is open, which means that a network operator could acquire Node Bs from one vendor and RNCs from another vendor. Together an RNC and the set of Node Bs that it supports are known as a *Radio Network Subsystem* (RNS).

Unlike in GSM where BSCs are not connected to each other, UTRAN contains an interface between RNCs. This is known as the Iur interface. The primary purpose of the Iur interface is to support inter-RNC mobility and a soft handover between Node Bs connected to different RNCs.

The user device is the UE. It comprises the *Mobile Equipment* (ME) and the *UTMTS Subscriber Identity Module* (USIM). UTRAN communicates with the UE over the Uu interface. The Uu interface is none other than the WCDMA air interface that we have already described in this chapter.

UTRAN communicates with the core network over the Iu interface. The Iu interface has two components—the Iu-CS interface, which supports circuit-switched services, and the Iu-PS interface, which supports

Figure 6-16
UTRAN Architecture.

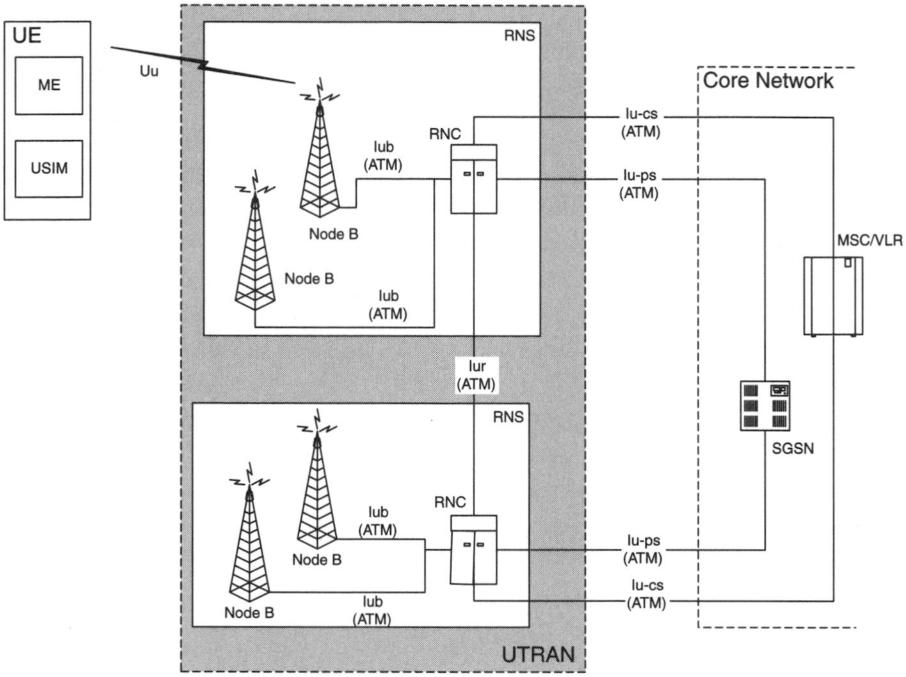

packet-switched services. The Iu-CS interface connects the RNC to an MSC and is similar to the GSM A-interface. The Iu-PS interface connects the RNC to an SGSN and is analogous to the GPRS Gb interface.

In 3GPP Release 1999, all of the interfaces within UTRAN, as well as the interfaces between UTRAN and the core network, use *Asynchronous Transfer Mode* (ATM) as the transport mechanism.

6.4.1 Functional Roles of the RNC

The RNC that controls a given Node B is known as the *Controlling RNC* (CRNC). The CRNC is responsible for the management of radio resources available at a Node B that it supports.

For a given connection between the UE and the core network, one RNC is in control. This is called the *Serving RNC* (SRNC). For the user in question, the SRNC controls the radio resources that the UE is using. In addition, the SRNC terminates the Iu interface to or from the core network for the services being used by the UE. In many cases, though not all, the SRNC is also the CRNC for a Node B that is serving the user.

As depicted in Figure 6-17, UTRAN supports soft handovers, which may occur between Node Bs controlled by different RNCs. During and after a soft handover between RNCs, one may find a situation where a UE is communicating with a Node B that is controlled by an RNC that is not the SRNC. Such an RNC is termed a *Drift RNC* (DRNC). The DRNC does not perform any processing of user data (beyond what is required for correct operation of the physical layer). Rather, data to or from the UE is controlled by the SRNC and is passed transparently through the DRNC.

As a UE moves further and further away from any Node B controlled by the SRNC, it will become clear that it is no longer appropriate for the same RNC to continue to act as the SRNC. In that case, UTRAN may make the decision to hand the control of the connection over to another RNC. This is known as *serving RNS* (SRNS) *relocation*. This action is invoked under the control of algorithms within the SRNC.

6.4.2 UTRAN Interfaces and Protocols

Figure 6-18 provides a generic model for the terrestrial interfaces used in UTRAN—the Iu-CS, Iu-PS, Iur, and Iub interfaces. Each interface has two main components—the radio network layer and the transport network layer. The radio network layer represents the application information to be carried—either user data or control information. This is the information that UTRAN actually cares about. The transport network layer represents the transport technology that the various interfaces use. In the case of 3GPP Release 1999, ATM transport is used, so the transport network layer represents an ATM-based transport. Another transport layer could be used instead. In such a case, the transport network layer would be different, but the radio network layer should not be.

Looking at Figure 6-18 in the vertical direction, we see three planes—the *control plane*, the *user plane*, and the *transport network user plane*. The control plane is used by UMTS-related control signaling. It includes the application protocol used on the interface in question. The control plane is responsible for the establishment of the bearers that transport user data, but the user data itself is not carried on the control plane. As seen from control plane, the user bearers established by the application protocol are generic bearers and are independent of the transport technology being used. If the application protocol were to view the bearers in

Figure 6-17 Soft Handover and SRNS Relocation.

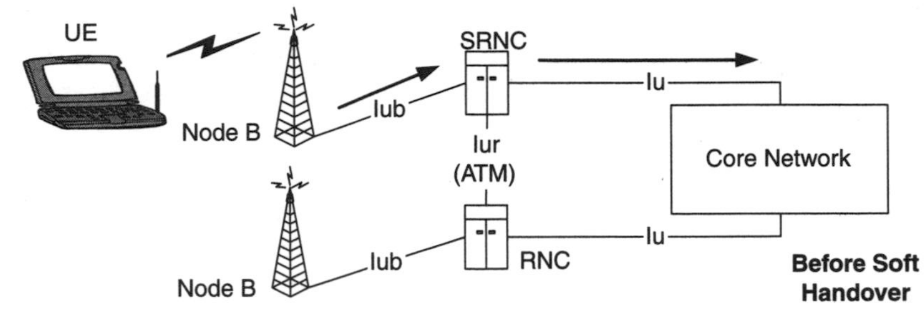

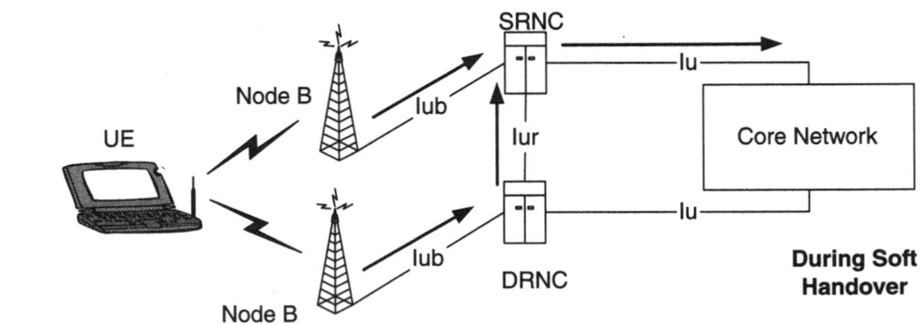

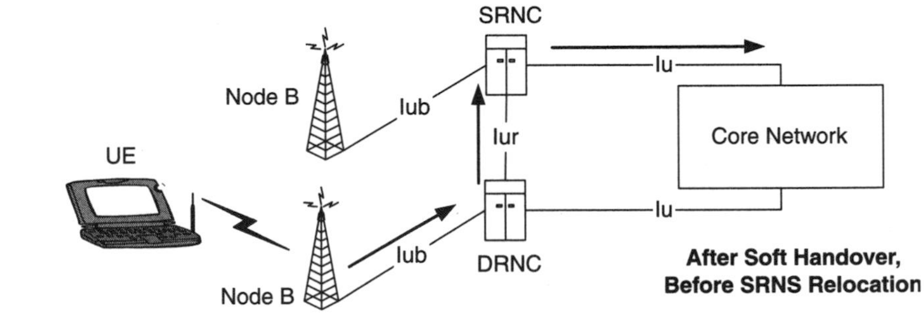

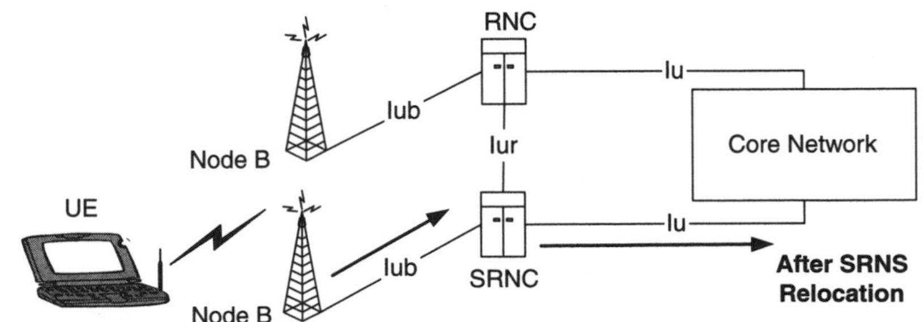

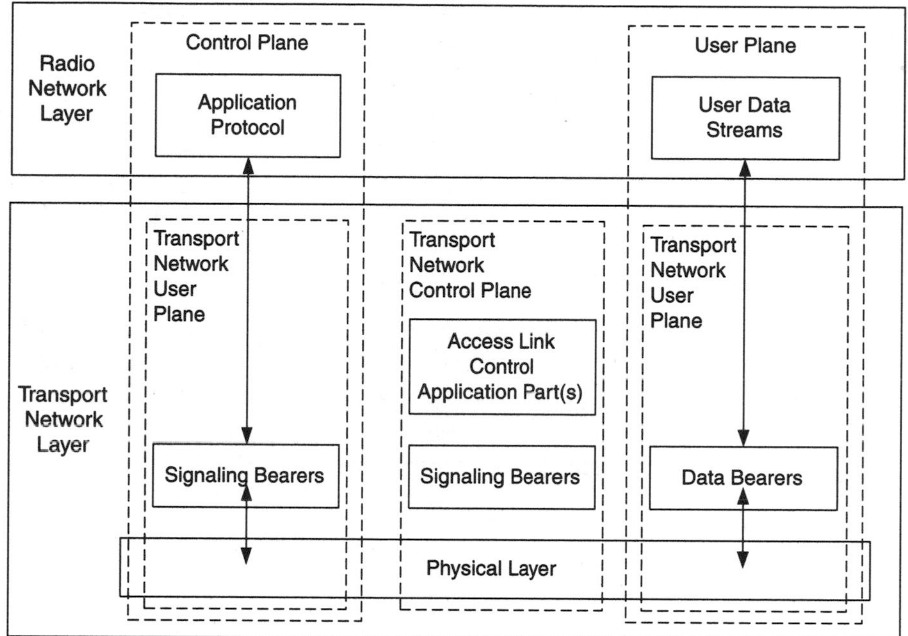

Figure 6-18 Generic Model for UTRAN Terrestrial Interfaces.

terms of a specific transport technology, then it would not be possible to cleanly separate the radio network layer from the transport network layer. In other words, the application protocol would have to be designed to suit a particular transport technology. The signaling bearers that carry the application signaling are established by O&M actions. These signaling bearers are analogous, for example, to the SS7 signaling links that are used between a BSC and a MSC in GSM.

The user plane is what carries the actual user data. This data could, for example, be data packets being sent or received by the UE as part of a data session. Each data stream carried in the user plane will have its own framing structure.

The transport network control plane contains functionality that is specific to the transport technology being used and is not visible to the radio network layer. If standard pre-configured bearers are to be used by the user plane and these are known to the control plane, then the transport network control plane is not needed. Otherwise, the transport network control plane is used. It involves the use of an *Access Link Control Application Part* (ALCAP). This is a generic term that describes a protocol or set of protocols used to set up a transport bearer. The ALCAP to be used is dependent on the user plane transport technology.

6.4.2.1 Iu-CS Interface If we apply this generic structure to the Iu-CS interface (RNC to MSC), then it appears as shown in Figure 6-19. The application protocol in the control plane is the *Radio Access Network Application Part* (RANAP). This provides functionality similar to that provided by BSSAP in GSM. Among the many functions supported by RANAP are the establishment of *radio access bearers* (RABs), paging, the direct transfer of signaling messages between the UE and core network, and SRNS relocation.

RANAP is carried over an ATM-based SS7 signaling bearer. This signaling bearer for the control plane is comprised of the *ATM Adaptation Layer 5* (AAL5), the *service-specific connection-oriented protocol* (SSCOP), the *service-specific coordination function at the network node interface* (SSCF-

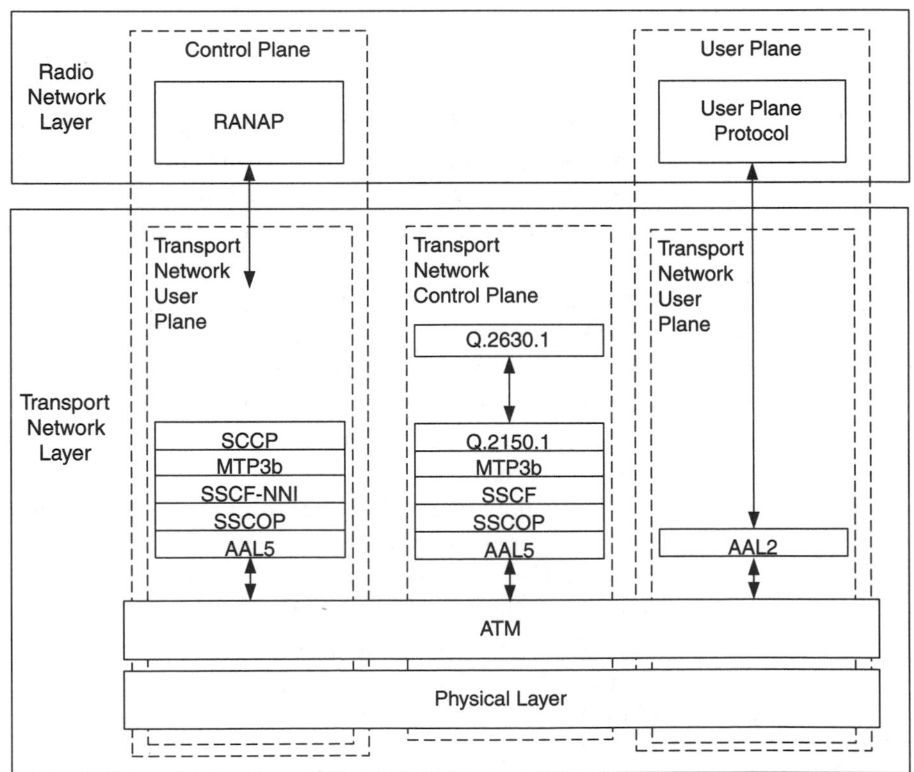

Figure 6-19 Iu-CS Protocol Sructure.

NNI), the layer *3 broadband message transfer part* (MTP3b), and the *Signaling Connection Control Part* (SCCP).

Different AAL layers may reside above the ATM layer depending on the type of service that needs to use ATM. In the case of signaling, it is normal to use AAL5, which supports variable bit rate services.

SSCOP provides mechanisms for the establishment and release of signaling connections. It also offers the reliable exchange of signaling information, including functions such as sequence integrity, error detection and message retransmission, and flow control. SSCF-NNI maps the requirements of the upper layer to the layer below. Together SSCOP and SSCF are known as the *signaling ATM adaptation layer* (S-AAL).

MTP3b is similar to standard MTP3, as used in standard SS7 networks, with some modifications to enable it to take advantage of the broadband transport that ATM can use. SCCP is the same SCCP as used in standard SS7 networks.

The same signaling stack is used for the transport network control plane — broadband SS7. Instead of SCCP, however, we find the Broadband ISDN ATM Adaptation Layer Signaling Transport Converter for the MTP3b (Q.2150.1). Above Q.2150.1, we have the ALCAP, which is AAL2 Signaling Protocol Capability Set 1 (Q.2630.1).

On the user side, things are much less complicated. We simply have the *ATM Adaptation Layer 2* (AAL2) as the user data bearer. This is an AAL specifically designed for the transport of short-length packets, such as those we find with packetized voice. One advantage of AAL2 is that it enables multiple user packets to be multiplexed within one cell to minimize ATM overhead. At the radio network layer, we have the User Plane Protocol. This is a simple protocol that provides either transparent or supported service. In transparent mode, data is simply passed onwards. In supported mode, the protocol takes care of functions such as data framing, time alignment, and rate control. Speech is an example of a service that would use supported mode.

6.4.2.2 Iu-PS Interface The protocol architecture for the Iu-PS interface is shown in Figure 6-20. We first notice that no transport network control protocol is involved. It is not needed because of the protocol that is used in the user plane. Specifically, in the user plane, we find that the *GPRS Tunneling Protocol* (GTP) tunnel extends to the RNC. This is different than standard GPRS where the tunnel ends at the SGSN and a special Gb interface is used from SGSN to BSC. The fact that the tunnel extends to the RNC means that only a tunnel identifier and IP addresses for each end are

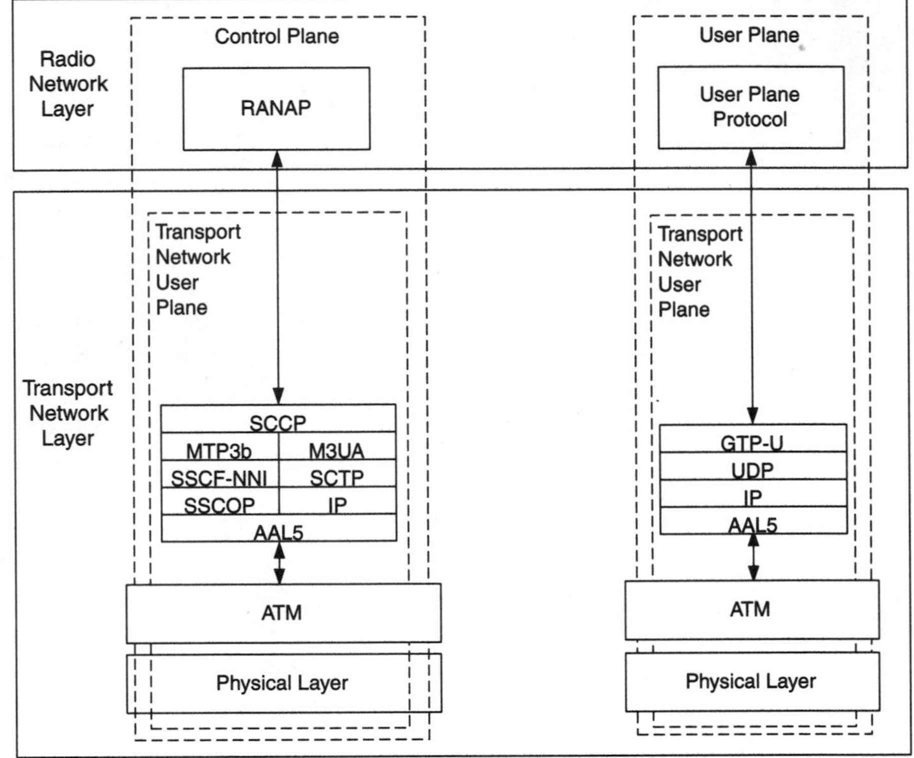

Figure 6-20 Iu-PS Protocol Sructure.

required for establishment of the bearer. These are included in the application messages used for establishment of the bearer, which means that no intermediate ALCAP is needed.

As mentioned, the user plane uses the GTP (GTP-U indicates a GTP user plane). This protocol uses the *User Datagram Protocol* (UDP) over IP. AAL5 over ATM is used as the transport. For a packet data transfer, the identification of individual user packets is supported within the GTP-U protocol. Consequently, it is not necessary to structure these user packets according to ATM cell boundaries. This means that multiple user packets can be multiplexed on a given ATM cell, thereby reducing ATM overhead.

In the control plane, we again find RANAP at the application layer. We have a choice of signaling bearer, however. One option is to use the standard ATM SS7 stack, as described previously, for the Iu-CS interface. Another option is to use SCCP over IP-based SS7 transport over ATM. For IP-based

SS7 transport, we use the *MTP3 User Adaptation* (M3UA) protocol over the *Stream Control Transmission Protocol* (SCTP). Both of these protocols are described in Chapter 8, "Voice over IP Technology."

6.4.2.3 Iub Interface The protocol architecture for the Iub interface is shown in Figure 6-21. This is the interface between an RNC and the Node B that it controls. In the protocol architecture, we again find the transport network control plane as was seen for the Iu-CS interface. In the control plane, we find the *Node B Application Part* (NBAP) as the application protocol. In the user plane, we find a number of frame protocols related to various types of transport channels previous described in this chapter. Basically, a specific framing protocol is applicable to each of the transport channels. Note that Figure 6-21 indicates the *Uplink Shared Channel* (USCH). This is a transport channel defined for TDD-mode only.

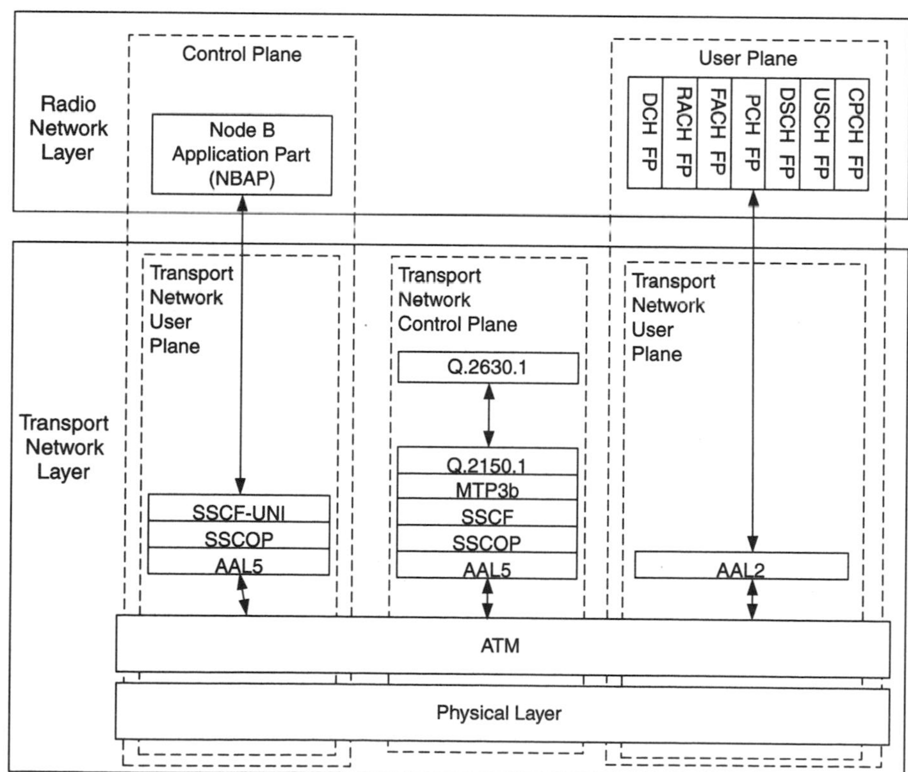

Figure 6-21 Iub Interface Protocol Sructure.

6.4.2.4 Iur Interface The interface between RNCs is the Iur interface. The primary purpose of this interface is to support inter-RNC mobility (SRNS relocation) and a soft handover between Node Bs connected to different RNCs. The protocol architecture for the Iur interface is shown in Figure 6-22. The controlling application protocol is known as the *Radio Network System Application Part* (RNSAP). Signaling between RNCs is SS7-based, whereby RNSAP uses the services of SCCP. As is the case for the Iu-PS interface, the signaling can be transported on a standard ATM SS7 transport or can use an IP-based transport over ATM. The same applies for the transport network control plane.

The user plane contains two frame protocols, one related to dedicated transport channels, the DCH FP, and one related to common transport channels, the CCH FP. These user protocols carry the actual user data and signaling between the SRNC and DRNC.

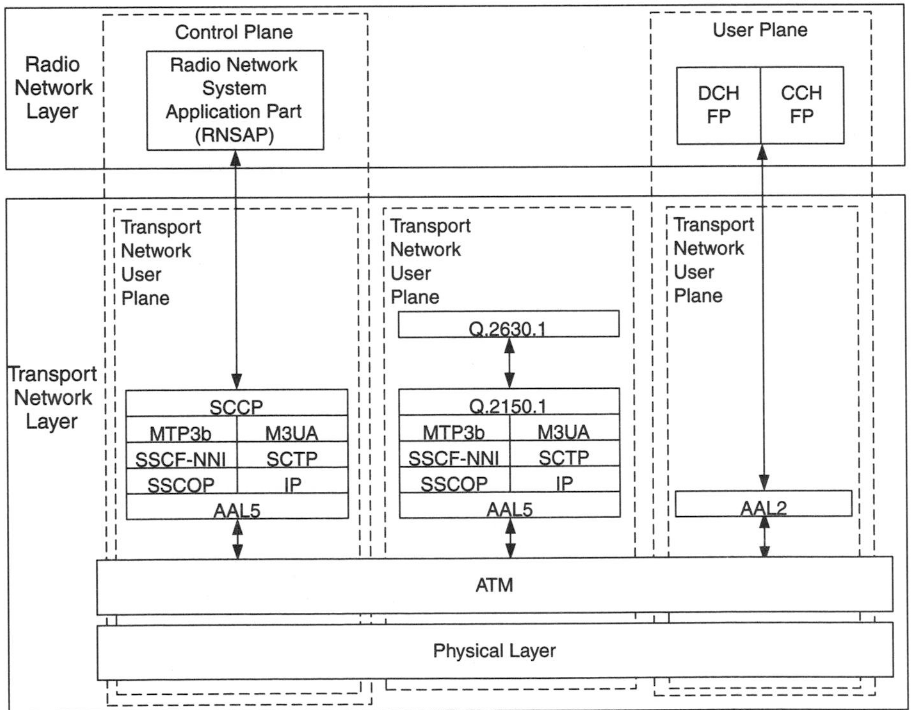

Figure 6-22 Iur Interface Protocol Sructure.

6.4.3 Establishment of a UMTS Speech Call

The procedure for the establishment of a basic speech call in UMTS is shown in Figure 6-23 (NBAP messaging has been omitted). The process begins with an access request from the UE. This access request is sent either on the RACH transport channel or the CPCH transport channel. The message sent is a request to establish an RRC connection, which must be done before signaling transactions or bearer establishment can take place. The RRC Connection Request includes an indication of the reason for the connection request.

The RNC responds with an RRC Connection Setup message. This message will be sent on the CCCH logical channel (typically mapped to the FACH transport channel). At the discretion of the RNC, the RRC Connection Setup message may or may not allocate a DCH transport channel to the UE. If a DCH transport channel is allocated, then the RRC Connection Setup message indicates the scrambling code to be used by the UE in the uplink. The channelization code is determined by the UE and is indicated on the uplink itself. Recall, for example, that a DPCCH is associated with a DPDCH. The DPDCH contains the TFCI that contains spreading factor information and enables the UTRAN to determine the channelization code for the DPDCH. If the RNC does not allocate a DCH, then further signaling is carried out on the FACH in the downlink and on the RACH or CPCH in the uplink.

The UE responds to the RNC with the message, RRC Connection Setup Complete. This message is carried on the uplink DCCH logical channel, which is mapped to the RACH, CPCH, or DCH transport channel. Next, the UE issues a message destined for the core network. This is sent in an RRC Initial Direct Transfer message. The payload of a direct transfer message is passed directly between the UE and the core network. In the case that a signaling relationship has not been established between the UE and core network, then the RRC message Initial Direct Transfer is used. This indicates to the RNC, and subsequently to the core network, that a new signaling relationship needs to be established between the UE and the core.

The RNC maps the Initial Direct Transfer message to the RANAP Initial UE message and sends the message to the core network. In this case, the message is passed to the MSC. The choice of MSC or SGSN is made based upon header information in the Initial Transfer message from the UE. The payload of the Initial Direct Transfer message is mapped to the payload of the RANAP Initial UE message to the MSC.

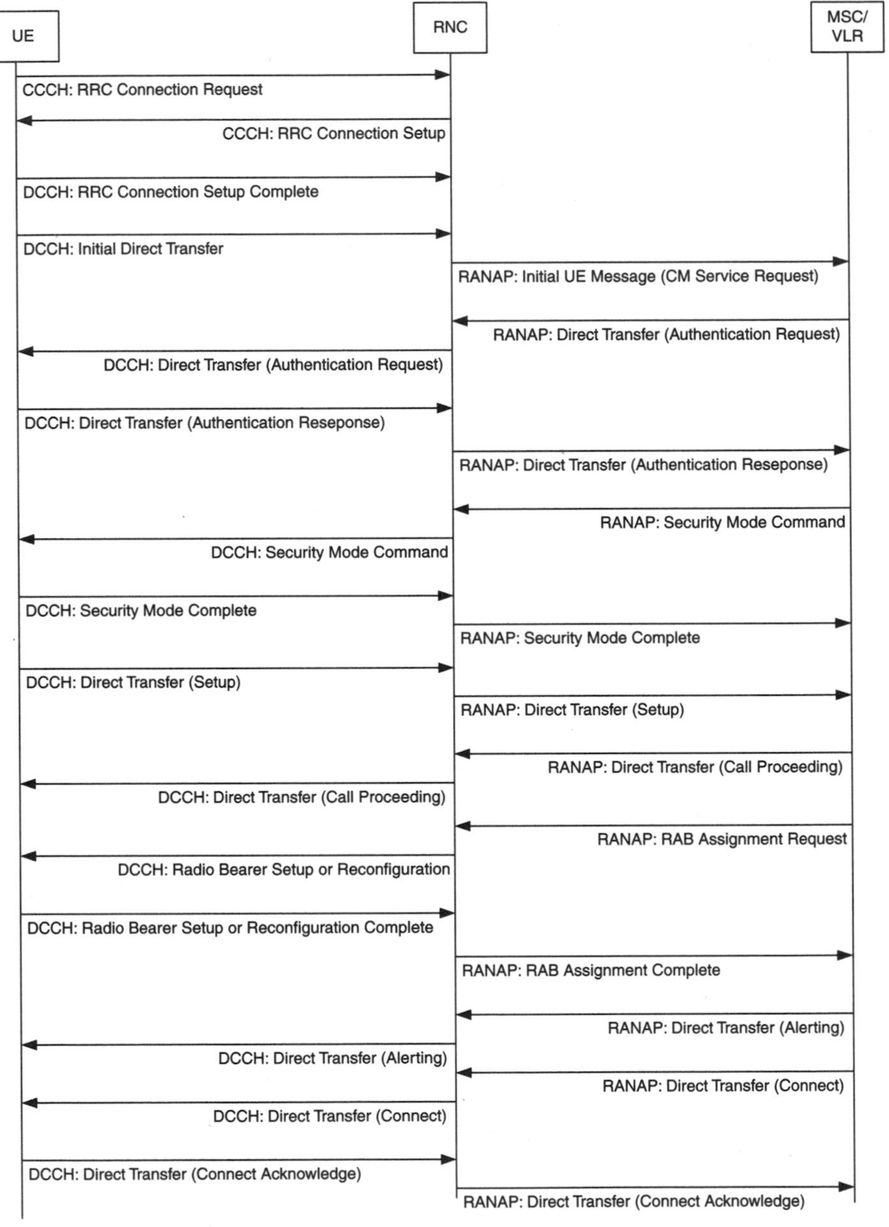

Figure 6-23 Establishing a Speech Call in UMTS.

Next, the MSC will initiate security procedures. This begins with authentication, which uses a challenge-response mechanism similar to that used in GSM. One difference, however, is that the UE and network authenticate each other. Not only does the network send a random number to the UE to which a correct response must be received, but it also sends a network *authentication token* (AUTN), which is calculated independently in the USIM and the HLR. The AUTN must match what the network is expecting. The authentication request is sent to the UE using the direct transfer messaging of RANAP and the RRC protocol.

Assuming that the AUTN is acceptable, the UE responds with an authentication response message, which contains a response that the MSC checks. This message is also carried using the direct transfer capabilities of RANAP and RRC.

Next, the core network will instigate encryption (ciphering) and integrity procedures. This is similar to the ciphering that is performed in GSM, with the addition that integrity assurance is also enabled. This capability enables the network or UE to verify that signaling messages from the other entity have not been maliciously altered. Ciphering and integrity procedures are initiated by the core network, but are executed between the UE and UTRAN. Therefore, the MSC sends the RANAP Security Mode Command message to the RNC. In turn, the RNC sends the RRC Security Mode Command message to the UE. The UE responds to the RNC with the RRC message, Security Mode Complete, and the RNC responds to the MSC with the RANAP message, Security Mode Complete.

At this point, the actual call establishment information such as the called party number data is sent in a Setup message from the UE to the MSC using direct transfer signaling. Provided that the call attempt can be processed, MSC responds with the Call Proceeding message, much like is done in GSM. Next, it is necessary to establish a *Radio Access Bearer* (RAB) for transport of the actual voice stream from the user.

A RAB is a bearer between the UE and the core network for the transport of user data, either speech or packet data. It is mapped to one or more radio bearers on the air interface. Each RAB has its own identifier that is used in signaling between the UE and the network. A RAB establishment is requested by the core network through a RANAP RAB Assignment Request message.

Based on the information in the RAB Assignment Request, the RNC may set up a new radio bearer for the UE to use, or it may reconfigure any existing bearer that the UE has active. The RNC uses either the RRC message Radio Bearer Setup or the Radio Bearer Reconfiguration to instruct the UE to use the new or reconfigured radio bearers. The UE responds with either Radio Bearer Setup Complete or Radio Bearer Reconfiguration Complete. The RNC, in turn, responds to the MSC with the RANAP message RAB Assignment Complete. Now a bearer path exists from the UE through to the MSC. Note that the establishment of the bearer path also requires the establishment of a terrestrial facility between the Node B and RNC and between the RNC and MSC. The details of this establishment have not been shown in Figure 6-23. Suffice it to say that the transport bearer (using AAL2) will be established through the transport user control plane and the ALCAP previously described.

The remainder of the call establishment is quite similar to call establishment in GSM. It involves Alerting, Connect, and Connect Acknowledge messages carried over direct transfer signaling.

It should be noted that speech service in the 3GPP Release 1999 architecture is still a circuit-switched service. Although the speech is actually packetized for transfer over the air and is also packetized as it is carried over the Iub and Iu interfaces, a dedicated bearer is established for the duration of a call, even when discontinuous transmission is active and no speech packets are being sent.

6.4.4 UMTS Packet Data Sessions

From a network perspective, packet data services in the 3GPP Release 1999 architecture use largely the same mechanisms as used for GPRS data, the big difference being the user data rates that can be supported. One notable difference is that the Gb interface of GPRS (between the SGSN and BSC) is replaced by the Iu-PS interface, which uses RANAP as the application protocol. This change includes the fact that IP over ATM is used between the SGSN and RNC. Thus, an IP network is set up from GGSN to SGSN to RNC. Consequently, the GTP-U tunnel can be relayed from the GGSN through the SGSN to the RNC, rather than terminating at the SGSN. The GTP-C tunnel, however, terminates at the SGSN, because the application protocol between RNC and SGSN is RANAP, rather than GTP. The establishment of the tunnel is still under the control of the SGSN. Figure 6-24 shows the Control Plane for packet data services in UMTS and Figure 6-25 shows the User Plane.

Packet data services are established in UMTS in largely the same manner as in GPRS—through the activation of a PDP context with an *Access Point Name* (APN), QoS criteria, and so on. One significant difference

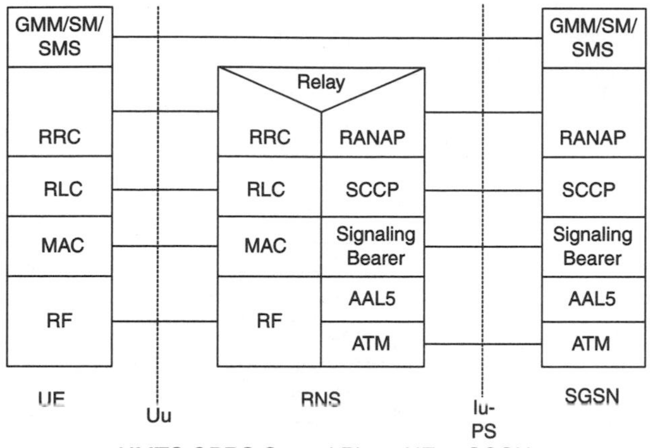

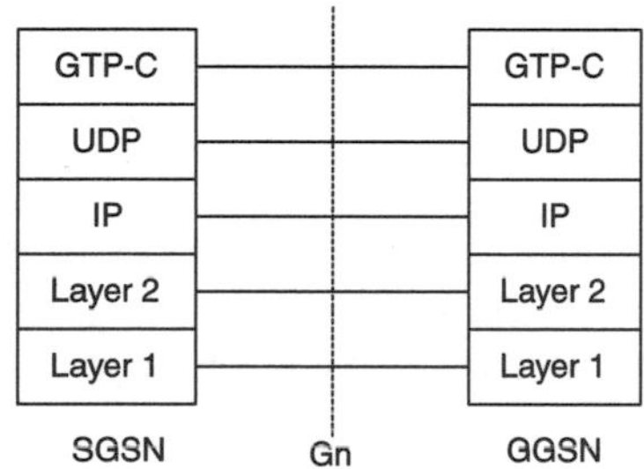

UMTS GPRS Control Plane SGSN to GGSN

Figure 6-24
UMTS GPRS Control Plane Protocol Stacks.

between UMTS and standard GPRS, however, involves SRNS relocation. Because of the fact that the GTP-U tunnel terminates at the RNC rather than the SGSN, relocation of the UE to another RNC may require the buffering of packets at the first RNC and a subsequent relay of those packets to the second RNC once relocation has taken place. That relay occurs via the SGSN. In case the two RNCs are connected to two different SGSNs, then the path for buffered packets is from RNC1 to SGSN1 to SGSN2 to RNC2.

From an air interface perspective, UMTS provides greater flexibility than GPRS in terms of how resources are allocated for packet data traffic. Not only does UMTS offer a greater range of speeds, but the WCDMA air

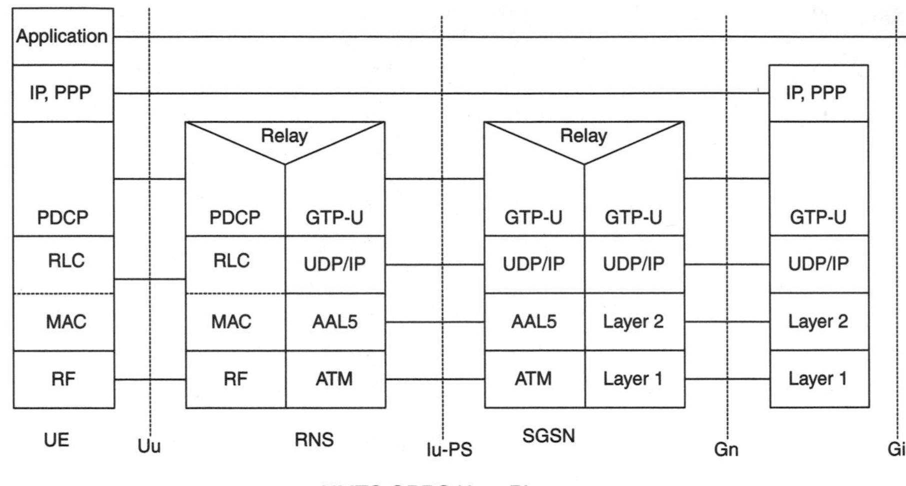

Figure 6-25 UMTS GPRS User Plane Protocol Stack.

interface has a selection of different channel types that can be used for packet data. In the uplink, the RACH, CPCH, and DCH are available. In the downlink, the DCH, FACH and DSCH are available. The choice of channel to be used is under the control of the RNC and is chosen depending on the characteristics of the session required by the user—e.g. high-volume streaming versus low-volume bursty traffic.

For a non-bursty service such as video streaming or for large file transfers, the DCH might be the best option as it has the greatest throughput capability. It has the disadvantage, however, of taking time to establish. For small amounts of bursty traffic, the RACH or CPCH in the uplink is likely to be more suitable. These are faster to establish, but cannot support rates as high as the DCH. In the case of the RACH, there is likely to be only one per cell (certainly no more than a few), whereas there can be many CPCH channels. Moreover, the CPCH can carry more data than the RACH.

In the downlink, the FACH is useful for small amounts of bursty user data. Like the RACH, however, the number of FACH channels is very limited. Another option in the downlink is the DSCH, which is a channel that is time-multiplexed among several users. It can support higher throughput than the FACH, through not as high as the DCH. It is, however, much more suited to bursty traffic than the DCH.

6.5 Handover

UMTS supports two main categories of handovers—soft handovers and hard handovers. A soft handover is make-before-break, whereby communication exists between the UE and more than one cell for a period of time. A hard handover is break-before-make, whereby communication with the first cell is terminated before establishing communication with the second cell.

A soft handover has two variants—soft handover and softer handover. These two situations are depicted in Figure 6-26. A soft handover occurs between two cells or sectors that are supported by different base stations. The UE is transmitting to and receiving from both base stations at the same time. The user information sent to the UE is sent from each base station simultaneously and is combined within the UE. In the uplink, the information sent from the UE is relayed from each base station to the RNC where the combining takes place. In the case of a soft handover, each base station is sending power control commands to the UE.

A softer handover occurs between two cells that are supported by the same base station. In this case, only one power control loop is active and is controlled by the base station that serves both cells. Depending on RF coverage, both a soft handover and a softer handover may occur at the same time for a given UE.

A hard handover can occur in several situations, such as from one cell to another where the two cells are using different carrier frequencies, or from one cell to another where the base stations are connected to different RNCs and no Iur interface exists between the RNCs. UMTS also supports a hard handover to and from GSM. This is a reasonable requirement as it takes time to roll out a UMTS network nationwide, and one would like UMTS subscribers to receive service from GSM in areas where holes occur in the UMTS coverage.

Regardless of the type of handover to take place, the decision when and how to invoke a handover is made at the serving RNC. This decision is based upon measurements reported by the UE. The set of cells for which measurement reports are to be generated is broadcast from the network on the BCH or FACH. If a neighboring cell uses a different frequency and the RNC requires reports related to that cell, then the UE needs time periodically to tune to the frequency in question. This means that the UE and UTRAN must operate in compressed mode. This mode means that in a given radio frame, not all 15 slots are used. The unused slots correspond to

Figure 6-26
Soft Handover and Softer Handover.

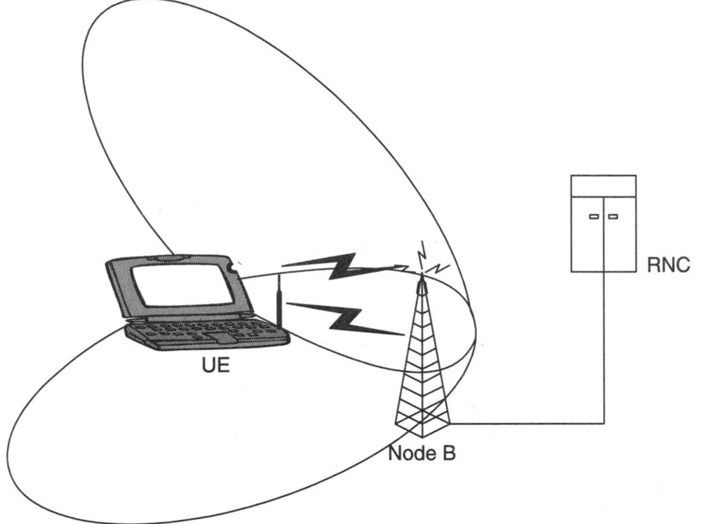

Softer Handover

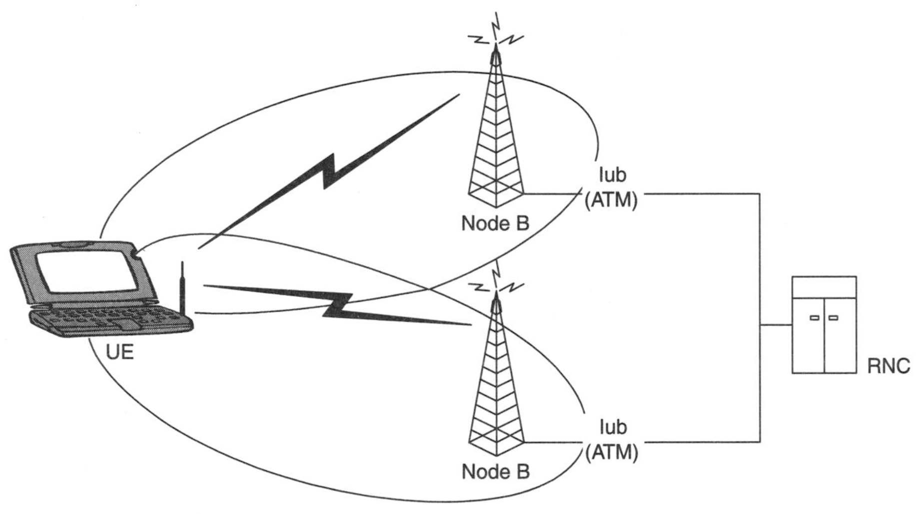

Soft Handover

durations where the UE can tune to another frequency to make the necessary measurements.

6.6 UMTS Core Network Evolution

The core network architecture for 3GPP Release 1999 is not greatly different than the core network architecture of GSM/GPRS. Clearly, the core network must be upgraded to support the new interfaces to the radio access network, but a completely new architecture is not needed. In 3GPP Release 4 and 3GPP Release 5, however, we find significant enhancements to the core network.

6.6.1 The 3GPP Release 4 Network Architecture

3GPP Release 4 introduces a significant enhancement to the core network architecture as it applies to the CS domain. Basically, the MSC is broken into constituent parts and it is allowed to be deployed in a distributed manner, as shown in Figure 6-27. Specifically, the MSC is divided into an MSC

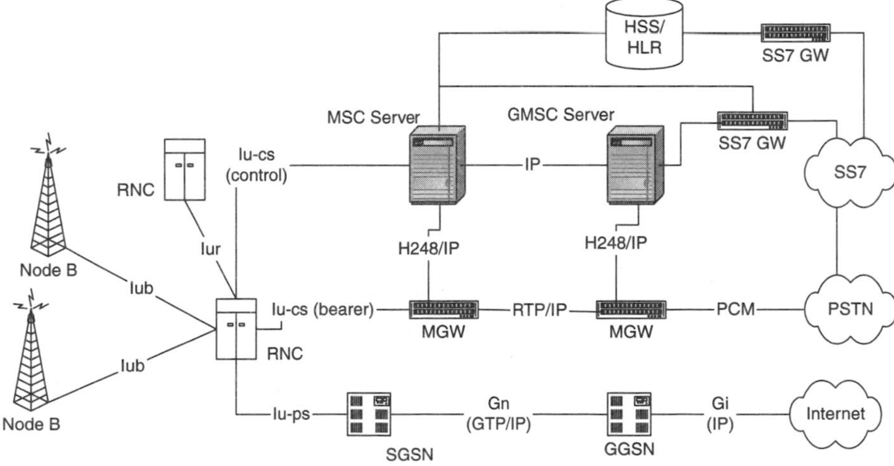

Figure 6-27 3GPP Release 4 Distributed Network Architecture.

server and a *media gateway* (MGW). The MSC server contains all of the mobility management and call control logic that would be contained in a standard MSC. It does not, however, reside in the media path. Rather, the media path is via one or more MGWs that establish, manipulate, and release media streams (voice streams) under the control of the MSC server.

Control signaling for circuit-switched calls is between the RNC and the MSC server. The media path for circuit-switched calls is between the RNC and the MGW. As far as the RNC is concerned, these two entities could be the same physical device, as would be the case when the RNC is communicating with a traditional MSC. Typically, an MGW takes calls from the RNC and routes those calls towards their destinations over a packet backbone. In many cases, the packet backbone will be IP-based, such that backbone traffic is *Voice over IP* (VoIP), as described in more detail in Chapter 8. Given that the PS domain also uses an IP backbone, then only one backbone network is needed, which can mean significant cost savings for the network operator.

At the remote end, where a call needs to be handed off to another network, such as the PSTN, another MGW is controlled by a *Gateway MSC server* (GMSC server). This MGW converts the packetized voice to standard PCM for delivery to the PSTN. It is only at this point that transcoding needs to take place. Thus, voice can be carried through the backbone at a far lower rate than 64 Kbps, with a step up to 64 Kbps only at the last point. This represents a lower bandwidth requirement in the backbone network and therefore a lower cost.

The control protocol between the MSC server or GMSC server and the MGW is the ITU H.248 protocol. This protocol is also known as MEGACO. The call control protocol between the MSC Server and the GMSC server can be any suitable call control protocol. The 3GPP standards suggest, but do not mandate, the *Bearer Independent Call Control* (BICC) protocol, which is based on the ITU-T recommendation Q. 1902.

Figure 6-28 shows an example of a voice call establishment using this architecture. For the sake of brevity, the messages are limited to those that relate to call establishment as seen from the core network. Thus, the RRC protocol messages between the UE and UTRAN have been omitted. These will be as shown in Figure 6-23. Figure 6-28 includes a number of H.248 messages. For details of the H.248 protocol, please refer to Chapter 8, "Voice-over-IP (VoIP) Technology."

When the Setup message arrives from the UE, the MSC server performs a call-routing determination. It then responds with a Call Proceeding message to the UE. Based on the call-routing determination, the MSC server chooses a MGW to handle the call. It instructs (Add Request) the MGW to

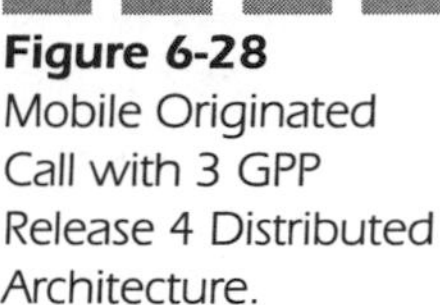

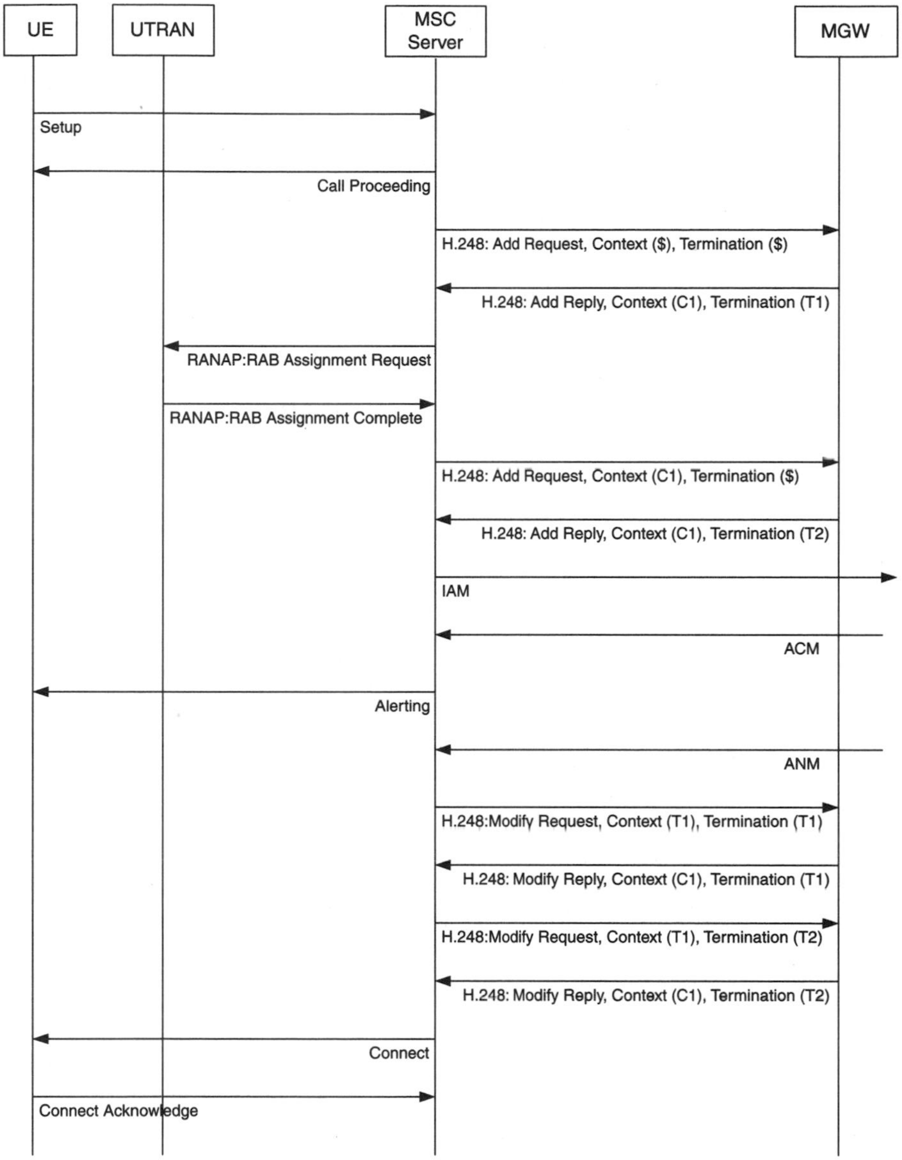

Figure 6-28 Mobile Originated Call with 3 GPP Release 4 Distributed Architecture.

establish a new context and places a termination in that context. The termination in question (T1) will be on the network side of the MGW. Once the new context is established by the MGW, the MSC server requests the RAN to establish a RAB to handle the call. Once the RAB has been assigned, the MSC server is in a position to establish a media connection between the RNC and the MGW. Therefore, it requests the MGW to add a new termination to the context that has just been established. This new termination (T2) will face towards the RNC. Because the new termination is in the same context as termination T1, a path is created from one side of the MGW to the other.

The MSC server then sends the ISUP *Initial Address Message* (IAM) to the called network (such as the PSTN). Upon receipt of an *Address Complete Message* (ACM) from the far end, the MSC server sends an Alerting message to the UE. Typically, the called user will answer, which causes an ISUP *Answer Message* (ANM) to be received at the MSC server. At this point, the MSC server may optionally modify the context established on the MGW. Specifically, when the terminations were established in the new context, they may have been configured to not provide a complete through-connection. For example, one or both of the terminations could have been configured to allow only a one-way media path (from the far end to UE for the purposes of receiving a ring-back tone). In such a case, the MSC server requests the MGW to modify the configuration so that a full two-way media path is established. Finally, the MSC server sends a Connect message to the UE and the UE responds with a Connect acknowledge.

Note that the distributed architecture just described does not rely on WCDMA-based UTRAN access. The core network architecture could just as well apply to standard GSM-based access, with BSCs instead of RNCs. In fact, as the distributed switching architecture is being deployed, it is likely that many deployments will simultaneously support both UTRAN and GSM access networks.

Note also that the signaling example of Figure 6-28 is just one possible sequence. Many different sequences are possible depending on the exact network configuration.

6.6.2 The 3GPP Release 5 IP Multimedia Domain

Figure 6-29 shows the network architecture for a new core network domain planned for 3GPP Release 5. This architecture has already been described in Chapter 4, "Third Generation (3G) Overview." It is important to note that this architecture represents an addition to the core network, rather than a change to the existing core. Instead, 3GPP Release 5 introduces a new core network domain in addition to the established CS and PS domains. This new domain is available for new user devices that have the capability and the call model logic needed to take advantage of the new domain. Thus, the UTRAN can now be connected to three different logical core network domains—the CS domain, the PS domain, and the *IP Multimedia* (IM) domain. When a terminal wants to use the services of the core network, it indicates which domain it wants to use. Existing (pre-Release 5) terminals will continue to request the services of the CS or PS domain. New terminals will also be able to request the services of the IM domain.

Note that while the IM domain is a new domain, it uses the services of the PS domain. All IM traffic is packet based and is transported using PS-domain nodes such as the SGSN and GGSN.

The IM domain is based on the *Session Initiation Protocol* (SIP), as described in Chapter 8. In fact, the *Call State Control Function* (CSCF) of Figure 6-29 is effectively a SIP proxy. The IM architecture enables voice and data calls to be handled in a uniform manner all the way from the UE to the destination. A complete convergence of voice and data takes place, such that voice is simply a type of data with specific QoS requirements. This convergence enables a number of new advanced services. Moreover, the use of SIP means that a great deal of service control can be placed in the UE rather

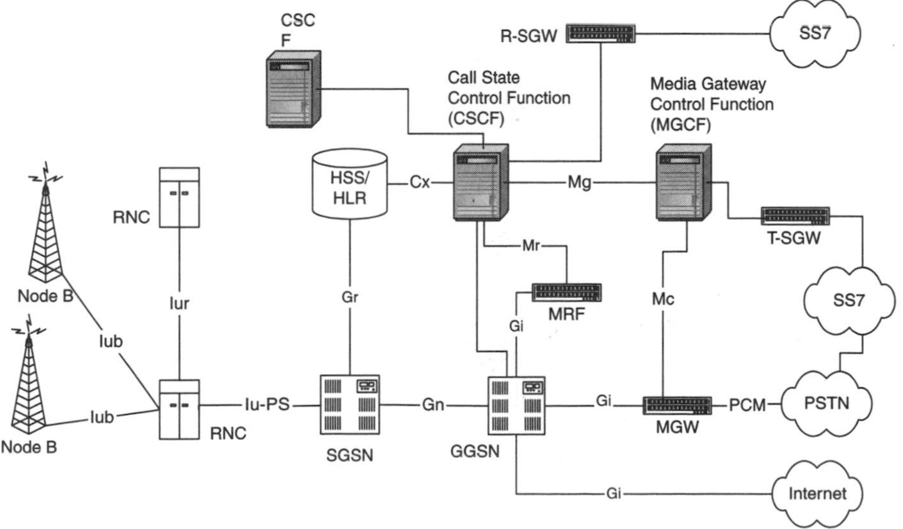

Figure 6-29
3GPP Release 5 IP Multimedia Network Architecture.

than the network, making it easier for the subscriber to customize services to meet his or her particular needs.

References

3GPP TR 23.930	Iu Principles
3GPP TR 25.931	UTRAN Functions, Examples of Signalling Procedures
3GPP TR 25.944	Channel Coding and Multiplexing Examples
3GPP TS 22.060	General Packet Radio Service (GPRS); Service Description, Stage 1
3GPP TS 23.002	Network Architecture (Release 1999)
3GPP TS 23.002	Network Architecture (Release 4)
3GPP TS 23.002	Network Architecture (Release 5)
3GPP TS 23.003	Numbering, Addressing, and Identification
3GPP TS 23.009	Handover Procedures
3GPP TS 23.018	Basic Call Handing; Technical Realization
3GPP TS 23.060	General Packet Radio Service (GPRS), Service Description, Stage 2
3GPP TS 23.060	General Packet Radio Service (GPRS); Service Description, Stage 2
3GPP TS 23.101	General UMTS Architecture (Release 1999)
3GPP TS 23.101	General UMTS Architecture (Release 4)
3GPP TS 23.107	QOS Concept and Architecture
3GPP TS 23.108	Mobile Radio Interface Layer 3 Specification, Core Network Protocols—Stage 2
3GPP TS 23.110	UMTS Access Stratum; Services and Functions
3GPP TS 23.205	Bearer Independent CS Core Network; Stage 2 (Release 4)
3GPP TS 24.002	GSM—UMTS Public Land Mobile Network (PLMN) Access Reference Configuration
3GPP TS 24.007	Mobile Radio Interface Signalling Layer 3; General Aspects

3GPP TS 24.008	Mobile Radio Interface Layer 3 Specification, Core Network Protocols—Stage 3
3GPP TS 25.101	UE Radio Transmission and Reception (FDD)
3GPP TS 25.104	UTRA (BS) FDD; Radio Transmission and Reception
3GPP TS 25.211	Physical Channels and Mapping of Transport Channels onto Physical Channels (FDD)
3GPP TS 25.212	Multiplexing and Channel Coding (FDD)
3GPP TS 25.213	Spreading and Modulation (FDD)
3GPP TS 25.214	Physical Layer Procedures (FDD)
3GPP TS 25.215	Physical Layer—Measurements (FDD)
3GPP TS 25.301	Radio Interface Protocol Architecture
3GPP TS 25.302	Services Provided by the Physical Layer
3GPP TS 25.304	UE Procedures in Idle Mode and Procedures for Cell Reselection in Connected Mode
3GPP TS 25.306	UE Radio Access Capabilities
3GPP TS 25.321	MAC Protocol Specification
3GPP TS 25.322	RLC Protocol Specification
3GPP TS 25.323	PDCP Protocol Specification
3GPP TS 25.331	RRC Protocol Specification
3GPP TS 25.401	UTRAN Overall Description
3GPP TS 25.410	UTRAN Iu Interface: General Aspects and Principles
3GPP TS 25.411	UTRAN Iu Interface: Layer 1
3GPP TS 25.412	UTRAN Iu Interface Signalling Transport
3GPP TS 25.413	UTRAN Iu Interface: RANAP Signalling
3GPP TS 25.414	Iu Interface Data Transport and Transport Signalling
3GPP TS 25.415	UTRAN Iu Interface User Plane Protocols
3GPP TS 29.060	General Packet Radio Service (GPRS); GPRS Tunnelling Protocol (GTP) across the Gn and Gp Interface
3GPP TS 33.102	Security Architecture
ITU-T H.248	Media Gateway Control Protocol
ITU-T Q.711	Functional Description of the Signalling Connection Control Part

CHAPTER 7

CDMA2000

CDMA2000 is a unique radio and network access system that is part of the IMT-2000 specification suite of access platforms that comprise what is known collectively as *third generation* (3G). The *International Mobile Telecommunications 2000* (IMT-2000) specification from the *International Telecommunication Union* (ITU) defines one of its platform standards that comprises the 3G suite of access platforms and is called IMT-2000-MC, or multi-carrier, called CDMA2000. CDMA2000 is unique in that, while supporting 3G services and bandwidth requirements, it enables a logical migration from the existing 2G platforms to 3G without forklifting the legacy system.

The IMT-2000 specification or vision for all the platforms supported has a common set of goals that all the standards are meant to achieve. The general specifications for the IMT-2000 are as follows:

- Support high-speed data services
- Global standard
- Worldwide common frequency band
- Flexibility for evolution
- Improved spectrum efficiency
- 2 Mbps for fixed environment
- 384 Kbps for pedestrian use
- 144 Kbps for vehicular uses

In reviewing this list, the underlying principal is that IMT-2000 is a high-speed packet data network designed for mobility using IP as the enabling protocol.

Some of the 3G applications that are envisioned to be enabled by CDMA2000 are as follows:

- Wireless Internet
- Wireless e-mail
- Wireless telecommuting
- Telemetry
- Wireless commerce
- Location-based services
- Longer standby battery life

CDMA2000 is standardized under the specification of IS-2000, which is backward-compatible with IS-95A and B, as well as with J-STD-008 specifications that collectively are called cdmaOne. The IS-95 and J-STD-008 specifications make up the existing CDMA mobility systems deployed currently in the world. CDMA2000, while being a 3G specification, is also backward-compatible with cdmaOne systems, allowing operators to make strategic deployment decisions in a graceful fashion.

Since CDMA2000 is backward-compatible with existing cdmaOne networks, upgrades or, rather, changes to the network from a fixed network aspect can be done in stages. More specifically, the upgrades or changes to the network involve the *Base Transceiver Stations* (BTSs) with Multimode Channel Element cards, the *Base Station Controller* (BSC) with IP-routing capabilities, and the introduction of the *Packet Data Server Network* (PDSN). The radio channel bandwidth is the same for CDMA2000-1X as it is for existing cdmaOne channels, leading to a graceful upgrade. Of course, the subscriber units and mobiles need to be capable of supporting the CDMA2000 specification, but this can be done in a more gradual fashion because the existing cdmaOne subscriber units can utilize the new network.

As indicated earlier, CDMA2000 is IMT2000-MC, which stipulates the use of more then one carrier. However, the initial introduction of the CDMA2000 will primarily utilize a single carrier even though CDMA2000 supports multiple carrier operation. Several terms are used to describe CDMA2000 for the different radio carrier platforms, some of which exist at present while others are in the development phase. However, the sequence of different CDMA2000 platforms or the migration path is as follows:

CDMA2000-1X (1xRTT)

1xEV-DO

1xEV-DV

CDMA2000-3X (3xRTT)

The 1xRTT utilizes a single carrier requiring 1.25 MHz of radio spectrum, which is the same as the existing cdmaOne system's channel bandwidth requirement. However, the 1xRTT platforms can utilize a different vocoder and more Walsh codes, 256/128 versus 64, allowing for higher data rates and more voice conversions than are possible over existing cdmaOne systems.

Under CDMA2000-1X, also called 1xRTT, three primary methods are used: 1x, 1xEV-DO, and 1xEV-DV, which are not mutually exclusive of each other. The term 1x is used to describe the first version of CDMA2000. 1xEV-DO means one carrier, which is data-only, while 1xEV-DV means one carrier that supports data and voice services.

However, when referring to CDMA2000-3X, the use of 3.75 MHz of the spectrum, or 3 × 1.25 MHz, is defined with a change in the modulation scheme as well as the vocoders to mention a few of the salient issues that come about with the introduction of this platform. The migration from 1X to 3X is talked about as being transparent but will likely involve the reallocation of the existing spectrum. The details of this will be covered in the design phase discussed later.

Another important aspect of CDMA2000 is that it supports not only IS-41 system connectivity, as does IS-95, but it also supports *Global System for Mobile communications-Mobile Application Part* (GSM-MAP) connectivity requirements. This can lead to the eventual harmonization or dual-system deployment in the same market by a wireless operator wanting to deploy both WCDMA and CDMA2000 concurrently.

Several key specifications are used to help define the particulars associated with a CDMA2000 system, as listed in Table 7-1.

Table 7-1
CDMA2000 Specifications

TIA	3GPP2	Description
IS-2000-1	C.S.0001	Cdma2000 Introduction
IS-2000-2	C.S.0002	Cdma2000 Physical Layer
IS-2000-3	C.S.0003	CDMA2000 MAC Layer
IS-2000-4	C.S.0004	CDMA2000 Layer 2 LAC
IS-2000-5	C.S.0005	CDMA2000 Layer 3
IS-2000-6	C.S.0006	CDMA2000 Analog
TIA/EIA-97	C.S.0010	Base Station Minimum Standard
TIA/EIA-98	C.S.0011	Mobile Station Minimum Performance
IS-127	C.S.0014	Enhanced Variable Rate Codec (EVRC)
TIA/EIA-637	C.S.0015	Short Message Service
TIA/EIA-683	C.S.0016	Over the Air service provisioning
TIA/EIA-707	C.S.0017	Data Services for Spread Spectrum Systems
TIA/EIA-733	C.S.0020	High Rate (13 Kbps) Speech SO
IS-801	C.S.0022	Location Services (Position Determination Service)
IS-95A		Mobile Station-Base Station Compatibility Standard for Dual-Mode Wideband Spread Spectrum Cellular System
IS-95B		Mobile Station-Base Station Compatibility Standard for Dual-Mode Wideband Spread Spectrum Cellular System
	A.S.0001	Access Network Interfaces Technical Specification

7.1 Radio and Network Components

CDMA2000, whether 1X or 3X, requires upgrades to the radio and network architecture of the existing system. It is important to note that the migration path for a CDMA2000 operator will be from 1X to 3X if the CDMA2000 platform is implemented in the near term.

To understand which radio and network components are required for the successful implementation of a CDMA2000 system, it is best to start with a simplified network layout for a cdmaOne system. Figure 7-1 is a standalone cdmaOne system employing several BTSs that are homed to two BSCs. The BSCs are shown not colocated with the MSC but in reality could be colocated depending on the specific interconnection requirements and commercial agreements arrived at. The *Home Location Register* (HLR) is shown, but many of the supporting systems are left out of the picture for simplification purposes. The backhaul from the BTSs to the BSC and from the BSC to the MSC could be via microwave links or fixed facilities.

What follows next is an example of a general CDMA2000 network, shown in Figure 7-2. The connectivity to other similar networks is not shown to keep the diagram less cluttered. Both Figures 7-1 and 7-2 identify the new platforms required to support the CDMA2000 network over a cdmaOne system.

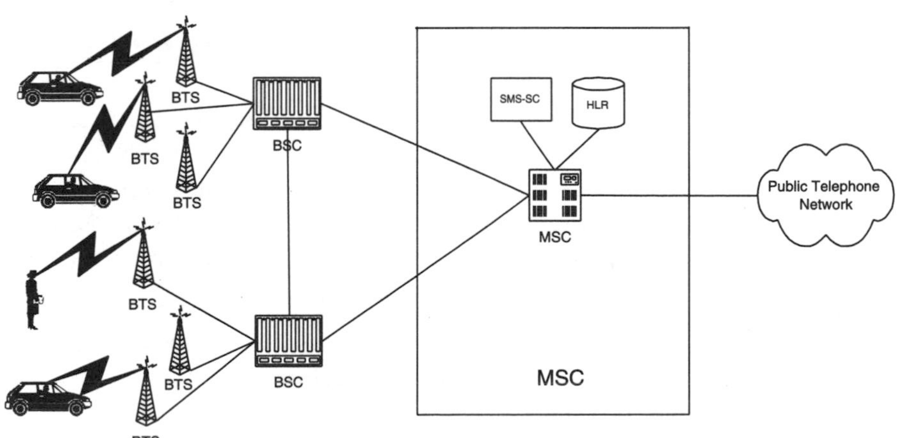

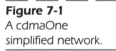

Figure 7-1 A cdmaOne simplified network.

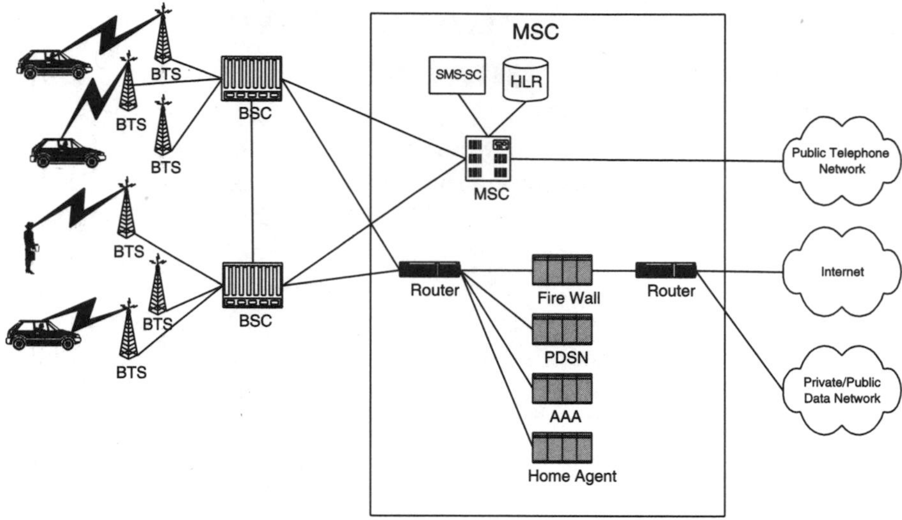

Figure 7-2 CDMA2000 system architecture.

What Figure 7-2 does not show are the platform upgrades needed. However, Figure 7-3 indicates the various major platforms that either have upgrades performed or are essentially new to the CDMA2000 network, as compared to a cdmaOne system.

The platform upgrades involve the BTS and BSC that can be facilitated by module additions or swaps, depending on the infrastructure vendor that is being used. Whether the system is new or upgrading from a cdmaOne system, the heart of the packet data services for a CDMA2000 network is the Packet Data Serving Node (PDSN).

7.1.1 Packet Data Serving Node (PDSN)

The PDSN is a new component associated with a CDMA2000 system, as compared to cdmaOne networks. The PDSN is an essential element in the treatment of packet data services that will be offered, and its location in the CDMA2000 network is shown in Figure 7-2. The purpose of the PDSN is to support packet data services and it performs the following major functions in the course of a packet data session:

- Establishes, maintains, and terminates *Point-to-Point Protocol* (PPP) sessions with the subscriber

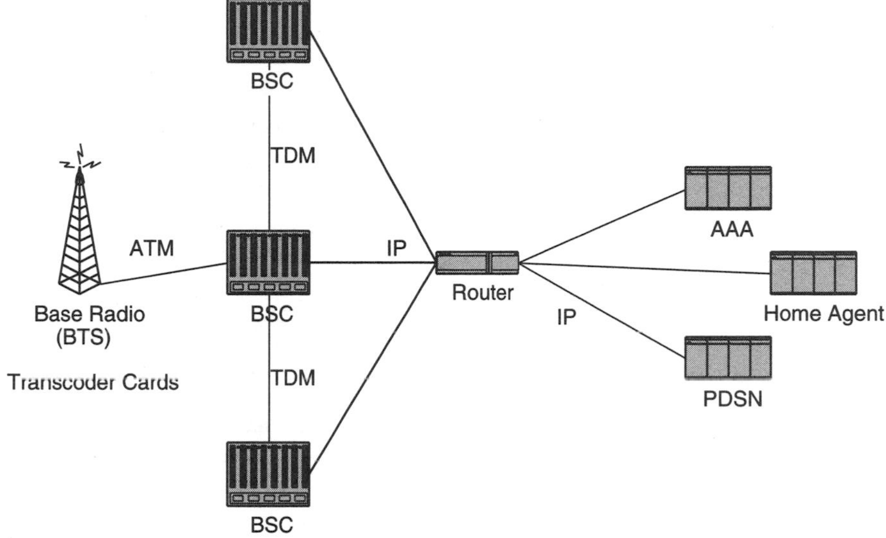

Figure 7-3
2.5G and 3G network element alterations.

- Supports both Simple and Mobile IP packet services
- Establishes, maintains, and terminates the logical links to the *Radio Network* (RN) across the *radio-paket* (R-P) interface
- Initiates *Authentication, Authorization, and Accounting* (AAA) for the mobile station client to the AAA server
- Receives service parameters for the mobile client from the AAA server
- Routes packets to and from the external packet data networks
- Collects usage data that is relayed to the AAA server

The overall capacity of the PDSN is determined by both the throughput and the number of PPP sessions that are being served. The specific capacity of the PDSN is, of course, dependant upon the infrastructure vendor used as well as the particular card population that is implemented. It is important to note that capacity is only one aspect of the dimensioning process and that the overall network reliability factor must be addressed in the dimensioning process.

7.1.2 Authentication, Authorization, and Accounting (AAA)

The AAA server is another new component associated with CDMA2000 deployment. The AAA provides, as its names implies, authentication, authorization, and accounting functions for the packet data network associated with CDMA2000 and utilizes the *Remote Access Dial-In User Service* (RADIUS) protocol.

The AAA server, as shown in Figure 7-2, communicates with the PSDN via IP and performs the following major functions in its role in a CDMA2000 network:

- Authentication associated with PPP and mobile IP connections
- Authorization (service profile and security key distribution and management)
- Accounting

7.1.3 Home Agent

The *Home Agent* (HA) is the third major component to the CDMA2000 packet data service network and should be compliant with IS-835, which is relevant to the HA functionality within a wireless network. The HA performs many tasks, some of which are tracking the location of the mobile IP subscriber as it moves from one packet zone to another. In tracking the mobile, the HA will ensure that the packets are forwarded the mobile itself.

7.1.4 Router

The *router* shown in Figures 7-2 and 7-3 has the function of routing packets to and from the various network elements within a CDMA2000 system. The router is also responsible for sending and receiving packets to and from the internal network to the offnet platforms. A firewall, not shown in the figures, is needed to ensure that security is maintained when connecting to offnet data applications.

7.1.5 Home Location Register (HLR)

The HLR used in existing IS-95 networks needs to store additional subscriber information associated with the introduction of packet data services. The HLR performs the same role for packet services as it currently does for voice services in that it stores the subscriber packet data service options and terminal capabilities along with the traditional voice platform needs. The service information from the HLR is downloaded in the *Visitor Location Register* (VLR) of the associated network switch, during the successful registration process. The same process as it is done in existing IS-95 systems and other 1G and 2G voice-oriented systems.

7.1.6 Base Transceiver Station (BTS)

The BTS is the official name of the cell site. It is responsible for allocating resources and both power and Walsh codes for consumption by the subscribers. The BTS also has the physical radio equipment that is used for transmitting and receiving the CDMA2000 signals.

The BTS controls the interface between the CDMA2000 network and the subscriber unit. The BTS also controls many aspects of the system that are directly related to the performance of the network. Some of the items the BTS controls are the multiple carriers that operate from the site, the forward power (allocated for traffic overhead and soft handoffs), and, of course, the assignment of the Walsh codes.

With CDMA2000 systems, the use of multiple carriers per sector, as with IS-95 systems, is possible. Therefore, when a new voice or packet session is initiated, the BTS must decide how to best assign the subscriber unit to meet the services being delivered. The BTS in the decision process not only examines the service requested, but also must consider the radio configuration, the subscriber type, and, of course, whether the service requested is voice or packet. Thus, the resources the BTS has to draw upon can be both physically and logically limited, depending on the particular situation involved.

BTS can perform a downgrade from a higher RC or spreading rate to a lower RC or spreading rate if

- The resource request is not a handoff
- The resource request is not available
- Alternative resources are available

The following are some of the physical and logical resources the BTS must allocate when assigning resources to a subscriber:

- The *Fundamental Channels* (FCHs) (the number of physical resources available)
- The FCH forward power (the power already allocated and that which is available)
- The Walsh codes required (and those available)

The physical resources the BTS draws upon also involve the management of the channel elements that are required for both voice and packet data services. Although discussed in more detail, handoffs are accepted or rejected on the basis of available power only.

Integral to the resource assignment scheme is Walsh code management, covered in another section in more detail. However, for CDMA2000, phase 1, whether 1x, 1xEV-DO, or 1xEV-DV is used, a total of 128 Walsh codes can be drawn upon. With the introduction of 3X, the Walsh codes are expanded to a total of 256.

For CDMA2000 1x, the voice and data distribution is handled by parameters set by the operator that involve

- Data resources (percent of available resources, which includes FCH and *supplemental channel* (SCH))
- FCH resources (percent of data resources)
- Voice resources (percent of total available resources)

This is best described by a brief example to help facilitate the issue of resource allocation, as shown in Table 7-2.

Obviously, the allocation of data/FCH resources directly controls the amount of simultaneous data users on a particular sector or cell site.

Table 7-2

Channel Resource Allocations Example

Topic	Percentage	Resources
Total Resources		64
Voice Resources	70%	44
Data Resources	30%	20
FCH Resources	40%	8

7.1.7 Base Station Controller (BSC)

The BSC is responsible for controlling all the BTSs under its domain. The BSC routes packets to and from the BTSs to the PDSN. In addition, the BSC routes *Time Division Multiplexing* (TDM) traffic to the circuit-switched platforms and it routes packet data to the PDSN.

7.2 Network Structure

The network structure for a CDMA2000 system that supports 2.5G and 3G has all the traditional voice elements associated with 2G wireless voice systems. However, the introduction of a packet network requires the additional network equipment to provide the connectivity between the radio access network and the data network, whether it is public or private.

Obviously, or maybe it's not obvious, numerous IP network configurations can, will, and are being utilized for the support of 2.5G and 3G. The reason for many different configurations lies in the fact that the information is packet-based and therefore can be shipped between different company networks or kept localized. Of course, the issue of the required throughput and the physical interfaces is also a location dependant.

The packet network is often called the IP network, the IP access network, or the carrier IP network, depending on your particular situation. However, the fundamental premise is that the packet network needs to support the transport and treatment of the packets in the chosen configuration.

Because numerous implementation methods are available for configuring the packet network, only three main variants will be covered. The main variants of a network configuration can then be modified to meet your particular requirements. For example, it might be best to send all Internet traffic to a local ISP, and *virtual private network* (VPN) applications, depending on the treatment required, can be brought to one centralized location for distribution on, say, an *Asynchronous Transfer Mode* (ATM) network when connection to a corporate *local area network* (LAN) is required.

The three main variants in configuring a CDMA2000 network are as follows:

- Distributed
- Regional
- Centralized

The regional and centralized variants are similar in concept, except that the centralized variant is an aggregation of several potential regional networks. Some of the determinations for deciding on which variant to implement is determined based on the following issues:

- Services supported
- Traffic volume
- Location of PDSN
- Commercial interconnection agreements
- Network reliability and availability

Regardless of which configuration is utilized, it will need a router for the backbone and, of course, a gateway for all of the offnet service delivery and reception.

The following sections contain simplified figures and descriptions of the three major network configuration variants that should be considered for a CDMA2000 system. In all likelihood, the network architecture, will be initially dictated by the existing 2G system that is in place. When reviewing the network figures, the traditional voice networks and the packet networks have implied different structures. The reason is that a packet network can and should be treated separately from that of the voice transport network unless *Voice over IP* (VoIP) is being utilized for internetwork transportation of voice-based services.

7.2.1 Distributed

The *distributed network*, also referred to as a localized network, involves establishing the network as being independent of other networks that the wireless company may have. The distributed network is ideal for a wireless company that has only a few markets that are geographically disbursed, such as in New York and San Francisco.

The advantages of a distributed network lie in its simple implementation. The distributed architecture can also be folded at a later time into a regional or centralized approach.

Of course, the disadvantage with the distributed approach lies in the issue of duplication and the lack of economies of scale for implementing and operating the network. Also, there is the distinct possibility that the networks will implement services differently and no commonality will exist between the data or even the voice networks unless standard practices and procedures for design and operation are implemented.

The typical implementation of a distributed packet network layout is shown in Figure 7-4.

7.2.2 Regional

The regional network depicted in Figure 7-5 has only two markets illustrated in order to simplify the concept. The regional approach could be utilized for a wireless carrier that, say, had multiple markets in the Northeastern and Southwestern United States. In this case, two separate networks would be established, one for the Northeast and the other for the Southeast. The distributed network shown involves several cities in the Northeastern United States, but the concept can easily be migrated to other areas and regions.

The advantage of the regional approach lies in the fact that it enables some economies of scale while at the same realizes the difficulty of man-

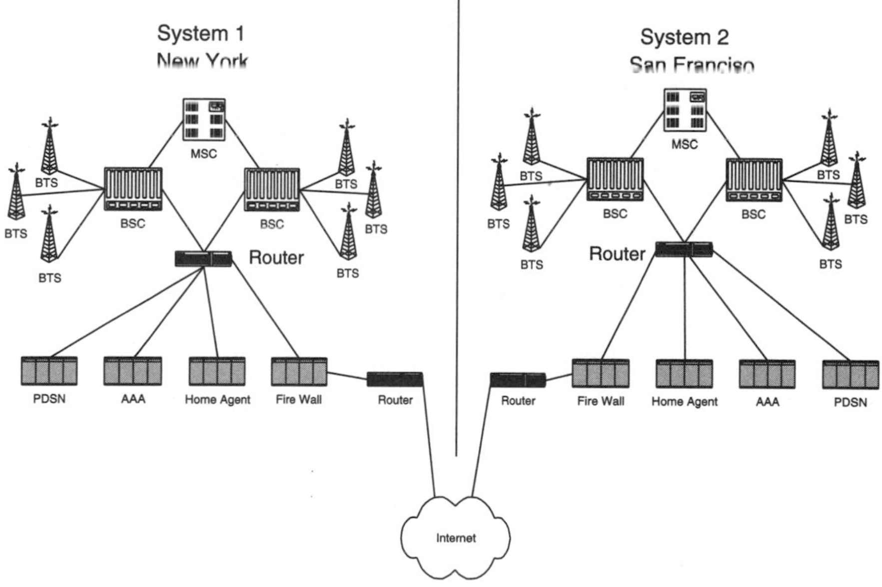

Figure 7-4 A distributed network.

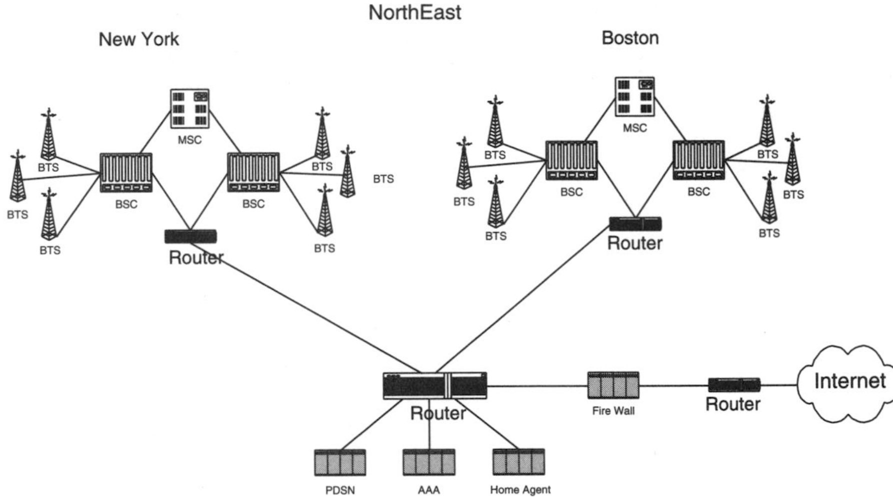

Figure 7-5
A regional network.

aging segmented markets effectively from one localized point. The configuration also enables expansion and service introductions to be expedited and uniform for the region. The regional configuration also enables the segregation of different vendor platforms from each other.

The disadvantage of using a regional configuration is that the networks may not be designed and managed the same way, leading to the classic issue of two networks run by the same company but having different design goals and performance. Again, as with a distributed network, the implementation of standard practices and procedures helps eliminate or mitigate most of the concerns mentioned.

7.2.3 Centralized

The third general configuration promoted for 2.5G and 3G implementation purposes is the *centralized approach*, shown in Figure 7-6. The centralized approach, as the name implies, facilitates the management of various markets and systems from a centralized location. This particular approach has the distinct advantage of providing economies of scale for platforms and a uniformity for service creation and treatment. The centralized approach could also be migrated easily from a regional structure-based system. The

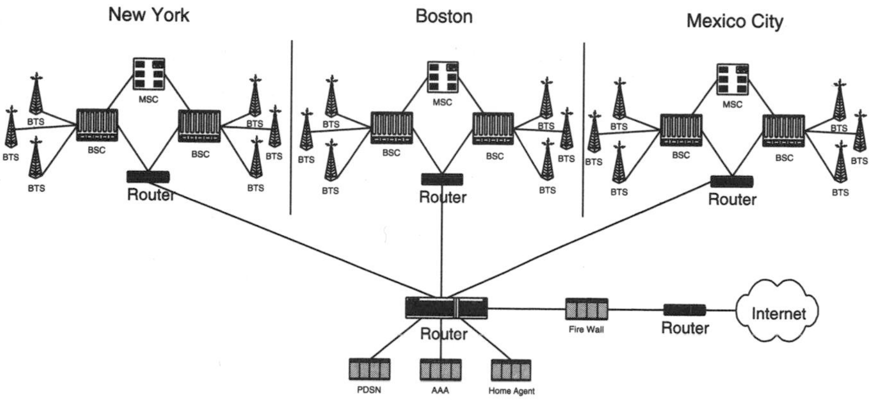

Figure 7-6 A centralized network.

centralized example shows New York, Boston, and Mexico City as being included in the configuration.

The chief disadvantage is that with the centralized approach local market flexibility is lost. In addition, the backbone transport size may become unwieldy because much of the traffic transported is destined for the Internet and it would be better to terminate it locally. Therefore, only the control of the system plus possibly the packet network should be centralized in reality.

7.3 Packet Data Transport Process Flow

CDMA2000 data services fall within two distinct categories: circuit-switched and packet. Circuit-switched data is handled effectively the same as a voice call. But for all packet data calls, a PDSN is used as the interface between the air interface data transport and the fixed network transport. The PDSN interfaces to the *base station* (BS) through a *Packet Control Function* (PCF), which can be colocated with the BS.

The CDMA2000 has three packet data service states that need to be understood in the process:

- **Active/connected** Here a physical traffic channel exists between the subscriber unit and the BS, with packet data being sent and received in a bidirectional fashion.

- **Dormant** No physical traffic channel exists, but a PPP link between the subscriber unit and the PDSN is maintained.
- **Null/inactive** Neither a traffic channel nor a PPP link is maintained or established.

The relationship between the three packet data states is best shown in the simplified state diagram in Figure 7-7.

CDMA2000 introduces to the mobility environment real packet data transport and treatment at speeds that meet or exceed the IMT-2000 system requirements. The voice call processing that is implemented by CDMA2000 is functionally the same as that of existing cdmaOne networks, with the exception that a vocoder change exists in the subscriber units. However, the key difference is that packet data can now be handled by the network.

The mobile initiates the decision as to whether the session will be a packet data session, voice session, or concurrent session, meaning voice and data. The network at this time cannot initiate a packet data session with the subscriber unit, with the exception of the *Standard Management System* (SMS), which does require a packet data session.

For call processing, the voice and data networks are segregated in general once the information, whether it is voice or data, leaves the radio environment at the BSC itself. Therefore, for packet data, the PDSN is central to all decisions. Figure 7-8 depicts a generalized network architecture.

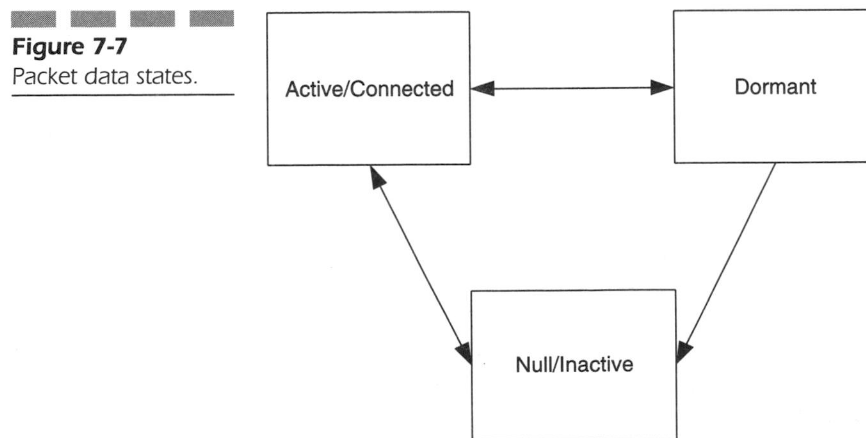

Figure 7-7
Packet data states.

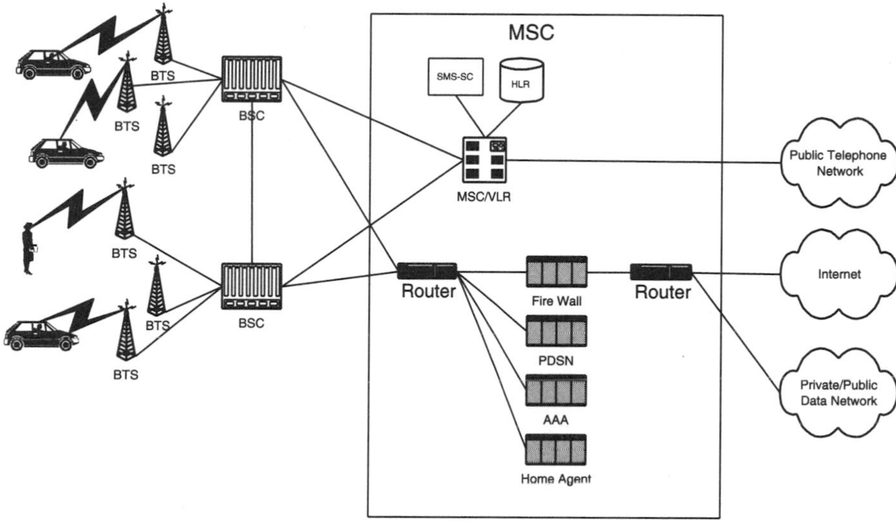

Figure 7-8 Generalized CDMA2000

The PDSN does not communicate directly with the voice network nodes like the HLR and VLR; instead it is done via the AAA. As discussed previously, the voice and data networks normally are segregated once they leave the radio environment at the BSC. Additionally, in a CDMA2000 network, the system utilizes PPP between the mobile and the PDSN for every type of packet data session that is transported and/or treated.

The PDSN is meant to provide several key packet data services, including *Simple IP* and *Mobile IP*. Also, several variants, to be discussed shortly, are relative to each of these services. However, the concepts behind Simple IP and Mobile IP need to be explored first.

Simple IP is a packet data service relative to CDMA2000 1xRTT and is where the subscriber is assigned a *Dynamic Host Configuration Protocol* (DHCP) address from the serving PDSN with its routing service provided by the local network. The specific IP address that the subscriber is assigned remains with the subscriber as long as it is served by the same radio network that maintains connectivity with the PDSN that issued the IP address. It is important to note that Simple IP does not provide for mobile terminations and therefore is an origination-based service only, that is, a PPP service using DHCP.

In Mobile IP, the public IP network provides the mobile's IP routing service. In this functionality, the mobile is assigned a static IP address that resides with the HA. A key advantage of Mobile IP over simple IP is that the mobile, due to the static IP address, can handoff between different radio networks that are served via different PDSNs, which resolves the ROAMing issues that are part of Simple IP. Mobile IP, due to the static IP, also enables the possibility for mobile terminations.

Now with mobility, whether the packet service is Simple or Mobile IP, the notion of mobility is fundamental to the concept of CDMA2000. Figure 7-9 illustrates some of the internetwork communication that needs to take place for establishing a packet data session.

It is important to note that the transport of the packets is not depicted in Figure 7-9. It shows just the elements in the network that need to communicate in order to establish which services the subscriber is allowed to have and how the network is going to meet the *Service-Level Agreement* (SLA) that is expected for the packet session.

The VLR, is normally colocated with the MSC, as shown in Figure 7-9. When a subscriber initiates a packet data session the BSC via the MSC/VLR to check the subscriber subscription information prior to the system granting the service request to the mobile subscriber. This will take place prior to the PDSN being involved with the packet session.

Elaborating on the various packet sessions available for use within a CDMA2000 network are, of course, Simple IP and Mobile IP. However, along with each of these packet session types are two variants where one uses a VPN and the other does not:

- Simple IP
- Simple IP with VPN
- Mobile IP
- Mobile IP with VPN

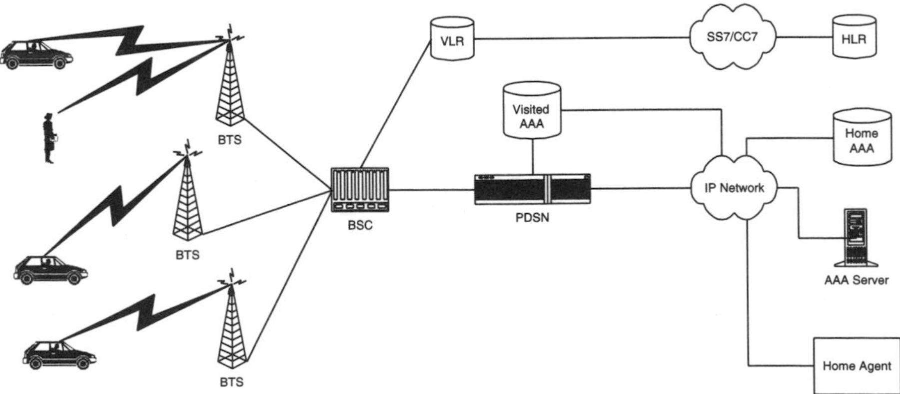

Figure 7-9 CDMA2000 packet network nonhome.

A more specific discussion of Simple IP and Mobile IP is covered in the next section.

7.3.1 Simple IP

Simple IP is similar to the dial-up Internet connections used by many people over standard landline facilities. Simple IP is where a PPP session is established between the mobile and the PDSN. The PDSN basically routes packets to and from the mobile in order to provide end-to-end connectivity between the mobile and the Internet. A diagram depicting Simple IP is shown in Figure 7-10.

When using Simple IP, the mobile must be connected to the same PDSN for the duration of the packet session. If the mobile, while in transit, moves to a coverage area whose BSC/BTSs are homed out of another PDSN, the Simple IP connection is lost and needs to be re-established. The loss of the existing packet session effectively is the same as when the Internet connection on the landline is terminated and you need to re-establish the connection.

Let's refer back to Figure 7-9, which is a simplified model of the simple IP implementation. Many of the details are left out, but the concept shows that the mobile is connected to the PDSN using a PPP connection in a best-effort data delivery method at the agreed-upon transfer rate. The transfer rate is determined by the subscribers profile, the radio resource availability, and the radio environment itself.

The IP address of a mobile is linked to the PDSN, which can be static or DHCP; for Simple IP, the choice is DHCP. A mobile with an active or dormant data call can transverse around the network, going from cell to cell, provided it stays within the PDSN's coverage area. Additionally, the PDSN

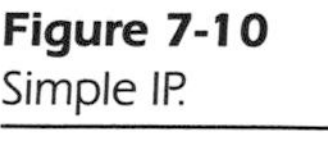

Figure 7-10
Simple IP.

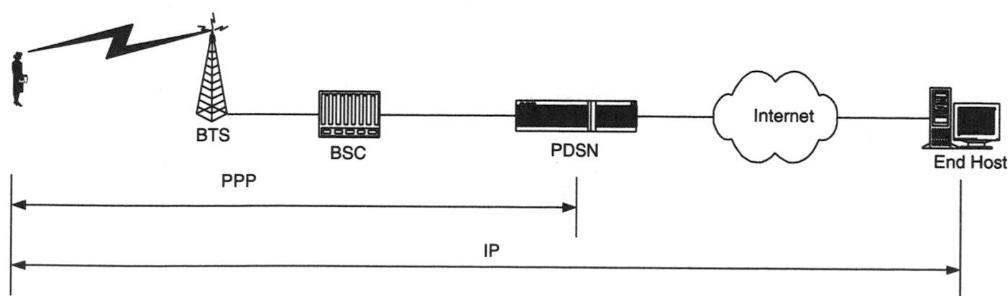

should support both the *Challenge Handshake Authentication Protocol* (CHAP) and the *Password Authentication Protocol* (PAP).

Simple IP, as indicated, does not enable the subscriber full mobility with packet data calls. When the subscriber exits the PDSN coverage area, it must negotiate for a new IP address from the new PDSN, which, of course, results in the termination of the existing packet session and requires a new session to begin.

Regarding the radio environment, the CDMA2000 radio network provides the mobile with a traffic channel that consists of a fundamental channel and possibly a supplemental channel for higher traffic speeds. To help explain the use of the Simple IP process, a call flow or a packet session flow chart is shown in Figure 7-11, which represents a subscriber operating in his or her home PDSN network. In Figure 7-12, the mobile is considered to be ROAMing.

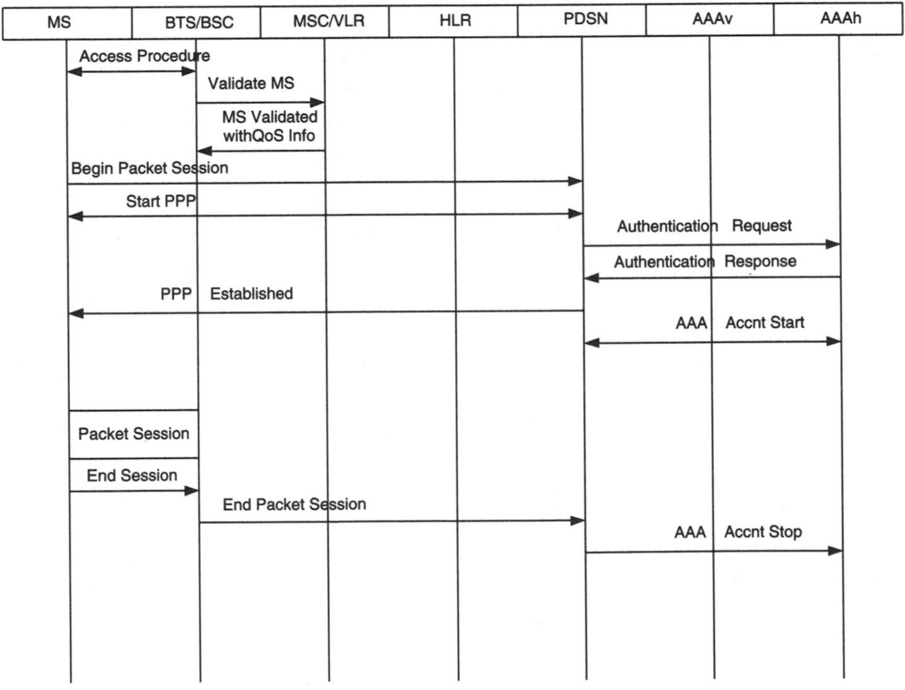

Figure 7-11 Simple IP flow chart.

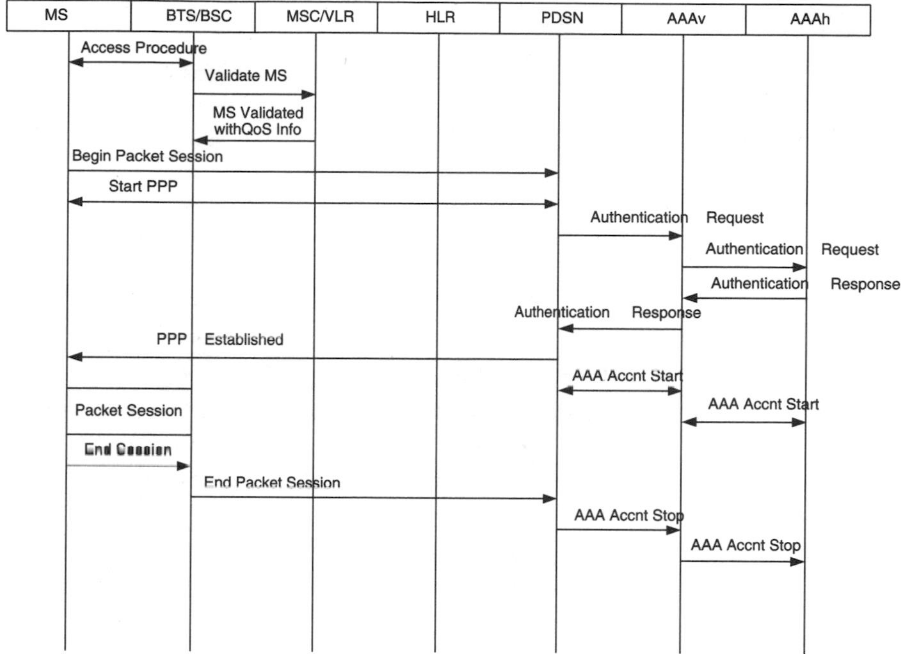

Figure 7-12 Simple IP ROAMing flow chart.

7.3.2 Simple IP with VPN

An enhancement to simple IP is the capability to introduce a VPN to the path for security and also to provide connectivity to a corporate LAN or other packet networks. With VPN, the mobile user should have the appearance of being connected directly to the corporate LAN.

The PDSN establishes a tunnel using the *Layer Two Tunneling Protocol* (L2TP) between the PDSN and the private data network. The mobile is effectively still using a PPP connection, but it is tunneled. The private network that the PDSN terminates to is responsible for assigning the IP address and, of course, authenticating the user beyond what the wireless system needs to perform for billing purposes.

Because of the specific termination and authentication that is performed by another network, the PDSN does not apply any IP services for the mobile and except for the predetermined speed of the connection that is all the system can provide.

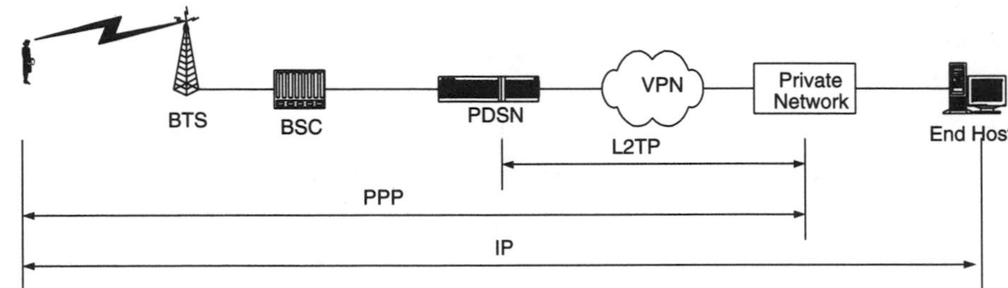

Figure 7-13
Simple IP with VPN.

Just as in the case of Simple IP, the mobile must still be connected to the same PDSN for the packet session. If the mobile moves to another area of the network, which is covered by a separate PDSN, the VPN is terminated and the mobile must reestablish the session. A simplified diagram is shown in Figure 7-13.

The packet session flowchart for Simple IP with VPN is shown in Figure 7-14 and assumes the subscriber is not ROAMing.

7.3.3 Mobile IP (3G)

Mobile IP, whereas a packet-transport method, is quite different than Simple IP in that it actually transports the data. Mobile IP utilizes a static IP address that can be assigned by the PDSN. The establishment of a static IP address facilitates ROAMing during the packet session, provided the static IP address scheme is unique enough for the subscriber unit to be uniquely identified.

With Mobile IP, the PDSN is the *Foreign Agent* (FA) and the *Home Agent* (HA) is set up as a virtual HA. The mobile needs to register each time it begins a packet data session, whether it is originating or terminating. Also, the PDSN on the visited network terminates the packet session using an IP-in-IP tunnel. The HA delivers the IP traffic to the FA through an IP tunnel.

The mobile is responsible for notifying the system that it has moved to another service area. Once the mobile has moved to another service area, it need to register with another FA. The FA assigns the mobile a *care of address* (COA).

The HA forwards the packets to the visited network for termination on the mobile. The HA encapsulates the original IP packet destined for the

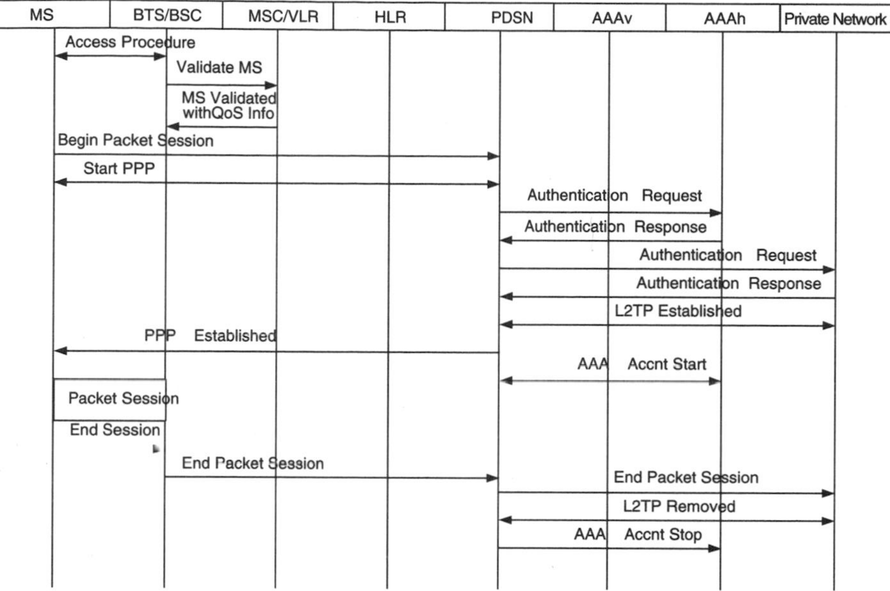

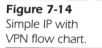

Figure 7-14 Simple IP with VPN flow chart.

mobile using the COA. The FA using IP-in-IP tunneling extracts the original packet and routes it to the mobile.

The IP address assignment is a done via DHCP and is mapped to the HA. However, PAP and CHAP are not used for Mobile IP as it is in Simple IP.

In the reverse direction, the routing of IP packets occurs the same as if on the home network and does not require an IP-in-IP tunnel unless the wireless operator decides to implement reverse IP tunneling.

In summary:

- The PDSN in the visited network always terminates the IP-in-IP tunnel.
- The HA delivers the IP traffic through the mobile IP tunnel to the FA.
- The FA performs the routing to the mobile and assigns the IP address using DHCP.

Figure 7-15 is a simplified depiction of Mobile IP.

Figure 7-16 is an example of a Mobile IP packet session flow.

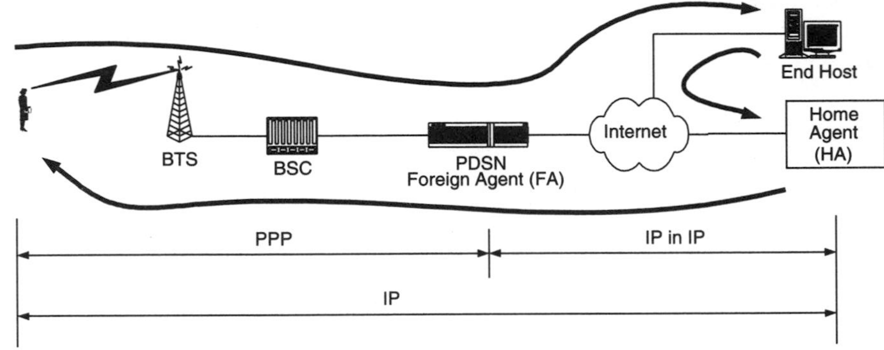

Figure 7-15 Mobile IP.

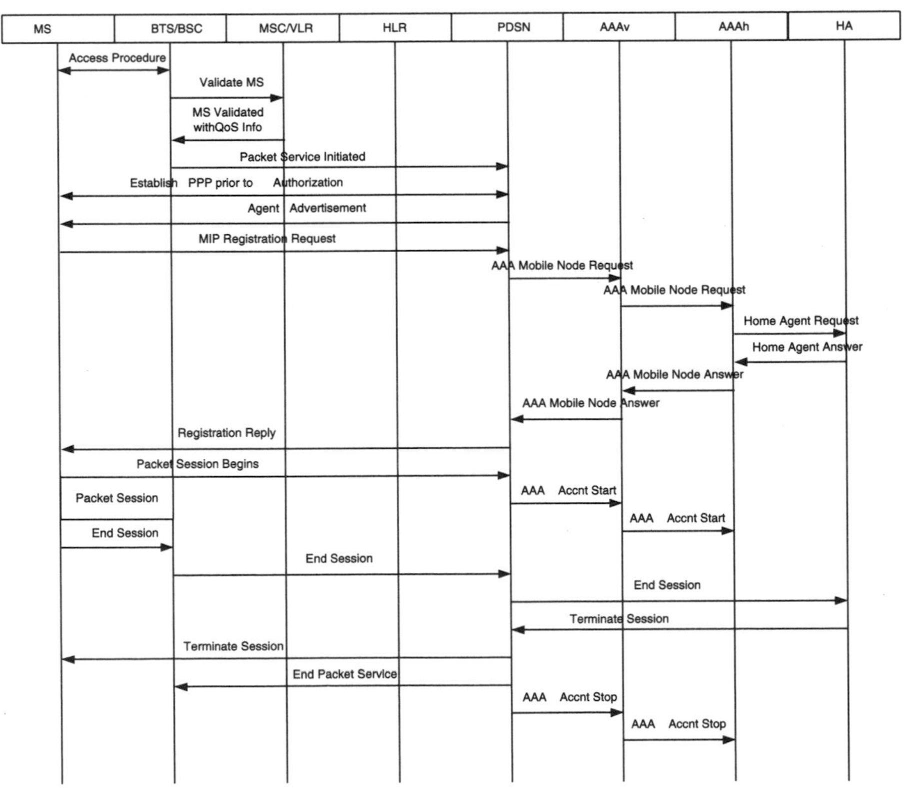

Figure 7-16 Mobile IP packet session flow.

7.3.4 Mobile IP with VPN

The second variant to Mobile IP is Mobile IP with VPN. Mobile IP with VPN affords greater mobility for the subscriber over Simple IP with VPN because it can maintain a session when it moves from one PDSN area to another. Like Mobile IP, the IP address assigned to the subscriber is static; however, the private network that the mobile is connected to provides the IP address, which needs to be drawn from a predefined IP scheme that is coordinated. The PDSN provides a COA when operating in a non-home PDSN for routing purposes. However, the IP packets in both directions flow between the HA and the FA using IP-in-IP encapsulation, and no treatment, with the exception of throughput speed allowed, is performed by the wireless network. Figure 7-17 depicts the general packet flow for Mobile IP with VPN.

7.4 Radio Network

The radio network for a CDMA2000 system has several enhancements over existing IS-95/J-STD-008 wireless systems. These enhancements involve better power control, diversity transmitting, modulation-scheme changes, new vocoders, uplink pilot channels, expansion of the existing Walsh codes, and channel-bandwidth changes. The CDMA2000 radio system, following the IS-2000 specification, is designed to provide an existing cdmaOne operator with a phased entrance into the 3G arena.

The CDMA2000 radio network for phase 1 implementation, also called CDMA2000 1xRTT, is the same as that defined for IS-95/J-STD-008 systems where the channel bandwidth is 1.25 MHz. However, a bandwidth change takes place with the introduction of CDMA2000 phase 2, which is

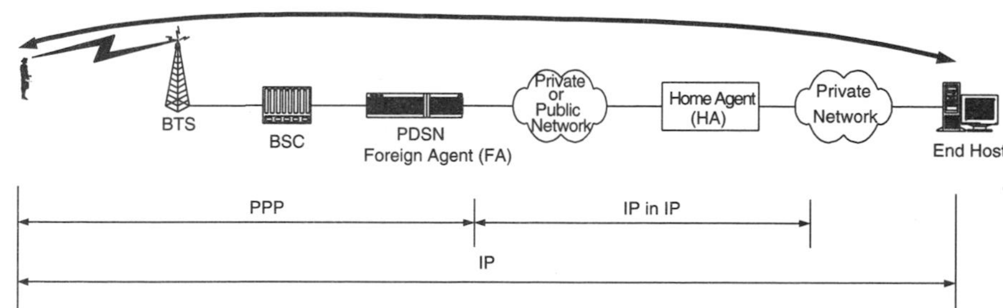

Figure 7-17 Mobile IP with VPN.

Figure 7-18 Radio channel bandwidth

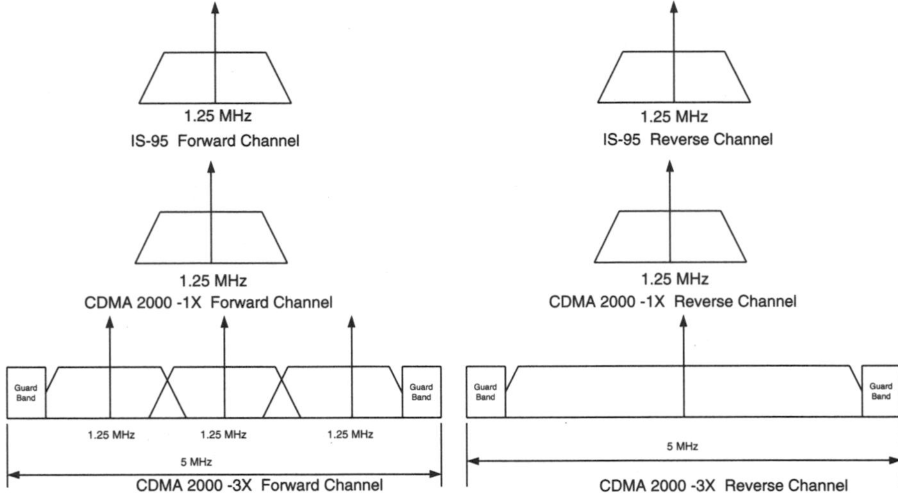

referred to as CMDA2000-3xRTT where multiple carriers are now used. A brief and simplified channel bandwidth diagram is shown in Figure 7-18, which illustrates the radio carrier differences between a CDMA IS-95, 1xRTT, and a 3xRTT system.

CMDA2000 introduces several new channel types for the radio access scheme. The new channel types are implemented in both the 1xRTT and 3xRTT schemes and are introduced to support high-speed data as well as enhanced paging functions. To accomplish the higher data rates, CDMA2000 uses a combination of expanded Walsh codes along with modulation and vocoder changes.

As depicted in Figure 7-18, a wireless operator can migrate to CDMA2000 from either the IS-95A or IS-95B platforms using the same amount of existing spectrum when transitioning to a 1xRTT format. The two common migration paths for implementing CDMA2000 are relative to operators utilizing CDMAOne (Is-95A/B) platforms:

- cdmaOne (IS-95A)—CDMA2000 (phase 1)—CDMA2000 (phase 2)
- cdmaOne (IS-95A)—cdmaOne (IS-95B)—CDMA2000 (phase 1)—CDMA2000 (phase 2)

The CDMA2000 radio access scheme has several enhancements over the existing IS-95 systems and they are as follows:

- Forward link:
  - Fast power control
  - *Quadrature Phase Shift* (QPS) keying modulation, rather than dual *Binary Phase Shift* (BPS) keying
- Reverse link:
  - Pilot signal, to enable coherent demodulation for the reverse link
  - *Hybrid Phase Shift* (HPS) keying spreading in the reverse link

Table 7-3 shows the various relationships between the IS-95 and CDMA2000 radio channels.

7.4.1 CDMA Channel Allocation

The CDMA2000 channel allocations, just as with IS-95, have preferred locations and methods for deploying which are envisioned at this time to help facilitate the migration from 1X to 3X in the future. Tables 7-4 and 7-5 are

Table 7-3
CDMA2000

Platform	Description
IS-95A	Primarily voice with circuit switch speeds of 9,600 bps or 14.4 Kbps 04 Walsh codes SR1 (1.2288 Mbps)
IS-95B	Primarily voice, data on forward link, improved handoff 64 Walsh codes SR1 (1.2288 Mbps)
CDMA2000—phase1 (1xRTT)	SR1 (1.2288 Mbps), Voice and data (packet data via separate channel) 128 Walsh codes closed loop power control
CDMA2000—phase2 (3xRTT)	SR3 (3.6864 Mbps) Data primarily Higher data rate 256 walsh codes

Table 7-4

Cellular CDMA2000-1x and 3x Carrier Assignment Scheme

Cellular System Carrier	Sequence	A	B
1	F1	283	384
2	F2	242*	425*
3	F3	201	466
4	F4	160	507
5	F5	119	548
6	F6	78	589
7	F7	37	630
8	F8 (Not advised)	691	777

Table 7-5

PCS CDMA2000-1x and 3x PCS Carrier Assignment Scheme

PCS System Carrier	A	B	C	D	E	F
1	25	425	925	325	725	825
2	50	450	950	350*	750*	850*
3	75*	475*	975*	375	775	875
4	100	500	1000	NA	NA	NA
5	125	525	1025	NA	NA	NA
6	150*	550*	1050*	NA	NA	NA
7	175	575	1075	NA	NA	NA
8	200	600	1100	NA	NA	NA
9	225*	625*	1125*	NA	NA	NA
10	250	650	1150	NA	NA	NA
11	275	675	1175	NA	NA	NA

for North America, but CS0002 has a channel plan for the conceivable band that this technology can be implemented into.

Please note that the channels defined in the tables are for 1.25-Mhz channel spacing. The asterisk denotes the locations where the first of three 1.25-MHz carriers is expected to be located for a 3X deployment. The first carrier is used in 3X for access and this is used to help steer the subscriber to the correct carrier(s) to support the services being requested.

7.4.2 Forward Channel

The forward link for a CDMA2000 channel, whether for 1X or 3X implementation, utilizes the structure shown in Figure 7-19.

Reviewing the channel structure, the base station transmits multiple common channels as well as several dedicated channels to the subscribers in their coverage area. Each CDMA2000 user is assigned a forward traffic channel that consists of the following combinations. An important point to note is that F-FCHs are used for voice, while F-SCHs are for data.

- **1** *Forward Fundamental Channel* (F-FCH)
- **0–7** *Forward Supplemental Code Channels* (F-SCHs) for both RC1 and RC2
- **0–2** *Forward Supplemental Code Channels* (F-SCHs) for both RC3 and RC9

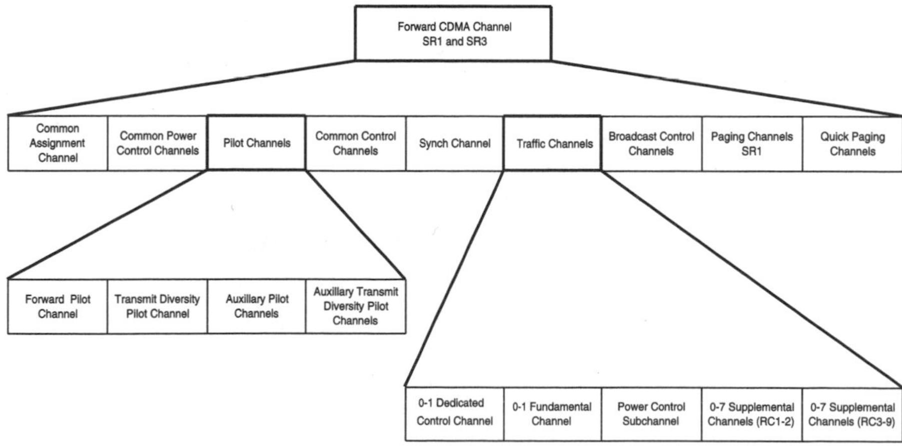

Figure 7-19 A forward CDMA channel transmitted by base station [33].

When the channel is associated with a 3XRTT implementation, the data for the subscriber is mapped to each of the three different carriers, enabling the high throughput. However, the Walsh codes are the same for each carrier, meaning they share the same throughput, distributing the traffic load evenly.

The CDMA2000 channel utilizes different modulation schemes depending on the radio configuration that is employed. The description of the radio configurations are shown later. However, the modulation scheme used for RC1 and RC2 is *Binary Phase Shift Keying* (BPSK), while *Quadrative Phase Shift Keying* (QPSK) is used for RC3-RC9. For RC3 through RC9, the data is converted into a two-bit-wide parallel data stream that initially would seem counterintuitive because it reduces the data rate for each stream by a factor of two. Each data stream, however, is then spread by a 128 Walsh code to get the spreading rate up to 1.2288 Mbps, which effectively doubles the processing gain, allowing for greater throughput at the same effective power level.

The following are some forward channel descriptions:

- **Forward Supplemental Channel (F-SCH)** Up to two F-SCHs can be assigned to a single mobile for high-speed data ranging from 9.6K to 153.6K in release 0 and 307.2 Kbps and 614.4 Kbps in release A. It is important to note that each F-SCH assigned can be assigned at different rates. The F = SCH must be assigned with a R-SCH when only one F-SCH is assigned.
- **Forward Quick Paging Channel (F-QPCH)** The quick paging channel enables the mobile battery life extension by reducing the amount of time the mobile spends parsing pages that are not meant for it. The mobile monitors the F-QPCH and when the flag is set, the mobile looks for the paging message. There are a total of three F-QPCH channels per sector.
- **Forward Dedicated Control Channel (F-DCCH)** This replaces the dim and burst and the blank and burst. It is used for messaging and control for data calls.
- **Forward Transmit Diversity Pilot Channel (F-TDPICH)** This is used to increase RF capacity.
- **Forward Common Control Channel (F-CCCH)** This is used to send paging, data messages, or signaling messages.

Table 7-6 helps to quantify the channel types and quantity of each for CDMA2000, both 1X and 3X.

Table 7-6

Forward and Reverse CDMA2000 Channel Descriptions

Channel Type (SR1)	Maximum Number
Forward Pilot Channel	1
Transmit Diversity Pilot Channel	1
Sync Channel	1
Paging Channel	7
Broadcast Control Channel	8
Quick Paging Channel	3
Common Power Control Channel	4
Common Assignment Channel	7
Forward Common Control Channel	7
Forward Dedicated Control Channel	1 per Fwd Traffic Channel
Forward Fundamental Channel	1 per Fwd Traffic Channel
Forward Supplemental Code Channel (RC1 and RC2 only)	7 per Fwd Traffic Channel
Forward Supplemental Channel (RC3, RC4 and RC5 only)	2 per Fwd Traffic Channel
Channel Type (SR3)	**Maximum Number**
Forward Pilot Channel	1
Sync Channel	1
Broadcast Control Channel	8
Quick Paging Channel	3
Common Power Control Channel	4
Common Assignment Channel	7
Forward Common Control Channel	7
Forward Dedicated Control Channel	1 per Fwd Traffic Channel
Forward Fundamental Channel	1 per Fwd Traffic Channel
Forward Supplemental Channel	2 per Fwd Traffic Channel

7.4.3 Reverse Channel

The reverse link or channel for CDMA2000 has many similar properties as the forward link and therefore differs significantly from that used in IS-95. One of the major differences or rather enhancements to CDMA2000 over IS-95 is the inclusion of a pilot on the reverse link. The structure of the reverse channel for CDMA2000 is shown in Figure 7-20.

Elaborating on the reverse channel, the subscriber, or mobile, is allowed to transmit more than one code channel to accommodate the high data rates. The minimum configuration consists of a *Reverse Pilot* (R-Pilot) channel to enable the base station to perform synchronous detection and a *Reverse Fundamental Channel* (R-FCH) for voice. The inclusion of additional channels, such as the *Reverse Supplemental Channels* (R-SCHs) and the *Reverse Dedicated Control Channel* (R-DCCH) can be used to send data or signaling information. The association between the radio configuration and the spreading rates is best shown in Table 7-9. It is important to note that the reverse channel for 3X is different than 1X in that it is a direct spread but can be overlaid over a 1X implementation. Depending on the subscribers operating in that sector, the appropriate SR and RC are then selected.

The following are some of the Reverse Link channel descriptions:

- **Reverse Supplemental Channel (R-SCH)** When data rates are greater than 9.6K, a R-SCH is required and also a R-FCH is also

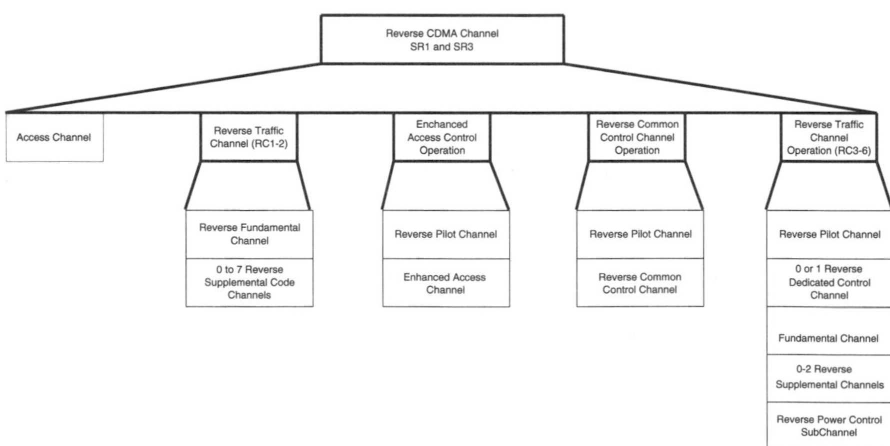

Figure 7-20 A reverse CDMA channel received at base station [33].

assigned for power control. A total of one or two R-SCHs can be assigned per mobile.

- **Reverse Pilot Channel (R-PICH)** The R-PICH provides pilot and power control information. The R-PICH enables the mobile to transmit at a lower power level and allows the mobile to inform the base station of the forward power levels being received, enabling the base station to reduce power.
- **Reverse Dedicated Control Channel R-DCCH** This replaces the dim and burst and the blank and burst. It is used for messaging and control for data calls.
- **Reverse Enhanced Access Channel (R-EACH)** This is meant to minimize the collisions and therefore reduce the access channel's power.

Table 7-7

CDMA2000 Channel Types

Channel Type (SR1)	Maximum Number
Reverse Pilot Channel	1
Access Channel	1
Enhanced Access Channel	1
Reverse Common Control Channel	1
Reverse Dedicated Control Channel	1
Reverse Fundamental Control Channel	1
Reverse Supplemental Code Channel (RC1 and RC2 only)	7
Reverse Supplemental Channel	2
Channel Type (SR3)	**Maximum Number**
Reverse Pilot Channel	1
Enhanced Access Channel	1
Reverse Common Control Channel	1
Reverse Dedicated Control Channel	1
Reverse Fundamental Control Channel	1
Reverse Supplemental Channel	2

- **Reverse Common Control Channel (R-CCCH)** Used by mobiles to send their data information after they have been granted access.

Table 7-7 helps to quantify the reverse channel types and the quantity of each for CDMA2000, both 1X and 3X.

7.4.4 SR and RC

CDMA2000 defines two spreading rates, referred to as *spreading rate 1* (SR1) and *spreading rate 3* (SR3). The SR1 spreading rate is utilized for IS-95A/B and CDMA2000 phase 1, 1xRTT implementations, whereas SR3 is destined for CDMA2000 Phase 2, 3xRTT.

For CDMA2000, the SR1 has a chip rate of 1.2288 Mbps and occupies the same bandwidth as CDMAOne signals. The SR1 is a direct spread method and follows the same concept as that used for IS-95 systems. However, for 3xRTT, a SR3 signal is introduced and has a rate of 3.6864 Mbps (3 × 1.2288 Mcps) and therefore occupies three times the bandwidth of a cdmaOne or 1xRTT channel. The SR3 system incorporates all the new coding implemented in a SR1 system while supporting even higher data rates. The 3xRTT channel scheme utilizes a multicarrier forward link and direct spread reverse link.

The IS-2000 specification also defines for both 1xRTT and 3xRTT radio access methods a total of nine forward and six reverse link radio configurations, as well as two different spreading rates. The radio configurations involve different modulations, coding, and vocoders, while the spreading rates address the usage amount of two different chip rates. The radio configurations are referred to as RC1 for radio configuration 1.

RC1 is backward-compatible with cdmaOne for 9.6-Kbps voice traffic and it supports circuit-switched data rates of 1.2 Kbps to 9.6 Kbps. RC3 is based on the 9.6-Kbps rate and supports variable voice rates from 1.2k to 9.6 Kbps, while also supporting packet data rates of 19.2, 38.4, 76.8, and 153.6 Kbps, but it operates using a SR1.

Tables 7-8 and 7-9 are meant to help illustrate the perturbations that exist with the different radio configurations and spreading rates. Table 7-8 is associated with the forward link, whereas Table 7-9 is associated with the reverse link.

Table 7-8

Forward Link RC and SR [16]

Forward RC	SR	Data Rates	Characteristics
1	1	1200,2400,4800,9600	R=1/2
2	1	1800,3600,7200,14400	R=1/2
3	1	1500,2700,4800,9600,38400,76800,153600	R=1/4
4	1	1500,2700,4800,9600,38400,76800,153600,307200	R=1/2
5	1	1800,3600,7200,14400,28800,57600,115200,230400	R=1/4
6	3	1500,2700,4800,9600,38400,76800,153600,307200	R=1/6
7	3	1500,2700,4800,9600,38400,76800,153600,307200, 614400	R=1/3
8	3	1800,3600,7200,14400,28800,57600,115200,230400, 460800	R=1/4 (20ms) R=1/3 (5ms)
9	3	1800,3600,7200,14400,28800,57600,115200,230400, 460800,1036800	R=1/2 (20ms) R=1/3 (5ms)

Table 7-9

Reverse Link RC and SR [16]

Reverse Link RC	SR	Data Rates	Characteristics
1	1	1200,2400,4800,9600	R=1/3
2	1	1800,3600,7200,14400	R=1/2
3*	1	1200,1350,1500,2400,2700,4800,9600,19200,38400, 76800,153600, 307200	R=1/4 R=1/2 for 307200
4*	1	1800,3600,7200,14400,28800,57600,115200,230400	R=1/4
5*	3	1200,1350,1500,2400,2700,4800,9600,19200,38400, 76800,153600, 307200,614400	R=1/4 R=1/2 for 307200 and 614400
6*	3	1800,3600,7200,14400,28800,57600,115200,230400, 460800,1036800	R=1/4 R=1/2 for 1036800

*Reverse pilot

7.4.5 Power Control

Power control is a major enhancement of CDMA2000 over IS-95, which enables higher data rates. The primary power control enhancement is with the fast-forward link power control.

As discovered through practical implementation issues, CDMA systems are interference-limited, and reducing the interference results in an improvement in system capacity.

Enabling better power control of both the forward and reverse links has several advantages:

- System capacity is enhanced or optimized.
- Mobile battery life is extended.
- Radio path impairments are properly or better compensated for.
- *Quality of Service* (QoS) at various bit rates can be maintained.

Obviously, with any wireless system that is interference-limited, it is important to ensure that all transmitters, whether mobile or located at a base station, transmit at the lowest power level while maintaining a good communication link.

To achieve this, CDMA2000 utilizes fast-response, closed-loop power control on the reverse link. In summary, the BTS measures the reverse link from the mobile and sends power control commands to increase or decrease the mobile's power level, which is similar to IS-95. It is important to note that the mobile can also operate autonomously and make power corrections based on the *Frame Error (Erasure) Rate* (FER) of the forward link. From that, it infers what it needs to do for the reverse link in terms of power control.

Also, a refinement to the closed-loop power control is located on the reverse link and that is where the base station performs an outer-loop power control, which is a refinement process for the inner power control process. Specifically, if the frame received from the mobile arrives without error, the base station instructs the mobile to power down, while on the other side if the frame arrives in error, the mobile is instructed to power up.

With CDMA2000, the use of power control on the forward channel is possible with the introduction of the reverse pilot channel. The reverse pilot channel for power control was introduced to help reduce the interference caused by forward energy. Effectively, the mobile measures the received power and compares it against a threshold that the mobile then feeds back to the base station. Upon receipt of the power information, the mobile is then instructed to power up or power down.

In addition, as with the reverse power link, an outer loop power control process dynamically adjusts the target *Energy per bit per Noise Ratio* (Eb/No). This is done by measuring the FER with a target FER, and if the FER is greater than the target, it is instructed to power up. If it is below the target FER, it is instructed to power down.

7.4.6 Walsh Codes

CDMA2000 introduces an increase in the number of Walsh codes, from 64 with IS-95 to a total of 256 with 3XRTT. As with IS-95, CDMA2000 utilizes PN long codes for both the forward and reverse directions. However, in CDMA2000, the introduction of variable-length Walsh codes is introduced to accommodate fast-packet data rates.

The Walsh code chosen by the system is determined by the type of reverse channel. The R-SCH also uses a reserve Walsh code. If only one R-SCH is used, it utilizes a two- or four-chip Walsh code, but when the second R-SCH is utilized, it uses a four- or eight-chip code. Therefore, in order

Table 7-10
Walsh Code Tree Table

		Walsh Codes						
	RC	256	128	64	32	16	8	4
SR1	1	Na	Na	9.6	Na	Na	Na	Na
	2	Na	Na	14.4				
	3	Na		9.6	19.2	38.4	76.8	153.6
	4	Na	9.6	19.2	38.4	76.8	153.6	307.2
	5	Na	Na	14.4	28.8	57.6	115.2	230.4
SR3	6		9.6	19.2	38.4	76.8	153.6	307.2
	7	9.6	19.2	38.4	76.8	153.6	307.2	614.4
	8		14.4	28.8	57.6	115.2	230.4	460.8
	9	14.4	28.8	57.6	115.2	230.4	460.8	1036.8

to maintain or obtain the higher data rates on the F-SCH, the Walsh code must be shorter in order to maintain the same spreading rate.

Table 7-10 shows the relationship between Walsh codes, the SR, the RC, and, of course, the data rates. One very important issue or, rather, effect with utilizing variable-length Walsh codes is that if a shorter Walsh code is being used, then it precludes the use of the longer Walsh codes that are derived from it.

Table 7-10 helps in establishing the relationship between which Walsh code length, which is associated with a particular data rate.

Table 7-11, a simplified table, shows the maximum number of simultaneous users for any data rate.

For an SR1 and RC1, a maximum number of users have individual Walsh codes equating to 64, a familiar number from IS-95A.

Looking at Table 7-11, if we had a total of 12 RC1 and RC2 mobiles under a sector, then one that would allow for three data users at 153.6K, 6 at 76.8 Kbps, 13 at 38.4 Kbps, 26 at 19.2 Kbps, or 104 at 9.6 Kbps. This relationship between the number of simultaneous users for a cdma channel is depicted in Table 7-12. Obviously, the negotiated mobile data rate complicates the determination for the total throughput of traffic levels. The real issue

Table 7-11

Simultaneous users with SR1 and SR3

		Simultaneous Users*						
	RC	**256**	**128**	**64**	**32**	**16**	**8**	**4**
SR1	1	Na	Na	9.6	Na	Na	Na	Na
	2	Na	Na	14.4				
	3	Na		9.6	19.2	38.4	76.8	153.6
	4	Na	9.6	19.2	38.4	76.8	153.6	307.2
	5	Na	Na	14.4	28.8	57.6	115.2	230.4
SR3	6		9.6	19.2	38.4	76.8	153.6	307.2
	7	9.6	19.2	38.4	76.8	153.6	307.2	614.4
	8		14.4	28.8	57.6	115.2	230.4	460.8
	9	14.4	28.8	57.6	115.2	230.4	460.8	1036.8

Table 7-12

Simultaneous Users for SR1 Only

	RC	Data Rates*						
		256	128	64	32	16	8	4
SR1	1 & 2	Na	Na	9.6/14.4	Na	Na	Na	Na
	3	Na	Na	9.6	19.2	38.4	76.8	153.6
		Simultaneous Users*						
SR1	1 & 2	Na	Na	12	Na	Na	Na	Na
	3	Na	Na	48	8	12	2	3

behind this is the type of data that will be allowed to be transported over the network, which has a direct impact on the available users.

It is important to note that the shorter Walsh codes inhibit the use of longer Walsh codes because of the orthogonality required. Also, all channel requests are allocated from the same Walsh code pool on a per-sector basis. In addition, to achieve the higher data rate, not only is the Walsh code implementation modified, but also the modulation scheme has been changed.

Also, if there was a need for high-speed data for interactive video with Phase 1 CDMA2000, the transport of 384 Kbps of data would not be feasible with a SR1 as indicated in Table 7-11.

References

3GPP2 A.R0003. "*Abis interface Technical Report for cdma2000 Spread Spectrum Systems*," Dec. 17, 1999.

3GPP2 A.S0001-A. "*Access Network Interfaces Interoperability Specification*," Nov. 30, 2000.

3GPP2 A.S0004. "*Tandem Free Operation Specification*," Nov. 8, 2000.

3GPP2 C.S0007-0. "*Direct Spread Specification for Spread Spectrum Systems on ANSI-41 (DS-41) (Upper Layers Air Interface)*," June 9, 2000.

3GPP2 C.S0008-0. "*Multi-carrier Specification for Spread Spectrum Systems on GSM MAP (MC-MAP) (Lower Layers Air Interface)*," June 9, 2000.

3GPP2 C.S0009-0. "*Speech Service Option Standard for Wideband Spread Spectrum Systems.*"

3GPP2 C.S0024. "*cdma2000 High Rate Packet Data Air Interface Specification*," Oct. 27, 2000.

3GPP2 C.S0025. "*Markov Service Option (MSO) for cdma2000 Spread Spectrum Systems*," Nov. 2000.

3GPP2 S.R0005-A. "*Network Reference Model for cdma2000 Spread Spectrum Systems*," Dec. 13, 1999.

3GPP2 S.R0021 Version 1.0. "*Video Streaming Services – Stage 1*," July 10, 2000.

3GPP2 S.R0022. " *Video Conferencing Services – Stage 1*," July 10, 2000.

3GPP2 S.R0023 Version 2.0 3. "*High-Speed Data Enhancements for cdma2000 1x – Data Only*," Dec. 5, 2000.

3GPP2 S.R0024 Version: 1.0. "*Wireless Local Loop*," Sept. 22, 2000.

3GPP2 S.R0026. "*High-Speed Data Enhancements for cdma2000 1x—Integrated Data and Voice*," Oct. 17, 2000.

3GPP2 S.R0027. "*Personal Mobility*," Dec. 8, 2000.

3GPP2 S.R0032. "*Enhanced Subscriber Authentication (ESA) and Enhanced Subscriber Privacy (ESP)*," Dec. 6, 2000.

3PP2 P.S0001-A-1 Version 1.0 3. "*Wireless IP Network Standard*," Dec. 15, 2000.

Agilent. "*Designing and Testing cdma2000 Base Stations*," Application Note 1357.

Agilent. "*Designing and Testing cdma2000 Mobile Stations*," Application Note 1358.

Andersen Consulting, Detecon, Telemate Mobile Consultants. "*The GSM-CDMA Economic Study*," Feb. 16, 1998.

Barron, Tim. "*Wireless Links for PCS and Cellular Networks*," Cellular Integration, Sept. 1995, pgs. 20–23.

Bates, Gregory. "*Voice and Data Communications Handbook*," Signature Ed., McGraw-Hill, 1998.

Black. "*TCP/IP and Related Protocols*," McGraw-Hill, 1992.

Brewster. "*Telecommunications Technology*," John Wiley & Sons, New York, NY, 1986.

Collins, Daniel. "*Carrier Grade Voice Over IP*," McGraw-Hill, 2001.

DeRose. "*The Wireless Data Handbook*," Quantum Publishing, Inc., Mendocino, CA, 1994.

Dixon. "*Spread Spectrum Systems*," 2nd Ed., John Wiley & Sons, New York, NY, 1984.

Held, Gil. "*Voice & Data Interworking*," 2nd Ed., McGraw-Hill, 2000.

McDysan, Spohn. "*ATM Theory and Applications Signature Edition*," McGraw-Hill, 1999.

Molisch, Andreas F. "*Wideband Wireless Digital Communications*," Prentice Hall, New Jersey, 2001.

Qualcomm. "*An Overview of the Application of Code Division Multiple Access (CDMA) to Digital Cellular Systems and Personal Cellular Networks*," Qualcomm, San Diego, CA, May 21, 1992.

Smith, Clint. "*Practical Cellular and PCS Design*," McGraw-Hill, 1997.

Smith, Clint. "*Wireless Telecom FAQ*," McGraw-Hill, 2000.

TIA/EIA-98-C. "*Recommended Minimum Performance Standards for Dual-Mode Spread Spectrum Mobile Stations (Revision of TIA/EIA-98-B)*," Nov., 1999.

TIA/EIA IS-127. "*Enhanced Variable Rate Codec, Speech Service Option 3 for Wideband Spread Spectrum Digital Systems*," Sept., 1999.

TIA/EIA/IS-683-A ."*Over-the-Air Service Provisioning of Mobile Stations in Spread Spectrum Systems*," May, 1998.

TIA/EIA IS-718. Minimum Performance Specification for the Enhanced Variable Rate Codec, Speech Service Option 3 for Spread Spectrum Digital Systems, July 1996.

TIA/EIR IS-733-1. "*High Rate Speech Service Option 17 for Wideband-Spread Spectrum Communication System*," Sept., 1999.

TIA/EIA IS-736-A. "*Recommended Minimum Performance Standard for the High-Rate Speech Service Option 17 for Spread Spectrum Communication Systems*," Sept. 6, 1999.

TIA/EIA IS-820. "*Removable User Identity Module (R-UIM) for cdma2000 Spread Spectrum Systems*," June 9, 2000.

TIA.EIA IS-2000-1. "*Introduction to cdma2000 Standards for Spread Spectrum Systems*," June 9, 2000.

TIA/EIR IS-2000-2. "*Physical Layer Standard for cdma2000 Spread Spectrum Systems*," Sept. 12, 2000.

TIA/EIA IS-2000-3. "*Medium Access Control (MAC) Standard for cdma2000 Spread Spectrum Systems*," Sept.12, 2000.

TIA/EIA IS-2000-4. "*Signaling Link Access Control (LAC) Specification for cdma2000 Spread Spectrum Systems*," Aug. 12, 2000.

TIA/EIA IS-2000-6. "*Analog Signaling Standard for cdma2000 Spread Spectrum Systems*," June 9, 2000.

Webb, William. "*CDMA for WLL*," Mobile Communications International, Jan. 1999, pg 61.

Willenegger, Serge. "*cdma2000 Physical Layer: An Overview*," Qualcomm 5775, San Diego, CA.

William, C.Y. Lee. "*Lee's Essentials of Wireless Communications*," McGraw-Hill, 2001.

William, C.Y. Lee. "*Mobile Cellular Telecommunications Systems*," 2nd Ed., McGraw-Hill, New York, NY, 1996.

www.fcc.gov

CHAPTER 8

Voice-over-IP (VoIP) Technology

8.1 Introduction

As we have seen in previous chapters, the development of wireless technology involves a migration from the circuit-switched solutions of *first-generation* (1G) and *second-generation* (2G) networks towards a completely packet-switched configuration for both voice and data. Although a number of packet-switching solutions could be leveraged, such as *Asynchronous Transfer Mode* (ATM) or Frame Relay, the ultimate goal is to use the *Internet Protocol* (IP). In fact, if we examine the migration from *Global System for Mobile communications* (GSM) to the *Universal Mobile Telecommunications Service* (UMTS) Release 5, we see the use of Frame Relay (the *General Packet Radio Service* [GPRS] Gb interface), followed by ATM, followed by IP.

Although the IP transport of data is well understood, the IP transport of voice is a relatively recent development. Given that *Voice over IP* (VoIP) will be used in *third-generation* (3G) networks, it is appropriate that we describe the solutions that make VoIP possible. Therefore, this chapter is devoted to a brief overview of VoIP technology. It should be noted, however, that the explanations provided in this chapter are at a relatively high level and are certainly not detailed enough to provide a complete understanding of all aspects of VoIP. This is, after all, a book about 3G wireless.

8.2 Why VoIP?

IP clearly has a number of advantages over circuit-switching. The most notable of these is the fact that it can leverage today's advanced voice coding techniques, such as the *Adaptive MultiRate* (AMR) coder used in *Enhanced Data Rates for Global Evolution* (EDGE) and UMTS networks. Thus, voice can be transported with far less bandwidth than the 64 Kbps used in traditional circuit-switched networks.

If we consider, for example, the network architecture of GSM, we find that speech from the MS must be transcoded to 64 Kbps before it enters the MSC. Thereafter, it is carried to the destination at 64 Kbps. If the voice were to be carried most of the way with a packet transport such as IP, then the transcoding up to 64 Kbps might not be needed at all, or might be needed only very close to the destination. The bandwidth savings enabled by such technology can be significant. Of course, IP is not the only technology that can enable such bandwidth savings. ATM, for example, can also transport voice at rates less than 64 Kbps. IP, however, has other advantages.

Perhaps the biggest advantage of IP over technologies such as ATM is the fact that IP is practically everywhere. Not only is it supported by every PC on the market today, it is also supported by handheld computers and personal organizers. ATM just does not have the same ubiquitous presence. Moreover, the availability of IP knowledge and experience is widespread, with numerous companies devoted to the development of IP-based applications. If IP resides in the handset and IP is used to carry both voice and data, then real voice-data convergence opportunities arise, offering the possibility of exciting new services.

8.3 The Basics of IP Transport

As shown in Figure 8-1, IP corresponds to layer 3 of the *Open Systems Interconnection* (OSI) seven-layer protocol stack. At its most basic level, IP simply passes a packet of data from one router to another through the network to the appropriate destination, as identified by the destination IP address in the IP packet header. This simple operation means that IP is inherently unreliable. IP provides no protection against a loss of packets, which might happen if congestion occurs along the path from the source to the destination. Moreover, in a given stream of packets from the source to the destination, it is quite possible that packets will take different routes through the

Figure 8-1 OSI and IP Protocol Stacks.

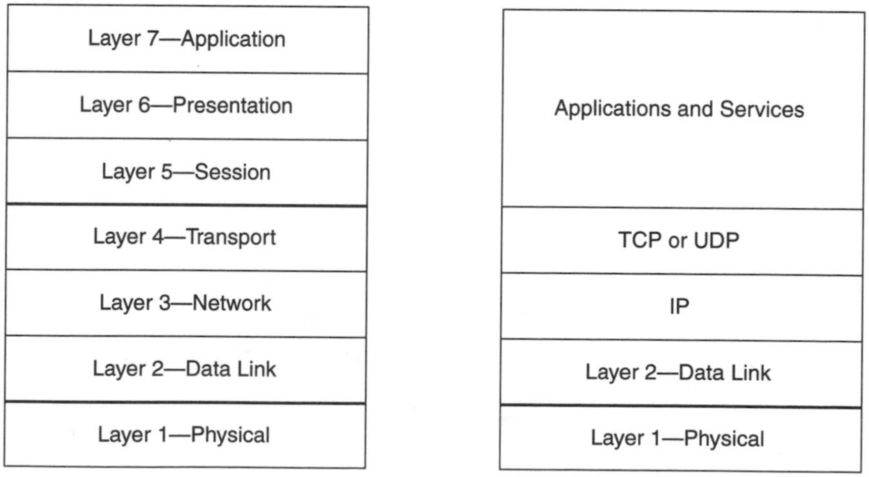

OSI	IP
Layer 7—Application	Applications and Services
Layer 6—Presentation	
Layer 5—Session	
Layer 4—Transport	TCP or UDP
Layer 3—Network	IP
Layer 2—Data Link	Layer 2—Data Link
Layer 1—Physical	Layer 1—Physical

network, meaning that different packets can have different delays and also that packets may arrive at the destination out of sequence.

In data networks, in order to ensure an error-free, in-sequence delivery of packets to the destination application, the *Transmission Control Protocol* (TCP) is used. This protocol resides on the layer above IP. When a session is to be set up between two applications, the application data is first passed to TCP where a TCP header is applied, the data is then passed to IP where an IP header is applied, and then it is forwarded through the network. The information contained in the TCP header includes, among other things, *source and destination port numbers*, which identify the applications at each end; *sequence numbers* and *acknowledgement numbers*, which enable the detection of lost packets; and a *checksum*, which enables the detection of corrupted packets. TCP uses these information elements to request retransmission of lost or corrupted packets and to deliver packets to the destination application in the correct order. In order to do all of this, TCP first establishes a connection between peer TCP instances at each end. This involves a sequence of messages between the TCP instances prior to the transfer of user data.

Instead of using TCP at layer 4 in the stack, the *User Datagram Protocol* (UDP) is another option. This is a simple protocol, which does little more than enable the identification of the source and destination applications. It does not support recovery from loss or error and does not ensure an in-sequence delivery of packets. It is meant for simple request-response types of transactions, rather than the sequential transfer of multiple packets. An application that would use UDP rather than TCP, for example, is the *Domain Name Service* (DNS), a classic one-shot request-response protocol.

8.4 VoIP Challenges

Good speech quality is a strong requirement of any commercial network, wireless or otherwise. Traditionally, this has been achieved through 64-Kbps (G.711) voice coding and the use of circuit-switching, which establishes a dedicated transmission path from the source to the destination. Nowadays, more advanced speech coding schemes can approach the quality of G.711 with a much lower bandwidth requirement. The GSM Enhanced Full-Rate Coder is one such advanced coding scheme and many others are available. Good speech coding is not the only requirement, however. Other requirements include low-transmission delay, low jitter (delay variation), and the requirement that everything transmitted at one end is received at the other (low loss).

These requirements are somewhat contradictory when viewed from an IP perspective. For example, the requirement for low loss could be achieved through the use of TCP at layer 4. That, however, would cause excessive delay, both at the start of the transfer when a TCP connection needs to be established and during the transfer when acknowledgements and retransmissions would cause a delay in the delivery of the voice packets. In order to minimize delay, one could use UDP at layer 4. UDP, however, offers no protection against packet loss.

Given the choice between UDP or TCP, the issue is whether we consider minimizing delay to be more important than eliminating packet loss. The answer is that, for speech, excessive delay and excessive jitter are far more disturbing than occasional packet loss. Obviously, excessive packet loss is unacceptable, but a limited amount (less than 5 percent) can be tolerated without noticeable speech quality degradation. Consequently, when transporting voice, UDP is chosen at layer 4, rather than TCP.

It is clear, however, that something more than UDP is required if VoIP is to offer reasonable voice quality. At a minimum, the destination application needs to know the coding scheme being used by the source application so that the voice packets can be decoded. The application also needs timing information so that packets can be played out to the user in a synchronized manner and help mitigate against delay in the network. Moreover, the application needs to know when packets are lost, so that a previous packet could be replayed to fill the gap if appropriate.

In order to fulfill these needs, a protocol known as the *Real-Time Transport Protocol* (RTP) has been developed. This protocol resides above UDP in the protocol stack. Whenever a packet of coded voice is to be sent, it is sent as the payload of an RTP packet. That packet contains an RTP header, which provides information such as the voice coding scheme being used, a sequence number, a timestamp for the instant at which the voice packet was sampled, and an identification for the source of the voice packet.

RTP has a companion protocol, the *RTP Control Protocol* (RTCP). RTCP does not carry coded voice packets. Rather, RTCP is a signaling protocol that includes a number of messages, which are exchanged between session users. These messages provide feedback regarding the quality of the session. The type of information includes such details as lost RTP packets, delay, and inter-arrival jitter.

Whenever an RTP session is opened, an RTCP session is also implicitly opened. This means that, when a UDP port number is assigned to an RTP session for the transfer of voice packets (or any other media packets, such as video), a separate port number is assigned for RTCP messages. An RTP port number will always be even, and the corresponding RTCP port number will be the next highest number, and hence odd. Thus, if we again consider the IP protocol stack for voice transport, it appears as shown in Figure 8-2.

It should be noted that RTP and RTCP do not guarantee minimal delays, low jitter, or low packet loss. In order to do that, other protocols are required. RTP and RTCP simply provide information to the applications at either end so that those applications can deal with loss, delay, or jitter with the least possible impact to the user.

8.5 H.323

In all telephony networks, specific signaling protocols are invoked before and during a call to communicate a desire to set up a call, to monitor call progress, and to gracefully bring a call to a conclusion. Perhaps the best example is the *ISDN User Part* (ISUP), a component of the *Signaling System 7* (SS7) signaling suite. In VoIP systems, signaling protocols also need to be used for exactly the same reasons. The first successful set of protocols for VoIP was developed by the *International Telecommunications Union* (ITU). This set is known as H.323 and has the title, "Packet-based Multi-

Figure 8-2
VoIP Protocol Layers

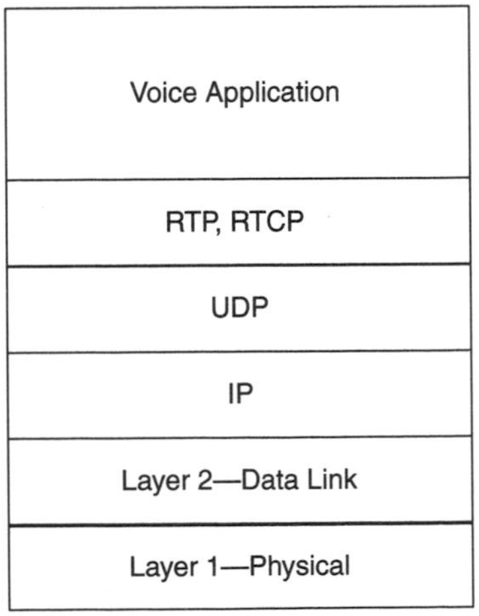

Voice Application
RTP, RTCP
UDP
IP
Layer 2—Data Link
Layer 1—Physical

media Communications Systems." Although recently new protocols have emerged, notably the *Session Initiation Protocol* (SIP), H.323 is still the most widely deployed VoIP signaling system.

8.5.1 H.323 Network Architecture

As is the case for most signaling systems, H.323 defines a specific network architecture, which is depicted in Figure 8-3. This architecture involves H.323 terminals, gateways, gatekeepers, and *multipoint controller units* (MCUs). The overall objective of H.323 is to enable the exchange of media streams between H.323 endpoints, where an H.323 endpoint is an H.323 terminal, a gateway, or an MCU.

An H.323 *terminal* is an endpoint that offers real-time communications with other H.323 endpoints. It is typically an end-user communications device. It supports at least one audio codec and may optionally support other audio codecs and/or video codecs.

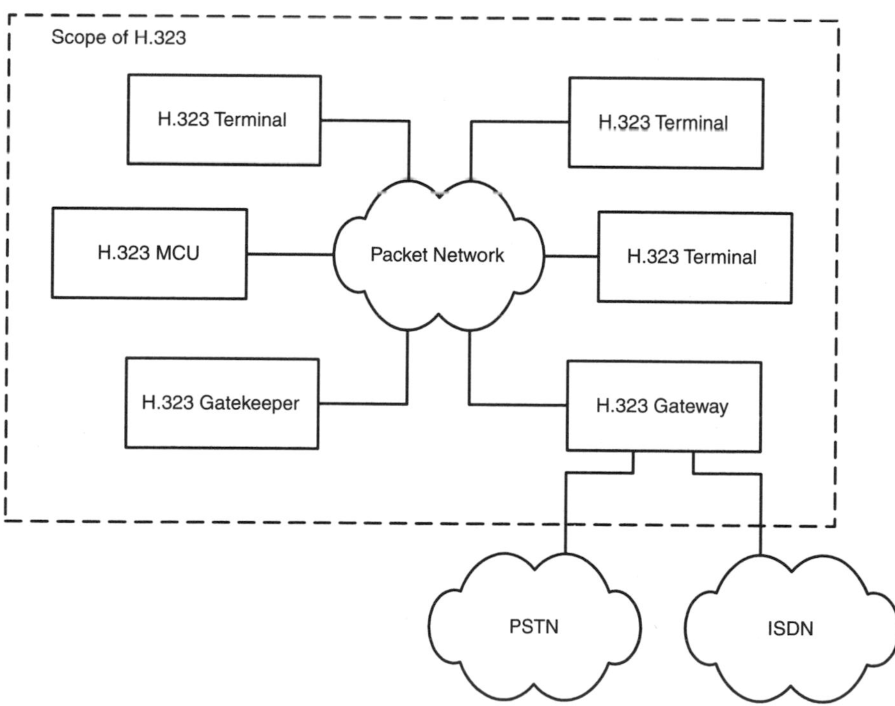

Figure 8-3 H.323 Network Architecture.

A *gateway* is an H.323 endpoint that provides translation services between the H.323 network and another type of network, such as an *Integrated Services Digital Network* (ISDN) or the *Public Switched Telephone Network* (PSTN). One side of the gateway supports H.323 signaling and terminates packet media according to the requirements of H.323. The other side of the gateway interfaces to a circuit-switched network and supports the transmission characteristics and signaling protocols of the circuit-switched network. On the H.323 side, the gateway has the characteristics of an H.323 terminal. On the circuit-switched side, it has the characteristics of a node in the circuit-switched network. A translation between the signaling protocols and media formats of one side and those of the other side is performed internally within the gateway. The translation is totally transparent to other nodes in the circuit-switched network and in the H.323 network. Gateways may also serve as a conduit for communications between H.323 terminals that are not on the same network, where the communication between the terminals needs to pass via an external network such as the PSTN.

A *gatekeeper* is an optional entity within an H.323 network. When present, it controls a number of H.323 terminals, gateways, and *multipoint controllers* (MCs). By control, we mean that it authorizes network access from one or more endpoints and may choose to permit or deny any given call from an endpoint within its control. It may offer bandwidth control services, which, if used in conjunction with bandwidth and/or resource management techniques, can help to ensure service quality. A gatekeeper also offers address translation services, enabling the use of aliases within the network. The set of terminals, gateways, and MCs controlled by a single gatekeeper is known as a *zone*. A zone can span multiple networks or subnetworks and it is not necessary that all entities within a zone be contiguous.

An MC is an H.323 endpoint that manages multipoint conferences between three or more terminals and/or gateways. For such conferences, it establishes the media that may be shared between entities by transmitting a capability set to the various participants, and an MC may change the capability set in the event that other endpoints join or leave the conference. An MC may reside within a separate MCU or may be incorporated within the same platform as a gateway, a gatekeeper, or an H.323 terminal.

Overview of H.323 Protocols

Figure 8-4 shows the H.323 protocol stack. Upon examination, we find a number of protocols already discussed, such as RTP, TCP, and UDP. It is clear from the figure that the exchange of media is performed using RTP over UDP and, of course, wherever there is RTP, there is also RTCP.

In Figure 8-4, we also find two protocols that have not yet been discussed —namely H.225.0 and H.245. These two protocols define the actual messages that are exchanged between H.323 endpoints. They are generic protocols in that they could be used in any number of network architectures. When it comes to the H.323 network architecture, the manner in which the H.225.0 and H.245 protocols are applied is specified by recommendation H.323.

H.225.0 is a two-part protocol. One part is effectively a variant of ITU-T recommendation Q.931, the ISDN layer 3 specification, and should be quite familiar to those with knowledge of ISDN. It is used for the establishment and tear-down of connections between H.323 endpoints. This type of signaling is known as call signaling or Q.931 signaling. The other part of H.225.0 is known as *Registration, Admission, and Status* (RAS) signaling. It is used between endpoints and gatekeepers and enables a gatekeeper to manage the endpoints within its zone. For example, RAS signaling is used by an endpoint to register with a gatekeeper and it is used by a gatekeeper to allow or deny endpoint access to network resources.

H.245 is a control protocol used between two or more endpoints. The main purpose of H.245 is to manage the media streams between H.323 session participants. To that end, it includes functions such as ensuring that the media to be sent by one entity is limited to the set of media that can be

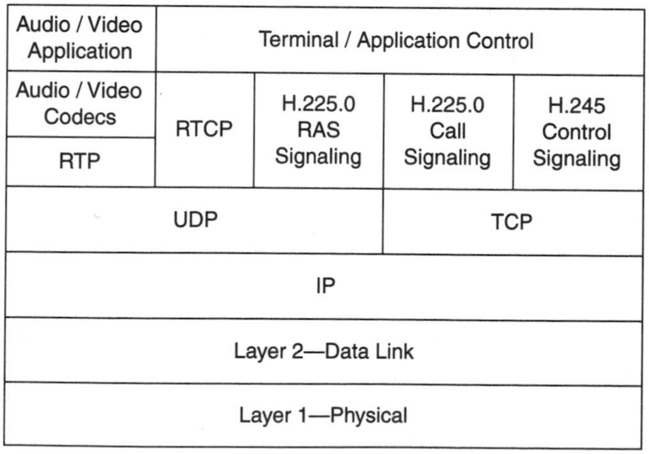

Figure 8-4
H.323 Protocol Layers.

Audio / Video Application	Terminal / Application Control			
Audio / Video Codecs	RTCP	H.225.0 RAS Signaling	H.225.0 Call Signaling	H.245 Control Signaling
RTP				
UDP			TCP	
IP				
Layer 2—Data Link				
Layer 1—Physical				

received and understood by another. H.245 operates by the establishment of one or more logical channels between endpoints. These logical channels carry the media streams between the participants and have a number of properties such as media type, bit rate, and so on.

All three signaling protocols—RAS, Q.931, and H.245—may be used in the establishment, maintenance, and tear-down of a call. The various messages may be interleaved. For example, consider an endpoint that wants to establish a call to another endpoint. Firstly, it may use RAS signaling to obtain permission from a gatekeeper. It may then use Q.931 signaling to establish communication with the other endpoint and set up the call. Finally, it may use H.245 control signaling to negotiate media parameters with the other endpoint and set up the media transfer. Figure 8-5 shows an example of the interaction between the different types of signaling.

8.5.3 H.323 Call Establishment

In theexample of Figure 8-5, two terminals (H.323 endpoints) need to establish a VoIP call between them, and different gatekeepers control the two terminals. As a first step, the calling terminal requests permission from its gatekeeper to establish the call. This is done with the *Admission Request* (ARQ) message. The terminal indicates the type of call in question (two-party or multi-party), the endpoint's own identifier, a call identifier (a unique string), a call reference value (an integer value also used in call signaling messages for the same call), and information regarding the other party or parties to participate in the call. The information regarding other parties to the call includes one or more aliases and/or signaling addresses. One of the most important mandatory parameters in the ARQ is the bandwidth parameter. This specifies the amount of bandwidth required in units of 100 bps.

Note that the endpoint should request the total media stream bandwidth needed, excluding overhead. Thus, if a two-party call is needed, with each party sending voice at 64 Kbps, then the bandwidth required is 128 Kbps, and the value carried in the bandwidth parameter is 1280. The purpose of the bandwidth parameter is to enable the gatekeeper to reserve resources for the call.

The gatekeeper indicates a successful admission by responding to the endpoint with an *AdmissionConfirm* (ACF) message. This includes many of the same parameters that are included in the ARQ. The difference is that, when a given parameter is used in the ARQ, it is simply a request from the

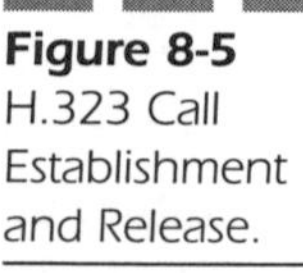

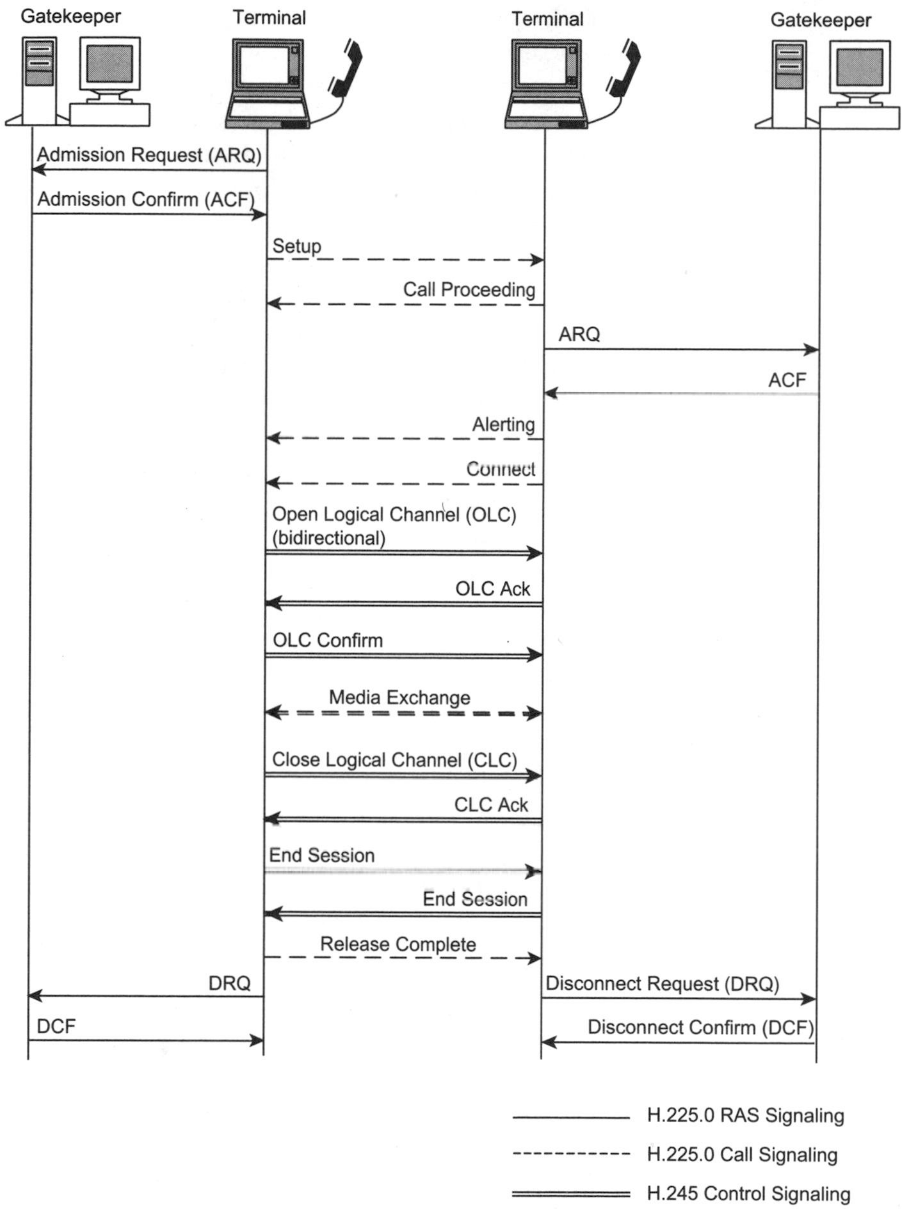

Figure 8-5 H.323 Call Establishment and Release.

endpoint, whereas a given parameter value in the ACF is a firm order from the gatekeeper. For example, the ACF includes the bandwidth parameter, which may be a lower value than that requested in the ARQ, in which case the endpoint must stay within the bandwidth limitations imposed by the gatekeeper.

Another parameter of particular interest in both the ARQ and the ACF is the *callModel parameter*, which is optional in the ARQ and mandatory in the ACF. In the ARQ, callModel indicates whether the endpoint wants to send call signaling directly to the other party, or prefers that call signaling be passed via the gatekeeper. In the ACF, it represents the gatekeeper's decision as to whether call signaling is to pass via the gatekeeper or directly between the terminals. In the example of Figure 8-5, the calling gatekeeper has chosen not to be in the path of the call signaling.

The *Setup message* is the first call-signaling message sent from one terminal to the other to establish the call. The message must contain the Q.931 Protocol Discriminator, a Call ReferenceSetup, a Bearer Capability, and the User-User information element. Although the Bearer Capability information element is mandatory, the concept of a bearer, as used in the circuit-switched world, does not map very well to an IP network. For example, no B-channel exists in IP and the actual agreement between endpoints regarding the bandwidth requirements is done as part of H.245 signaling, where RTP information such as the payload type is exchanged. Consequently, many of the fields in the Bearer Capability information element, as defined in Q.931, are not used in H.225.0. Of those fields that are used in H.225.0, many are used only when the call has originated from outside the H.323 network and has been received at a gateway, where the gateway performs a mapping from the signaling received to the appropriate H.225.0 messages.

A number of parameters are included within the mandatory User-to-User information element. These include the call identifier, the call type, a conference identifier, and information about the originating endpoint. Among the optional parameters, we may find a source alias, a destination alias, an H.245 address for subsequent H.245 messages, and a destination call-signaling address. The User-to-User information element is included in all H.225.0 call-signaling messages. It is the inclusion of this information element that enables Q.931 messages, originally designed for ISDN, to be adapted for use with H.323.

The *Call Proceeding message* may optionally be sent by the recipient of a Setup message to indicate that the Setup message has been received and that call establishment procedures are underway. When sent, it usually precedes the *Alerting message*, which indicates that the called device is "ringing." Strictly speaking, the Alerting message is optional.

In addition to Call Proceeding and Alert, we may also find the optional Progress message (not shown). Ultimately, when the called party answers, the called terminal returns a Connect message. Although some of the messages from the called party to the calling party, such as Call Proceeding and Alerting, are optional, the Connect message must be sent if the call is to be completed. The User-User information element contains the same set of parameters as defined for the Call Proceeding, Progress, and Alert messages, with the addition of the Conference Identifier. These parameters are also used in a Setup message and their use in the Connect message is to correlate this conference with that indicated in a Setup. Any H.245 address sent in a Connect message should match that sent in any earlier Call Proceeding, Alerting, or Progress messages. In fact, the called terminal must include at least an H.245 signaling address to which H.245 messages must be sent because H.245 messages are used to establish the media (that is, voice) flow between the parties.

In the example of Figure 8-5, the H.245 message exchange begins after the Connect message is returned. This message exchange could, in fact, occur earlier than the Connect message. It is important to note that H.245 is not responsible for carrying the actual media. For example, there is no such thing as an H.245 packet containing a sample of coded voice. That is the job of RTP. Instead, H.245 is a control protocol that manages the establishment and release of media sessions. H.245 does this through messaging that enables the establishment of logical channels, where a logical channel is a unidirectional RTP stream from one party to the other.

A logical channel is opened by sending an *Open Logical Channel* (OLC) request message. This message contains a mandatory parameter called forwardLogicalChannelParameters, which relates to the media to be sent in the forward direction, that is, from the endpoint issuing this command. It contains information such as the type of data to be sent (e.g. AMR-coded audio), an RTP session ID, an RTP payload type, and an indication as to whether silence suppression is to be used. If the recipient of the message wants to accept the media to be sent, then it will return an OpenLogicalChannelAck message containing the same logical channel number as received in the request and a transport address to which the media stream should be sent.

Strictly speaking, a logical channel is unidirectional. Therefore, in order to establish a two-way conversation, two logical channels must be opened—one in each direction. According to the description just presented, this requires four messages, which is rather cumbersome. Consequently, H.323 defines a bidirectional logical channel. This is a means of establishing two logical channels, one in each direction, in a slightly more efficient manner. Basically, a bidirectional logical channel really means two logical channels that are associated with each other. The establishment of these two channels can be achieved with just three H.245 messages rather than four. In order to do so, the initial OLC message not only contains information regarding the media that the calling endpoint wants to send, but it also contains reverse logical channel parameters. These indicate the type of media that the endpoint is willing to receive and to where that media should be sent.

Upon receipt of the request, the far endpoint may send an Open Logical Channel Ack message containing the same logical channel number for the forward logical channel, a logical channel number for the reverse logical channel, and descriptions related to the media formats that it is willing to send. These media formats should be chosen from the options originally received in the request, thereby ensuring that the called end will only send media that the calling end supports.

Upon receipt of the Open Logical Channel Ack, the originating endpoint responds with an Open Logical Channel Confirm message to indicate that all is well. RTP streams and RTCP messages can now flow in each direction.

8.5.4 H.323 Call Release

Figure 8-5 also shows the disconnection of a call after media have been exchanged (a conversation has taken place). The first step in the process involves closing the logical channels that have been created by H.245 signaling—closing the RTP streams between the users.

Closing a logical channel involves the sending of a CloseLogicalChannel message. In the case of a successful closure, the far end should send the response message CloseLogicalChannelAck. In general, a logical channel can be closed only by the entity that created it in the first place. For example, in the case of a unidirectional channel, only the sending entity can close the channel. However, the receiving endpoint in a unidirectional channel can humbly request the sending endpoint to close the channel. It does so by sending the RequestChannelClose message, indicating the channel that the endpoint would like to have closed. If the sending entity is willing to grant the request, then it responds with a positive acknowledgment and then proceeds to close the channel. When an entity closes the forward logical channel of a bidirectional logical channel, then it also closes the reverse logical channel.

Once all logical channels in a session are closed, then the session itself is terminated when an endpoint sends an EndSession command message. The receiving endpoint responds with an EndSession command message. Once an entity has sent this message, it must not send any more H.245 messages related to the session.

At this point, the call signaling comes to a close with the issuance of a Release Complete message. Unlike standard Q.931 ISDN signaling, no Release message is sent—just the Release Complete message, which is all that is needed to end call signaling.

Finally, each endpoint uses the *Disconnect Request* (DRQ) message to request permission from its gatekeeper to disconnect. The gatekeeper responds with the *Disconnect Confirm* (DCF) message.

8.5.5 The H.323 Fast Connect Procedure

It is clear from Figure 8-5 that H.323 call establishment and release can be quite cumbersome. Moreover, it is possible for a given gatekeeper to choose to be in the path of all call signaling and H.245 signaling in addition to RAS signaling. In such a scenario, the number of messages exchanged can be very large, which can extend the call setup time beyond acceptable limits. In order to speed things up, H.323 includes a procedure known as *Fast Connect*, a method that can significantly reduce the amount of call establishment signaling. The Fast Connect Procedure is depicted in Figure 8-6.

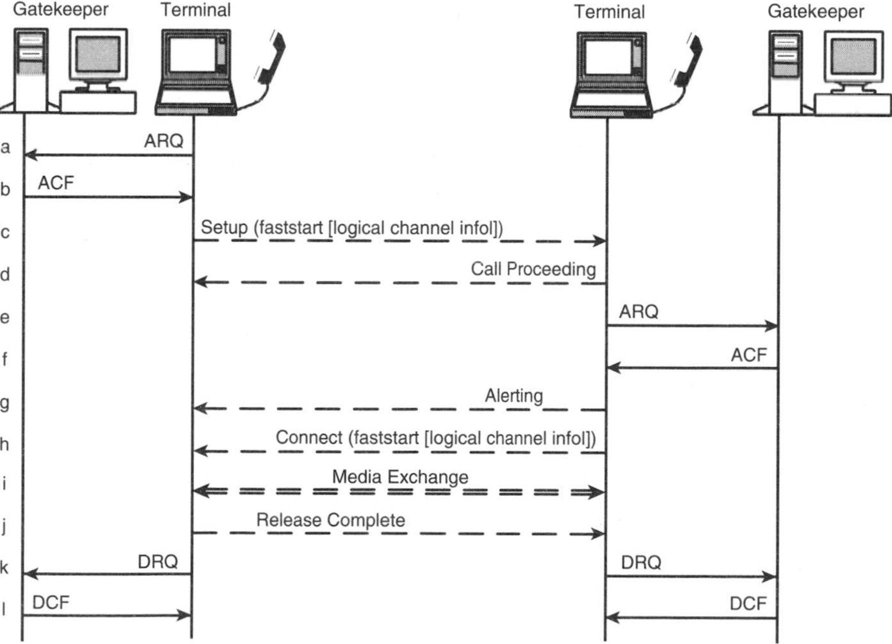

Figure 8-6 H.323 Fast Connect Procedure.

The Fast Connect procedure involves the setting up of media streams as quickly as possible. To achieve this, the Setup message can contain a faststart element within the User-User information element. The faststart element is actually one or more Open Logical Channel request messages containing all the information that would normally be contained in such requests. It includes reverse logical channel parameters if the calling endpoint expects to receive media from the called endpoint.

If the called endpoint also supports the procedure, then it can return a faststart element in one of the Call Proceeding, Alerting, Progress, or Connect messages. That faststart element is basically another OLC message, which appears like a request to open a bidirectional logical channel. The included choices of media formats to send and receive are chosen from those offered in the faststart element of the incoming Setup message. The calling endpoint has effectively offered the called endpoint a number of choices for forward and reverse logical channels, and the called endpoint has indicated those choices that it prefers. The logical channels are now considered open as if they had been opened according to the procedures of H.245.

Note that a faststart element from the called party to the calling party may be sent in any message up to and including the Connect message. If it has not been included in any of the messages, then the calling endpoint shall assume that the called endpoint either cannot or does not want to support faststart. In such a case, the standard H.245 methods must be used.

The use of the Fast Connect procedure means that H.245 information is carried within the call signaling messages and no separate H.245 control channel exists. Therefore, bringing a call to a conclusion is also faster. The call is released simply by the sending of the call-signaling Release Complete message. When used with the fast connect procedure, this has the effect of closing all of the logical channels associated with the call and is equivalent to using the procedures of H.245 to close the logical channels.

8.6 The Session Initiation Protocol (SIP)

The Session Initiation Protocol (SIP) is considered by many to be a powerful alternative to H.323. It is considered to be a more flexible solution, simpler than H.323, easier to implement, better suited to the support of intelligent user devices, and better suited to the implementation of advanced features. Although H.323 may still have a larger installed base than SIP, most people in the VoIP community believe that the future of VoIP revolves around SIP. In fact, 3GPP has endorsed SIP as the session management protocol of choice for 3GPP Release 5, albeit with some enhancements.

Like H.323, SIP is simply a signaling protocol and does not carry the voice packets itself. Rather, it makes use of the services of RTP for the transport of the voice packets (the media stream).

8.6.1 The SIP Network Architecture

SIP defines two basic classes of network entities—clients and servers. Strictly speaking, a *client*, also known as a user agent client, is an application program that sends SIP requests. A *server* is an entity that responds to those requests. Thus, SIP is a client-server protocol. VoIP calls using SIP are originated by a client and terminated at a server. A client may be found within a user's device, which could be, for example, a SIP phone. Clients may also be found within the same platform as a server. For example, SIP enables the use of proxies, which act as both clients and servers.

Four different types of servers are available—*proxy servers*, *redirect servers*, *user agent servers*, and *registrars*. A proxy server acts similarly to a proxy server used for Web access from a corporate local area network (LAN). Clients send requests to the proxy, which either handles those requests itself or forwards them on to other servers, perhaps after performing some translation. To those other servers, it appears as though the message is coming from the proxy rather than some entity hidden behind it. Given that a proxy both receives requests and sends requests, it incorporates both server and client functionality. Figure 8-7 shows an example of the operation of a proxy server. It does not take much imagination to realize how this type of functionality can be used for call forwarding/follow-me services.

A redirect server is a server that accepts SIP requests, maps the destination address to zero or more new addresses, and returns the translated address to the originator of the request. Thereafter, the originator of the request may send requests to the address(es) returned by the redirect server. A redirect server does not initiate any SIP requests of its own.

Figure 8-8 shows an example of the operation of a redirect server. This can be another means of providing the call forwarding/follow-me service that can be provided by a proxy server. This difference is that, in the case of

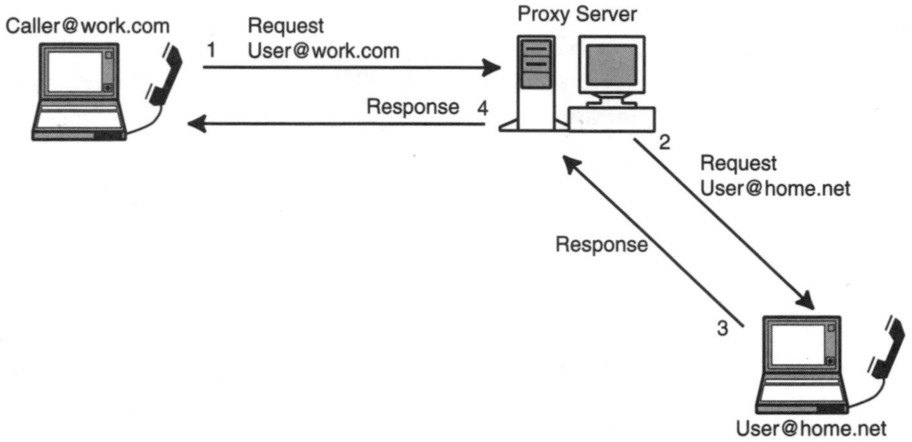

Figure 8-7
SIP Proxy Server.

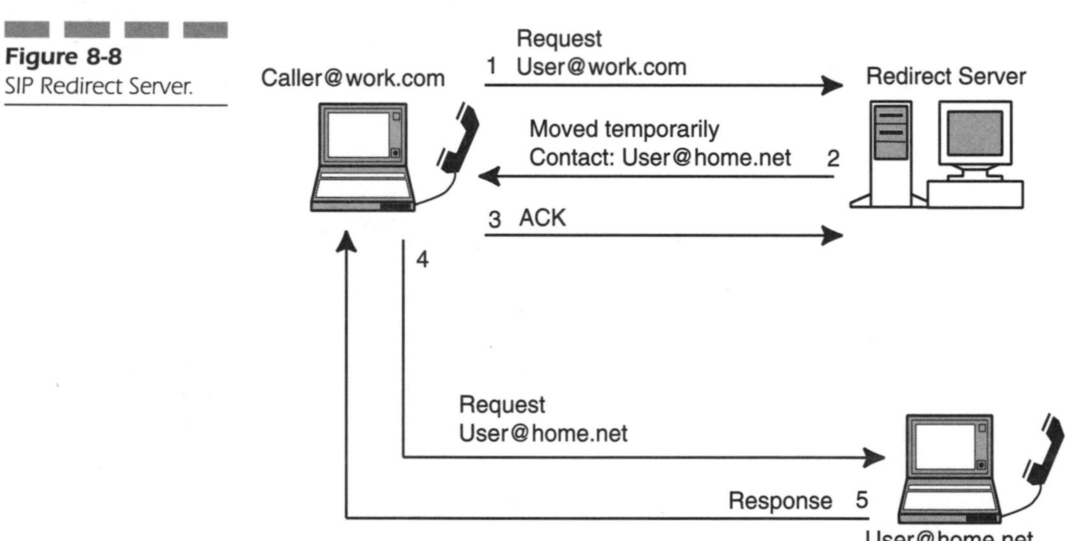

Figure 8-8
SIP Redirect Server.

a redirect sever, the originating client does the actual forwarding of the call. The redirect server simply provides the information necessary to enable the originating client to do so, after which the redirect server is no longer involved.

A user-agent server accepts SIP requests and contacts the user. A response from the user to the user-agent server results in a SIP response on behalf of the user. In reality, a SIP device, such as a SIP-enabled phone, will function as both a user-agent client and a user-agent server. Acting as a user-agent client, it is able to initiate SIP requests. Acting as a user-agent server, it can receive and respond to SIP requests. In practical terms, this means that it is able to initiate calls and receive calls. This enables SIP, a client-server protocol, to be used for peer-to-peer communication.

A registrar is a server that accepts SIP REGISTER requests. SIP includes the concept of user registration, whereby a user signals to the network that it is available at a particular address. Such registration is performed by the issuance of a REGISTER request from the user to the registrar. Typically, a registrar will be combined with a proxy or redirect server. Registration in SIP serves a similar purpose to location updating in a GSM network; it is a means by which a user can signal to the network that he or she is available at a particular location.

Given that practical implementations involve the combination of a user-agent client and a user-agent server and the combining of registrars with either proxy servers or redirection servers, a real network may well involve only user agents and the redirection or proxy servers.

8.6.2 SIP Call Establishment

At a high level, SIP call establishment is very simple, as shown in Figure 8-9. The process starts with a SIP INVITE message, which is used from the calling party to the called party. The message invites the called party to participate in a session—a call. Included with the INVITE message is a session description— a description of the media that the calling party wants to use. This description includes the voice-coding scheme that the caller wants to use, plus an IP address and a port number that the called party should use for sending media back to the caller.

A number of interim responses to the INVITE may be sent, prior to the called party accepting the call. For example, the caller might be informed that the call is queued and/or that the called party is being alerted; that is, the phone is ringing. Subsequently, the called party answers the calls, which generates an OK response back to the caller. The OK response is actually indicated by the status code value of 200 in the response. In the example of Figure 8-9, the 200 (OK) response contains a session description, indicating the media that the caller wants to use plus an IP address and port number to which the caller should send packets.

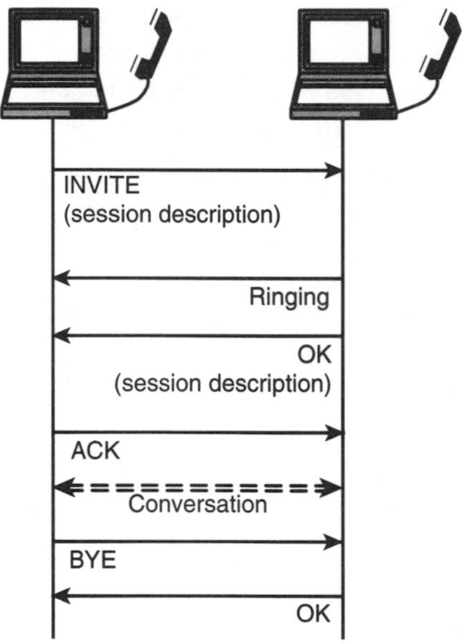

Figure 8-9 SIP Basic Call Establishment and Release,

Upon receipt of the 200 (OK) response, the caller responds with ACK to confirm that the OK response has been received. At this point, media are exchanged. These media will most often be coded speech, but could also be other media such as video. Finally, one of the parties hangs up, which causes a BYE message to be sent. The party receiving the BYE message sends 200 (OK) to confirm receipt of the message. At that point, the call is over.

All in all, SIP call establishment is quite a simple process. Of course, the signaling could well pass via one or more proxy servers, in which case the process becomes somewhat more complex. Nonetheless, it is clear that SIP call establishment is much simpler than the equivalent H.323 process.

8.6.3 Information in SIP Messages

Obviously, there is more to SIP signaling than the messages outlined in Figure 8-9. To start with, each SIP request or response contains addresses for the calling and called parties. Each such address is known as a SIP *uniform resource locator* (URL) and has the format "SIP:user@domain." This is somewhat similar to an e-mail URL, which has the format mailto:user@domain. A SIP user might well want to have the same values for user and domain in his or her SIP and e-mail addresses, which would make it very easy to know how to contact a SIP user—much easier than having to remember a telephone number.

Several requests and many responses can be sent between SIP entities. For example, if, in the example of Figure 8-9, the called user were not available, then the response "Temporarily Unavailable" (status code 480) could have been returned, rather than the 200 (OK).

Not only are there several requests and many responses, many information elements can be contained in those requests and responses. In SIP, these information elements are known as header fields. For example, when sending an INVITE, the message contains not only a session description and the to and from addresses (contained in the To and From header fields), but it can also contain a Subject header field. This field indicates the reason for the call and can be presented to the called user, who may choose to accept or reject the call based on the subject in question. One can easily imagine this capability being used to filter out unwanted telemarketing calls.

Other header fields include, for example, Call ID, Date, Timestamp, In-reply-to, Retry-after, and Priority. The Retry-after header could be used, for example, with the 480 (Temporarily unavailable) response to indicate when the caller should try the call again (if ever). One of the most important header fields is Content-type, which indicates the type of additional information included in the message. For example, when a user issues an INVITE message, the message includes a session description. The Content-type field indicates how that session description is coded so that the receiver of the message can understand whether or not that type of session can be supported.

8.6.4 The Session Description Protocol (SDP)

Clearly, SIP is used to establish sessions between users, which requires that the users agree on the type and coding of the information to be shared. For example, the two users must agree on the voice-coding scheme to be used, which requires that they share session descriptions. These session descriptions are coded according to the *Session Description Protocol* (SDP).

SDP is simply a language for describing sessions. It contains information regarding the parties to be involved in the session, the date and time when the session is to take place, the types of media streams to be shared, and addresses and port numbers to be used. It is perfectly possible that a session description could refer to multiple media streams, such as in a video conference where one media stream relates to coded voice and another media stream relates to coded video. Consequently, SDP is structured so that it can describe information related to the session as a whole (e.g. the name of the session), plus information associated with each individual stream (e.g. the media format and the applicable port number). Some of the information included in an SDP session description will also be included in the SIP message that carries the SDP description. This overlap is due to the fact that SDP is designed to be used by a range of other protocols, not just SIP.

Perhaps the best way to describe the combined usage of SIP and SDP is by example. Consider Figure 8-10, which is a more detailed version of the call establishment scenario presented in Figure 8-9. In this case, we see a call from User1@work.com, who is logged in at station1.work.com, to User2@work.com, who is logged in at station2.work.com.

As with any SIP session establishment, the call begins with an INVITE, which is indicated in the first line of the request. The first line also indicates the address of the entity to which the message is being sent, known as the request *uniform resource indicator* (URI). In this case, the message is being sent directly to User2. If, however, there happens to be a proxy server between User1 and User2, then the request would first go to the proxy, in which case the request URI would indicate the proxy.

The Via header field is inserted by each entity in the chain from the source of a message to the destination. This is to ensure that the response can follow the same path back through the network, as was taken by the original request. The From and To header fields indicate the initiator of the request and the recipient of the request. The Call ID is a globally unique identification. To ensure uniqueness, the Call ID should take the form indicated in the figure. The CSeq field refers to the command sequence. The CSeq contains an integer and an indication of the type of request. The purpose of the CSeq header is to enable the initiator of a request to correlate a response with the request that generated the response.

Finally, we have two headers that provide information about the message body. The first, Content-Length, indicates the length of the message body. The second, Content-Type, indicates the type of message body. Strictly speaking, the message body could be any *Multipurpose Internet Mail Extension* (MIME coded) type, such as text. In our example, the message body contains a session description code according to SDP.

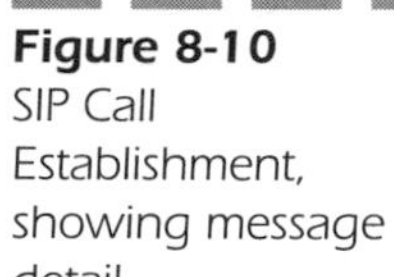

User1@work.com
User2@work.com
INVITE sip:User2@work.com SIP/2.0
Via: SIP/2.0/UDP station1.work.com
From: sip:User1@work.com
To: sip:User2@work.com
Call-ID: 123456@station1.work.com
CSeq: 1 INVITE
Content-Length: 168
Content-Type: application/sdp

v=0
o=User1 123456 001 IN IP4 station1.work.com
s=vacation
c=IN IP4 station1.work.com
t=0 0
m=audio 4444 RTP/AVP 98
a=rtpmap 98 AMR/8000

SIP/2.0 180 Ringing
Via: SIP/2.0/UDP station1.work.com
From: sip:User1@work.com
To: sip:User2@work.com
Call-ID: 123456@station1.work.com
CSeq: 1 INVITE
Content-Length: 0

SIP/2.0 200 OK
Via: SIP/2.0/UDP station1.work.com
From: sip:User1@work.com
To: sip:User2@work.com
Call-ID: 123456@station1.work.com
CSeq: 1 INVITE
Content-Length: 167
Content-Type: application/sdp

v=0
o=user2 45678 001 IN IP4 station2.work.com
s=vacation
c = IN IP4 station2.work.com
t=0 0
m=audio 6666 RTP/AVP 98
a=rtpmap 98 AMR/8000

ACK sip:User2@work.com SIP/2.0
Via: SIP/2.0/UDP station1.work.com
From: sip:User1@work.com
To: sip:User2@work.com
Call-ID: 123456@station1.work.com
CSeq: 1 ACK
Content-Length: 0

Conversation

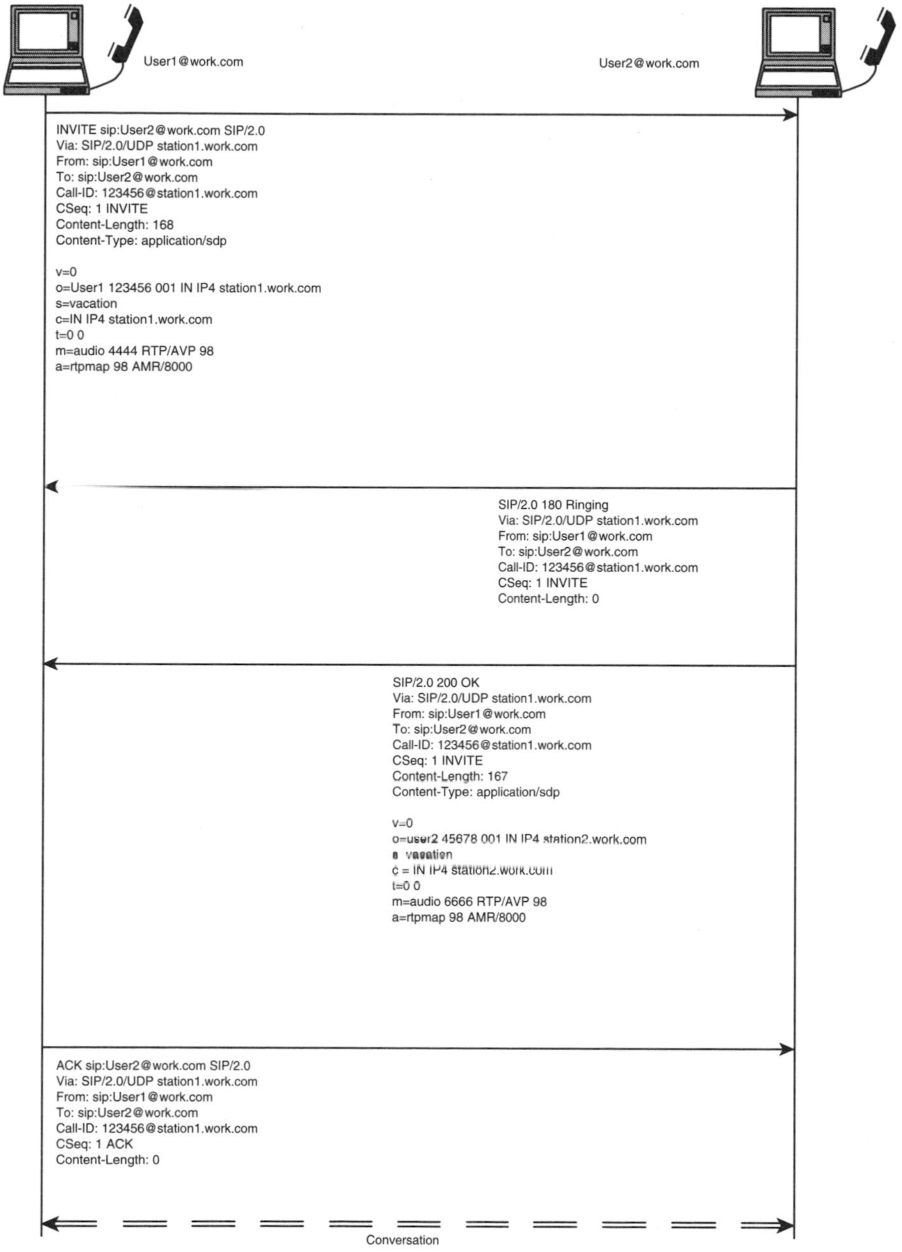

Figure 8-10 SIP Call Establishment, showing message detail.

The message body is separated from the SIP headers by a blank line. In SDP, it starts with a version identifier, which is version 0. Next, we find the *Origin* (o) field, which indicates the user name (User1), a session ID (123456 in our case) that does not have to match the SIP Call ID, a version for the session (001 in our example), the type of network (IN indicates Internet), the type of addressing used (IP4 indicates IP version 4), and an address for the machine that initiated the session (station1.work.com in our example).

After the Origin field, we find the optional *Subject* (s) field and the *Connection* (c) field. The connection field provides information regarding where the user would like the media to be sent. In our case, it indicates that the type of network is *Internet* (IN), that the addressing uses *IP version 4* (IP4), and the address to which media should be sent. This address could be different from the address of the machine that created the session.

After the Connection field, we find the *Time* (t) field, which indicates the start and stop times for the session. In our example, these are both set to 0, which means that the session does not have any set start or stop time.

Next, we find the *Media* (m) field, which provides information about the media to be used and the port to which the media should be sent. In our example, the type of media is audio, and it should be received at port number 4444 (i.e. the far end should sent the media to port number 4444). The media field also indicates the type of RTP *audio/video profile* (AVP) to be used, which is 98 in our example.

In RTP, certain types of media stream codings are assigned specific values of payload types and are known as static payload types. For example, payload type 0 indicates G.711 coded voice. Thus, if the media field of the SDP description indicated RTP/AVP value 0, then the far end would know that G.711 coded voice is required. RTP also includes the concept of a dynamic payload type, where the payload type value is significant only within one session. Therefore, it is necessary to indicate additional attributes in order for the far end to understand the meaning of the payload type chosen. In our example, the attribute "a=rtpmap AMR/8000" indicates that the payload type is adaptive multirate and sampled at 8000 Hz.

Figure 8-10 shows that the first response includes status code 180, indicating that the user is being alerted. It contains the same Via, To, From, Cseq, and Call ID header fields as the original request and they enable the sender of the request to match the response with the request. This response does not contain any session description.

When the called user answers, a 200 (OK) response is generated. The SIP header fields are the same as the original INVITE request, with the exception of the content-length and the message body itself. This is because the called device has included a session description of its own. This is quite similar to the session description in the INVITE request, but indicates a different address and port number. This makes sense, as the address and port number indicate where User2 expects to receive the media stream.

Once User1 has received the 200 (OK) response, it sends an ACK message. The header fields in this message are identical to those of the original INVITE, with the exception of the CSeq field, which now indicates the ACK request. At this point, media can flow between the two parties and a conversation can take place.

8.7 Distributed Architecture and Media Gateway Control

The foregoing discussions regarding SIP and H.323 have focused primarily on the signaling needed to establish media streams between session participants. Although not clearly stated, it is implied that the entities generating the signaling are the same entities that will generate the actual media streams. In other words, we have not described a clear separation of media from call control.

If one looks carefully at the description of SDP, however, one sees that it is possible to indicate different addresses for the entity that sends a session description and the entity that actually terminates the media stream. This indicates that the separation of media from call control is possible. Moreover, we have seen from the architectures of 3GPP Release 4 and 3GPP Release 5 that the separation of media and call control is not only possible, but is often desirable.

If we physically separate a call control entity from an entity that handles media streams (such as a gateway that performs voice coding), then we need a protocol between those two types of entities so that the call control entity can manage the media entity for the setup and tear-down of calls. Provided we have such a protocol, then there is no reason why one call control device could not manage multiple media-handling devices. It would simply be a question of the processing power of the call control device. In such a scenario, we can envisage an architecture such as that shown in Figure 8-11, where a single controller manages multiple media-handling devices such as *media gateways* (MGs). In some quarters, this separation between call control and media is known as the softswitch architecture.

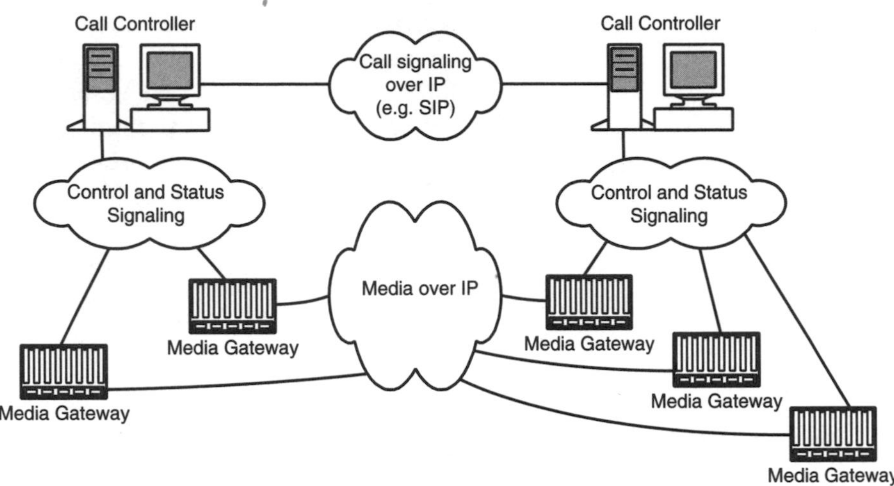

Figure 8-11 Separation off Bearer and Call Control.

The advantages of such an approach are that MGs can be placed as close as possible to the source or sink of the media stream, which can be of great significance if voice is being carried on one side of the gateway at 64 Kbps while it is carried at a much lower bandwidth over the IP network. Though we may place the MGs close to the edge of the network, we can centralize the call control and network intelligence. Depending on the processing power of the controller, the required size of the various gateways, and the cost of each type of node, it is possible to design a network that is very cost-efficient both from a capital cost and operating cost perspective. Of course, the critical requirement is that there be a fast, robust, and scalable control protocol between the controllers and the media devices.

As it happens, several such protocols exist. The control protocol most widely deployed in VoIP networks today is the *Media Gateway Control Protocol* (MGCP), which was developed within the *Internet Engineering Task Force* (IETF). This protocol, however, has been superseded by a protocol known as MEGACO/H.248, which was jointly developed by the IETF and the ITU. In fact, it is known as MEGACO in the IETF community and as H.248 within the ITU; the terms MEGACO and H.248 are interchangeable. MEGACO has been endorsed by 3GPP as the protocol of choice for gateway control in 3GPP Release 4 and 3GPP Release 5.

8.7.1 The MEGACO Protocol

The architecture associated with MEGACO defines MGs, which perform the conversion of media from the format required in one network to the format required in another. The architecture also defines *media gateway controllers* (MGCs), which control call establishment and tear-down within MGs.

The MEGACO protocol involves a series of transactions between MGCs and MGs. Each transaction involves the sending of a transaction request by the initiator of the transaction and the sending of a transaction reply by the responder. A *transaction request* comprises a number of commands and the *transaction reply* comprises a corresponding number of responses. For the most part, transactions are requested by an MGC and the corresponding actions are executed within an MG. However, a number of cases occur where an MG initiates the transaction request.

MEGACO defines *terminations*, which are logical entities on an MG that act as sources or sinks of media streams. Certain terminations are physical. They have a semi-permanent existence and may be associated with external physical facilities or resources. These would include a termination connected to an analog line or a termination connected to a DS0 channel, or perhaps an ATM virtual circuit. Such terminations exist as long as they are provisioned within the MG. Other terminations have a more transient existence and only exist for the duration of a call or media flow. These are known as *ephemeral terminations* and represent media flows such as a stream of RTP packets. These terminations are created as a result of a MEGACO Add command and they are destroyed by means of the Subtract command.

Terminations have specific properties and the properties of a given termination will vary according to the type of termination. It is clear, for example, that a termination connected to an analog line will have different characteristics than a termination connected to a TDM channel such as a DS0. The properties associated with a termination are grouped into a set of descriptors. These descriptors are included in MEGACO commands, thereby enabling termination properties to be changed according to instructions from MGC to MG.

A termination is referenced by a *termination ID*. This is an identifier chosen by the MG. MEGACO enables the use of the wildcards "all" (*) and "any," or "choose" ($). A special termination ID is available called "Root." This termination ID is used to refer to the gateway as a whole rather than to any specific terminations within the gateway.

MEGACO also defines *contexts*, where a context is an association between a number of terminations for the purposes of sharing media between those terminations. Terminations may be added to contexts, removed from contexts, or moved from one context to another. A termination may exist in only one context at any time, and terminations in a given gateway may only exchange media if they are in the same context.

A termination is added to a context through the use of the Add command. If the Add command does not specify a context to which the termination should be added, then a new context is created as a result of the execution of the Add command. This is the only mechanism for creating a new context. A termination is moved from one context to another through the use of the Move command and is removed from a context through the use of the Subtract command. If the execution of a Subtract command results in the removal of the last termination from a given context, then that context is deleted.

The relationship between terminations and contexts is illustrated in Figure 8-12, where a gateway is depicted with four active contexts. In context C1, we see a simple two-way call across the MG. In Context C2, we see a three-way call across the MG. In contexts C3 and C4, we see a possible implementation of call waiting. In context C3, terminations T6 and T7 are involved in a call. Another call arrives from termination T8 for termination T7. If the user wants to accept this waiting call and place the existing call on hold, then this could be achieved by moving termination T7 from context C3 to context C4.

The existence of several terminations within the same context means that they have the potential to exchange media. However, the existence of terminations in the same context does not necessarily mean that they can all send data to each other and receive data from each other at any given time. The context itself has certain attributes. These include the *topology*, which indicates the flow of media between terminations (which terminations may send media to others/receive media from others). Also, the *priority attribute* indicates the precedence applied to a context when an MGC must handle many contexts simultaneously. An *emergency attribute* is used to give preferential handling to emergency calls.

A context is identified by a context ID, which is assigned by the MG and is unique within a single MG. As is the case for terminations, MECAGO enables wildcarding when referring to contexts, such that the all (*) and any, or choose ($) wildcards may be used. The all wildcard may be used by an MGC to refer to every context on a gateway. The choose ($) wildcard is used when an MGC requires the MG to create a new context.

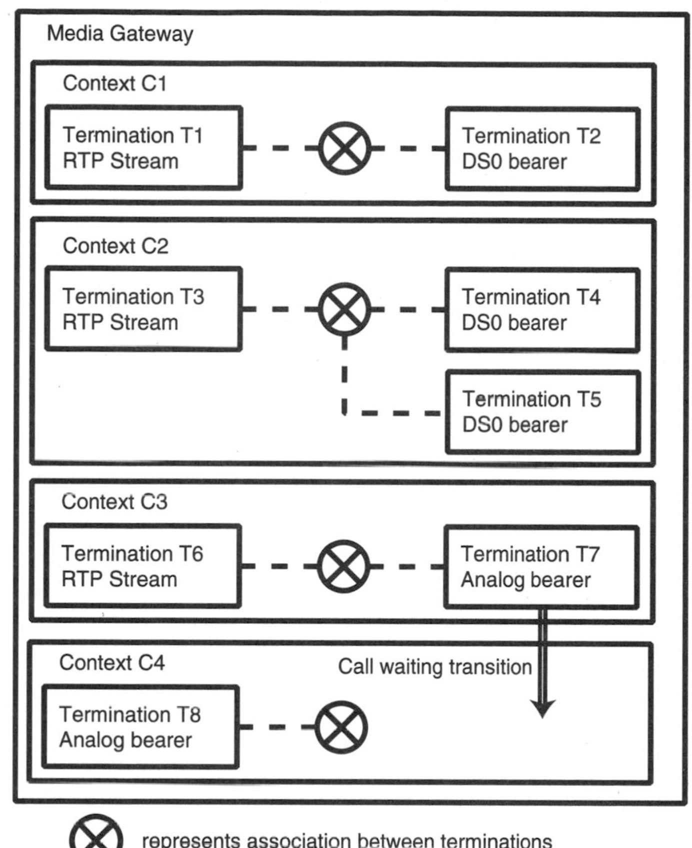

Figure 8-12 Contexts and Terminations.

A special context, known as the null context, also exists. This contains all terminations that are not associated with any other termination—that is, all terminations that do not exist in any other context. Idle terminations normally exist in the null context. The context ID for the null context is simply "-."

8.7.1.1 MEGACO Transactions MEGACO transactions involve the passing of commands and the responses to those commands. Commands are directed towards terminations within contexts. In other words, every command specifies a context ID and one or more termination IDs to which the command applies. This is the case even for a command that requires some action by an idle termination that does not exist in any specific context. In such a case, the null context is applicable.

Multiple commands may be grouped together in a transaction structure whereby a set of commands related to one context may be followed by a set of commands related to another context. The grouped commands are sent together in a single transaction request. This can be represented as

```
Transaction Request (Transaction ID {
     ContextID1 {Command, Command, . . . Command},
     ContextID2 {Command, Command, . . . Command},
     ContextID3 {Command, Command, . . . Command} } )
```

No requirement specifies that a transaction request contain commands for more than one context or even contain more than one command. It is perfectly valid for a transaction request to contain just a single command for a single context.

Upon receipt of a transaction request, the recipient executes the enclosed commands. The commands are executed sequentially in the order specified in the transaction request. Upon completed execution of the commands, a transaction reply is issued. This has a similar structure to the transaction request in that it contains a number of responses for a number of contexts. A transaction reply may be represented as

```
TransactionReply (TransactionID {
     ContextID1 {Response, Response, . . . Response},
     ContextID2 {Response, Response, . . . Response},
     ContextID3 {Response, Response, . . . Response} } )
```

8.7.1.2 MEGACO Commands MEGACO defines the following eight commands. Most of the commands are sent from an MGC to an MG. The exceptions are the Notify command, which is always sent from an MG to an MGC, and the ServiceChange command, which can be sent from either an MG or an MGC.

Add The *Add command* adds a termination to a context. If the command does not specify a particular context to add the termination to, then a new context is created. If the command does not indicate a specific TerminationID, but instead uses the choose ($) wildcard, the MG will create a new ephemeral termination and add it to the context.

Modify The *Modify command* is used to change the property values of a termination, to instruct the termination to issue one or more signals, or to instruct the termination to detect and report specific events.

Subtract The *Subtract command* is used to remove a termination from a context. The response to the command is used to provide statistics related to the termination's participation in the context. These statistics depend upon the type of termination in question. For an RTP termination, the statistics may include items such as packets sent, packets received, and jitter. If the result of a Subtract command is the removal of the last termination from a context, then the context itself is deleted.

Move The *Move command* is used to move a termination from one context to another. It should not be used to move a termination from or to the null context, as these operations must be performed with the Add and Subtract commands respectively. The capability to move a termination from one context to another provides a useful tool for accomplishing the call-waiting service.

Audit Value The *Audit Value command* is used by the MGC to retrieve current values for properties, events, and signals associated with one or more terminations.

Audit Capabilities The *Audit Capabilities command* is used by an MGC to retrieve the possible values of properties, signals, and events associated with one or more terminations. At first glance, this command may appear very similar to the Audit Value command. The difference between them is that the Audit Value command is used to determine the current status of a termination, whereas the Audit Capabilities command is used to determine the possible statuses that a termination might assume. For example, Audit Value would indicate any signals that are currently being applied by a termination, while Audit Capabilities would indicate all of the possible signals that the termination could apply if required.

Notify The *Notify command* is issued by an MG to inform the MGC of events that have occurred within the MG. The events to be reported will have previously been requested as part of a command from the MGC to the MG, such as a Modify command. The events reported will be accompanied by a RequestID parameter to enable the MGC to correlate reported events with previous requests.

Service Change The *Service Change command* is used by an MG to inform an MGC that a group of terminations is about to be taken out of service or is being returned to service. The command is also used in a situation where an MGC is handing over control of an MG to another MGC. In that case, the command is first issued from the controlling MGC to the MG to instigate the transfer of control. Subsequently, the MG issues the Service Change command to the new MGC as a means of establishing the new relationship.

8.7.1.3 MEGACO Descriptors Associated with each command and response are a number of *descriptors*. These descriptors are effectively the parameters or information elements associated with each command or response. The content of a given descriptor will depend on the termination in question.

Many such descriptors exist, but one in particular is worth noting. This is the *media descriptor*, which describes media streams. It contains two components—the termination state descriptor and the stream descriptor. The stream descriptor is comprised of three components—the local control descriptor, the local descriptor and the remote descriptor. This structure can be represented as follows:

- Media descriptor
  - Termination state descriptor
  - Stream descriptor
    - Local control descriptor
    - Local descriptor
    - Remote descriptor

The *termination state descriptor* indicates whether the termination is currently in service, out of service, or in test. It also provides information about how events detected by the termination are to be handled.

The *stream descriptor* is identified by a stream ID. Stream ID values are used between an MG and an MGC to indicate which media streams are interconnected. Within a given context, streams with the same stream ID are connected. A stream is created by specifying a new stream ID on a particular termination in a context.

The *local control descriptor* is used to indicate the current mode of the termination, such as send-only, receive-only, or send-receive, where these terms refer to the direction from the context to the outside world. Thus, the term receive-only means that a termination can receive media from outside the context and pass it to other terminations in the context, but it cannot send media to anywhere outside the context.

The *local descriptor* and *remote descriptor* are basically SDP session descriptions related to the local end of a media stream and the far end of a media stream respectively. Imagine, for example, a VoIP gateway (gateway A) that is communicating with another VoIP gateway (gateway B) across an IP network. The local descriptor for gateway A specifies the media formats that gateway A wants to receive and the address and port number to which that media (that is, the RTP stream) should be sent. The remote descriptor for gateway A indicates the media formats that gateway B wants to receive and the address and port to which that media should be sent.

8.7.1.4 Call Establishment Using MEGACO Based on the foregoing high-level descriptions, we are in a position to describe how a basic call can be established using MEGACO. Figure 8-13 shows a scenario where a call is to be established between two terminations, T1 and T4, which reside on two different MGs. In this example, the two MGs are controlled by the same MGC.

The MGC has determined, through call control signaling (not shown), that a call needs to be established between termination T1 on MG-A and termination T4 on MG-B. It first requests MG-A to add T1 to a new context. The fact that it is a new context is indicated by the $ wildcard. It also requests that the MG create a new ephemeral termination (indicated by the wildcard associated with the second Add command) and add that termination to the same context. The MGC specifies that the new termination should be able to receive media from the far end, but not send media to the far end. This is reasonable, because the MG has not yet received any information as to where the media should be sent.

The MGC also makes a suggestion as to the media coding that the new termination should use. This can be seen in the local descriptor, which contains an SDP description. In this example, the suggestion is that the new termination use audio coded according to AMR and use the dynamic RTP payload type value of 98. Note that the MGC has not specified any IP address or port number, as these are associated with a termination that MG-A has yet to create. Note also that the session description provided by the MGC is merely a suggestion. The MGC is not required to suggest any format. If it does suggest a format, then the MG should comply with that suggestion if possible, but it does not have to.

MG-A responds to the MGC using the same transaction ID. It indicates that it has created a new context with ContextID = 1001. It has added

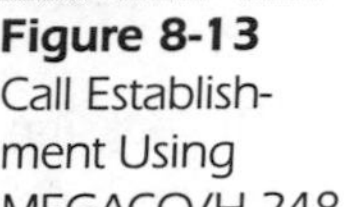

Figure 8-13 Call Establishment Using MEGACO/H.248

MG - B
322.322.1.1
T4 T3

MGC
333.333.1.1

MG - A
311.311.1.1
T2 T1

MGC → MG - A:

```
Transaction = 1 {
Context = $ {
Add = T1,
Add = $ { Media {
  Stream = 1 {
  LocalControl { Mode = receiveonly }
  Local {
  v=0
  C=IN IP4 $
  m= audio $ RTP/AVP 98
  a= rtpmap AMR/8000 } } } } } }
```

MG - A → MGC:

```
Reply = 1 { Context = 1001 {
Add = T1,
Add = T2 { Media {
  Stream = 1 {
  Local {
  v=0
  C=IN IP4 311.311.1.1
  m= audio 1199 RTP/AVP 98
  a= rtpmap AMR/8000 } } } } } }
```

MGC → MG - B:

```
Transaction = 2 {
Context = $ {
Add = T4,
Add = $ { Media {
  Stream = 2 {
  LocalControl { Mode = sendreceive }
  Local {
  v=0
  C=IN IP4 $
  m= audio $ RTP/AVP 98
  a= rtpmap AMR/8000 } ,
  Remote {
  v=0
  C=IN IP4 311.311.1.1
  m= audio 1199 RTP/AVP 98
  a= rtpmap AMR/8000 } } } } } }
```

MG - B → MGC:

```
Reply = 2 { Context = 2002 {
Add = T4,
Add = T3 { Media {
  Stream = 2 {
  Local {
  v=0
  C=IN IP4 322.322.1.1
  m= audio 2299 RTP/AVP 98
  a= rtpmap AMR/8000 } } } } } }
```

MGC → MG - A:

```
Transaction = 3 {
Context = 1001 {
Modify = T2 {
  Media {
  Stream = 1 {
  LocalControl { sendreceive }
  Remote {
  v=0
  C=IN IP4 322.322.1.1
  m= audio 2299 RTP/AVP 98
  a= rtpmap AMR/8000 } } } } } }
```

MG - A → MGC:

```
Reply=3 { Context = 1001 {
modify = T2 } }
```

termination T1 to the context as requested. It has also created termination T2 and added it to the same context. Associated with termination T2 is an SDP session description. Unlike the suggested session description received from the MGC, this session description (included in the local descriptor) includes an IP address (311.311.1.1) and port number (1199) at which termination T2 expects to receive the RTP stream.

The MGC now requests MG-B to set up a new context and to add two terminations to that context—termination T4 and a new termination that MG-B must create. For the new termination, the MGC makes a suggestion as to the content of the local descriptor. It also specifies the exact content of the remote descriptor. Although the information for the local descriptor is simply a suggestion, the information in the remote descriptor is what the new termination must use. The content of the remote descriptor is, after all, the content of the local descriptor for termination T2 on MG-A. In other words, the local descriptor specifies which media format the new termination should send and where it should send it.

Note the use of the local control descriptor. In this case, the MGC specifies that the mode should be send-receive. This is because the far end is ready to receive RTP packets and will soon know where to send them, even though it does not know that quite yet.

MG-B creates the new context (Context ID = 2002) and adds termination T4 to that context. It also creates termination T3 and adds it to the context. For the new termination, it specifies a local descriptor, which includes the media format it wants to receive, and the address and port number to which the packets should be sent.

The MGC takes the local descriptor information related to termination T3 and, using the Modify command, sends it to MG-A as a remote descriptor for termination T2. It also specifies the mode for termination T2 to be send-receive. Termination T2 now knows where to send RTP packets and has permission to send them.

The chain is now complete. Terminations T1 and T2 are in the same context, so a path exists between them across MG-A. Equally, terminations T3 and T4 are in the same context, so a path is created between them across MG-B. Finally, T2 and T3 have established a bidirectional RTP stream between them. Thus, a path is available from T1 to T4, as originally intended.

8.7.1.5 MEGACO and SIP Interworking Imagine the case where the two MGs of Figure 8-13 happen to be controlled by separate MGCs. In that case, a protocol needs to be used between the two MGCs. The obvious choice for that protocol is SIP. Once a local descriptor is available at the gateway where the call is being originated, this can be passed in a SIP INVITE as a SIP message body. Upon acceptance of the call at the far end, the corresponding session description is carried back as a SIP message body within the SIP 200 (OK) response and is passed to the originating side gateway. Once the gateway on the originating side has acknowledged receipt of the remote session description, then the MGC can send a SIP ACK to complete the SIP call setup.

8.8 VoIP and SS7

Although new signaling solutions, such as H.323 and SIP, exist for VoIP networks, the standard in traditional telephony and in mobile networks is SS7. Therefore, if a VoIP-based network is to communicate with any traditional network, not only must it interwork at the media level through media gateways, it must also interwork with SS7. To support this, the IETF has developed a set of protocols known as Sigtran.

In order to understand Sigtran, it is worth considering the type of interworking that needs to occur. Imagine, for example, an MGC that controls one or more media gateways. The MGC is a call control entity in the network and, as such, uses call control signaling to and from other call control entities. If other call control entities use SS7, then the MGC must use SS7, at least to the extent that the other call control entities can communicate freely with it. This means that the MGC does not necessarily need to support the whole SS7 stack—just the necessary application protocols.

Consider Figure 8-14, which shows the SS7 stack. The bottom three layers are called the *Message Transfer Part* (MTP). This is a set of protocols responsible for getting a particular SS7 message from the source signaling point to the destination signaling point. Above the MTP, we find either the *Signaling Connection Control Part* (SCCP) or the *ISDN User Part* (ISUP). ISUP is generally used for the establishment of regular phone calls. SCCP can also be used in the establishment of regular phone calls, but it is more often used for the transport of higher-layer applications, such as the GSM *Mobile Application Part* (MAP) or the *Intelligent Network Application Part* (INAP). In fact, most such applications use the services of the *Transaction Capabilities Application Part* (TCAP), which in turn, uses the services of SCCP.

SCCP provides an enhanced addressing mechanism to enable signaling between entities even when those entities do not know each other's signal-

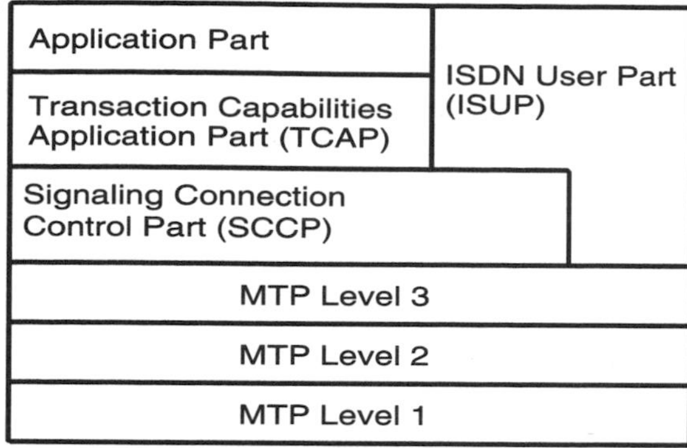

Figure 8-14 Signaling System 7 Protocol Stack.

ing addresses (known as point codes). This addressing is known as *global title addressing*. Basically, it is a means whereby some other address, such as a telephone number, can be mapped to a point code, either at the node that initiated the message or some other node between the originator and destination of the message.

Figure 8-15 provides some examples of communication between different SS7 entities. Consider scenario A. In this case, the two entities, represented by point code 1 and point code 4, communicate at layer 1. At each layer, a peer-to-peer relationship exists between the two entities. Scenario B has a peer-to-peer relationship at layer 1, layer 2, and layer 3 between point codes 1 and 2, 2 and 3, and 3 and 4. At the SCCP layer, a peer-to-peer relationship exists between point codes 1 and 2 and between point codes 2 and 4.

At the TCAP and Application layers, a peer-to-peer relationship can only take place between point codes 1 and 4. In other words, the application at point code 1 is only aware of the TCAP layer at point code 1 and the application layer at point code 4. Similarly, the TCAP layer at point code 1 is aware only of the application layer above it, the SCCP layer below it, and the corresponding TCAP layer at point code 4. It is not aware of any of the MTP layers. Equally, if we consider communication between point code 2 and point code 4, the SCCP layer at each point code knows only about the layer above (TCAP), the layer below (MTP3), and the corresponding SCCP peer. As far as the SCCP layers are concerned, nothing else exists. Therefore, SCCP neither knows nor cares that point code 3 exists.

Consider Scenario C, where point code 3 is replaced by a gateway that supports standard SS7 on one side and an IP-based MTP emulation on the

Figure 8-15 Example SS7 Communication Scenarios.

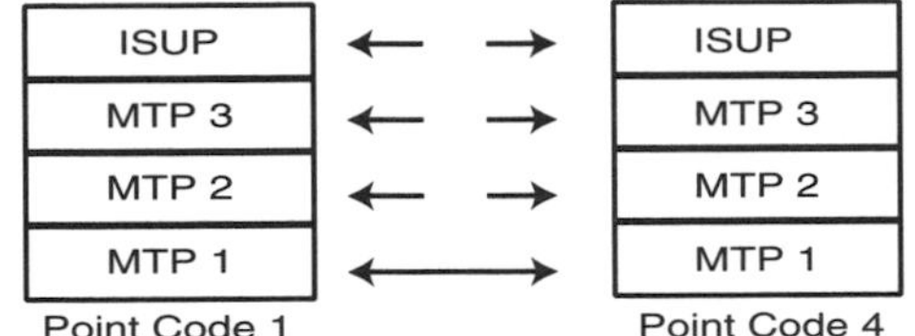

Scenario A—Communication Between Adjacent Signaling Points

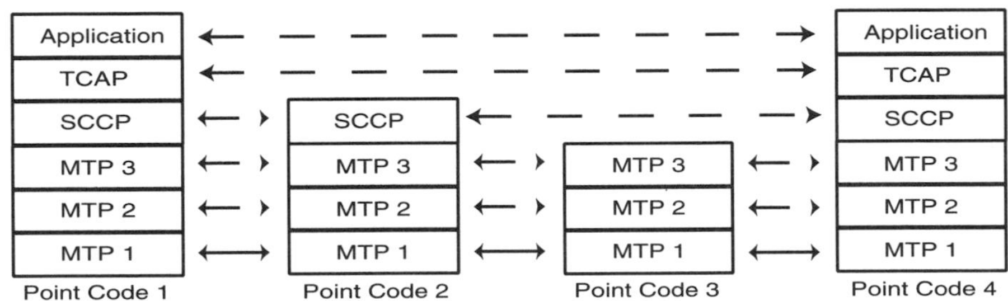

Scenario B—Communication Between non-Adjacent Signaling Points

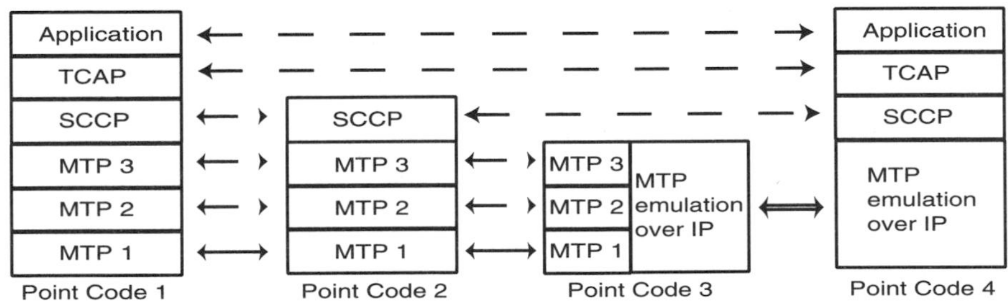

Scenario C—Communication Between SS7-based and IP-based Applications

other side. Point code 4 does not support the lower SS7 layers at all—just an MTP emulation over IP. Provided that the MTP emulation at point code 4 appears to the SCCP layer as standard MTP, then the SCCP layer does not care, nor do any of the layers above SCCP. Equally, the SCCP layers at point code 1 and 2 do not care. Consequently, it is possible to implement SS7-based applications at point code 4 without implementing the whole SS7 stack. This is the concept behind the Sigtran protocol suite.

8.8.1 The Sigtran Protocol Suite

Figure 8-16 shows the Sigtran protocol suite and the relationship between the Sigtran protocols and standard SS7 protocols. Above IP, we find a protocol known as the *Stream Control Transmission Protocol* (SCTP). The primary motivation behind the development of SCTP is the fact that neither UDP nor TCP offer both the speed and reliability required of a transport protocol used to carry signaling. The design of SCTP is an attempt to make such reliability and speed available to the users of SCTP.

In the SCTP specification, such a user is known as an *Upper Layer Protocol* (ULP). A ULP can be any of the protocols directly above the SCTP layer, as illustrated in Figure 8-16. Each of the protocols above SCTP is an adaptation layer. For example, M3UA is the *MTP3 User Adaptation Layer*. Thus, we could have ISUP over M3UA over SCTP. Each of the adaptation layers uses the same primitives to and from the layer above, as are used by the equivalent SS7 layer. Thus, the layer above does not see any difference between the adaptation layer and its SS7 equivalent. Thus, if we have ISUP over M3UA, the ISUP layer believes the M3UA to be standard MTP3 and does not know that the transport is IP-based.

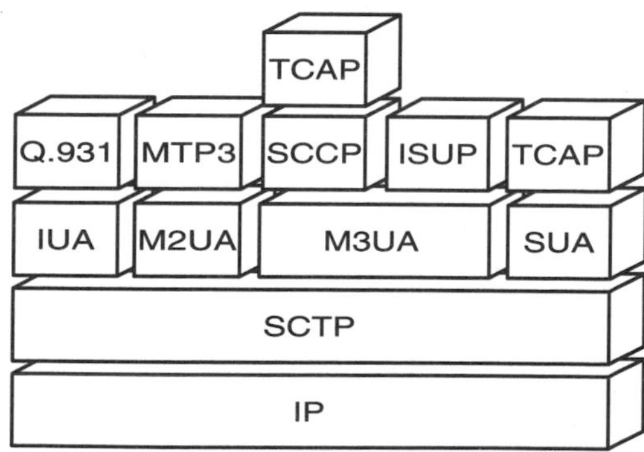

Figure 8-16 Sigtran Protocol Suite.

The following applications layers are defined:

- SS7 *MTP2-User Adaptation Layer* (M2UA) provides adaptation between MTP3 and SCTP. It provides an interface between MTP3 and SCTP such that standard MTP3 may be used in the IP network, without the MTP3 application software realizing that messages are being transported over SCTP and IP, instead of MTP2. For example, a standard MTP3 application implemented at an MGC could exchange MTP3 signaling network management messages with the external SS7 network. In the same manner that MTP2 provides services to MTP3 in the SS7 network, M2UA provides services to MTP3 in the IP network.
- SS7 *MTP3-User Adaptation Layer* (M3UA) provides an interface between SCTP and those applications that typically use the services of MTP3, such as ISUP and SCCP. M3UA and SCTP enable seamless peer-to-peer communication between MTP3 user applications in the IP network and identical applications in the SS7 network. The application in the IP network does not realize that SCTP over IP transport is used instead of typical SS7. In the same manner that MTP3 provides services to applications such as ISUP in the SS7 network, M3UA offers equivalent services to applications in the IP network.
- SS7 *SCCP-User Adaptation Layer* (SUA) provides an interface between SCCP user applications and SCTP. Applications such as TCAP use the services of SUA in the same way that they use the services of SCCP in the SS7 network. In fact, those applications do not know that the underlying transport is different in any way. Hence, transparent peer-to-peer communication can take place between applications in the SS7 network and applications in the IP network.
- *ISDN Q.921-User Adaptation Layer* (IUA) is the Sigtran equivalent of the Q.921 Data-link layer which is used to carry Q.931 ISDN signaling. Thus, Q.931 messages may be passed from the ISDN to the IP network, with identical Q.931 implementations in each network, and neither of them recognize any difference in the underlying transport.

8.8.1.1 Stream Control Transmission Protocol (SCTP) SCTP provides for the reliable and fast delivery of signaling messages. It is reliable because it includes mechanisms for the detection and recovery of lost or corrupted messages. It is faster than TCP, however, because it avoids head-of-line blocking, which can occur with TCP, and it also has more efficient retransmission mechanisms than TCP.

Head-of-line blocking is avoided in SCTP through the use of streams. A *stream* is a logical channel between SCTP endpoints. It may also be thought of as a sequence of user messages between two SCTP users. When an *association* is established between endpoints, part of the establishment of the association involves each endpoint specifying how many inbound streams and how many outbound streams are to be supported. If we think of a given association as a one-way highway between endpoints, then the individual streams are analogous to the individual traffic lanes on that highway. The advantage with the stream concept is that resources (or queues) are allocated individually to each stream, rather than to the complete set of packets that might pass between two endpoints. Consequently, a message from one stream does not have to wait in a queue behind a message from another stream.

Retransmission in SCTP is based on the fact that SCTP packets carrying user data (known as chunks) include a *transmission sequence number* (TSN). The receiver of the chuncks checks to make sure that all chunks have been received by ensuring that no gap exists in the TSNs. If a gap is found, then SCTP enables the receiver to specify which TSNs are missing and it is only those TSNs that need to be retransmitted, which is more efficient than TCP.

Consider, for example, the situation depicted in Figure 8-17. Chunks with TSNs 1 to 4 have been received correctly, the chunk with TSN 5 is missing, the chunk with TSN 9 is missing, and the chunks with TSNs 8 and 11 have been received twice. If TCP were to deal with this situation, then all chunks from 5 onwards would be retransmitted. SCTP, however, has the means for the receiver to clearly specify to the sender what is missing and what is duplicated so that the minimum retransmission takes place.

Not only does SCTP support fast transmission and efficient retransmission, it also supports congestion avoidance and it supports network-level redundancy. Congestion avoidance is achieved through the use of a parameter in SCTP messages called the Advertised Receiver Credit Window. This parameter indicates to the far end how much buffer space the receiver has for the receipt of new messages. This helps to avoid flooding a receiver with more messages than it can handle. Redundancy is achieved through the fact that a given endpoint can be logically distributed across multiple platforms with multiple IP addresses. If a given platform fails, then another platform can take over. SCTP includes messages for monitoring the reachability of a given endpoint and failover messages for one endpoint to indicate to another that a different IP address should be used for future messages.

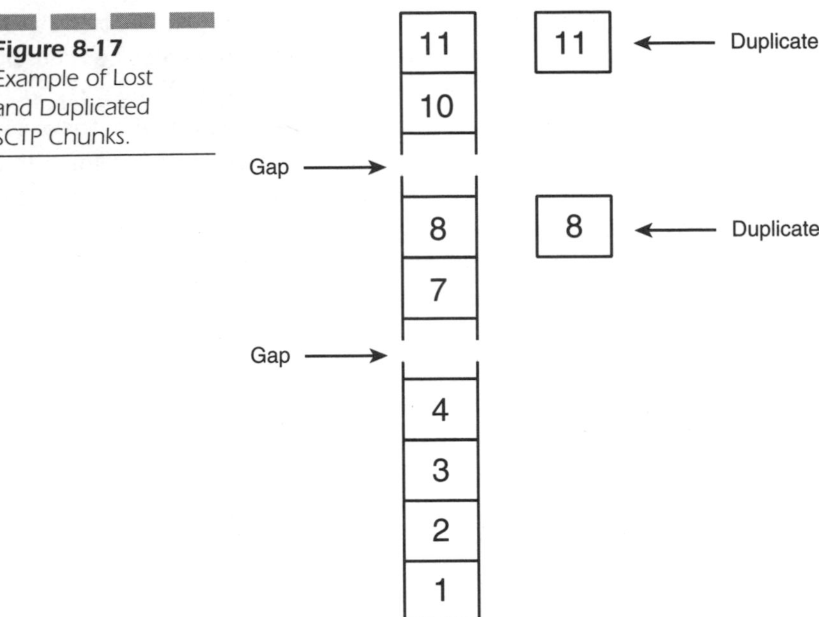

Figure 8-17 Example of Lost and Duplicated SCTP Chunks.

8.8.2 Example of Sigtran Usage

Figure 8-18 provides an example of how IP and SS7 networks can interwork using Sigtran. The IP-to-SS7 connectivity diagram shows how a SIP device could be connected to an MG and an MGC such that it can communicate with a standard telephone in the PSTN. The IP-to-SS7 protocol interworking diagram shows how the protocol interworking can take place via a *signaling gateway* (SG). The net effect is that the nodes, such as a PSTN switch, in the SS7 network can communicate with the SIP terminal via the SG and MGC and MG without realizing that the SIP terminal is not a standard telephone connected to a standard SS7-enabled switch.

Of course, the MGC must be able to translate SIP messages to ISUP messages and vice versa. Although these two protocols are different, the messages of SIP and those of ISUP do serve similar functions, and it is possible to map from one protocol to the other. For example, the ISUP *Initial Address Message* (IAM) maps quite well to the SIP INVITE. The SIP 183

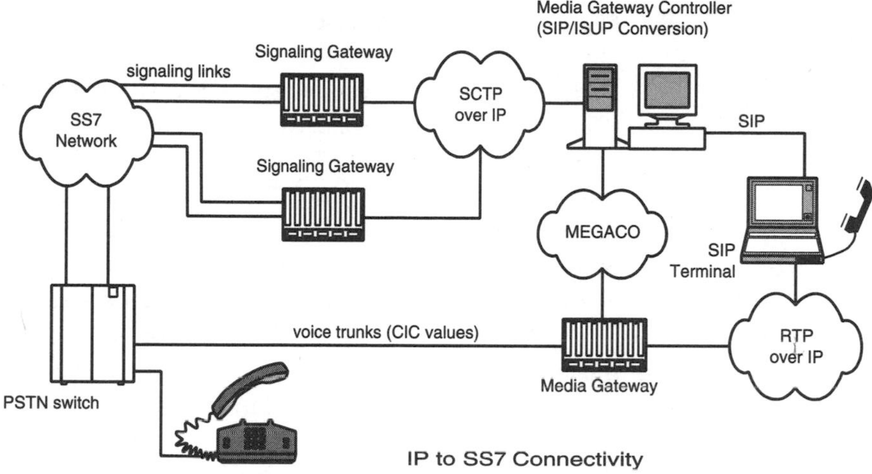

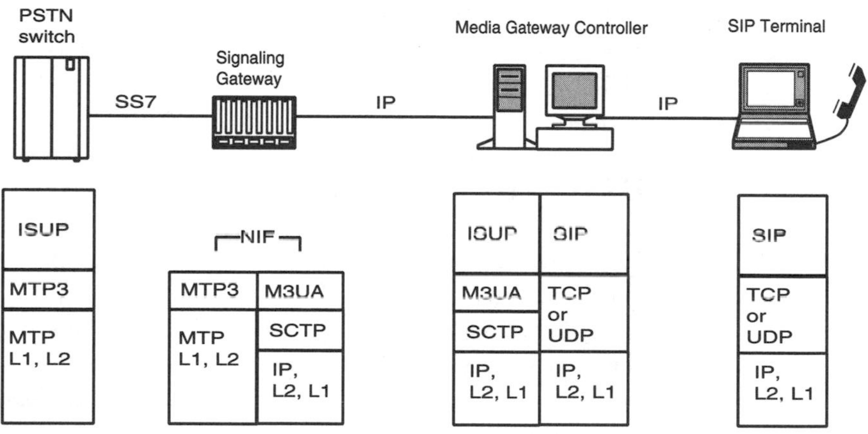

Figure 8-18 IP/SS7 Interworking Example.

(Session Progress) response, an extension to the original SIP specification, maps to the ISUP *Address Complete message* (ACM). The SIP 200 (OK) response maps to the ISUP *Answer* (ANS) message.

8.9 VoIP Quality of Service

Perhaps the biggest issue with VoIP is ensuring that the *Quality of Service* (QoS) is comparable to the QoS achieved in traditional circuit-switched telephony. As we have seen, IP and UDP provide no quality guarantees whatsoever. Although RTP and RTCP provide QoS-related information (such as jitter, number of lost packets, and so on), they do not provide any assurance of quality. In order to ensure that VoIP is not a low-quality service, specific solutions must be implemented in the network.

One way to help ensure that VoIP offers high quality is to ensure that more than enough bandwidth is available—both in terms of throughput on transmission facilities and in terms of processing power within routers. By overprovisioning the network, one can reduce the likelihood of congestion and thereby improve quality. This, however, is an expensive option that leaves much of the network capacity unused much of the time. Moreover, it does not guarantee quality. Thus, one needs technical solutions within the network.

The following sections provide a brief overview of some QoS techniques. For more detailed explanations, the reader is referred to the applicable IETF specifications.

8.9.1 The Resource Reservation Protocol

Resource reservation techniques for IP networks are specified in RFC 2205, the *Resource Reservation Protocol* (RSVP), which is part of the IETF integrated services suite. It is a protocol that enables resources to be reserved for a given session or sessions prior to any attempt to exchange media between the participants. Of the solutions available, it is the most complex, but is also the solution that comes closest to circuit emulation within the IP network. It provides strong QoS guarantees, a significant granularity of resource allocation, and significant feedback to applications and users.

RSVP currently offers two levels of service. The first is guaranteed, which comes as close as possible to circuit emulation. The second is controlled load, which is equivalent to the service that would be provided in a best-effort network under no-load conditions.

Basically, RSVP works as depicted in Figure 8-19. A sender first issues a PATH message to the far end via a number of routers. The PATH message contains a *traffic specification* (TSpec), which provides details of the data that the sender expects to send, in terms of the bandwidth requirement and packet size. Each RSVP-enabled router along the way establishes a "path state" that includes the previous source address of the PATH message (that is, the next hop back towards the sender). The receiver of the PATH message responds with a *reservation request* (RESV) that includes a flowspec. The flowspec includes a Tspec and information about the type of reservation service requested, such as controlled-load service or guaranteed service.

The RESV message travels back to the sender along the same route that the PATH message took (in reverse). At each router, the requested resources are allocated, assuming that they are available and that the receiver has the authority to make the request. Finally, the RESV message reaches the sender with a confirmation that resources have been reserved.

One interesting point about RSVP is that reservations are made by the receiver, not by the sender of data. This is done in order to accommodate multicast transports, where there may be large numbers of receivers and only one sender.

Note that RSVP is a control protocol that does not carry user data. The user data (e.g. voice) is transported later using RTP. This occurs only after the reservation procedures have been performed. The reservations that RSVP makes are soft, which means that they need to be refreshed on a regular basis by the receivers(s).

Figure 8-19
Resource Reservation.

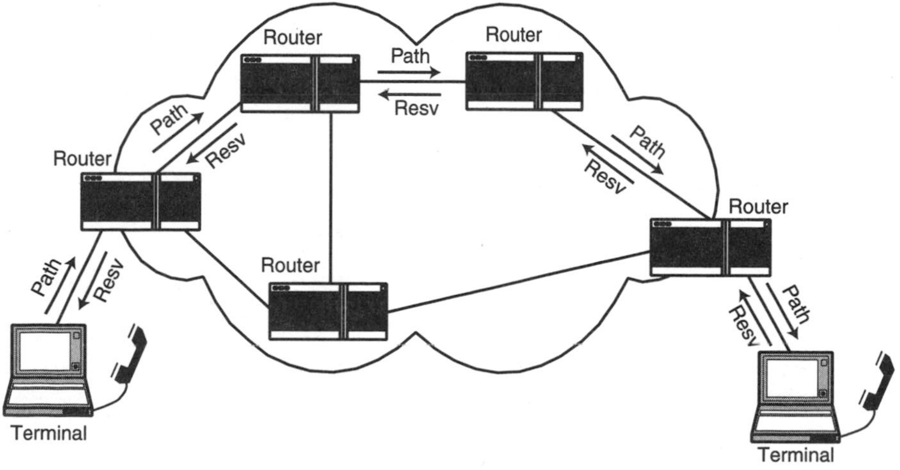

8.9.2 Differentiated Service (DiffServ)

Differentiated Service (DiffServ) is a relatively simple means for prioritizing different types of traffic. The DiffServ protocol is described in RFC 2475, An Architecture for Differentiated Services. Basically, DiffServ makes use of the IPv4 *Type of Service* (TOS) field, contained in the IPv4 header and the equivalent IPv6 Traffic Class field. The portion of the TOS/Traffic Class field used by DiffServ is known as the DS field. The field is used in specific ways to mark a given stream as requiring a particular type of forwarding. The type of forwarding to be applied is known as *per-hop behavior* (PHB), of which DiffServ defines two types. These are *expedited forwarding* (EF) and *assured forwarding* (AF).

EF is specified in RFC 2598. It is a service whereby a given traffic stream is assigned a minimum departure rate from a given node, one that is greater than the arrival rate at the same node, provided that the arrival rate does not exceed a pre-agreed maximum. This ensures that queuing delays are removed. Since queuing delays are a major cause of end-to-end delay and are the main cause of jitter, this ensures that delay and jitter are minimized. In fact, EF can provide a service that is equivalent to a virtual leased line.

AF is defined in RFC 2597. This is a service whereby packets from a given source are forwarded with a high probability, provided that the traffic from that source does not exceed some pre-agreed maximum. AF defines four classes, with each class allocated a certain amount of resources (buffer space and bandwidth) within a router. Within each class, a given packet may have one of three drop rates. At a given router, if congestion occurs within the resources allocated to a given AF class, then the packets with the highest drop rate values will be discarded first so that packets with a lower drop rate value receive some protection. In order to work well, it is necessary that the incoming traffic does not have packets with a high percentage of low drop rates. After all, the purpose is to ensure that the highest-priority packets get through in the case of congestion, and that cannot happen if all the packets have the highest priority.

8.9.3 MultiProtocol Label Switching (MPLS)

Label switching is something that has seen significant interest from the Internet community, and significant effort has been made to define a protocol called *Multi-Protocol Label Switching* (MPLS). In some ways, it is similar to DiffServ in that it marks traffic at the entrance to the network. However, the primary function of the marking is not to allocate a priority within a router, but to determine the next router in the path from the source to the destination.

MPLS involves the attachment of a short label to a packet in front of the IP header. This effectively is like inserting a new layer between the IP layer and the underlying link layer of the OSI model. The label contains all the information that a router needs to forward a packet. The value of a label may be used to look up the next hop in the path and forward to the next router. The difference between this and standard IP routing is that the match is an exact one and is not a case of looking for the longest match (that is, the match with the longest subnet mask). This enables faster routing decisions within routers.

The label identifies something called a *Forwarding Equivalence Class* (FEC). This term is chosen because it means that all packets of a given FEC are treated equally for the purposes of forwarding. All packets in a given stream of data, such as a voice call, will have the same FEC and will receive the same forwarding treatment. It is therefore possible to ensure that the forwarding treatment applied to a given stream can be set up such that all packets from A to B follow the same path. If that stream has a particular bandwidth requirement, then that bandwidth can be allocated at the start of the session. This can ensure that a given stream has the bandwidth that it needs and the packets that make up the stream arrive in the same order as transmitted. Hence, a higher QoS is provided.

In many ways, MPLS is as much of a traffic engineering protocol as it is a QoS protocol. It is somewhat analogous to the establishment of virtual circuits in ATM and can lead to similar QoS benefits. It helps to provide QoS by helping to better manage traffic. Whether it should be called a traffic engineering protocol or a QoS protocol hardly matters if the end result is better QoS.

References

IETF Draft — SS7 MTP2-User Adaptation Layer, work in progress

IETF Draft — SS7 MTP3-User Adaptation Layer (M3UA), work in progress

IETF Draft — SS7 SCCP-User Adaptation Layer (SUA), work in progress

IETF RFC 768	User Datagram Protocol (STD 6)
IETF RFC 791	Internet Protocol (STD 5)
IETF RFC 793	Transmission Control Protocol (STD 7)
IETF RFC 1889	RTP: A Transport Protocol for Real-Time Applications
IETF RFC 1890	RTP Profile for Audio and Video Conferences with Minimal Control
IETF RFC 2205	Resource ReSerVation Protocol (RSVP)—Version 1 Functional Specification
IETF RFC 2327	SDP: Session Description Protocol
IETF RFC 2475	An Architecture for Differentiated Services
IETF RFC 2543	Session Initiation Protocol (SIP)
IETF RFC 2597	Assured Forwarding PHB
IETF RFC 2598	An Expedited Forwarding PHB
IETF RFC 2701	Media Gateway Control Protocol (MGCP) Version 1.0
IETF RFC 2719	Architectural Framework for Signaling Transport
IETF RFC 2805	Media Gateway Control Protocol Architecture and Requirements
IETF RFC 2960	Stream Control Transmission Protocol
IETF RFC 3031	Multiprotocol Label Switching Architecture
IETF RFC 3051	MEGACO Protocol
ITU-T H.225.0	Call-Signalling Protocols and Media Stream Packetization for Packet-Based Multimedia Communication Systems
ITU-T H.245	Control Protocol for Multimedia Communication
ITU-T H.323	Packet-Based Multimedia Communications Systems
ITU-T Q.931	ISDN User-Network Interface Layer 3 Specification for Basic Call Control
ITU-T H.248	Media Gateway Control Protocol

CHAPTER 9

3G System RF Design Considerations

The *radio frequency* (RF) design criteria is a set of rules or parameters that are used by the RF Engineering department to not only design the network and the new components that are added, such as cell sites, but also for improving the performance of the network. The values that are included for each of the design criteria topics is driven by the desire to offer the best service within the monetary and technological constraints.

Therefore, the design criteria for the radio access part of a 3G system is extremely important to establish at the onset of the design whether it is for a new system, migrating to an new platform, or expanding an existing system. Many aspects are associated with an RF design and surprisingly they are common, in concept, with any radio access platform that is being utilized by a wireless operator.

This chapter will try and consolidate many of the most important issues concerning the generation and execution of a design criteria associated with the radio access portion of a system. The topics that will be discussed in this chapter are as follows:

- RF system design procedures
- Methodology
- Propagation models
- Link budget
- Tower top amplifiers
- Cell site design
- RF design report

The chapter concludes with a recommended format for presenting the design criteria in a formalized report that will list the design criteria, assumptions, and other key issues.

In summary, the RF design process for a wireless network is an ongoing process of refinements and adjustments based on a multitude of variables, most of which are not under the control of the engineering department. The RF system design process involves both RF and network engineering efforts with implementation, operations, customer care, marketing, and, of course, operations. However, it is important to note that although many issues are outside the control of the technical services group of a wireless company, the need to stipulate a design and its associated linkages is essential if there is any desire to obtain an operating system that meets the system objective of fulfilling the customer's requirements.

Therefore, the RF system design process that should be followed is listed here in summary form. The process can be used for an existing system or a new system because the material needs to be revisited for each of the topics when any system design takes place:

- Marketing requirements
- Methodology
- Technology decision
- Defining the types of cell sites
- Establishing a link budget
- Defining coverage requirements
- Defining capacity requirements
- Completing RF system design
- Issuing a search area
- *Site qualification test* (SQT)
- Site acceptance/site rejection
- Land use entitlement process
- Integration
- Handover to operations

It is important to note that the design process or guidelines involve not only the establishment of the criteria, but also the realization of the design itself.

The information needed for a system design varies from market to market and, of course, nuances can be noticed between the different technology platforms. However, commonality exists between markets and also technology platforms. The following is a brief listing of the most important pieces of information needed for a system design:

- Time frames for the report to be based on
- Subscriber growth projections (current and future by quarter)
- Subscriber voice usage projection (current and forecasted by quarter)
- Subscriber packet usage projection (current and forecasted by quarter)
- Subscriber types (mobile, portable, packet capable, blend)
- New features and services offered
- Design criteria (technology-specific issues)
- Baseline system numbers for building on the growth study
- Cell site construction expectations (ideal and with land-use entitlement issues factored in)

- *Fixed Network Equipmemt* (FNE) ordering intervals
- New technology deployment and time frames
- Budget constraints
- Due date for design
- Maximum and minimum off loading for cell sites when new cells are added to a design

Of course, many sources and types of information are required for an RF design. The basic inputs usually obtained from the Marketing and Sales organization within a wireless network are listed in this chapter. The output from the RF design process will determine the requirements and fundamental structure of the radio access aspects of a wireless system. A simplified radio access structure is shown in Figure 9-1 but can apply to any situation with the expansion of the individual components relative to the different technology platforms utilized.

In order to design either a new system or establish the migration path for a system, the RF design is relegated to determining the specific access method that the subscriber will have with the wireless system. The subscriber and base stations (*Base Transceiver Stations* [BTSs]) both have a transmitter and receiver incorporated in their fundamental architecture. Figure 9-2 is an illustration of the various components that need to be fac-

Figure 9-1 Generic radio access system.

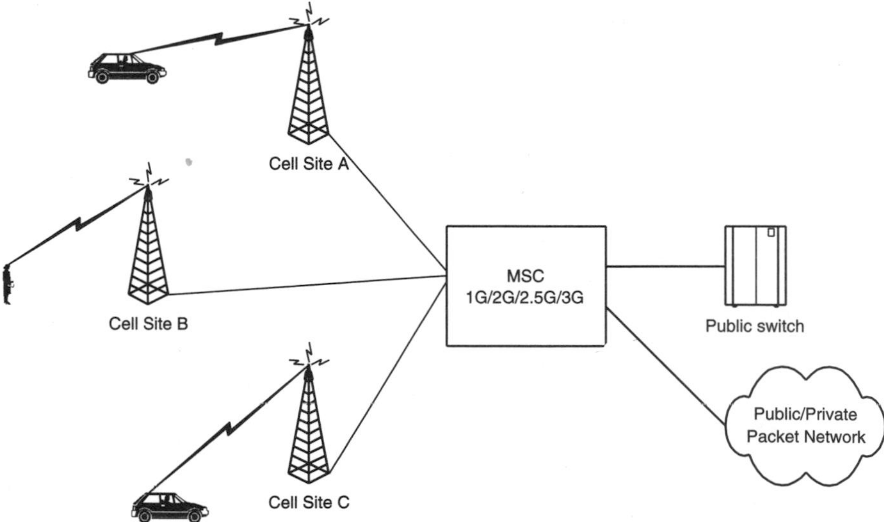

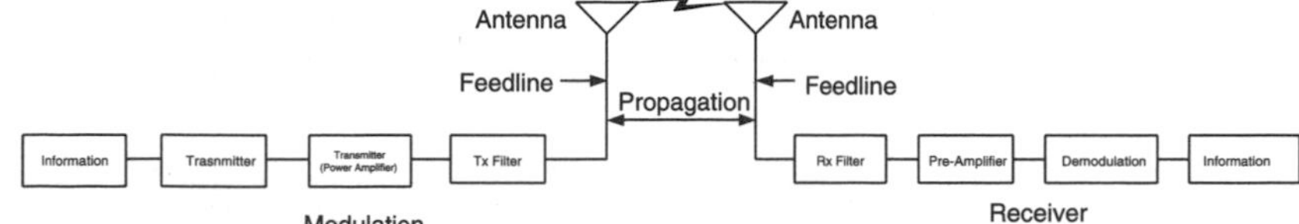

Figure 9-2
Generic radio system.

tored into the design of a system. Figure 9-2 can be used for any technology platform and the specifics with the various network elements make up part of the propagation analysis where *Carrier to Interferer* (C/I) or *Energy per bit per noise* (Eb/No) values are used to determine the performance criteria necessary for the successful transmission and reception of the information being delivered.

9.1 RF System Design Procedures

The RF system design procedures associated with a *third-generation* (3G) system design are similar to those followed for a *second-generation* (2G) or even *first-generation* (1G) wireless system. Amazing similarities exist between implementing 2.5/3 G into an existing system, as was the case when 2G was introduced into cellular systems.

Fundamentally, a wireless communication system has three possible system designs:

- Existing system expansion
- New system design
- Introduction of a new technology platform to an existing system

The radio system design needs to factor in to the process all the components that comprise the path the radio signal takes, as well as how the individual base stations are integrated into a larger system. The specific procedures that need to be followed vary depending on the market, the individual technology platform being installed, and the type of legacy system that is in place, if any. However, basic procedures should be followed and they are listed in this section. It is important to restate that you need to know what your objective is from the onset of the design process, and that objective needs to be linked to the business and marketing plans for the company. Following the direction of design discovery (we will build it and they will come) has seen some very negative consequences in the wireless industry to date.

With that said, this chapter is a brief list of the general design procedures that need to be performed whether the system is for a new or existing 2.5 or 3G system. If, as in most cases, you first migrate from a 2G to a 2.5 platform, and then from a 2.5 to a 3G, the design procedure to follow is that of introducing new technology for both scenarios.

9.1.1 New Wireless System Procedure

The RF design process for a new 2.5G or 3G system is basically the same as that followed for a new 2G or even 1G wireless system. However, the subscriber usage needs to factor in both voice and packet data usage. The steps to do this are as follows:

1. Obtain a marketing plan and objectives.
2. Establish a system coverage area.
3. Establish system on air projections.
4. Establish technology platform decisions.
5. Determine the maximum radius per cell (link budget).
6. Establish environmental corrections.
7. Determine the desired signal level.
8. Establish the maximum number of cells to cover the area.
9. Generate the coverage propagation plot for the system.
10. Determine subscriber usage.
11. Determine usage/sq km (voice and packet).
12. Determine the maximum number of cells for capacity.
13. Determine if the system is capacity- or coverage-driven.
14. Establish the total number of cells required for coverage and capacity.
15. Generate the coverage plot, incorporating coverage and capacity cell sites (if different).
16. Re-evaluate the results and make assumption corrections.
17. Determine the revised (if applicable) number of cells required for coverage and capacity.

18. Check the number of sites against the budget objective; if it exceeds the number of sites, reevaluate the design.
19. Using a known database of sites overlay this onto the system design and check of matches or close match (<0.2R).
20. Adjust system design using site-specific parameters from known database matches.
21. Generate propagation and usage plots for system design.
22. Evaluate the design objective with time frame and budgetary constraints and readjust if necessary.
23. Issue search rings.

9.1.2 2.5G or 3G Migration RF Design Procedure

The process for introducing a 2.5G or 3G platform into an existing wireless system needs to account for the impact the reallocation of the spectrum will have on the legacy system. Also, the design needs to address the new platforms and the modifications to the existing platforms, which are needed to facilitate the introduction of the new system. Therefore, the following is a brief summary of the main issues that need to be addressed when integrating a new platform into an existing system:

1. Obtain a marketing plan.
2. Establish a technology platform introduction time table.
3. Determine new technology implementation tradeoffs.
4. Determine a new technology implementation methodology (footprint and 1:1 or 1:N).
5. Identify coverage problem areas.
6. Determine the maximum radius per cell (link budget for each technology platform).
7. Establish environmental corrections.
8. Determine the desired signal level (for each technology platform).
9. Establish the maximum number of cells to cover an area(s).
10. Generate the coverage propagation plot for the system and the areas, showing before and after coverage.

11. Determine the subscriber usage (existing and new; packet and voice).
12. Determine the subscriber usage by platform type.
13. Allocate the percentage of system usage to each cell.
14. Adjust the cells' maximum capacity by spectrum reallocation method (if applicable).
15. Determine the maximum number of cells for capacity (technology-dependent).
16. Establish which cells need capacity relief.
17. Determine the new cells needed for capacity relief.
18. Establish the total number of cells required for coverage and capacity.
19. Generate the coverage plot, incorporating the coverage and capacity cell sites (if different).
20. Re-evaluate the results and make assumption corrections.
21. Determine the revised (if applicable) number of cells required for coverage and capacity.
22. Check the number of sites against the budget objective; if it exceeds the number of sites, reevaluate the design.
23. Using the known database of sites, overlay on the system design and check the matches or close matches (<0.2R).
24. Adjust the system design using site-specific parameters from known database matches.
25. Generate the propagation and usage plots for the system design.
26. Evaluate the design objective with time frame and budgetary constraints and readjust if necessary.
27. Issue search rings.

The previous two procedures can be easily crafted into a checklist to follow for the design team. Obviously, the lists are generic and need to be tailored to the specific situation. However, when using the lists, whether it is for a new system or the migration from 2G to 2.5G or 3G, it should provide sufficient guidance in order to organize the process for a successful design that will meet the customer and business objectives for the wireless company.

9.2 Methodology

Although more subjective at times, the methodology that is followed for the RF design process is essential in establishing an RF design that correlates with the business plans of the wireless company. The methodology that is followed for the RF design process involves a look at which services need to be supported, where they will be supported, and how they will be supported. Answering the four fundamental questions of what, where, when, and how will determine the methodology for the design:

- *What* defines what you are trying to accomplish with the design.
- *Where* clarifies the issue of where the service will be introduced.
- *When* defines a time frame to follow.
- *How* clarifies the concept of how this will be realized.

More specifically, if the plan is to offer high-speed packet data services for fixed applications only and medium-speed packet services for mobility, then the methodology for implementing the packet data services will have a direct impact on the RF system design. Another issue is a decision to bifurcate the system by introducing two new platforms. Yet another issue is whether the choice is to plan for a 3G implementation or just a 2.5G platform with the concept that migration to a full 3G platform will not be considered.

Regarding the implementation of the technology, a decision needs to be made as to how it will be introduced. Two possible alternatives involve a direct overlay on the existing system for a 1:1 overlay or to proceed with only introducing the new technology with a limited footprint, say, in the core of the network.

Therefore, the methodology chosen determines the fundamental direction of the RF design itself. The methodology obviously should not be left to the purview of the technical services group, but needs to have direct involvement with senior management from marketing, sales, customer service, new technology, operations, implementation, network engineering, and, of course, the RF engineering group.

9.3 Link Budget

The establishment of a *link budget* is one of the first tasks that the RF engineer needs to perform when beginning the design process. The establishment of the link budget can only be done after a decision has been made as to which technology platform(s) to use. When introducing, say, a 2.5G platform into a 2G system, it will be necessary to have a link budget established for each of the individual technology platforms involved. In addition, with the introduction of packet data, the higher data rates have a direct influence on the range of the site and/or its capacity.

What exactly is a link budget? The link budget is a power budget that is one of the fundamental elements of a radio system design. The link budget is the part of the RF system design where all the issues associated with propagation are included. Simply put, the link budget can either be forward- or reverse-oriented; it must account for all the gains and losses that the radio wave will experience as it goes from the transmitter to the receiver.

The link budget, as it is commonly called, is the primary method that an RF engineer must first determine in order to ascertain if a valid communication link can and does exist between the sender and the recipient of the information content. The link budget, however, incorporates many elements of the communication path. Unless the actual path loss is measured empirically, the RF engineer has to estimate or rather predict just how well the RF path itself will perform. The many elements involved in the communication path incorporate assumptions made regarding various path impairments.

Figure 9-3 shows which part of the radio communication path the link budget tries to account for. The link budget has two paths: *up-link* and *down-link*. The up-link path is the path from the subscriber unit to the base station. The down-link path is the path from the base station to the sub-

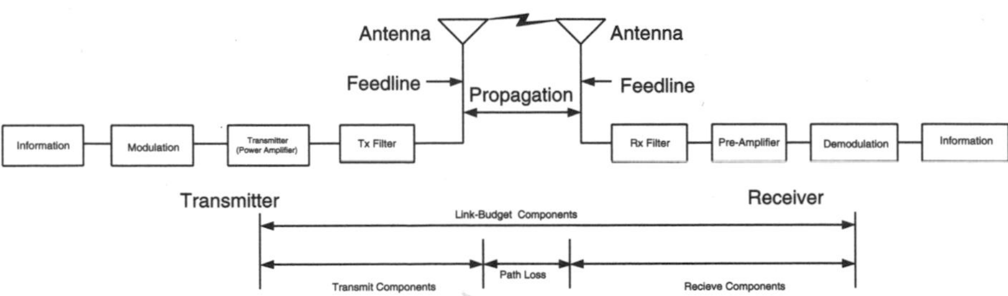

Figure 9-3 Radio path components.

scriber unit. Both the up-link path and the down-link path are reciprocal, provided they are close enough in frequency. However, the actual paths should be the same, with the exception of a few key elements that are hardware-related. The actual path loss associated with the path the radio wave transverses from antenna to antenna is the same whether it is up-link- or down-link-directed.

The maximum path loss, or limiting path, for any communication system used determines the effective range of the system. Table 9-1 involves a simplistic calculation of a link budget associated with a 1G system and is used for determining which path is the limiting case to design from. In this example, the receiver sensitivity value has the thermal noise, bandwidth, and noise figures factored into the final value presented.

The uplink path, defined as mobile to base, is the limiting path case. As shown in Table 9-1, the talk-back path is 6 dB less than the talk-out path. The limiting path loss is then used to determine the range for the site using the propagation model for the network.

However, with the introduction of a 2.5G and/or 3G platform into the wireless system, the issue of the components with a link budget becomes more complicated. The complications arise due to differing modulation techniques, bandwidth as well as process gain, and finally the Eb/No or *Carrier to noise ratio* (C/N) values required for a proper *Bit error rate* (BER) or *Frame error rate* (FER) rate. The link budget for both UMTS and CDMA2000 is included in their respective chapters, Chapter 12, "UMTS System Design," and Chapter 13, "CDMA2000 System Design." Therefore, the individual link budgets for each technology platform will not be referenced here; instead, because many issues are associated with either

Table 9-1

1G Link Budget

	1G Link Budget	
	Downlink	**Uplink**
Transmit (ERP)	50 dBm	36 dBm
Rx Antenna Gain	3 dBd	12 dBd
Cable Loss	2 dB	3 dB
Rx Sensitivity	−116 dBm	−116 dBm
C/N Ratio	17 dB	17 dB
Max Pathloss	150 dB	144 dB

CDMA2000 or UMTS and their associated legacy systems, the fundamental issues for a link budget will be discussed.

When putting together a link budget for a system, it will be common to have more than one link budget based on the morphology and, of course, the technology platform used. However, the morphology variation in the actual link budget is included in the propagation analysis for the particular site and is a varying value depending the local particulars. The link budget itself is an establishment of the maximum path loss; either in the uplink or downlink, that the signal can attenuate while still meeting the system design requirements for a quality signal.

When calculating the actual link budget, the items in Tables 9-2 and 9-3 are recommended the be included in the calculation. The items listed should have more items included in them than what may be utilized in the physical system being installed. However, the inclusion or exclusion of any of the items that can impact the link budget is included for reference. It is also highly possible that other devices can be added in the path to either enhance or potentially degrade the performance of the network.

Tables 9-2 and 9-3 help define the forward and reverse radio path components that comprise the forward and reverse link budgets. One final note on the path is that certain wireless access technologies utilize different modulation formats on both the uplink and downlink paths. If this is the case, then some of the reciprocity may not be applicable.

Because the link budget is such an integral part of the RF design process, the link budget used for the system design needs to be documented and made available for the design community to utilize.

9.4 Propagation Models

The use of propagation modeling is a requirement in the RF design process. The propagation modeling techniques used are meant to determine the attenuation of the radio wave as it transverses from the transmitter antenna to that of the receiver's antenna. The propagation model therefore is meant to characterize the radio path shown in Figure 9-4.

As with all aspects of radio design, numerous methods are used in the course of arriving at the desired result, that is, how much attenuation did the signal experience and does it exhaust the values defined in the link budget.

Some of the most popular propagation models used are Hata, Carey, Elgi, Longley-Rice, Bullington, Lee, and Cost 231, to mention a few. Each of these

Table 9-2

Generic Downlink Link Budget

Downlink Path		Units
Base Station Parameters		
	Tx PA Output Power	dBm
	Tx Combiner Loss	dB
	Tx Duplexer Loss /Filter	dB
	Jumper and Connector Loss	dB
	Lightening Arrestor Loss	dB
	Feedline Loss	dB
	Jumper and Connector Loss	dB
	Tower Top Amp Tx Gain or Loss	dB
	Antenna Gain	dBd or dBi
	Total Power Transmitted (ERP/EIRP)	**W or dBm**
Environmental Margins	Tx Diversity Gain	dB
	Fading Margin	dB
	Environmental Attenuation (building,car, pedestrian)	dB
	Cell Overlap	dB
	Total Environmental Margin	**dB**
Subscriber Unit Parameters		
	Antenna Gain	dBd or dBi
	Rx Diversity Gain	dB
	Processing Gain	dB
	Antenna Cable Loss	dB
	C/I or Eb/No	dB
	Rx Sensitivity	dB
	Effective Subscriber Sensitivity	**dBm**

models has advantages and disadvantages associated with each of them. Specifically, some baseline assumptions are used with any propagation model and need to be understood prior to utilizing them. Most cellular operators use a version of the Hata model for conducting propagation characterization. The Carey model, however, is used for submitting information to the FCC with regards to cell site filing information. Cellular and *Personal Communication Services* (PCS) operators utilize either Hata or Cost231 as

Table 9-3
Generic Uplink Link Budget

Uplink		Units
Subscriber Unit Parameters		
	Tx PA ouput	dBm
	Cable and jumper loss	dB
	Antenna Gain	dBd or dBi
	Subscriber Unit Total Tx Power (ERP, EIRP)	**W or dBm**
Environmental Margins	Tx Diversity Gain	dB
	Fading Margin	dB
	Environmental Attenuation (building,car,pedestrian)	dB
	Total Environmental Margin	**dB**
Base Station Parameters		
	Rx Antenna Gain	dBd or dBi
	Tower Top Amp Net Gain	dB
	Jumper and connector loss	dB
	Feedline Loss	dB
	Lightening Arrestor Loss	dB
	Jumper and connector loss	dB
	Duplexer /Rx Filter Loss	dB
	Rx Diversity Gain	dB
	C/I Eb/No	dB
	Processing Gain	dB
	Rx Sensitivity	dBm
	Base Station Effective Sensitivity	**dBm**

Figure 9-4
Propagation path.

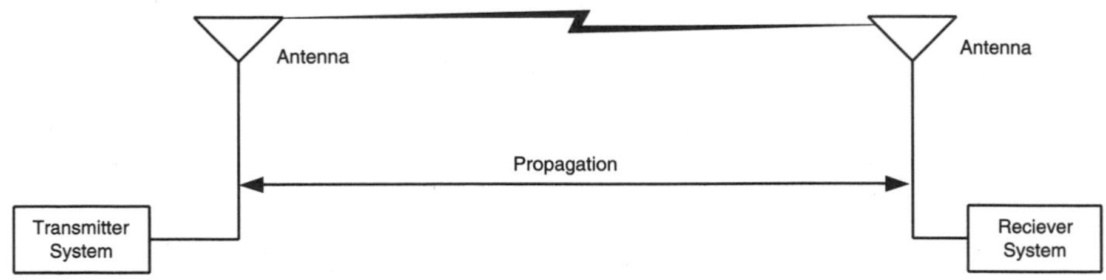

their primary method for determining path loss. With the introduction of 3G, the use of Cost231 is the model of choice to use that can be applied to any of the spectrum allocations defined by the ITU.

Regardless of the frequency band of operation, the model used for predicting coverage needs to factor into it a large amount of variables that directly impact the actual RF coverage prediction of the site. The positive attributes affecting coverage are the receiver sensitivity, transmit power, antenna gain, and the antenna height above average terrain. The negative factors affecting coverage involve line loss, terrain loss, tree loss, building loss, electrical noise, natural noise, antenna pattern distortion, and antenna inefficiency, to mention a few.

With the proliferation of cell sites, the need to theoretically predict the actual path loss experienced in the communication link is becoming more and more critical. To date, no overall theoretical model has been established that explains all the variations encountered in the real world. However, as the cellular and PCS communication systems continue to grow, a growing reliance is placed on the propagation prediction tools. The reliance on the propagation tool is intertwined in the daily operation of the wireless communication system. The propagation model employed by the cellular and PCS operator has a direct impact on the capital build program of the company for determining the budgetary requirements for the next few fiscal years. Therefore, it is essential that the model utilized for the propagation prediction tool be understood. The model should be understood in terms of what it can actually predict and what it cannot predict.

Over the years, numerous articles have been written with respect to propagation modeling in the cellular communications environment. With the introduction of PCS, there has been an increased focus on refining the propagation models to assist in planning out the networks. However, no one model can predict every variation that will take place in the environment. To overcome this obstacle, some operators have resorted to utilizing a combination of models, depending on the environmental conditions relevant to the situation.

In addition to which model would be the best to utilize, other perturbations to the model need to be considered. One of the most basic considerations is determining the morphology that the model will be applied to. Morphologies are normally defined in four categories: dense urban, urban, suburban, and rural. The selection of which morphology to utilize at times is more of an art than a direct science and this often leads to gross assumptions being made for a geographic area. The morphologies are generally defined using a rough set of criteria:

- **Dense urban** This is normally the dense business district for a metropolitan area. The buildings for the area generally are 10 to 20 stories or above, consisting of skyscrapers and high-rise apartments.
- **Urban** This type of morphology usually consists of building structures that are from 5 to 10 stories in height.
- **Suburban** This morphology is a mix of residential and business with the buildings ranging from one to five stories, but mainly consisting of one- to two-story structures.
- **Rural** This morphology, as the name applies, generally consists of open areas with structures not exceeding two stories and that are sparsely populated.

From these morphologies, it may seem obvious that classifying an area is rather ambiguous because the geographic size of the area is left to the engineer to define.

As mentioned before, several propagation models are currently utilized throughout the industry, and each of the models has pros and cons. It is through understanding the advantages and disadvantages of each of the models that a better engineering design can actually take place in a network.

9.4.1 Free Space

Free space path loss is usually the reference point for all the path loss models employed. Each propagation model points out that it more accurately predicts the attenuation experienced by the signal over that of free space. The equation that is used for determining free space path loss is based on a 20 dB/decade path loss. The free space equation is as follows:

$$L_f = 32.4 + 20\ log\ (R) + 20\ log\ (f)$$

$$where\ R = km,\ f = MHz\ and\ L_f = dB$$

The free space path loss equation has a constant value that is used for the air interface loss, a distance and frequency adjustment. Using some basic values, the different path loss values can be determined for comparison with other models that will be discussed.

9.4.2 Hata

The most prolific path loss model employed in cellular presently is the empirical model developed by Hata or some variant of it. The *Hata model* is an empirical model derived from the technical report made by Okumura, so the results could be used in a computational model. The Okumura report is a series of charts that are instrumental in radio communication modeling. The Hata model is as follows:

$$L_H = 69.55 + 26.26\,Log\,(f) - 13.87log\,(h_b) - a(h_m) + (44.9 - 6.55log\,h_b)log\,R$$

$$Where\ f = 150\text{–}1500\ MHz - Frequency$$

$$h_b = 20\text{–}200m - Height\ of\ base\ station\ above\ ground\ level$$

$$h_m = 1\text{–}10\ m - Height\ of\ receive\ antenna\ above\ ground$$

$$R = 1\text{–}20\ km = Distance\ from\ the\ site$$

$$L_H = dB$$

It should be noted that some additional conditions are applied when using the Hata model as compared to the free space equation. The values utilized are dependent upon the range over which the equation is valid. If the equation is used with parameters outside the values, the equation is defined for the results and will be suspect to error.

Therefore, the Hata model should not be employed when trying to predict a path loss less than 1 km from the cell site or if the site is less than 30 meters in height. This is an interesting point to note since cellular sites are being placed less than 1 km apart and often below the 30-meter height.

In the Hata model, the value h_m is used to correct the mobile antenna height. The interesting point is that if you assume a height of 1.5 meters for the mobile, that value nulls out or becomes 0 of the equation.

A critical point to mention here is that the Hata model employs three correction factors based on the environmental conditions that path loss prediction is evaluated over. The three environmental conditions are urban, suburban, and open.

The environmental correction values are easily calculated but vary for different values of mobile height. For the following values, a mobile height of 1.5 meters has been assumed.

Urban:	0 dB
Suburban:	−9.88 dB
Open:	−28.41 dB

9.4.3 Cost231 Walfisch/Ikegami

The Cost231 Walfish/Ikegami propagation model is used for estimating the path loss in an urban environment for wireless communication systems. The Cost231 model is a combination of empirical and deterministic modeling for estimating the path loss in an urban environment over the frequency range of 800 to 2000 MHz. The Cost231 model is used primarily in Europe for GSM modeling and in some propagation models used for cellular in the United States.

The Cost231 model is composed of three basic components:

1. Free space loss
2. Roof-to-street diffraction loss and scatter loss
3. Multiscreen loss

The equations which comprise Cost231 are listed next where

$$L_c = \begin{cases} L_f + L_{RTS} + L_{ms} \\ L_f \text{ where } L_{RTS} + L_{ms} \leq 0 \end{cases}$$

$$L_f = \text{Free space loss}$$

$$L_{RTS} = \text{Rooftop-to-street diffraction and scatter loss}$$

$$L_{ms} = \text{Multi-screen loss}$$

$$L_f = 32.4 + 20 \log R + 20 \log_{10} f_c$$

where R = km
f_c = MHz

$$L_{RTS} = -16.9 - 10 \log_{10} W + 10 \log f_c + 20 \log \Delta h_m + L_o$$

where ω = street width, m
$\Delta hm = h_r - h_m$

$$L_o = \begin{cases} -10 + 0.354\phi & 0 \leq \phi \geq 35 \\ 2.75 + .075(\phi\text{-}35) & 35 \geq \phi \geq 55 \\ 4.0 - 0.114(\phi - 55) & 55 \leq \phi \geq 90 \end{cases}$$

where ϕ = the incident angle relative to the street.

$$L_{ms} = L_{bsh} + k_a + k_d \log R + k_f \log f - 9\log b$$

where b = the distance between buildings along the radio path.

$$L_{bsh} = -18 \log (1 - \Delta h_b) \text{ when } h_b > h_r$$
$$= 0 \text{ when } h_b < h_r$$

and

$$k_a = 54 \qquad \text{when } h_b > h_r$$
$$= 54 - 0.8\, h_b \qquad \text{when } d >= 500m \text{ and } h_b <= h_r$$
$$= 54 - 1.6\, h_b \times R \qquad \text{when } d < 500m \text{ and } h_b <= h_r$$

Both L_{bsh} and k_a increase the path loss with a lower base station antenna. And for the final k factor,

$$k_f = 4 + 0.7(f/925 - 1) \text{ for a midsized city and suburban area with moderate tree density}$$
$$= 4 + 1.5(f/925 - 1) \text{ for a metropolitan center}$$

As with the Hata equation, the equation is designed to operate within a useful range, which is shown here:

$$f = 800\text{–}2000\ Mhz$$
$$h_b = 4\text{–}50\ m$$
$$h_m = 1\text{–}3\ m$$
$$R = 0.02\text{–}5\ km$$

Some additional default values apply to the Cost231 model when specific values are not known. The default values recommended are listed in the following section. The default values can and will significantly alter the path loss values arrived at.

$$B = 20\text{-}50m$$

$$W = b/2$$

$$h_r = 3 \times (\#\ floors) + roof$$

$$roof = 3\ m\ for\ pitched\ and\ 0\ for\ a\ flat\ roof$$

$$\phi = 90\ degrees$$

Figure 9-5 helps bring the Cost231 equation variables into perspective.

In the previous equations that comprise the Cost231 model, it is important, as always, to know what the valid ranges are for the model.

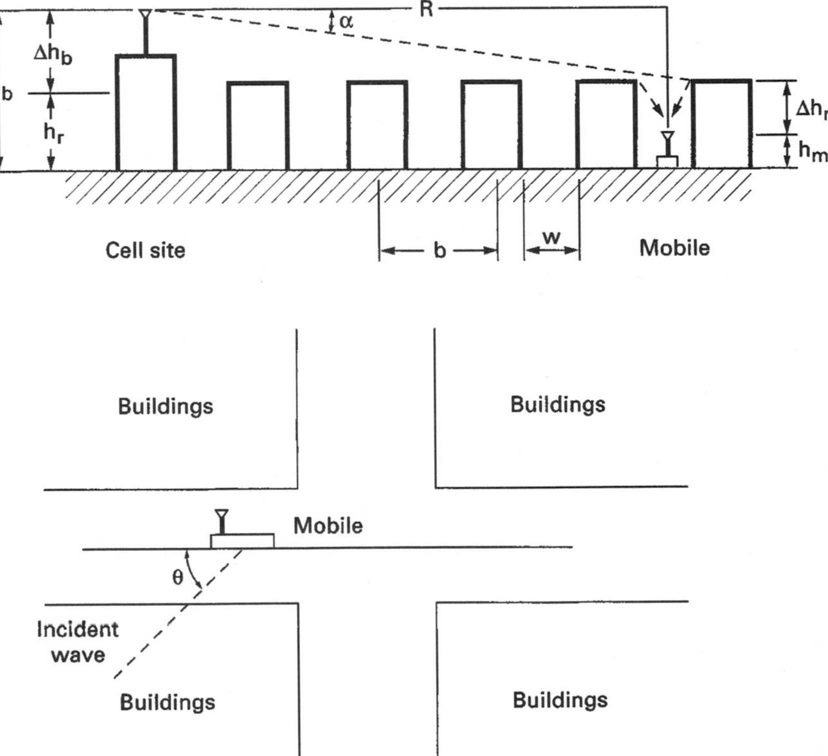

Figure 9-5 Cost231 Parameter Diagram.

9.4.4 Cost 231—Hata

The Cost231 Hata model has been tailored for the PCS 1900-MHz environment and is being used by many of the PCS operators in establishing their system design. The equation utilized for Cost 231 Hata is shown here. The Cost231 Hata model is similar to the Hata model with the exception of frequency and correction factors added based on the morphology that the model is applied to.

$$L_{CH} = 46.3 + 33.9\,Log\,(f) - 13.82 log\,(h_b) + (44.9 - 6.55 log\,h_b) log\,d + c$$

$$\begin{aligned} \text{Where c} &= 13 \text{ db dense urban} \\ &= 0\ urban \\ &= -12\ suburban \\ &= -27\ rural \end{aligned}$$

9.4.5 Quick

The *Quick model* is a down and dirty estimate that can be used to estimate the general propagation expectations for the area. The model is rather simplistic and straightforward. The advantage with this model is its quickness for use in roughly estimating the situation at hand. The disadvantage is it lacks the refinement of the other models.

The Quick method should be used when conducting some generalized approaches to a cell design and a rough answer is needed.

The Quick method utilizes two equations one for cellular, 880 MHz, and another for PCS, 1900 MHz.

$$\begin{aligned} 880\ Mhz\ PL &= 121 + 36 log\,(km) \\ 1900\ Mhz\ PL &= 130 + 40 log\,(km) \end{aligned}$$

The Quick method gives a reasonable approximation for a propagation prediction over a variety of morphologies and can be used when details regarding the particular environment may not be readily available.

Regardless of which model is used for your analysis, the propagation model or models employed by your organization must be chosen with extreme care and undergo a continuous vigil to ensure they are truly being a benefit to the company as a whole. The propagation model employed by the engineering department not only determines the capital build program, but also plays a direct factor in the performance of the network. The RF design is directly affected by the propagation model chosen and particularly by the underlying assumptions that accompany the use of the particular model.

The propagation model is used to determine how many sites are needed to provide a particular coverage requirement for the network. In addition, the coverage requirement is coupled into the traffic-loading requirements. These traffic-loading requirements rely on the propagation model chosen to determine the traffic distribution, or off-loading, from an existing site to new sites as part of the capacity relief program. The propagation model helps determine where the sites should be placed in order to achieve an optimal position in the network. If the propagation model used is not effective in helping place sites correctly, the probability of incorrectly justifying and deploying a site into the network is high.

Reiterating the point that, although no model can account for all the perturbations experienced in the real world, it is essential that you utilize one or several propagation models for determining the path loss of your network.

9.5 Tower-Top Amplifiers

The use of tower-top amplifiers has been deployed in numerous communication sites and is anticipated to be used in the introduction of 2.5G and 3G also. The use of a tower-top amplifier has occasionally been misapplied in that the gain exhibited from the tower-top amplifier is added directly to the link budget. However, the purpose of the tower-top amplifier is to improve the noise figure for the receive system.

The noise figure is improved by having the first amplification stage placed as close as possible to the antenna itself, thereby eliminating the loss experienced due to the feedline that connects the antenna to the rest of the receive system. The location of the tower top amplifier is shown in Figure 9-6

The tower-top amplifier has to have a minimal gain of 10 dB and, because the feedline usually has 2 to 4 dB of loss, the additional gain needs to be attenuated by the insertion of a resistive pad, shown in Figure 9-6.

The typical improvement in the receive path due to the introduction of the tower-top amplifier is equal to the line loss that would have been attributed to the feedline, nothing else. The negative issues with tower-top ampli-

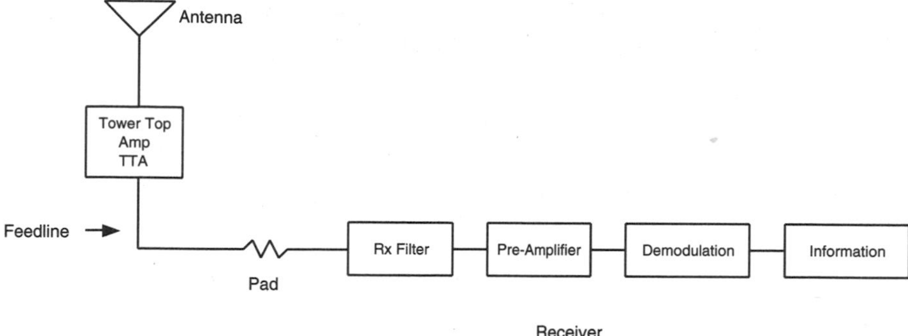

Figure 9-6
Tower-top amplifier.

fiers include the requirement for power, in DC, to be supplied to the unit and increased maintenance issues in the event of a failure. Another problem is the increased system noise due to the amplifier having a less-than-optimal receive filter due to the size and weight restrictions imposed with installing the unit on a tower.

9.6 RF Design Guidelines

No true RF engineering can take place without some RF design guidelines, whether formal or informal. However, with the level of complexity introduced when integrating a 2.5G or 3G platform into an existing system, the need for a clear definitive set of design guidelines is paramount for success. Although this concept seems straightforward and simple, many wireless engineering departments when pushed have a difficult time defining what exactly their design guidelines are.

The actual format, or method of how it is conducted, should be structured in such a fashion as to facilitate ease, documentation, and minimization for formal meetings. For most of the design reviews, a formal overhead presentation is not required; instead, a meeting with the manager of the department is the level of review that is needed. It is also important that another qualified member of the engineering staff reviews the material in order to prevent the common or simple mistakes from taking place. Ensuring that a design review process is in place does not eliminate the chances of mistakes occurring. Design reviews ensure that when mistakes do take place, the how, why, and when issues needed to expedite the restoration process are already in place.

It is highly recommended that the department's RF design guidelines are reasonably documented and updated on a predetermined basis, yearly at a minimum. The use of design guidelines will facilitate the design review process and establish a clear set of directions for the engineering department to follow. The RF design guidelines will also ensure that a consistent approach is maintained for designing and operating the capital infrastructure that has or will be put into place within the network.

The actual design guidelines that should be utilized by the RF engineers need to be well documented and distributed. The design guidelines, however, do not need to consist of voluminous amounts of data. The design guidelines should consist of a few pages of information that can be used a quick reference sheet by the engineering staff. The design guideline sheet has to be based on the system design goals and objectives set forth in the RF system design.

The actual content of the design guideline can and will vary from operator to operator. However, it is essential that a list of design guidelines be put together and distributed. The publication and distribution of RF design guidelines will ensure a minimum level of RF design specifications exist in the network.

The proposed RF design guideline is shown in Table 9-4 and is a generic wireless system. The guideline can easily be crafted to reflect the particular design guidelines utilized for the market where it will be applied. In addition, a need exists to have RF design criteria for each technology platform with links to each other to ensure that one platform, by mistake, is not factored over another.

9.7 Cell Site Design

Although this is not necessarily the first step in any design process, it is one of the most important for the RF Engineering department. The reason the cell site design is critical lies in the fact it is where the bulk of the capital is spent. The cell site design guidelines can be utilized directly or modified to meet your own particular requirements.

The use of a defined set of criteria will help facilitate the cell site build program by improving interdepartmental coordination and provide the proper documentation for any new engineer to review and understand the entire process with ease. Often when a new engineer comes onto a project, all the previous work done by the last engineer is reinvented,

Table 9-4

RF Design Guidelines (Per Radio Access Platform)

RF Design Guideline

System Name

Date:

	RSSI	**ERP**	**Cell Area**	**Antenna Type**	
Urban	−80 dBm	16W	3.14 km/sq	12 dBd	90H/14E
Suburban	−85 dBm	40W	19.5 km/sq	12 dBd	90H/14E
Rural	−90 dBm	100W	78.5 km/sq	10 dBd	110H/18E

Eb/No	7 dB (90th percentile)
Frequency Reuse or N=1	
Maximum # carriers per secrtor	
Maximum # traffic channels per carrier	
Maximum Kbps per sector	
Sector Cell Orientation	0, 120, 240
Antenna Height:	100 feet or 30 Meters
Antenna Pass Band	XXX-XXX MHz
Antenna Feedline Loss	2 dB
Antenna System return Loss	20–25dB
Diversity Spacing	d=h/11 (d = Receive antenna spacing, h=antenna AGL)
Receive Antennas per Sector	2
Transmit Antennas per Sector	1
Roof Height Offset	h=x/5 (h=height of antenna from roof, x=distance from roof edge)

Performance Criteria

Lost Call Rate	<2%
Attempt Failure	< 2%
RF Blocking (voice)	1%><2%
FER	<1%

primarily due to a lack of documentation and/or design guidelines from which to operate from.

The cell site design process takes on many facets and each company's internal processes are different. However, no matter what the internal process you have, the following items are needed as a minimum:

- Search area
- Site qualification test (SQT)
- Site acceptance
- Site rejection
- FCC guidelines
- FAA guidelines
- *Electro Magnetic Force* (EMF) compliance

9.7.1 Search Area

The definition of a *search area* and the information content provided is a critical first step in the cell site design process. The search area request is a key source document that is used by the real estate acquisition department of the company. The selection and form of the material presented should not be taken lightly because more times than not the RF engineers rely heavily upon the real estate group to find a suitable location for the communication facility to exist. If the search area definition is not done properly in the initial phase, it should not be a surprise when the selection of candidate properties is poor.

The search areas issued need to follow the design objectives for the area following the RF system design objectives. The search area should be put together by the RF engineer responsible for the site's design. The final paper needs to be reviewed and signed by the appropriate reviewing process, usually the department manager, to ensure that checks and balances are used in the process. The specifications for the search area document need to not only meet the RF Engineering department's requirements, but also the real estate and construction groups' needs. Therefore, the proposed form needs to be approved by the various groups, but be issued by the RF Engineering department. It is imperative that the search area request undergoes a design review prior to its issuance.

9.7.2 Site Qualification Test (SQT)

The Site Qualification Test (SQT) is an integral part of any RF system design. Even in the age of massive computer modeling, it is still essential that every system has some form of transmitter or site qualification test conducted. The fundamental reason behind requiring a test is to assure that the site is a viable candidate before a large amount of company capital is spent on building the site. This test is also required to make sure the site will operate well within the network. The financial implications associated with accepting or rejecting a transmitter necessitates a few thousand dollars expended in the front end of the build process. If a site is accepted that will not perform its intended mission statement, additional capital will need to be spent to accomplish it.

Based on the volume of sites required within a specified time frame, it may not be possible to physically test every cell site candidate. Therefore, it is essential that a goal be defined as to many sites should be physically tested. The establishment of a goal for physically testing or using a propagation model evaluation will help establish the risk factors associated with the building of the network.

Regardless of whether a site is to be physically tested or evaluated through a computer simulation, several stages need to be done in this process. It is very important that the SQT be performed properly since this will determine the cost of the potential facility, which could range from $500,000 to $1.

It is strongly recommended that the RF engineer responsible for the final site design visit the location prior to any SQT taking place. This site visit will facilitate several factors. First, the engineer will now have a better idea of the potential usefulness of the site and its capability to be built. He or she can also provide more accurate instructions to the testing team.

It is strongly recommended that the RF engineer does not design the test on the fly by telling the testing team where to place the transmitter and which routes to drive. The desired approach is to have the engineer determine where to place the transmitter, either as part of the tower or rooftop, and the location for the crane. The RF engineer then puts together his or her test plan, identifying the location of the transmitter antenna, the ERP, the drive routes, and any particular variations. The test plan is then submitted to the manager of the department for approval and is then passed to the SQT team.

9.7.3 Site Acceptance (SA)

Once a site has been tested for its potential use in the network, it is determined to either be acceptable or not acceptable. For this section, the assumption will be that the site is acceptable for use by the RF Engineering department as a communication facility. It is imperative that the desires of the RF Engineering department be properly communicated to all the departments within the company in a timely fashion. The method of communication can be done verbally at first, based on time constraints, but a level of documentation must follow that will ensure that the design objectives are properly communicated.

The forms outlined later in the chapter are meant to be general guides and might need to be modified based on your particular requirements. Before the *site acceptance* (SA) is released, it is imperative that it go through the design review process to ensure that nothing is overlooked. The SA will be used to communicate the RF Engineering department's intention for the site and will be a key source document used by Real Estate, Construction, Operations, and the various subgroups within the Engineering department itself.

The SA will also need to be given a document control number to ensure that changes in personnel during the project are as transparent as possible.

9.7.4 Site Rejection (SR)

In the unfortunate event that a potential site has been tested and is determined not to be suitable for a potential use in the network, a *site rejection* (SR) form needs to be filled out. The issuance of an SR form may seem trivial until a change of personnel occurs and the site is tested again at a later date. The SR form serves several purposes. The first purpose is that it formally lets the Real Estate Acquisition team know that the site is not acceptable for engineering to use, and they need to pursue an alternative location. The second purpose is that this process identifies why the site does not qualify as a potential communication site. The third purpose ties into future use when the SQT data is stored, and when the site might be more favorable for the network.

It is recommended that the SR process include a design review with a sign-off by the manager. This is to ensure that the reasons for rejecting the site are truly valid and the issues are properly communicated. The form proposed in the SR needs to be distributed to the same parties that the SA would be sent to. The reason is that if a site does not meet the design criteria specified at this time, it does not mean it will always be unsuitable. Therefore, it is imperative that the SQT information collected for this site be stored in the search area's master file. The storing of the SQT information in the central file will assist later design efforts that could involve the capacity or a relocation of existing sites to reduce lease costs.

9.7.5 Site Activation

The *activation* of a cell site into the network is exciting. It is at this point that the determination is made for how effective the design of the cell site is in resolving the problem area. Numerous steps must be taken after the site acceptance process. The degree of involvement with each of these steps is largely dependent upon the company resources available and the interaction required to take place between the Engineering and Construction departments.

At a minimum, these two groups should perform site visits together. These site visits involve the group responsible for the cell site's architectural drawings and the overall design of the site's structure. Regardless of the interaction between the groups, when it comes to show time, it is imperative to have a plan of action to implement.

9.7.6 FAA Guidelines

Federal Aeronautics Administration (FAA) compliance is mandatory for all the sites within a system. The verification of whether the site is within FAA compliance should be covered during the design review process. If a site does not conform within the FAA guidelines, then a potential redesign might be in order to ensure FAA compliance.

The overall key elements that need to be followed for compliance are as follows:

- Height
- Glide slope
- Alarming
- Marking and lighting

The verification of the height and glide slope calculations is needed for every site. It is recommended that every site have the FAA compliance checked and included in the master site reference document. If no documented record had been made for a site regarding FAA compliance, it is strongly recommended that this be done immediately. The time and effort required to check FAA compliance is not long and could be done within a week for a several-hundred-cell system.

9.7.8 EMF Compliance

EMF compliance needs to be factored into the design process and the continued operation of the communication facility. The use of a EMF budget is strongly recommendedand can ensure personnel safety and government compliance. A simple source for the EMF compliance issue should be the company's EMF policy.

The establishment of an EMF power budget should be incorporated into the master source documents for the site and be stored on the site itself, identifying the transmitters used, the power, who calculated the numbers, and when it was last done. As a regular part of the preventative maintenance process, the site should be checked for compliance and changes to the fundamental budget calculation.

The method for calculating the compliance issue is included in the IEEE C95.1-1991 specification with measurement techniques included in C95.3. Both cellular and PCS utilize the same C95.1-1991 standard. Currently, different guidelines are used for different wireless services. It is recommended that the C95.1-1991 specification be used for all the wireless services. However, be sure that the license you are operating under complies with the applicable EMF standard.

9.8 RF Design Report

As with any good effort, the results need to be documented and communicated to the respective parts of the wireless organization. The following is a proposed guideline that can be used to help construct such a report. The report is not inclusive of the circuit-switched and packet-fixed networks; it just applies to the RF portion of the system.

9.8.1 Cover Sheet

This is the cover sheet for the report and should include the following items:

- The system it is meant for (such as "New York Metro")
- Date of issuance (7/1/2001)

- Revision number
- Who or which group issued the report
- Confidentiality statement (this should be on every page of the document, usually in the footer)

9.8.2 Executive Summary

This is a one- or two-page summary that includes the findings from the report and is meant to serve as a base from which most management decisions are made.

9.8.3 Revision

This is meant to document which version of the report this particular version is. The sign-off section that is included is meant to ensure that the version that is under scrutiny is the current one and has undergone a design review.

The format of the revision page should be as follows:

Date	Originator	Reviewed by	Comments	Rev. Number

9.8.4 Table of Contents

The Table of Contents section is meant to serve as a simple reference point so that anyone picking the package up can quickly find a particular section without having to read the whole document.

The suggested format for the Table of Contents is shown here:

- Page
- Introduction
- Revision
- Coverage objectives

9.8.5 Introduction

This is where the description of the objective is meant to be discussed. Specifically, the topics here should cover the market, the general types of equipment, and who this document is intended for. Also included with the Introduction is the time frame this report is meant to cover. Specifically, if this is a new system, then the time frame for the validity of this report could be one year. However, if this is an existing system, or one that is particularly built out, then the time frame may also be one year but should really be two years as a minimum.

9.8.6 Design Criteria

The design criteria used for the establishment of the design should be listed here. The inclusion of the link budget and propagation modeling assumptions need to be listed here as well.

9.8.7 Existing System Overview

This section is meant to describe which areas the system incorporates. A map showing the physical boundaries will also be necessary for this section. The key elements that need to be included here include the technology used and the changes envisioned in the future.

9.8.8 Coverage Objectives

This section is meant to describe the coverage objectives for the system. The following are suggested points that need to be covered in this section:

- What is the current coverage of the system?
- What are the coverage requirements?
- Which areas need coverage?

This information should be derived from the Marketing, Operations, RF, and System Performance Engineering departments.

This section should include a map of the geographic area that encompasses the system. The map should include which type of coverage objective is desired and its approximate differentiation on a map. The map could either be of the existing system or of an area that is currently being contemplated for building a system.

If multiple phases are associated with the build program, then this should also be reflected in the coverage objective section. In the event of multiple phases, a map showing the overall plan differentiating the different phases should be included also. Such a map will be used as the foundation for the deployment of resources that will be tied into the overall system design document.

If the system is currently existing, then a coverage map should also be included with this section. The coverage map should convey where the current system coverage problems are.

9.8.9 Coverage Quality

This section is meant to describe the coverage quality requirements for the system. The coverage quality is a series of parameters that will be used to clearly define the link budget requirements for the system and the geographic areas within the network.

The coverage quality is meant to define the different morphology requirements that will be used in determining how much of an area will need to be satisfied by the coverage requirements. The coverage quality could also include not only the cell edge coverage requirements, but also the overall coverage requirements for the cell itself, depending on the morphology that the cell is referenced to.

9.8.10 Inter System Coverage

This section of the report would include the coverage requirements needed to provide contiguous coverage into another market. Specifically, this would be applicable for an area of the system that, say, interfaces to another *Basic Trading Area* (BTA), *Metropolitan Trading Area* (MTA), *Cellular Geographic Service Area* (CGSA), or *Rural Service Area* (RSA) and the capability to handle roaming traffic was desired.

The coverage objective here should define what the overlap should be in terms of dB and where the coverage objective should be. A map indicating the desired geographic areas would be directly applicable here. Also, any comments with regards to the other system's build program and coverage objectives should be listed here also.

9.8.11 Link Budget

The calculations and assumptions that comprise the link budget need to be included in this section. The link budget format shown in Tables 9-2 and 9-3 should be utilized for this section.

9.8.12 Analysis

This is where the analysis conducted is put into the report. Issues that are included pertain to the spectrum utilization, channels selected, and migration strategy.

9.8.13 Summary of Requirements

This section is the end result of the design work and should include a summary table that indicates the amount of capital, either in product and dollars or just in product.

References

American Radio Relay League. "*The ARRL 1986 Handbook*," 63rd Ed., The American Radio Relay League, Newington, Conn., 1986.

American Radio Relay League. "*The ARRL Antenna Handbook*," 14th Ed,. The American Radio Relay League, Newington, Conn., 1984.

AT&T. "*Engineering and Operations in the Bell System*," 2nd Ed., AT&T Bell Laboratories, Murry Hill, N.J. 1983.

Carr, J.J. "*Practical Antenna Handbook*," Tab Books, McGraw-Hill, Blue Ridge Summit, PA,1989.

Fink, Beaty. "*Standard Handbook for Electrical Engineers*," 13th Ed., McGraw Hill, 1995.

Fink, Donald and Donald Christiansen. "*Electronics Engineers Handbook*," McGraw-Hill, 3rd Ed., New York, NY, 1989.

Lathi. "*Modern Digital and Analog Communication Systems*," CBS College Printing, New York, NY, 1983.

Lynch, Dick. "*Developing a Cellular/PCS National Seamless Network*," Cellular Integration, Sept. 1995, pgs. 24–26.

Jakes, W.C. "*Microwave Mobile Communications*," IEEE Press, New York, NY, 1974.

Johnson, R.C., and H. Jasik. "*Antenna Engineering Handbook*,", 2nd Ed., McGraw-Hill, New York, NY, 1984.

Kaufman, M., and A.H. Seidman, "*Handbook of Electronics Calculations*," 2nd Ed., McGraw-Hill, New York, NY, 1988.

MacDonald. "*The Cellular Concept*," Bell Systems Technical Journal, Vol. 58, No. 1, 1979.

Miller, Nathan, "*Desktop Encyclopedia of Telecommunications*," McGraw-Hill, 1998.

Pautet, Mouly. "*The GSM System for Mobile Communications*," Mouly Pautet, 1992.

"*Reference Data for Radio Engineers*," Sams, 6th Ed., 1983.

Schwartz, Bennett, Stein. "*Communication Systems and Technologies*," IEEE, New York, NY, 1996.

Smith, Clint. "*Practical Cellular and PCS Design*," McGraw-Hill, 1997.

Smith, Clint. "*Wireless Telecom FAQ*," McGraw-Hill, 2000.

Smith, Gervelis. "*Cellular System Design and Optimization*," McGraw-Hill, 1996.

Steele. "*Mobile Radio Communications*," IEEE, 1992.

Stimson. "*Introduction to Airborne Radar*," Hughes Aircraft Company, El Segundo, CA, 1983.

Webb, Hanzo. "*Modern Amplitude Modulations*," IEEE, 1994.

Webb, William. "*Introduction to Wireless Local Loop, Second Editions: Broadband and Narrowband Systems*," Artech House, Boston, MA, 2000.

White, Duff. "*Electromagnetic Interference and Compatibility*," Interference Control Technologies, Inc., Gainesville, GA, 1972.

William, C.Y Lee. "*Mobile Cellular Telecommunications Systems*," 2nd Ed., McGraw-Hill, New York, NY, 1996.

Williams, Taylor. "*Electronic Filter Design Handbook*," McGraw-Hill, 3rd Ed., 1995.

Winch, Robert. "*Telecommunication Transmission Systems*," 2nd Ed., McGraw-Hill, 1998.

www.fcc.gov

Yarborough. "*Electrical Engineering Reference Manual*," 5th Ed., Professional Publications, Inc., Belmont, CA, 1990.

CHAPTER 10

Network Design Considerations

10.1 Introduction

Chapter 9, "3G System RF Design Considerations," addressed the RF design issues related to the implementation of a 3G network. This chapter focuses on the design of the non-RF aspects of the network. Thus, we consider issues such as placement and dimensioning of *Mobile Switching Centers* (MSCs), *Base Station Controller* (BSC), *Serving GPRS Support Node* (SGSN), *Packet Data Serving Node* (PDSN), and so on. We also address the connectivity and transport requirements between the various network elements.

In general, the design of the core network involves striking a balance between three requirements—meeting or exceeding the capacity needed to handle the projected demand; minimizing the capital and operational cost of the network; and ensuring high network reliability/availability. In short, we can refer to these three issues as cost, capacity, and quality. Of course, meeting one or more of the requirements often means making sacrifices elsewhere such that it is impossible to divorce one network design consideration from any of the others. For example, a lower cost might well mean a lower network capacity or a lower network quality. Thus, we will never get a network that is remarkably cheap to implement and operate while still offering high capacity and high quality. Instead, we must aim to establish some "happy medium" where we satisfy at least the most important criteria.

10.2 Traffic Forecasts

Obviously, we need to design a network that will support the projected traffic demand. Consequently, projecting subscriber usage is a critical first step in the network design process. This projection often involves a certain amount of up-front guesswork, particularly if this is the first network of a given type in a given market. If one is building a network to compete with someone else's established network, then one can forecast subscriber growth based on the competitor's subscriber numbers, which are often publicly available. If, however, one is building the first network in a given market or one is building a network very soon after a competing network has been launched, then less data is available. In such a situation, one needs to make educated estimates based upon factors such as average household income, existing penetration of mobile voice service (such as 2G service), average Internet usage in the market, and similar data.

Traffic forecasts need to address several considerations, which include the total subscriber numbers, per-subscriber voice usage, per-subscriber data usage, and signaling demand.

10.2.1 Subscriber Forecast

For a given market, an estimate of total subscriber population is needed. Ideally this should be broken down on a monthly basis so that we have an understanding of how subscriber numbers will grow over time. This is necessary as the design of the network will involve a certain amount of build-ahead.

If there are to be a number of different commercial offerings, then there may well be a number of different subscriber categories, in which case forecasts are needed for each type of category. For example, a network operator may choose to offer some combination of services involving voice-only, voice and data, or data-only. Moreover, the data services may be further broken down depending on commercial offerings and subscriber devices. For example, one data offering might be limited to Web browsing, another might include Web browsing, e-mail service provided by the network operator, plus some other service such as Web space. Yet another data service might be aimed at telematic devices. Forecasts are needed for each type of user category.

10.2.2 Voice Usage Forecast

A voice usage forecast involves an estimation of the amount of voice traffic generated by the average voice user. Ideally this should also be provided on a monthly basis. The voice profile should include the distribution of traffic in terms of mobile-to-land, land-to-mobile, mobile-to-mobile, and mobile-to-voice mail. For the mobile-to-land aspect, there should also be a breakdown of what percentage is local and what percentage is long distance. Ideally, the voice usage profile information should include the average number of calls per subscriber in the busy hour and the *mean holding time* (MHT) per call. Quite often, however, marketing organizations are likely to simply provide information in terms of *minutes of use* (MoUs) per subscriber per month. In that case, it is up to the engineering organization to derive the busy-hour usage. The following is an example of how this can be done.

Example: Average user has 400 MoUs per month.

Assume for example that 90 percent of traffic occurs during work days (that is, only 10 percent on weekends).

Assume 21 work days per month.

Assume that in a given day, 10 percent of voice traffic occurs during the busy hour.

Then the average busy-hour usage (in MoUs) per subscriber is given by

$$(MoUs\ per\ month) \times (fraction\ during\ work\ days)$$
$$\times\ (percentage\ in\ busy\ hour)/(work\ days\ per\ month).$$

Thus, in our example, we get $400 \times 0.9 \times 0.10/21 = 1.71$ MoU/sub/busy hour.

Dividing by 60 gives the number of Erlangs = 0.0286 in our example = 28.6 milliErlangs.

If we multiply this number by the total number of subscribers, then we can determine the total busy-hour Erlang demand, which is a critical network dimensioning factor. What we also need, however, is the total number of call attempts as some network elements are limited more by the processing effort involved in call establishment rather than the total throughput. If we assume that most calls are completed (which is often the case in today's world of voice mail), then determining the number of *busy-hour call attempts* (BHCAs) is done by the following formula:

$$BHCA = (Traffic\ in\ Erlang) \times (3600)/(MHT\ in\ seconds)$$

If in our example we assume a MHT of 120 seconds, then we get a per-subscriber BHCA of

$$0.0286 \times 3600/120 = 0.86.$$

Thus, the average subscriber makes 0.86 call attempts in the busy hour.

10.2.3 Data Usage Forecast

As mentioned, we need to address the various categories of data users and forecast for each user type and the amount of data throughput. We also need to forecast where the throughput begins and ends. If for example a given user has Web browsing service plus operator provided e-mail, then a certain amount of traffic will terminate on an e-mail server within the operator's network, while a certain amount of traffic will be sent to and from the Internet. The dimensioning of the interfaces to the e-mail system and to the Internet will depend on the amount of traffic related to those services. Moreover, the e-mail system will need to be dimensioned to meet requirements for total number of users, total storage, and total traffic in and out.

For each type of user and data service, we perform a similar analysis to determine busy-hour usage. We assume, for example, a certain amount of usage during work days and a certain percentage in the busy hour. From this we calculate the average throughput per user and per type of service in the busy hour. This throughput should be calculated in bps, and the uplink/downlink split should be specified. For most services, we will find that the downlink traffic is far greater than the uplink traffic with an 80 percent/20 percent split being common. Once we have determined the busy-hour usage, we need to add some buffer to allow for burstiness or peaks within the busy hour. The amount of buffer to be added will depend on the amount of build-ahead factored into the network design. If for example there is already a build-ahead of 12 months, then the system will be purposely over dimensioned at the beginning, in which case a further buffer would be wasteful. On the other hand, if little build-ahead has been factored in, then a 25 percent buffer for data traffic peaks could well be appropriate.

It should be noted that the busy hour for voice traffic and the busy hour for data traffic might not coincide. Given that for many network technologies, different core network nodes are used for voice and data, whether the two busy hours happen to be the same will often not be an issue for network node dimensioning. For example, in 3GPP Release 1999, voice traffic is handled by an MSC and data traffic by an SGSN. Similarly, in CDMA 2000, voice traffic is handled by an MSC and data traffic is handled by a PDSN. Therefore, for dimensioning of an SGSN or PDSN, whether the voice busy hour is coincident with the data busy hour is of no consequence.

The same cannot be said for the access network, however. Nor can the same necessarily be said for the backbone transport network. On the access network, for example, the capacity of a BSC or *Radio Network Controller* (RNC) will be determined both by the voice usage and the data usage. In the core network backbone, we may wish to use VoIP (such as with 3GPP Release 4), in which case the backbone network will carry both voice and packet data, in which the issue of coincident busy hours is important. Until experience tells us otherwise, it is wise to assume the worst case—that is, that the voice busy hour and data busy hour are coincident.

10.3 Build-Ahead

It makes no sense to design a network to support the traffic demand that we expect today. This means that we must return tomorrow to enhance the network capacity. Instead, we need to design the network to support the demand that we expect at some point in the future so that we are not enhancing network capacity on a daily basis. Moreover, a reasonable build-ahead provides extra capacity so that the network is prepared to handle extra traffic in case subscriber growth is greater than projected. Build-ahead also provides a buffer in case of a sudden change in marketing tactics. For business reasons it may be necessary to introduce new pricing plans or incentives, which can significantly change subscriber numbers or usage patterns. It is wise to have the network prepared in advance for such eventualities.

So how much build-ahead is reasonable? Typically, it is wise to design the network to support the traffic demand expected 6 to 12 months in the future. If for example we launch a network in December of 2001, then a 12-month build-ahead would mean that we use the subscriber forecasts and usage projections applicable to December 2002 as input to the network design process. In general, the build-ahead can be larger at the beginning and be reduced over time. If for example we include a 12-month build-ahead at the beginning, we might want to reduce this to a 6-month build ahead after 2 years as we will have a better understanding of traffic growth patterns, and usage forecasts (assuming they are updated on a regular basis) will be more dependable.

10.4 Network Node Dimensioning

In order to determine the number of nodes of each type in the network, we must first understand the dimensioning rules associated with each type of node. If we understand the capacity limits of a given node type, then we can determine the minimum required number of nodes of that type. For a number of reasons, it is likely that we will deploy greater than the minimum number, but we must at least know the starting point.

Of course, for a given node type, the dimensioning rules and capacity limits will vary from vendor to vendor. Any examples provided in the following sections should be considered examples only and do not necessarily reflect the characteristics of any given vendor's implementation.

10.4.1 BSC Dimensioning

Typically a BSC will have a number of capacity limitations. The following types of limitations are typical:

- Maximum number of *transceivers* (TRXs) (such as 256 or 512)
- Maximum number of base stations (such as 128, 256, or 512)
- Maximum number of cells (that is, sectors) (such as, 256 or 512)
- Maximum number of packet data channels (such as 2000)
- Maximum number of physical interfaces (such as 128)
- In many cases, a BSC from a given vendor has a fixed capacity based on a combination of the previous limitations. In determining the number of BSCs required, one analyzes the RF design in the market and calculates the number of BSCs based on which limitation imposes the greatest restraint. Imagine for example that a given market has 200 sites, each with 3 sectors and 1 TRX per sector. Consider a BSC model that can support up to 256 sites, 256 sectors, and 512 TRXs. Then based on the site counts, we need two BSCs; based on the sector count, we need three BSCs, and based on the TRX count, we need two BSCs. Therefore, it is necessary to deploy at least three BSCs.

10.4.2 UMTS RNC Dimensioning

Although a BSC is generally limited by the number of RF network elements (such as sites, sectors, and TRXs) that can be supported, the capacity of an RNC tends to be traffic or throughput limited. This is because of the fact that an RNC can be involved in traffic handling for base stations that it does not directly control. For example, an RNC can act as a serving RNC or drift RNC during soft handover. In such cases the RNC may be handling traffic to/from a base station that it does not control. Thus, the number of controlled base stations becomes less important and the amount of traffic handled is of greater significance to the capacity of the RNC.

The capacity of an RNC is typically limited by a combination of the following factors:

- Total Erlangs
- Total BHCA
- Total voice subscribers

- Total data subscribers
- Total Iub interface capacity (Mbps)
- Total Iur interface capacity (Mbps)
- Total Iu interface capacity (Mbps)
- Total switching capacity (Mbps)
- Total number of controlled base stations
- Total number of RF carriers

The determination of the number of RNCs required in a given market will be based on which of these limitations is the most restrictive.

Unlike the situation for BSCs, it is more common for RNCs from a given vendor to be offered in a variety of configurations. For example, the Iu interface might be offered using different transmission interface capacities (such as E1, T1, or STM-1). Moreover, a given vendor's RNC might come in several multi-cabinet configurations, where one can start with a small configuration and expand capacity by adding additional cabinets.

The determination of the number of required RNCs is more complex than the equivalent determination of the number of required BSCs. In particular, the effect of soft handover needs to be considered. Imagine for example that there are two RNCs supporting a number of base stations. One RNC limitation will be the total switching capacity. If there is a great deal of inter-RNC soft handover, then switching capacity is consumed on both RNCs. In fact, switching capacity can be consumed on both RNCs even after the soft handover is finished if SRNS relocation has not yet taken place. Thus, the determination of the number of RNCs needs to consider not just the RF elements and not just the overall traffic load, it must also consider the effects of soft handover. For this reason, the calculation of the number of required RNCs should be done is close cooperation with the RF design effort.

In most cases, we find, however, that the most limiting factor is the Iub interface capacity.

10.4.3 MSC Dimensioning

In the case where the network technology involves BSCs (or RNCs) that are separated from the MSC, then the MSC capacity generally has two limitations—maximum BHCA and maximum Erlang. In networks where the BSC functionality is included within the same machine as the MSC, then there will also be limitations in termed of RF elements supported (such as sites, sectors, and TRXs).

The separation of BSCs or RNCs from the MSC is the most common configuration in 3G networks. Consequently, the MSC capacity is generally not limited by RF-specific factors. Thus, the capacity is Erlang or BHCA limited. (There may be a total cell limit, but this is generally sufficiently large that it is not a limiting factor.)

Although we say that the capacity of an MSC is typically Erlang or BHCA limited, the reality is that the BHCA limit is the real bottleneck. Although BHCA and Erlangs are closely related, Erlangs reflect the switching capacity and port capacity of the MSC, whereas BHCA reflects the processing power of the MSC. In general, the number of supported Erlangs can be increased by the addition of extra MSC hardware, while the maximum BHCA for a given release of MSC is typically fixed. Thus, it is usually possible to add hardware and increase the supported Erlangs until such time as the BHCA limit is reached. Adding extra hardware after this point provides no extra capacity. Thus, when determining the number of MSCs required to support a given market, the calculation is BHCA-based.

When we come to distributed architectures, such as the MSC Server—Media Gateway architecture of 3GPP Release 4, many of the same dimensioning rules will still apply. In this case, the MSC Server is most likely to be BHCA limited, while the Media Gateway is likely to be Erlang limited.

Today's MSCs typically have BHCA limitations in the order of 300,000 to 500,000 BHCA. As technology advances, these numbers will increase, and capacities of up to 1,000,000 BHCA will be common in the next few years.

For most vendors, the configuration of a given MSC in a given market is a custom configuration. In other words, the size of the switching matrix and the numbers and types of ports are custom designed to meet the specific market requirements. If one is building a limited number of markets at a given time, this is the optimum approach. On the other hand, if one is attempting to build a large network (such as a nationwide deployment) with many MSCs, custom design of the hardware configuration for each MSC may be overly time consuming and may jeopardize a timely launch. In such a situation, it is often wise to work with the MSC vendor to define a number of network-specific standard configurations, such as small, medium, and large configurations, depending on the types of markets to be supported. Thus, in a large metropolitan city, one might need to deploy two large MSCs, although is a smaller city, one might need only a single medium-sized MSC or a single small-sized MSC. Although this approach is not optimal from a hardware perspective, judicious determination of the different configurations is likely to ensure that there is not a great deal of over dimensioning. The resulting ease of cookie-cutter design and easier ordering and delivery may well result in savings in the design effort and more rapid deployment.

10.4.4 SGSN and GGSN Dimensioning

In UMTS we continue to use SGSNs and *Gateway GPRS Support Node* (GGSN) largely as they are used in standard *General Packet Radio Service* (GPRS) and the dimensioning rules that apply in UMTS are similar to those that apply in GPRS.

The dimensioning limits applicable to an SGSN are generally as follows:

- Total number of simultaneously attached subscribers
- Total number active PDP contexts
- Total number of Gb or Iu-PS interfaces
- Total number of routing areas
- Total throughput

It is common to find that the real bottlenecks will involve the total number of attached subscribers or the total throughput. Of course, the limitations will vary from vendor to vendor, but typical values will range from 25,000 to 150,000 attached subscribers. As with any technology, these limits tend to increase over time, so that much higher capacities will be available in one or two year's time.

For a GGSN, the typical limitations are the total throughput and the number of simultaneous PDP contexts. Typical systems of today have limitations in the order of 100,000 simultaneous PDP contexts, but we can expect significant capacity enhancements over the coming years.

10.4.5 PDSN and Home Agent Dimensioning

Typically, the capacity of a PDSN is limited by the total throughput it can support and the total number of simultaneous PPP sessions. One is likely to find that the PPP session limit is reached before the throughput limitation. With today's technology, limits in the order of 50,000 PPP sessions are common.

For a home agent, the most limiting factor is often the number of supported Mobile IP binding records. Values of 100,000 to 200,000 binding records are common. Note that, in some implementations, the PDSN and home agent may be combined within one physical machine.

10.4.6 Dimensioning of Other Network Elements

In the previous descriptions, we have provided the basic dimensioning information for a number of central nodes in a 3G network. There are, however, many other network elements that need to be sized correctly. These are nodes, such as *Home Location Registers* (HLRs), voice mail systems, SMSCs, and others. Each such node type has it own dimensioning limitations. For example, an HLR is typically limited by the number of subscriber records it can support. A voice mail system is often limited both by the number of subscriber mailboxes (of a given size) that it can support, plus the number of message deposits or retrievals in the busy hour. An SMSC is typically limited by the number of messages per second that can be supported. In the case of a *Global System for Mobile Communication* (GSM) or UMTS network, it should be noted that SMS is used as the delivery mechanism for voice mail notifications, which means that the number of short messages supported by an SMSC may well be greater than those supported in a CDMAOne or CDMA2000 network.

For all of the other network elements that need to be deployed—such as, an *Equipment Identity Register* (EIR), an *Intelligent Network* (IN) *Service Control Point* (SCP), an e-mail system, an HTTP gateway, a WAP gateway, a AAA server, and so on, one needs to acquire from the particular vendor the specific dimensioning rules and capacity limitations.

10.5 Interface Design and Transmission Network Considerations

In general the determination of the amount of bandwidth required for a given interface is a relatively straightforward process. For example if we expect a given RNC to support a given number of Erlangs, then we can easily determine the required bandwidth on the Iu-CS interface. Similarly if we size some RNCs to support a given amount of data traffic in the busy-hour, then we can determine the bandwidth required for Iu-PS interface, and so on. Thus, once we have used traffic forecasts and dimensioning rules for the access network elements, we can determine the bandwidth requirements from the access network to the core network.

For example if a given RNC is expected to carry 2,000 Erlangs of voice traffic in the busy hour and if we dimension the Iu-CS interface at a 0.1 percent grade of service, then Erlang B tables tell us that we need approximately 2,100 circuits. Assuming that traffic between the RNC and the MSC is carried at 16 Kbps, we need 2100/4 DS0s, which equates to 525 DS0s, or approximately 22 T1s. This bandwidth is carried over ATM, so we must add approximately 201 additional overhead for ATM.

Similarly, if a given RNC is expected to carry 50 Mbps of user data, then we can directly determine the Iu-PS bandwidth requirements. It will typically be about 120 percent to 130 percent of the user data bandwidth to enable for GTP overhead. Thus, for 50 Mbps of user data, we would need a bandwidth of 65 Mbps on the Iu-PS interface. Again, ATM overhead must be added.

Dimensioning of interfaces between RNCs (or between BSCs) will depend on the specifics of the radio network. RF design input regarding the amount of soft handover traffic is critical. For example, if RF designers estimate that 20 percent of all voice calls will involve inter-RNC soft handover, then we can use that information to determine the bandwidth requirement. For example, we can assume that 20 percent of the voice traffic will be carried across a given Iur interface.

Overall, the dimensioning of bandwidth requirements for individual interfaces is not overly complicated provided that we have determined the number of network elements and have established detailed traffic demand estimates. The next step, however, is the design of a transport network that supports those bandwidth requirements in an efficient but reliable manner. That design effort can involve more complex issues and involves a greater degree of network design expertise. Consider, for example, a scenario such as is shown in Figure 10-1. In this example there is a large local market and a remote medium-sized market. It has been determined that one MSC and three SGSNs should be placed in the large market. These are connected to four co-located RNCs that serve the local market. In addition, there are two RNCs placed in the remote market. Thus, we need Iu-CS and Iu-PS connections from the remote market to the local market. We may also need one or more Iur interfaces between the local market and the remote market, particularly if the RF coverage of an RNC in one market borders the RF coverage of an RNC in another market, as might be case along a highway between the two cities.

In addition, in North America at least, there will need to be connections from the MSC in the local market to the *Public Switched Telephone Network* (PSTN) in the remote market for support of PSTN calls to or from subscribers whose dialable numbers belong in the remote market. Imagine for

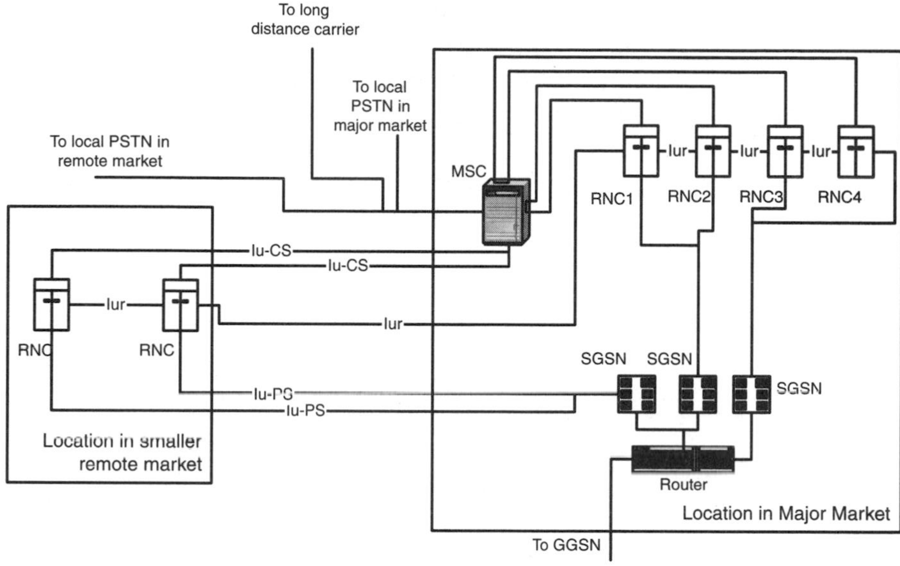

Figure 10-1
3GPP Release 1999 Example Network.

example a subscriber in the remote market who makes a local call. That call is first carried to the MSC and then must be carried back to the PSTN in the remote market. This can be done through direct trunks to that PSTN carrier as shown in Figure 10-1, or the calls can be handed over to a long distance carrier.

Finally, of course, there needs to be hundreds of Iub interfaces from the RNCs to the base stations in each city. All of these interfaces and the associated bandwidth requirements need to be supported by an integrated transmission design that provides the necessary bandwidth and reliable transport.

In most cases, the MSC will be placed on a fiber ring. The total capacity of the ring will depend on the total transmission in and out of the MSC site, but it will be divided into a number of discreet capacities such as a number of DS3s or OC-3s. Typically, the ring will have a number of nodes, including hubbing nodes that belong to the ring provider. The individual links from the base stations in the local market will be connected to such points on the ring where they will be multiplexed onto DS3s or OC-3s for transport to the RNCs at the MSC site.

At the remote market, there will also be a significant number of Iub interfaces from base stations. Depending on the number of such interfaces, the availability of transport facilities, and the cost of those facilities, the remote RNC site might also be placed on a ring. In fact, if the distance is not too great, the remote RNC location might be a node on the same ring as the MSC site. In many cases, however, the distance between the cities may be too great to justify the cost of extending the ring to the remote city. In any event, the traffic from the remote RNC location to the MSC location will need to be protected. This will generally mean that there are diverse transmission facilities between the remote city and the MSC location. These diverse facilities must be sized to support the Iu-CS, Iu-PS, Iur, and PSTN connectivity requirements. This diversity will involve extra cost. That extra cost, however, will generally be justifiable. After all, the traffic demand in the remote city will be significant or it would not have made sense to dedicate RNCs in that remote city.

Often, the availability of transport facilities is a major factor in the timely deployment of a network. In the United States, most transport uses transmission facilities leased from local- and long-distance carriers. Depending on the carriers, it might take six months before a ring can be installed at an MSC location. Moreover, it may be necessary to wait until the ring is installed before ordering individual circuits on that ring. Consequently, the earlier the transmission network requirements can be established the better.

10.6 Placement of Network Nodes and Overall Network Topology

The example of Figure 10-1 assumes that the number of nodes in each city was already established and that the transmission network design was based on that established network element distribution. Often, however, one must consider a multitude of factors before making the decision to place equipment in a given location.

10.6.1 Cost Optimization

Among other considerations, one should not determine the placement of network elements without considering the transmission requirements and the likely transmission cost. For example in Figure 10-1, one could equally have determined that it would be better to place an MSC (and perhaps some SGSNs) at the remote market in addition to just RNCs. This would greatly reduce the transport requirements between the two cities. On the other hand, however, there would be greater capital cost involved in placing an MSC in the remote city. Alternatively, one could have decided that it would be better to completely serve the remote city from equipment housed in the larger local city. This would likely reduce the total RNC cost and would avoid the need for a suitably conditioned building in the remote city to house RNC equipment. The capital cost reduction in such a situation could be considerable. On the other hand, the additional transport required between the remote city and the MSC site could be very great and could mean a large cost. (After all, there will likely be at least a T1 from every site to the serving RNC regardless of how heavily used that site happens to be. On the other hand, the Iu-Cs and Iu-PS interfaces are sized based upon utilization only.)

Having said that, there will need to be a certain amount of transport from the MSC to the PSTN in the remote market in any case. It may well be that the size of that transmission facility is such that extra capacity is available "for free" or that additional capacity can be added at a reasonably low cost. Thus, the cost structure for transmission bandwidth must also be considered. For example, although a DS3 supports 28 DS1s, the cost of a DS3 is approximately 8 to 10 times that of a DS1. Thus, if one needs 12 DS1s, one is better off to lease a DS3 and get up to 20 DS1s "for free." Similarly, an OC-3 costs less than 2 DS3s, even though it supports up to 3 DS3s.

Finally, one must consider future technology evolution and the expected costs and capacities of future network elements. If one were not anticipating an upgrade to 3GPP Release 4, then the capital cost of an MSC in the remote city might be justified if it could be depreciated over a seven- or ten-year period and the effective cost compare with the transmission cost of placing just BSCs or RNCs in the market. Imagine, however, that one is deploying 3GPP Release 1999 and expecting to upgrade the network to 3GPP Release 4 within a two-year timeframe. In that case, one could delay the deployment of switching equipment in the remote city until such time as media gateways are available, provided of course that those media gateways are sufficiently scalable and sufficiently inexpensive compared to a traditional MSC. It might make more financial sense to absorb the cost of transmission between the two cities until the more efficient architecture is available.

Similar issues need to be considered in the placement of other network nodes such as SGSNs or PDSNs, GGSNs, and so on. Let us take a UMTS example. An SGSN is at the same level as an MSC in the network hierarchy. Consequently, it generally makes sense for SGSNs and MSCs to be co-located. What about the placement of GGSNs? Well, that question comes down to the types of data services that the network operator wishes to offer and therelative use of those services. If for example a great deal of user traffic goes to and from the Internet, then it would make sense to place GGSNs at or close to the SGSNs and connect to the Internet relatively close to the user. That can save bandwidth. On the other hand, if one expects that subscribers will make a lot of use of operator-provided services, such as e-mail, where those services are housed in a limited number of centralized locations, then it can make sense to place the GGSNs nearer to those centralized locations. Although that approach can mean greater transmission overhead (because of the tunneling overhead between SGSN and GGSN), it may also mean a net fewer number of GGSNs in the network. Given that a GGSN or cluster of GGSNs needs to have other associated equipment, such as DHCP servers and firewalls, a reduction in the number of GGSNs or the number of GGSN locations may mean a considerable reduction in capital cost. Again, we are faced with the issue of striking a balance between capital expense and operating expense.

In the case of the placement of data nodes, there may also be special cases that need to be considered. Imagine for example that a given network operator establishes a relationship with a large corporate customer in a given city. The individual subscribers from that customer may have a totally different usage profile from other subscribers. They might, for example, use the wireless data service exclusively for access to the corporate network. In such a case it could be appropriate to dedicate one or more GGSNs in a specific location for the use of those subscribers.

10.6.2 Considerations for All—IP Networks

As we move towards all-IP network architectures, then we will find the situation where we can establish just a single IP-based backbone network for the support of voice, data, and signaling. This amalgamation can mean a more efficient and cost-effective network. It is important to remember, however, that different *quality of service* (QoS) requirements will apply to such categories of service. In fact there will be different QOS requirements for different data services. Consequently it is not sufficient to simply size the core network backbone network to meet the expected bandwidth requirements. We must also ensure that QoS mechanisms are built into the network so that quality is not jeopardized. Specifically one wishes to ensure that each service is provided with the required quality without adversely impacting any other service. The first step in doing this is ensuring that the backbone network has sufficient bandwidth. This does not only mean ensuring sufficient bandwidth in transmission facilities. It also means that core network routers have the switching capacity to handle the routing and switching of millions of packets.

Once we have established that we have the necessary bandwidth and packet switching capacity in place, we must then make sure that each service and perhaps each service user can have reasonable access to that capacity in accordance with desired QoS objectives. This means that no one service can hog capacity at the expense of others; it may mean that one service can pre-empt another, and it may mean traffic shaping at the edge of the network to ensure that the traffic entering the core network is in accordance with an agreed profile. A number of QoS solutions are available. To begin with, *Asynchronous Transfer Mode* (ATM) has the capability to provide QoS guarantees. A number of other techniques are also available, such as the Resource Reservation Protocol (RSVP) and *Multi-Protocol Label Switching* (MPLS). The various techniques each have different advantages and disadvantages. For example ATM is a layer 2 protocol. If one decides to use ATM at layer 2 in the network, then one can take advantage of the QoS mechanisms it can provide. On the other hand, one may not wish to be forced to choose ATM at layer 2, particularly if other options exist (such as packet over SONET) and QoS guarantees can be achieved in other ways. RSVP can provide strong QoS guarantees and comes very close to circuit emulation. It has the disadvantage, however, of being processing-intensive and does not scale well to support very large networks. For many, MPLS holds the greatest promise. It offers strong QoS capabilities, can scale better than RSVP and can be used with any layer 2 protocol, including ATM. MPLS is likely to be the most flexible solution for most large networks. Further discussion of IP QoS techniques is provided in Chapter 8, "Voice-over-IP Technology."

10.6.3 Network Reliability Considerations

Clearly, one would like to build a network that supports the expected demand and do so at the lowest overall cost—including both capital and operating costs. Reducing capital cost often involves a centralized design where equipment is deployed in fewer locations, thereby taking advantage of efficiencies of scale. This also helps to reduce some aspects of operating costs as it reduces the number of locations and the number of network elements that need to be managed. On the other hand, a centralized design can lead to greater operational costs caused by increased transmission requirements, particularly if transmission facilities are leased and paid for on a monthly basis.

Cost is not the only important factor, however. Network reliability also plays a big role. The fewer the number of locations used to support a given subscriber base, the greater the impact if one of those locations becomes inoperative due to some major catastrophe. If for example a single city is served by a single location and that location suffers some catastrophe, such as an earthquake or tornado, it is likely that service to the city will be degraded if not completely halted. If, however, that location also serves a number of other cities, then service in those cities will suffer equally. Thus in parts of California, one might want to limit the number of markets served by a particular location so that damage is somewhat contained in the event of an earthquake. Similar considerations might apply in parts of the eastern United States that are subject to hurricanes.

We understand that the foregoing discussion does not provide any real rules for determining the placement of network elements and for establishing the overall network topology. In reality, there are no hard and fast rules that can be applied to any network. Therefore, we have attempted to provide a description of the issues that need to be considered in the network design exercise. Each network or operator will vary in terms of geographical service area, quality and reliability objectives, capital and operating expenditure limitations, offered services, service packaging, equipment capacities, technology roadmap, and so on. All of these aspects must be considered in determining the initial network design and how that design should evolve over time. It is possible to include some of these factors in software-based network design models, to which network design experience should be added in developing the optimum design.

CHAPTER 11

Antenna System Selection

This chapter will briefly discuss some of the more important issues associated with an antenna system regarding 3G applications. The selection of the antenna type to utilize for base station whether it is for a macro, micro, or pico cell is similar for all the technology platforms. There are of course some differences related to the different design issues associated with a 1G, 2G, 2.5G, or 3G system. The key difference in the antenna design issues lies in the desire to keep the systems either separate or unified depending on the underlying technology platform the new system is being overlaid upon.

The antenna system for any radio communication platform utilized is one of the most critical and least understood parts of the system. The antenna system is the interface between the radio system and the external environment. The antenna system can consist of a single antenna at the base station and one at the mobile or receiving station. Primarily the antenna is used by the base station site and the mobile for establishing and maintaining the communication link.

There are many types of antennas available, all of which perform specific functions depending on the application at hand. The type of antenna used by a system operator can be a collinear, log periodic, folded dipole, or yagi to mention a few. Coupled with the type of antenna is the notion of an active or passive antenna. The active antenna usually has some level of electronics associated with it to enhance its performance. The passive antenna is more of the classical type where no electronics are associated with its use and it simply consists entirely of passive elements.

Along with the type of antenna there is the relative pattern of the antenna indicating in what direction the energy emitted or received from it will be directed. There are two primary classifications of antennas associated with directivity for a system, and they are omni and directional. The omni antennas are used when the desire is to obtain a 360 degree radiation pattern. The directional antennas are used when a more refined pattern is desired. The directional pattern is usually needed to facilitate system growth through frequency reuse or to shape the system's contour.

The choice of which antenna to use will directly impact the performance of either the cell or the overall network. The radio engineer is primarily concerned in the design phase with the base station antenna because this is the fixed location, and there is some degree of control over the performance criteria that the engineer can exert on the location.

The correct antenna for the design can overcome coverage problems or other issues that are trying to be prevented or resolved. The antenna chosen for the application must take into account a multitude of design issues. Some of the issues that must be taken into account in the design phase involve the antennas gain, its antenna pattern, the interface or matching to the transmitter, the receiver utilized for the site, the bandwidth and frequency range over which the signals desired to be sent will be applicable, its power handling capabilities, and its IMD performance. Ultimately the antenna you use for a network needs to match the system RF design objectives.

11.1 Base Station Antennas

There are a multitude of antennas that can be used at a base station. However the specifics of what comprise a base station antenna, or rather an antenna system, is determined by the design objectives for the site coupled with real world installation issues. For 3G radio systems as well as the 2.5G radio systems, most, if not all, of the antenna design decisions are determined by the type of base station they will be employed at. For instance the antenna system for a macro cell will most likely be different than that used for a micro and definitely different for a pico cell.

Base station antennas are either omni directional, referred to as omni, or directional antennas. The antenna selected for the application should be one that meets the following major points as a minimum:

- Elevation and azimuth patterns meet requirements.
- The antenna exhibits the proper gain desired.
- The antenna is available from common stock and company inventory.
- The antenna can be mounted properly at location, that is, it can be physically mounted at the desired location.
- Antennas will not adversely affect the tower, wind and ice loading for the installation,
- Visual impact, negative, has been minimized in the design and selection phase.
- Antenna meets the desired performance specifications required.

However this section will restrict itself to collinear, log periodic, folded dipole, yagi, and microstrip antennas with respect to passive, that is, no active electronics in the antenna system itself. Of the antennas classifications mentioned, two are more common for use in 1G and 2G communication systems for base stations and will be used for 2.5G and 3G also. The two types of antennas used for base stations are collinear and log periodic antennas.

11.2 Performance Criteria

The performance or performance criteria for an antenna is not restricted to its gain characteristics and physical attributes, that is, maintenance. With the introduction of 2.5G and 3G platforms, the performance criteria associated with new or existing antennas needs to be reviewed. There are many parameters that must be taken into account when looking at an antennas performance. The parameters that define the performance of an antenna can be referred to as the *figures of merit* (FOM) that apply to any antenna that is selected to use in a communication system:

- Antenna pattern
- Main lobe
- Side lobe suppression
- Input impedance
- Radiation efficiency
- Horizontal beamwidth
- Vertical beamwidth
- Directivity
- Gain
- Antenna polarization
- Antenna bandwidth
- Front-to-back ratio
- Power dissipation
- Intermodulation suppression (PIM)
- Construction
- Cost

The performance of an antenna is not restricted to its gain characteristics and physical attributes, that is, maintenance. There are many parameters that must be taken into account when looking at an antenna's performance.

Just because an antenna is performing or appears to be performing properly in a 2G system the introduction of a 2.5G or 3G platform may require the alteration of the existing antenna system. The antenna system alteration could involve the replacement or addition of more antennas in order to meet the design and performance criteria of the new system.

There are many parameters and FOM that characterize the performance of an antenna system. The following is a partial list of the FOM for an antenna that should be quantified by the manufacturer of the antennas you are using. The trade offs that need to be made with an antenna chosen involve all the FOM issues discussed in the following.

The antenna pattern of course is one of the key criteria the design engineer utilizes for directing the radio energy either in the desired area or to keep it out of another. The antenna pattern is typically represented by a graphical representation of the elevation and azimuth patterns.

The *antenna pattern* chosen should match the coverage requirements for the base station. For example if the desire is to utilize a directional antenna for a particular sector of a cell site, 120 degrees, then choosing an antenna pattern that covers 360 degrees in azimuth would be incorrect. Care must also be taken in looking for electrical downtilt that may or may not be referenced in the literature.

The *side lobes* are important to consider because they can and do create potential problems with generating interference. Ideally there would be no side lobes for the antenna pattern. For downtilting the sidelobes are important to note because they can create secondary sources of interference.

The *radiation efficiency* for an antenna is often not referenced but should be considered in that it is a ratio of total power radiated by an antenna to the net power accepted by an antenna from the transmitter. The equation is as follows where e = Power Radiated ÷ (Power Radiated Power Lost).

The antenna would be 100 percent efficient if the power lost in the antenna were zero. This number indicates how much energy is lost in the antenna itself, assuming an ideal match with the feedline and the input impedance. Using the efficiency equation, if the antenna absorbed 50 percent of the available power then it would only have 50 percent of the power for radiating and thus the effective gain of the antenna would be reduced.

The *beamwidth* of the antenna, either elevation or azimuth, is important to consider. The beamwidth is the angular separation between two directions in which radiation interest is identical. The 1/2 power point for the beamwidth is usually the angular separation where there is a 3 dB reduction off the main lobe. Why this is important to note is that the wider the beamwidth, the lower the gain of the antenna is normally. A simple rule of thumb is for every doubling of the amount of the elements associated with an antenna, a gain of 3 dB is realized. However this gain comes at the expense of beamwidth. The beamwidth reduction for a 3 dB increase in gain is about 1/2 the initial beamwidth, so if an antenna has a 12 degree beamwidth and has an increase in gain of 3 dB, then its beamwidth now is six degrees.

The *gain* of any antenna is a very important FOM. The gain is the ratio of the radiation intensity in a given direction to that of an isotropically radiated signal. The equation for antenna gain is as follows. G Maximum radiation intensity from antennas/maximum radiation from an isotopic antennas. The gain of the antenna can also be described as

$G = e \times G(D)$ If the antenna were without loss, $e = 1$, than $G = G(D)$.

Polarization is important to note for an antenna because wireless mobility systems utilize vertical polarization, with some exception notes, when the use of X-pole antenna is in play.

The *bandwidth* is a critical performance criteria to examine because the bandwidth defines the operating range of the frequencies for the antenna. The *Standing Wave Ratio* (SWR) is usually how this is represented besides the frequencies range it is constant over. A typical bandwidth that is referenced is the 1:1.5 SWR for the band of interest. Antennas are now being manufactured that exceed this, having a SWR value of 1:1.2 at the band edges.

The antenna's bandwidth must be selected with extreme care to not only account for current but also future configuration options with the same cell site. For example an antenna that is selected for use as the receive antenna at a cell site should also operate with the same performance in the transmit band and vise versa. The rational behind this dual purpose use is in the event of a transmit antenna failure a receive antenna can be switched internally in the cell for use as a transmit antenna.

The *front to back ratio* is a ratio that is with respect to how much energy is directed in the exact opposite direction of the main lobe of the antenna. The front to back ratio is a loosely defined term. The IEEE Std 145-1983 references the front to back ratio as the ratio of maximum directivity of an antenna to its directivity in a specified rearward direction. A front to back ratio is only applicable to a directional antenna because obviously with a omni directional antenna there is no rearward direction.

Many manufacturers reference high front to back ratios but care must be taken in knowing just how the number was computed. In addition if installation is say on a building and the antenna will be mounted on a wall, then the front to back ratio is not as important a FOM. However if the antenna is mounted so there are no obstructions between it and the reusing cell, then the front to back ratio can be an important FOM. Specifically in the latter case the front to back ratio should be at least the C/I level required for operation in the system.

The power dissipation needs to be looked at when integrating a new platform. The power dissipation is a measure of the total power the antenna can accept at its input terminals is its power dissipation. This is important to note because receive antennas may not need to handle much power but the transmit antenna might have to handle 1500 watts of peak power. The antenna chosen should be able to handle the maximum envisioned power load without damaging the antenna.

The amount of *intermodulation* which the antenna will introduce to the network in the presence of strong signals as referenced from the manufacturer needs to be considered in the antenna selection. The intermodulation that is referenced should be checked against how the test was run. For instance some manufacturers reference the IMD to two tones although some reference it to three or multiple tones. The point here is that the overall signal level that the IMD is generated at needs to be known in addition to how many tones were used, their frequency of operation, bandwidth, and of course, the power levels that they were at that caused the IMD level.

The *construction* attributes associated with its physical dimensions, mounting requirements, materials used, wind loading, connectors, and color constitute this FOM. For instance one of the items that needs to be factored into the construction FOM is the use of materials, whether the elements are soldered together or bolted. In addition the type of metals that are used in the antenna and the associated hardware needs to be evaluated with respect to the environment that the antenna will be deployed in. For instance if you install antennas near the ocean, or an aircon unit that uses salt water for cooling, then it will be imperative that the material chosen will not corrode in the presence of salt water.

How much the antenna *costs* is a critical FOM. No matter how well an antenna will perform in the system, the cost associated with the antenna will need to be factored into the decision. For example if the antenna chosen met or exceeded the design requirements for the system but cost twice as much as another antenna that met the requirements, the choice here would seem obvious; pick the antenna that meets the requirement at the lowest cost.

Another example of cost implications would involve selecting a new antenna type to be deployed in the network. The spares and stocking issues need to be factored into the antenna selection process. If the RF department designs every site's antenna requirements too uniquely, then it is possible to have a plethora of antenna types deployed in the network leading to a multitude of additional stocking issues for replacements. Therefore, it is important to select a specific number of antennas that should meet most if not all the design requirements for a system and utilize only those antennas.

11.3 Diversity

Diversity, as it applies to an antenna system, refers to a system used in wireless communication as a method for comparing signal fading in the environment. Diversity gain is based on the gain that over what fading would have taken place in the event that a diversity technique was not used. In the case of a two branch diversity system, if the received signal into both antennas is not of an equal signal strength, then there cannot be any diversity gain. This is an interesting point considering most link budget calculations incorporate diversity gain as a positive attribute. The only way diversity gain can be incorporated into a link budget is if a fade margin is included in the link budget and the diversity scheme chosen attempts to improve or reduce the fade margin that is included there.

There are several types of diversity that need to be accounted for in both the legacy systems as well as 2.5G and 3G platforms. When discussing diversity, the concept for a radio engineer is usually focused on the receive path, uplink from the mobile to the base station. With the introduction of 2.5G and 3G platforms, transmit diversity has been introduced but is implemented in a fashion where the subscriber does not need a second antenna.

The type of antenna diversity used can and is often augmented with another type of diversity that is accomplished at the radio level:

- Spacial
- Horizontal
- Vertical
- Polarization
- Frequency
- Time
- Angle

For most, 2G systems and 2.5G and 3G systems use two antennas separated by a physical distance, that is horizontal. Some 2.5G platforms like iDEN utilize a three branch diversity receive scheme but they are the exception, and the usual method is to deploy only two antennas per sector for diversity reception.

The spacing is associated with the antennas located in the same sector is normally a design requirement that is stipulated from RF engineering. Diversity spacing is a physical separation between the receive antennas that is needed to ensure that the proper fade margin protection is designed into the system. As mentioned earlier horizontal space diversity is the most common type of diversity scheme that is used in wireless communication systems.

The following is a brief rule of thumb used to determine the required horizontal diversity requirements for a site and is shown in Figure 11-1.

$n = h/d = 11$ where h = height (ft)

d = distance between antennas (ft)

The equation used was derived for cellular systems operating in the 800 MHz band but has been successfully applied for the other wireless bands in 1800 and 1900.

With the introduction of both CDMA2000 as well as *Universal Mobile Telecommunications Service* (UMTS), the application for transmit diversity needs to be factored into the antenna design. Two different transmit

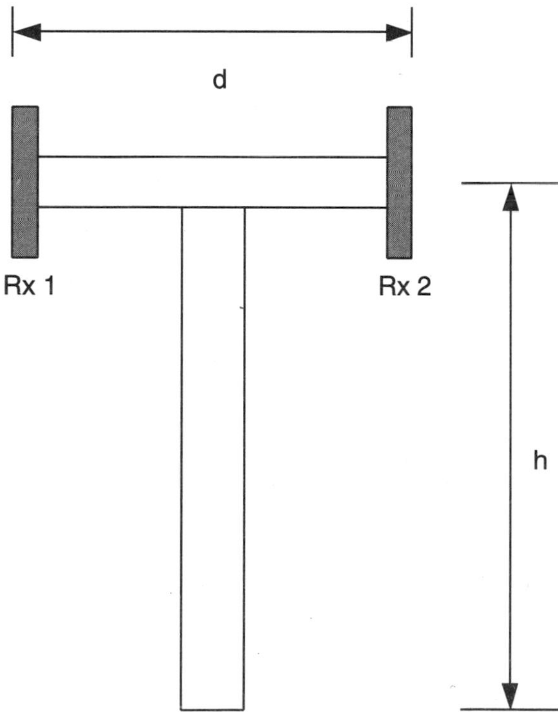

Figure 11-1 Two branch diversity spacing.

diversity schemes are possible with these platforms. The use of two different transmit diversity schemes is driven by the practical issue of antenna installation concerns. The two transmit diversity schemes are STD and OTD. STD is *space transmit diversity* while OTD is *orthogonal transmit diversity*. The preferred transmit diversity scheme, when implemented, is the STD method.

What follows is a simplified diagram for both the STD and OTD transmit diversity schemes. Figure 11-2 shows the STD transmit diversity scheme for a single channel when there are two antennas available on a sector. The two antennas could also be separate ports on a X-pole antenna. The important issue is that when integrating a second carrier, either more antennas need to be added or additional *transmit* (Tx) combing losses will ensue.

Figure 11-3 shows a configuration recommended for the rest of the network that involves using OTD transmit diversity. In examining the differences between Figure 11-2 and 11-3, one immediate observation is that a second carrier is introduced with the same amount of physical antennas.

One immediate observation with the use of Tx diversity for the new radio platforms is the issue with what happened to the legacy systems. That will be covered shortly.

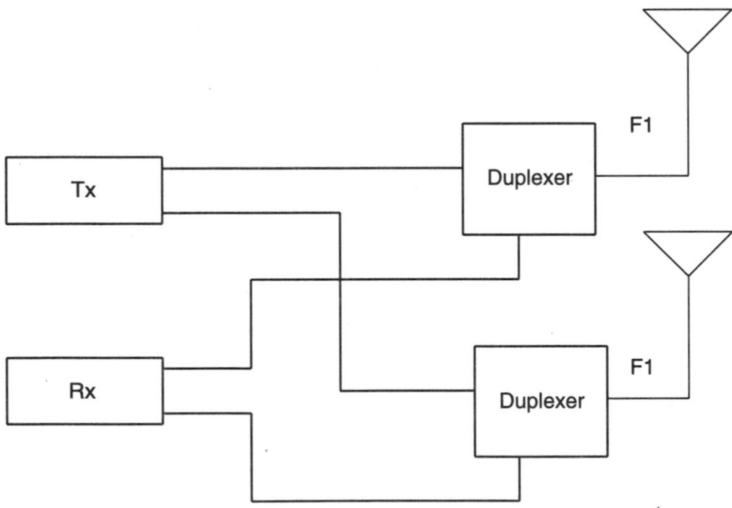

Figure 11-2 Sector STD transmit diversity scheme.

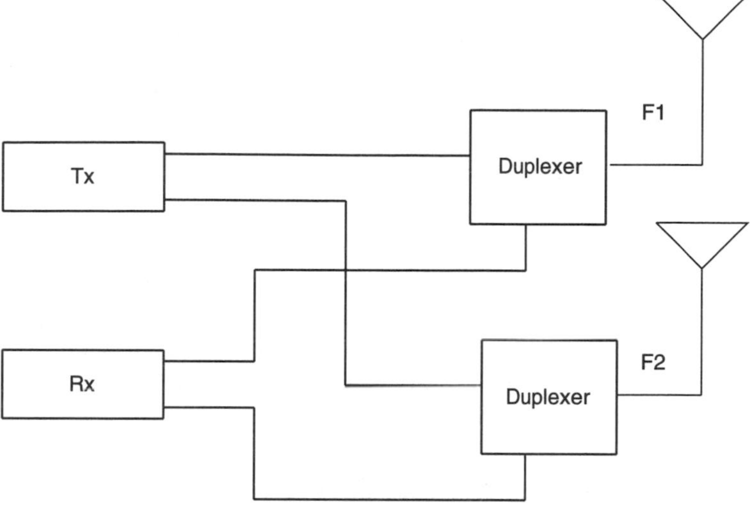

Figure 11-3 Sector OTD transmit diversity scheme.

11.4 Installation Issues

In any wireless communication system there are always a host of antenna installation problems ranging from space restrictions to tower compression or shearing loading factors or even the physical ports available to be used. However, the installations more common are associated with physically mounting the antennas. The introduction of 2.5G and 3G systems have guaranteed that the installation issues of the past will continue.

One of the prevalent issues associated with 2.5G and 3G is the lack of the number of physical antennas that will be available from which to utilize. As with the introduction of any new radio access platform, each technology has its own special issues.

For the CDMA2000 antenna systems there are some different considerations to take into account when migrating from an IS-95 system to a CDMA2000 system if it is an AMPS or PCS spectrum. The desire is to utilize transmit diversity, and this will be achieved either by a STD or OTD method. However, the STD method is the preferred version.

Figure 11-4 shows a typical situation where there are two or three antennas per sector available for use. Sometimes there is only one antenna if it is a cross pole antenna. With an AMPS system as the underlying legacy system, the use of a STD transmit diversity scheme is possible with a configuration shown in Figure 11-4 with the exception that only one carrier is used for CDMA. If a second carrier is added, then OTD diversity is utilized and the configuration shown in Figure 11-4 is used. Now, if the operator has been able to secure more antennas per sector, that is, five, then the configuration shown in in the figure is the desired method where the AMPS and CDMA systems are bifurcated. The use of STD or OTD is again dependant upon the number of carriers required at the site.

Regarding the deployment of GPRS into an existing Global System for Mobile Communications (GSM) network, the migration is rather straightforward from an antenna aspect because the carriers and fundamental infrastructure issues remain the same. The only difference lies in the amount of antennas that may need to be added due to transmitter combing losses. However, there is no unique antenna configuration issues that need to be adhered to other than standard GSM deployment schemes.

However, when implementing a GPRS system over a IS-136 system or migrating from GPRS to WCDMA, there are antenna issues that need to be thought about prior to acquiring the cell site or installing antennas. What needs to be thought about is the fundamental problem that GPRS or IS-136 relies on a different modulation scheme and therefore has different performance parameters and design guidelines.

A lesson learned with IS-95 deployment into an Advanced Mobile Phone System (AMPS) environment is that for performance and optimization reasons, a set of separate antennas should be sought were possible. It is not that the technologies cannot share the same antenna but that the optimization techniques for same GPRS or IS-136 is different than that envisioned for Wideband CDMA (WCDMA). Therefore, if the antenna system is not separated, performance compromises will be experienced in both the WCDMA and the legacy systems.

Figure 11-5 shows a typical situation where there are two or three antennas per sector available for use. Sometimes there is only one antenna but it is a cross pole antenna. With a GSM or IS-136 system as the underlying legacy system, the use of a STD transmit diversity scheme is possible with a configuration shown in Figure 11-5 with the exception that only one carrier is used for WCDMA. If a second carrier is added, then OTD diversity is utilized and the configuration shown in Figure 11-5 is used. Now if the operator has been able to secure more antennas per sector, that is, five, then

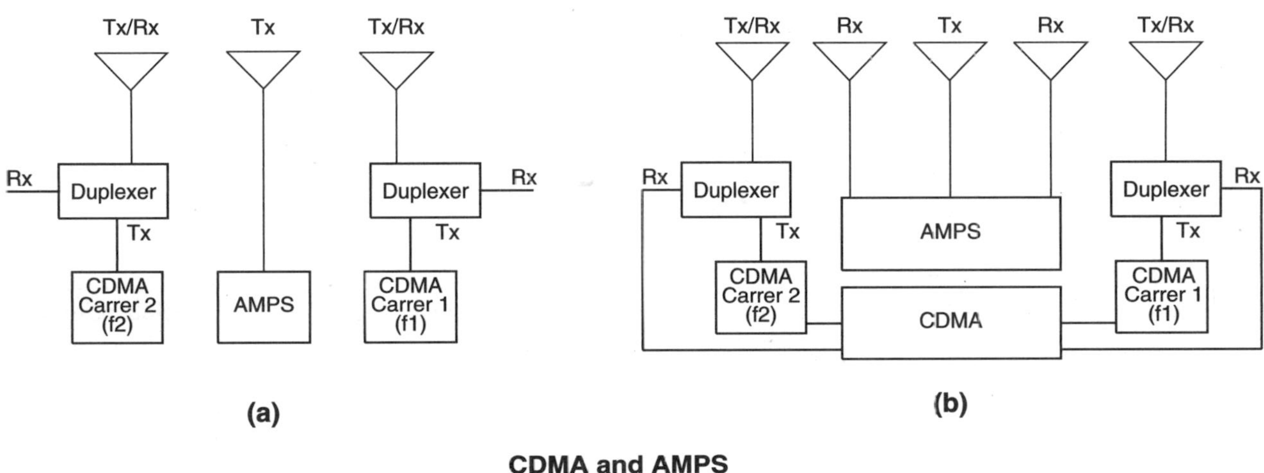

Figure 11-4 CDMA2000/IS-95 and AMPS systems sector antenna configurations.

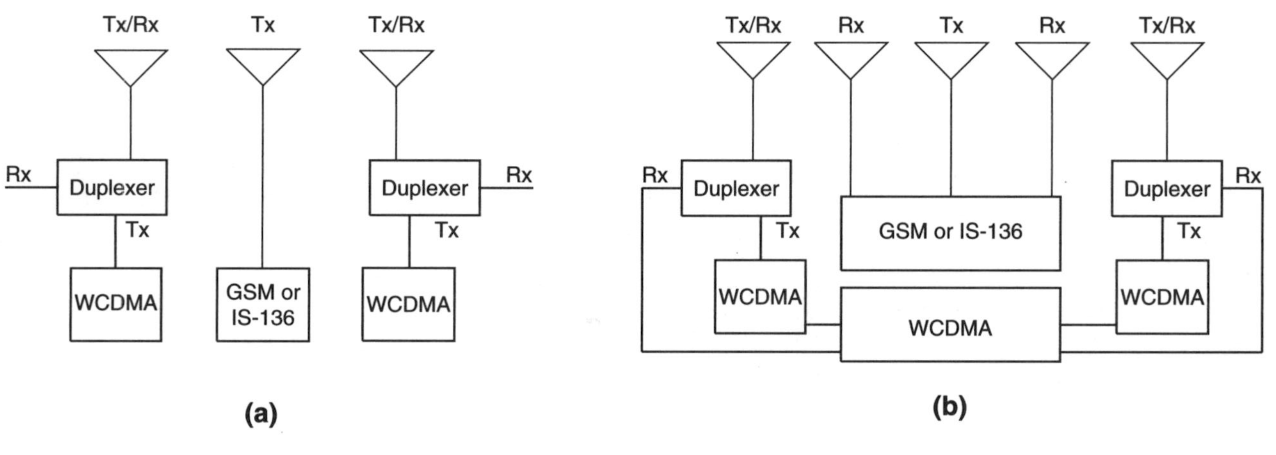

Figure 11-5 WCDMA sector antenna configuration.

the configuration shown in the figure is the desired method where the GSM/IS-136 and WCDMA systems are bifurcated. The use of STD or OTD is again dependant upon the number of carriers required at the site.

Both WCDMA and CDMA2000 systems are all envisioned to be deployed as three sectors only. Although there is the capability to deploy more sectors, the practicality of the situation favors three sectors.

The determination of where to place antennas or the methodology that is used to place the antennas is often encountered when not utilizing a monopole or tower installation. Figure 11-6 shows an example of an omni site or single sector involving a transmit and two receive antennas. The transmit antenna is installed in the center while the receive antennas are aligned as best as reasonably possible to provide maximal diversity reception for the major road shown in the figure. Obviously the example is more relevant for the small stretch of highway and different installation schemes can be implemented.

The diagram shown in Figure 11-7 is a slight modification from that shown in Figure 11-6. The change addresses the issue of when the antennas cannot be installed at the edge of the building's roof and needs to be installed on the penthouse of the building.

When installing on a penthouse or any building installation where the antennas are not installed at the edge of the building for either visual or structural reasons, a setback rule needs to be followed. The setback rule

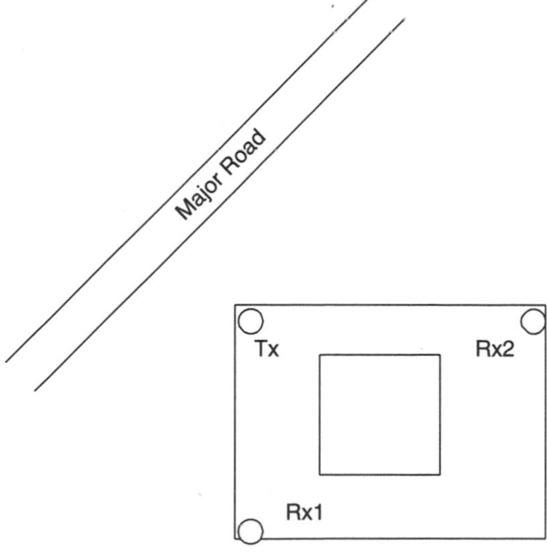

Figure 11-6 Antenna installation example.

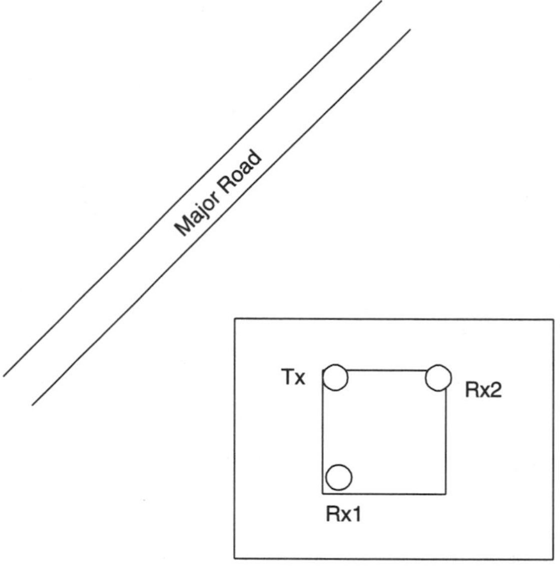

Figure 11-7 Penthouse antenna installation.

involves the relationship between the antennas installation point, its height above the roof top, and of course the distance between the antenna and the roof edge.

Figure 11-8 shows the relationship in a simplistic drawing of the antenna placement to the roof edge when installing on a roof. The concept is to avoid violating the first fresnel zone for the antenna, however, because each antenna has a different pattern, and there are different operating frequencies. The relationship shown next will provide the necessary clearance.

$$a = 5 \times b$$

Examination of the equation draws the conclusion that the farther the antenna is from the roof edge, the higher it will need to be installed to obtain the necessary clearance.

11.4.1 Wall Mounting

For many building installations it may not be possible to install the antennas above the penthouse or other structures for the building. Often it is necessary to install the antennas onto the penthouse or water tank of an

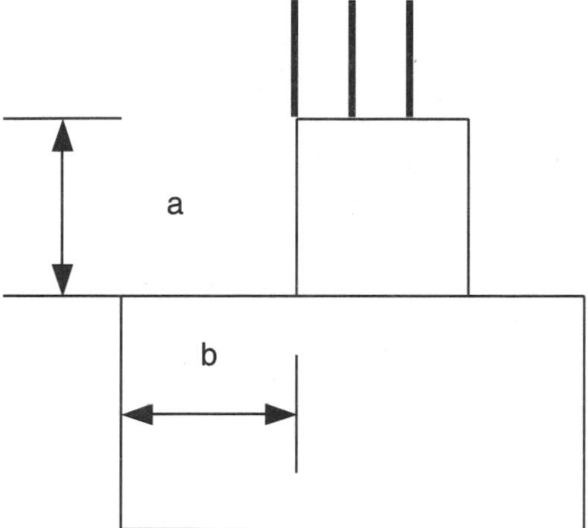

Figure 11-8
Building installation.

existing building. When installing antennas onto an existing structure, rarely has the building architect factored into the potential installation of antennas at the onset of the building design. Therefore, as shown in Figure 11-9, the building walls may meet one orientation needed for the system but rarely all three for a three sector configuration.

Therefore, it is necessary to determine what the offset from the wall of the building structure needs to be. Figure 11-10 illustrates the wall mounting offset that is required to ensure proper orientation for each sector.

Figure 11-10 shows that in order to obtain the directionality of the sector, a structure is installed that will meet that requirement. The method for determining the wall offset is shown in the following equation.

$$\alpha = \mathrm{d} \times \sin(\phi)$$

where
ϕ is the angle from wall
α is the distance from the wall
d = diversity separation for a two-branch system

11.4.2 Antenna Installation Tolerances

When designing or even installing antenna systems for a wireless communication facility, the installation will have some variance to the design. Just

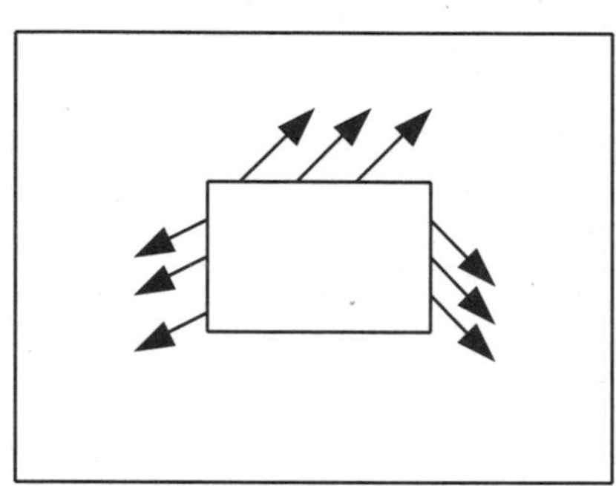

Figure 11-9 Sector Building installation.

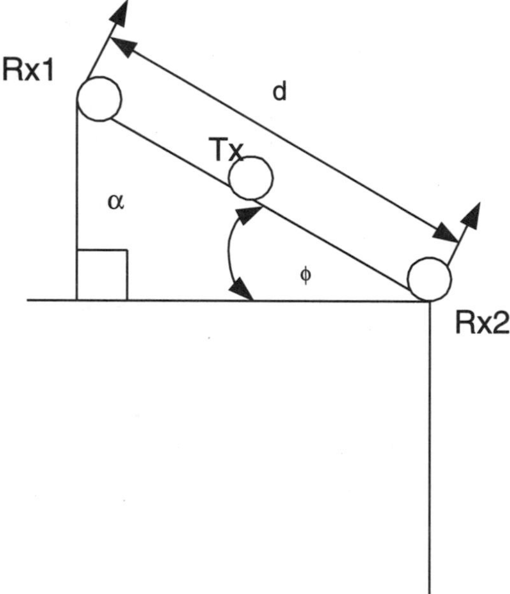

Figure 11-10 Wall offset.

what variances are enabled needs to be stipulated from the onset of the design process. The antenna installation tolerances apply directly to the physical orientation and plumbness of the antenna installation itself. There are usually two separate requirements: how accurate should the antenna orientation be and how plumb should the antenna installation be. The obvious issue here is not only the design requirements from engineering but also the practical implementation of the antennas for cost reasons.

Therefore, the following guidelines in Table 11-1 should be used.

The antenna orientation tolerance is a function of the antenna pattern and can be unique for each type of cell site. Obviously, for an omni cell site, there are no orientation requirements because the site is meant to cover 360 degrees. However, for a sector or directional cell site, the orientation tolerance becomes a critical issue. The orientation tolerance should be specified from Radio Frequency (RF) engineering but in the absence of this, the guideline is to be within 5 percent of the antennas horizontal pattern. Table 11-2 will help illustrate the issue by using some of the more standard types of antenna patterns used in the industry.

The obvious goal however is to have no error associated with the orientation of the antenna, but this is rather impractical.

Therefore, as the antenna pattern becomes more tight, the tolerance for the orientation error is reduced. The objective defined here is +/− 5 percent, but the number can be either relaxed or tightened depending on your particular system requirements. The 5 percent number should also factor

Table 11-1

Antenna Installation Tolerances

Type	Tolerance
Orientation	+/− 5% or antennas horizonatal pattern
Plumbness	+/1 1 degree (critical)

Table 11-2

Horizontal Antenna Tolerances

Antenna Horizontal Pattern	Tolerance from Boresite
110 degrees	+/− 5.5 degrees
92 degrees	+/− 4.6 degrees
90 degrees	+/− 4.5 degrees
60 degrees	+/− 3.0 degrees
40 degrees	+/− 2.0 degrees

into any potential building sway that can and does occur, usually a nonissue due to the height of the buildings used for wireless installations.

11.5 dBi and dBd

All too often the calculated value that you have is not in the right scale that is required for the form or questioner. Therefore, it is necessary to convert from either dBi to dBd or from dBd to dBi.

To convert a value in dBi to the equivalent dBd value, the following equation is utilized:

$$dBd = dBi - 2.14$$

Therefore, Table 11-3 can be used to help reinforce the conversion process that shows the calculated values along with the nearest approximate value found.

To convert a value in dBd to the equivalent dBi value, the following equation is utilized:

$$dBi = 2.14 + dBd$$

Therefore, Table 11-4 can be used to help reinforce the conversion process that shows the calculated values along with the nearest approximate value found.

Table 11-3

dBi to dBd

dBi	dBd
5	2.86 (3dBd)
10	7.86 (8dBd)
12	9.86 (10dBd)
14	11.86 (12dBd)
18	15.86 (16dBd)
21	18.86 (19dBd)

Table 11-4

dBd to dBi

dBd	dBi
3	5.14
10	12.14
12	14.14
14	16.14
18	20.14
21	23.14

11.6 Intelligent Antennas

Intelligent antenna systems are being introduced to commercial wireless communication systems. The concepts and implementation for intelligent antenna systems have been utilized in other industries for some time, primarily the military.

Intelligent antenna systems can be configured for either receive only or full duplex operations. The configuration of the intelligent antenna systems can be arranged as either in an omni or sector cell site depending on the application at hand.

With CDMA2000 and WCDMA the use of intelligent antenna systems are supported directly, unlike 1G and 2G systems, with the use of auxiliary and dedicated pilot channels.

Intelligent antennas were initially promoted as providing an increase to the S/N of a sector by reducing the amount of N, noise and interference, and possibly increasing the S, serving signal, in the same process. All the technologies referenced are based on the principle that narrower radiation beam patterns will provide increased gain and can be directed toward the subscriber and at the same time offer less gain to interfering signals that will arrive at an off axis angle due to the reduced beam width size.

Intelligent antennas are now promoted as not only being able to improve the S/N of a sector or system but also more uniformly balance traffic between sectors and cells and improve on the system performance through reducing softhandoffs for IS-95 systems.

Figure 11-11 illustrates three types of intelligent antenna systems, each has positive and negative attributes.

All the illustrations shown can be either receive only or full duplex. The difference between the receive only and the full duplex systems involves the amount of antennas and potential number of transmitting elements in the cell site itself.

The beam switching antenna arrangement shown is the simplest to implement. It normally involves four standard antennas of narrow azimuth beam width, 30 degrees for a 120 degree sector, and based on the receive signal received, the appropriate antenna will be selected by the base station controller for use in the receive path.

The multiple beam array shown involves utilizing an antenna matrix to accomplish the beam switching.

The beam steering array, however, utilizes phase shifting to direct the beam toward the subscriber unit. However, the direction that is chosen by the system for directing the beam will affect the entire sector. Normally amplifiers for transmit and receive are located in conjunction with the antenna itself. In addition the phase shifters are located directly behind each antenna element. The objective of placing the electronics in the mast head is to maximize the receive sensitivity and exploit the maximum transmit power for the site.

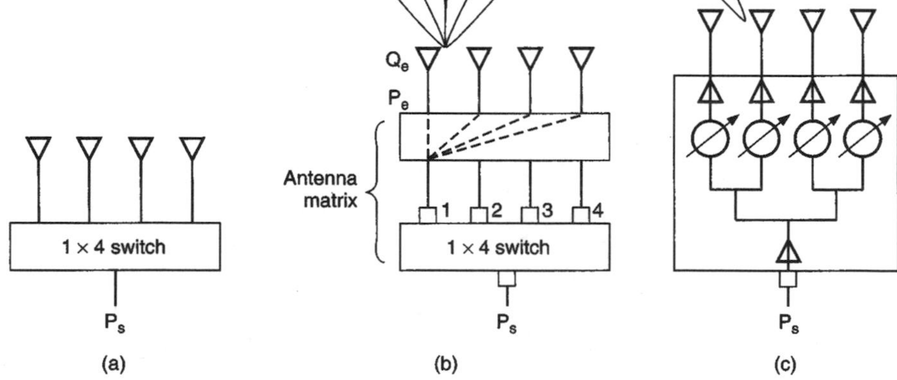

Figure 11-11 Intellient Antenna Systems: (a) Switched Antennas (b) Multiple Beam Array (c) Steered-Beam Array.

References

3GPP2 C.S0008-0. "*Multi-carrier Specification for Spread Spectrum Systems on GSM MAP (MC-MAP) (Lower Layers Air Interface)*," June 9, 2000.

American Radio Relay League. "*The ARRL Antenna Handbook*," 14th Ed., The American Radio Relay League, Newington, Conn., 1984.

Carr, J.J. "*Practical Antenna Handbook*," Tab Books, McGraw-Hill, Blue Ridge Summit, PA, 1989.

Fink, Donald, and Donald Christiansen. "*Electronics Engineers Handbook*," McGraw-Hill, 3rd Ed., New York, NY, 1989.

Fink, Beaty. "*Standard Handbook for Electrical Engineers*," 13th Ed., McGraw-Hill, NY, 1995.

Jakes, W.C. "*Microwave Mobile Communications*," IEEE Press, New York, NY, 1974.

Johnson, R.C., and H. Jasik. "*Antenna Engineering Handbook*," 2nd Ed., McGraw-Hill, New York, NY, 1984.

Kaufman, M., and A.H. Seidman. "*Handbook of Electronics Calculations*," 2nd Ed., McGraw-Hill, New York, NY, 1988.

Miller, Nathan. "*Desktop Encyclopedia of Telecommunications*," McGraw-Hill, 1998.

Mouly, Pautet. "*The GSM System for Mobile Communications*," 1992.

"*Reference Data for Radio Engineers*." Sams, 6th Ed., 1983.

Schwartz, Bennett, Stein. "*Communication Systems and Technologies*," IEEE, New York, NY, 1996.

Smith, Clint. "*Practical Cellular and PCS Design*," McGraw-Hill, 1997.

Smith, Clint. "*Wireless Telecom FAQ*," McGraw-Hill, 2000.

Steele. "*Mobile Radio Communications*," IEEE, 1992.

Stimson. "*Introduction to Airborne Radar*," Hughes Aircraft Company, El Segundo, CA, 1983.

TIA.EIA IS-2000-1. "*Introduction to cdma2000 Standards for Spread Spectrum Systems*," June 9, 2000.

TIA/EIR IS-2000-2. "*Physical Layer Standard for cdma2000 Spread Spectrum Systems*," Sept. 12, 2000.

TIA/EIA IS-2000-3. "*Medium Access Control (MAC) Standard for cdma2000 Spread Spectrum Systems*," Sept. 12, 2000.

TIA/EIA IS-2000-4. " *Signaling Link Access Control (LAC) Specification for cdma2000 Spread Spectrum Systems*," Aug. 12, 2000.

TIA/EIA-98-C. "*Recommended Minimum Performance Standards for Dual-Mode Spread Spectrum Mobile Stations (Revision of TIA/EIA-98-B)*," Nov. 1999.

Webb, Hanzo. "*Modern Amplitude Modulations*," IEEE, 1994.

White, Duff. "*Electromagnetic Interference and Compatibility*," Interference Control Technologies, Inc., Gainesville, GA, 1972.

CHAPTER 12

UMTS System Design

In previous chapters, we have described *universal mobile telecommunications service* (UMTS) from a pure technology perspective. In this chapter, we aim to address the design criteria and methodologies that apply to deploying UMTS technology in a real network. There are numerous inter-related considerations that must be addressed in the design and deployment of such a network. While some of these considerations are common to any wireless network design, a number are specific to the technology in question. Regardless, because of the multiple issues involved, it is very important that a well understood methodology is in place so that the network design can proceed from the initial establishment of requirements to the final deployed network.

12.1 Network Design Principles

Figure 12-1 shows the overall network design and deployment process at a very high level. To begin with, we must specify a number of criteria regarding the set of services that we wish to provide and the estimated demand for those services. We must establish exactly where we wish to offer those services, and we must establish any limiting factors that might constrain our ability to meet all objectives—such as spectrum limitations.

Based on the established input requirements, a number of network design activities take place. These can largely be broken into two main areas—*radio frequency* (RF) network design and core network design. Of course, within each of these areas there is a myriad of individual design efforts.

Once designs are established, implementation is undertaken. This largely involves the RF network implementation, core network implementation, integration, and optimization. Quite often during the implementation phase, one finds that it is not possible or optimal to deploy the system exactly as designed, in which case the design itself needs to be modified. There are many reasons why designs might need to be changed—such as an inability to acquire a Node B site in the exactly desired location, coverage or quality problems discovered during integration or optimization, and so on.

Finally, statistics and measurements generated during the performance of the operational network should be fed back to those who generate the design inputs and also to those who are responsible for the system design. This enables revised requirements related to design modifications, expanded coverage, capacity, or service demand to be based upon real experience.

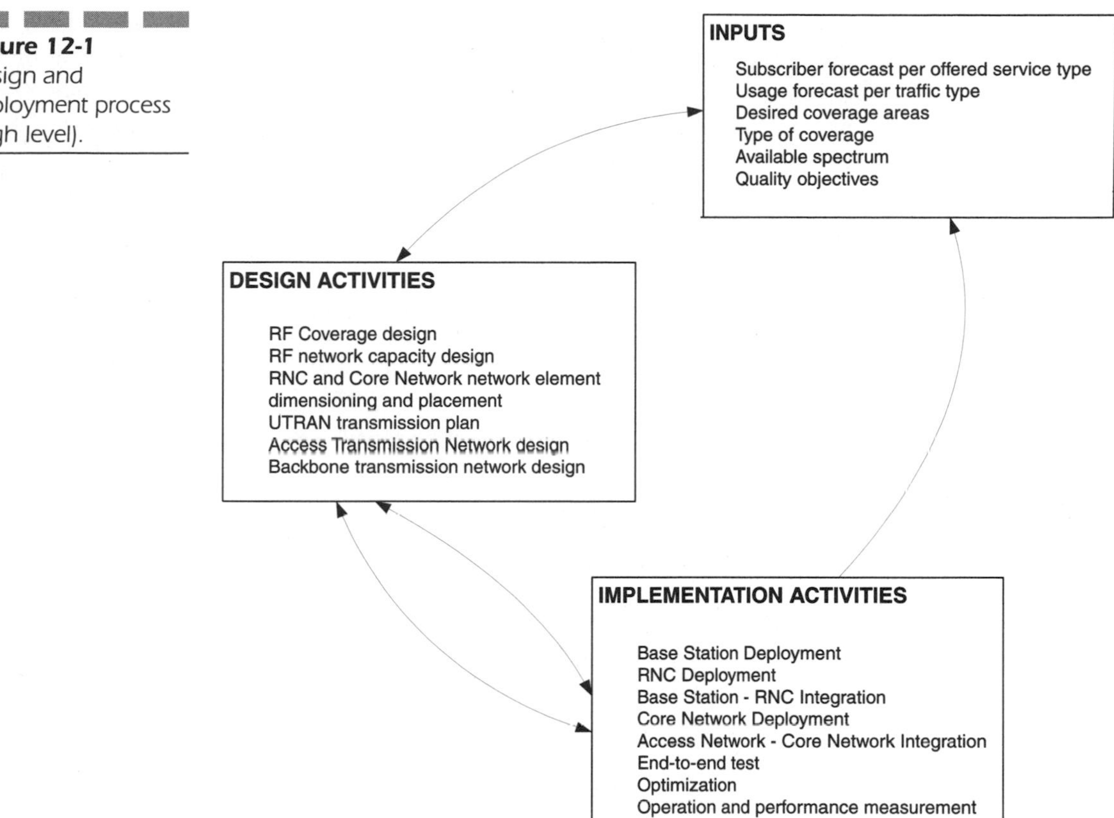

Figure 12-1 Design and deployment process (high level).

12.2 RF Coverage Analysis

Looking again at Figure 12-1, we now delve into a little more detail in each of the areas of concern. We start with the input requirements, specifically the RF coverage requirements, and consider the issues related to designing an RF network to meet those coverage requirements.

As described in Chapter 9, "3G System RF Design Considerations," a good deal of detail should be specified regarding where coverage is to be provided and the type of coverage to be provided in those areas. It is not sufficient to simply state that we wish to provide coverage in a given market. Often it is necessary and desirable to provide coverage only in certain areas

—such as commercial areas, areas with significant population density, and major highways. Therefore, we must obtain a good understanding of the market to be covered, which will require a great deal of map-based information specifying population densities; what areas are urban, suburban, rural; what areas are primarily commercial, residential, industrial, parkland; and so on. Figure 12-2 provides a simple example of a population density map. An understanding of these factors is important for two reasons: First we wish to make sure that we provide sufficient capacity in those

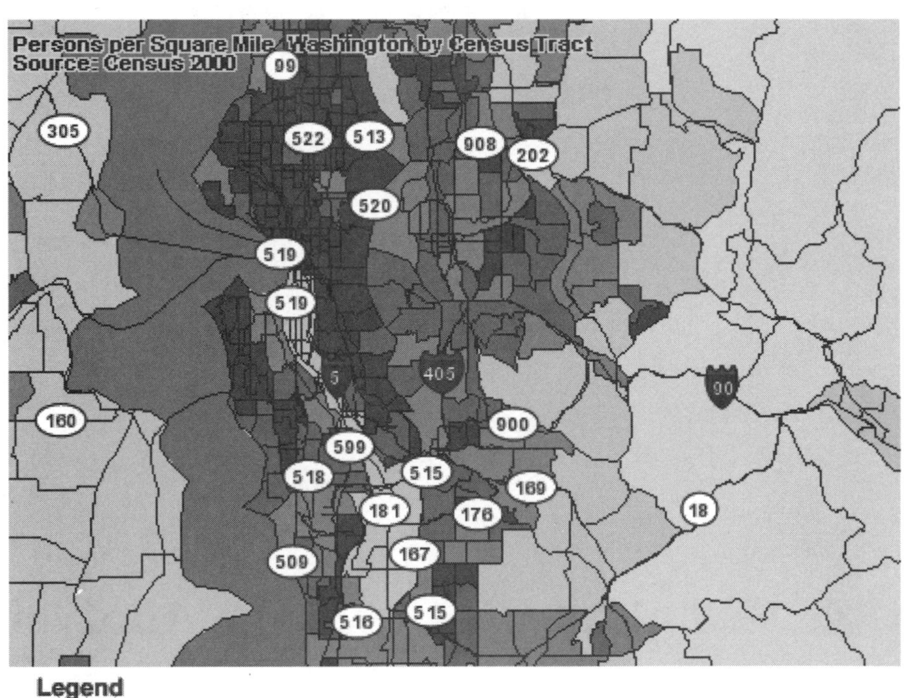

Figure 12-2 Example population density map.

areas where we expect the greatest traffic. Second, the type of environment will have a direct impact on propagation modeling, where we consider issues such as modeling correction factors and in-building penetration losses.

Depending on the type of environment, we may wish to provide different levels of coverage. For example, in urban and suburban areas we may wish to provide in-building coverage. On highways, however, we will be interested only in in-vehicle coverage. In other areas, such as parkland, we will likely want to provide only outdoor coverage. In systems such as *Global System for Mobile Communications* (GSM), a good understanding of these criteria may well be sufficient to start the design process. With UMTS, however, a further consideration is required—what types of services should be available in a given area. For example, in a given area, should a subscriber have access to data rates of up to, say 480 Kbps, or is a lower rate sufficient, or is speech-only service acceptable? These issues are important because, as depicted in Figure 12-3, the effective footprint of a cell is influenced by the data rate to be supported. The higher the throughput, the smaller the effective cell radius.

Once we have a solid understanding of the coverage requirements, then we can use that information in the preparation of an initial RF coverage plan. Before generating that plan, however, another critical input is required—link budgets.

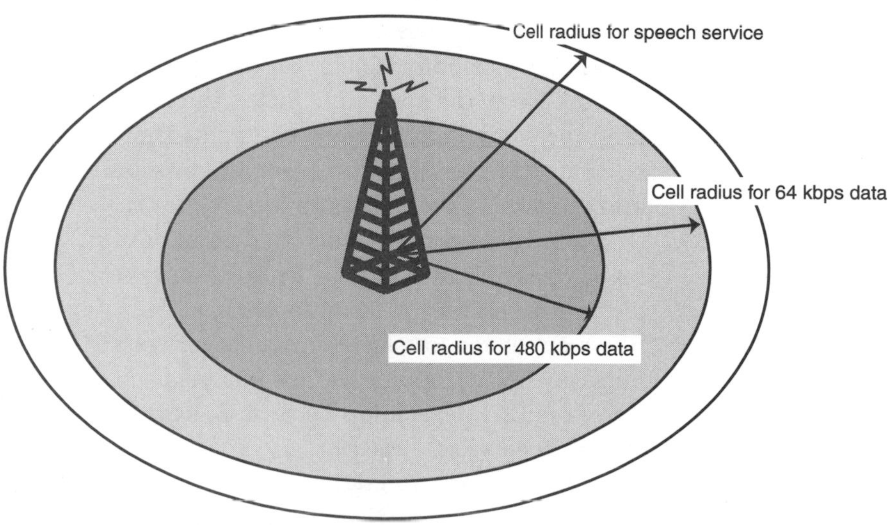

Figure 12-3 Relative cell footprints for different user data rates.

12.2.1 Link Budgets

A link budget is a calculation of the amount of power received at a given receiver based on the output power from a given transmitter. The link budget accounts for all of the gains and losses that a radio wave experiences along the path from transmitter to receiver. For a given transmitter power, we determine the maximum path loss that the signal can experience in order for the signal to be recoverable at the receiver. Given that the base station must be able to "hear" the mobile and the mobile must be able to "hear" the base station, we need to perform the calculation in both directions—from mobile to base station and from base station to mobile. We determine the maximum allowable path loss in each direction and the lesser of the two corresponds to the coverage limit for the cell and service in question. For example, if the maximum allowable path loss in the uplink is, say 130 dB, and the maximum allowable path loss in the downlink is, say 135 dB, then we should not exceed 130 dB, and we are said to be uplink limited.

The link budget needs to include a margin (that is, a buffer) to enable fading of the signal. In other words, we design the system such that service will still be supported even in the case of where the signal fades significantly. The greater the fade margin, the greater the reliability of the service. Moreover, because a *Wideband CDMA* (W-CDMA) system is interference limited, we also need to include an interference margin. As described later in this chapter, the size of that margin is load dependent.

As mentioned, the effective cell coverage is dependent upon the service to be provided. One reason for this is the fact that the higher the spreading factor (corresponding to a lower data rate), the higher the processing gain, and the lower the spreading factor, the lower the processing gain. Because the processing gain is one of the gains that needs to be included in a link budget, it follows that the lower the processing gain, the lower the maximum allowable path loss and the smaller the effective radius of the cell.

From a pure radio propagation point of view, we will generally find that coverage is uplink limited, if for no other reason than that the output power of the base station is far greater than that of the mobile. As we shall see, however, cell loading also impacts coverage, so the consideration of cell load must be considered in coverage analysis.

Tables 12-1, 12-2, and 12-3 provide example uplink link budgets for three WCDMA services—speech service at 12.2 Kbps outdoors, data service at 128 Kbps indoors, and data service at 384 Kbps indoors.

Table 12-1

Example Link Budget for Speech, Outdoor Pedestrian Service

Transmitter (mobile)		
Mobile TX power (dBm)	21	
Antenna gain (dBi)	0	
Body loss (dB)	3.0	
EIRP (dBm)	18	Equivalent Isotropic Radiated Power
Receiver (base station)		
Thermal noise density (dBm/Hz)	−174.0	Note 1
Receiver noise figure (dB)	5.0	Equipment/vendor-dependent
Receiver noise power (dBm), calculated for 3.84 Mcps	−103.2	= Thermal noise density + receiver noise figure + $10\log(3.84 \times 10^6)$
Interference margin (dB)	4	Cell load-dependent
Total noise + interference (dBm)	−99.2	
Processing gain (dB)	25.0	= 10 log(3,840,000/12,200)
Required E_b/N_o (dB)	4	Service-dependent
Effective receiver sensitivity (dBm)	−120.2	= Total noise + interference minus processing gain + E_b/N_o
Base station antenna gain (dBi)	18	
Base station feeder and connector losses (dB)	2	
Fast fading margin (dB)	4	Enables room for closed loop power control
Log normal fade margin (dB)	7.5	Enables for greater cell-area reliability (Note 2)
Building penetration loss (dB)	0	
Soft handover gain (dB)	2	
Maximum allowable path loss (dB)	144.7	

Table 12-2

Example Link Budget for 128 Kbps Data, Indoor Service

Transmitter (mobile)		
Mobile TX power (dBm)	24	
Antenna gain (dBi)	0	
Body loss (dB)	0	
EIRP (dBm)	24	Equivalent Isotropic Radiated Power
Receiver (base station)		
Thermal noise density (dBm/Hz)	−174.0	Thermal noise floor
Receiver noise figure (dB)	5.0	
Receiver noise power (dBm), calculated for 3.84Mcps	−103.2	= Thermal noise density + receiver noise figure + 10log(3.84×10^6)
Interference margin (dB)	4	Cell load-dependent
Total noise + interference (dBm)	−99.2	
Processing gain (dB)	14.8	=10 log(3,840,000/128,000)
Required E_b/N_o (dB)	2	Service-dependent
Effective receiver sensitivity (dBm)	−112.0	= Total noise + interference minus processing gain + E_b/N_o
Base station antenna gain (dBi)	18	
Base station feeder and connector losses (dB)	2	
Fast fading margin (dB)	4	Enables room for closed loop power control
Log normal fade margin (dB)	7.5	Enables for greater cell-area reliability (Note 2)
Building penetration loss (dB)	15	Typical value for suburban building
Soft handover gain (dB)	2	
Maximum allowable path loss (dB)	127.5	

Table 12-3

Example Link Budget for 384 Kbps Data, Indoor Service

Transmitter (mobile)		
Mobile TX power (dBm)	24	
Antenna gain (dBi)	2	
Body loss (dB)	0	
EIRP (dBm)	26	Equivalent Isotropic Radiated Power
Receiver (base station)		
Thermal noise density (dBm/Hz)	−174.0	Thermal noise floor
Receiver noise figure (dB)	5.0	
Receiver noise power (dBm), calculated for 3.84Mcps	−103.2	= Thermal noise density + receiver noise figure + 10log(3.84×10^6)
Interference margin (dB)	4	Cell load-dependent
Total noise + interference (dBm)	−99.2	
Processing gain (dB)	10.0	=10 log(3,840,000/384,000)
Required E_b/N_o (dB)	1	Service-dependent
Effective receiver sensitivity (dBm)	−108.2	= Total noise + interference minus processing gain + E_b/N_o
Base station antenna gain (dBi)	18	
Base station feeder and connector losses (dB)	2	
Fast fading margin (dB)	4	Enables room for closed loop power control
Log normal fade margin (dB)	7.5	Enables for greater cell-area reliability (Note 2)
Building penetration loss (dB)	15	Typical value for suburban building
Soft handover gain (dB)	2	
Maximum allowable path loss (dB)	125.7	

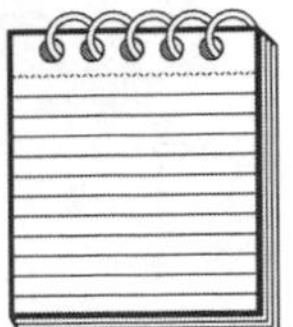

NOTE: *Thermal noise density = kT, where k = Boltzmann's constant and T is temperature in Kelvin. T is usually assumed to be 293K.*
NOTE2: *The log normal fade margin is a design value and depends upon the required level of signal reliability over the cell area. 7.5 dB corresponds to 93.4% coverage probability.*

In Table 12-1, we have a link budget that would apply to outdoor (non-vehicular) speech service. The user device has a nominal power output of 0.125 W (21 dBm). Thus, it is likely to be a Power Class 4 device (max power of 21 dBm, ± 2 dB) or a Power Class 3 device (max power of 24 dBm, +1/−3 dB). We assume that there is no antenna gain for the device, and we assume a 3 dB body loss as the device is likely to be close to user, and the signal will have to pass through the user.

At the receiving side, we assume a receiver noise figure of 5 dB and an interference margin of 4 dB. The interference margin accounts for the fact that there will be interference at the base station caused by multiple users. The greater the number of users, the greater the interference and the greater the required interference margin. Also at the receiving side, we specify the processing gain and the E_b/N_o required for the service. As we will describe shortly, the required E_b/N_o can vary according to the service in question.

If we include a typical antenna gain value for the base station antenna and typical losses for cables and connectors, then simple addition gives the maximum path loss. In reality, however, we need to add some additional considerations to account for real-world situations.

First, we need to add a fast fading margin. This is a buffer to enable for the mobile to adjust power according to closed loop power control. If the link budget in Table 12-1 were prepared for a mobile moving at a fast speed (such as 60 mph), then closed loop power control would be unlikely to be fast enough to change the transmitted power in response to the rapid changes in pass loss as the mobile moves. Thus, for a high-speed vehicular service, one would set the fast fading margin to zero.

We also need to add a log-normal fading margin, with a value that is determined by the desired cell area (or cell edge) coverage reliability. The higher the desired coverage reliability, the higher the log normal fading margin. Finally, we can add a gain that results from soft handover. Basically, if a subscriber is being covered by more than one cell and is in a soft-handover situation, then the signal from the handset is being received by two base stations (or perhaps by two cells at the same base station site).

This is the equivalent to an extra level of receiver diversity and offers a similar gain.

In Table 12-2, we have a link budget that would apply to an indoor data service at 128 Kbps. In this case, the service is assumed to be provided by a base station located outside of the building in question. The user device has a nominal power output of 0.25 W (24 dBm). Thus, it is likely to be a Power Class 3 device (max power of 24 dBm, +1/−3 dB) or a Power Class 2 device (max power of 27 dBm, +1/−3 dB). We assume that there is no antenna gain for the device. We further assume that, unlike the case for a speech service, the device is less likely to be very close to the user (that is, not against the user's head). Therefore, we do not allow for any body loss.

At the receiving side, many of the parameters are the same as for the example of Table 12-1. The required E_b/N_0 in this case, however, is 2 dB, and the processing gain is lower (due to the higher data rate). The other margins, gains, and losses are the same as for Table 12-1, with the exception of the building penetration loss, which we assume to be 15 dB. This figure is highly dependent on the area to be covered. In a dense urban environment, for example, the building penetration loss could be significantly higher.

In Table 12-3, we have a link budget that would apply to an indoor data service at 384 Kbps. We assume that the service is to be provided by a base station located outside of the building in question. The user device has a nominal power output of 0.25 W (24 dBm). Given that this is likely to be a specialized data device with an external antenna, we assume an antenna gain of 2 dBi for the device. We also assume that there is no body loss.

At the receiving side, many of the parameters are the same as for the example of Table 12-2. The required E_b/N_0 in this case, however, is 1 dB, and the processing gain is lower (due to the higher data rate). The other margins, gains, and losses are the same as for Table 12-2. We assume the same building penetration loss as in Table 12-2 because we are assuming that the service is provided from a base station outside of the building. If we were to assume an in-building base station, then the penetration loss would be much lower—just enough to accommodate for losses in internal walls within the building.

In reviewing the three example link budgets, we see that the maximum allowed path loss decreases as the required data rate increases. Thus, the higher the data rate to be offered over a given area, the greater the required density of base stations.

There is not an exact "apples-to-apples" comparison between the different services in our example link budgets as we have made different assumptions regarding mobile output power, antenna gains, and building penetration losses. If, for comparison purposes, we were to assume that these quantities were the same in each scenario, then we would have a clear picture of how the cell coverage reduces as the data rate increases (all other things being equal). The reduction in cell coverage is due to the reduced processing gain. It is noticeable, however, that while there is a reduced processing gain for higher data rates, this is somewhat counterbalanced by a lower E_b/N_o requirement for higher data rates.

The required E_b/N_o is dependent on many factors including the mobile speed, data rate, and multipath profile. Why should the E_b/N_o decrease as the data rate increases? The answer is the fact that higher bit rates mean greater power output from the mobile. There is greater output power for both the *Dedicated Physical Control Channel* (DPCCH) and the *Dedicated Physical Data Channel* (DPDCH) as the data rate increases. The pilot symbols on the DPCCH are used for channel estimation and received *Signal-to-Interface Ratio* (SIR) estimation. As the DPCCH power increases, the better the channel estimation, which means that a lower E_b/N_o can be accommodated. Of course, as the data rate increases, the DPDCH power also increases and, in fact, the relative power of the DPCCH versus the DPDCH decreases. In other words, as the data rate increases, a greater proportion of the total power is allocated to DPDCH rather than DPCCH. But the fact that the overall power increases with increasing data rate means that the total DPCCH power increases (albeit not as much as the DPDCH power). It is the absolute DPCCH power that is important in channel estimation, and because the absolute DPCCH power increases, the required E_b/N_o decreases.

12.3 RF Capacity Analysis

Based onlink budgets and using an appropriate propagation model as described in Chapter 9, we can perform an initial RF coverage plan. This is typically done using a software-based planning tool. This will only be an initial plan, however. The next step requires that we validate the plan to ensure that it will support the expected load. Recall that the link budget includes an interference margin that is based upon the loading expected on the cell. The greater the expected load, the greater the interference margin needs to be. Suppose, for example, that we perform an initial coverage analysis based on a nominal interference margin of, say 3 dB, equivalent to approximately a 50 percent cell loading. Therefore, we must validate the initial coverage-based plan to ensure, based upon the provided coverage and the expected traffic forecast in the covered area, that the interference margin chosen will be sufficient to support the expected load. Table 12-4 shows the required interference margins as a function of uplink cell load.

The reason for the interference margin is to account for the interference that will be caused by other users. That interference is effectively additional noise over and above thermal noise. In other words, the greater the cell load, the greater the noise, and we need to include a greater margin to account for that noise. This increase in noise is known as the *noise rise*, and the margin we include in the link budget matches the noise rise generated by the expected cell load.

From Table 12-4 we can see that the noise rise tends toward infinity as the cell load tends toward 100 percent. In other words, 100 percent cell load is not achievable. Moreover, the greater the cell load, the greater the noise rise and the smaller the effective cell coverage area.

We cannot achieve 100 percent cell load, but we can readily achieve a cell load of, say 60 percent. We must, of course, be able to translate that percentage into some measure of subscriber usage—such as total number of subscribers for a given service or total throughput. This will allow us to verify whether the cell coverage we expect (assuming a particular interference margin) will be sufficient to support the offered load. Imagine, for example, that we have a nominal cell plan based on a link budget for a particular service (such as 128 Kbps data) and with a particular interference margin (such as 4 dB or about 60 percent uplink load). That plan will mean that a given cell has a particular footprint. We then consider that footprint and

Table 12-4

Required Interference Margins as a Function of Uplink Cell Load

Uplink Cell Load	Required Interference Margin
0 %	0 dB
10 %	0.46 dB
20 %	1 dB
50 %	3 dB
75 %	6 dB
90 %	10 dB
95 %	13 dB
99 %	20 dB

determine whether the load expected within that footprint will be less than the loading for which the plan was designed in the first place. Clearly, this is an iterative process. If we find that the plan does not support the expected load in some areas, then we need to modify the plan, perhaps by the addition of extra base stations.

In order to determine whether a given cell can accommodate the expected load, we need to quantify that load. As described in Chapter 10, "Network Design Considerations," we should first determine the expected demand in the busy hour so that we can make sure that we design the system to accommodate peak demand. That peak demand needs to be specified for the various services we wish to offer—both voice and data at various rates. We then determine the cell capacity and check to make sure that it can support the expected demand. The expected demand should be specified both for the uplink and the downlink and the cell capacity calculation should also be performed for both directions. This is of particular importance because UMTS services can be asymmetrical.

12.3.1 Calculating Uplink Cell Load

The load placed on a cell is described in terms of load factor, which is some fraction of the maximum theoretical load. In other words, a load factor of 0.5 equates to 50 percent cell loading. The load placed on the cell can be viewed as the sum of the loads generated by all users of the cell. Alternatively, the total load factor is the sum of the load factors contributed by each user as shown in Equation 12-1:

$$\text{Load factor} = \sum_{j=1}^{N} L_j \qquad \text{(Equation 12-1)}$$

where L_j is the load factor of a single user (j) and we assume N users in the cell. L_j is simply the fraction of the total power at the base station that user j generates. Thus,

$$L_j = S_j / S_{total} \qquad \text{(Equation 12-2)}$$

Alternatively

$$S_j = L_j \cdot S_{total} \qquad \text{(Equation 12-3)}$$

The signal power (S_j) for a given user needs to be such that the E_b/N_o requirement is met for the service that the user wishes to obtain. Moreover, the E_b/N_o is a function of the total interference in the cell. E_b/N_o can be expressed as follows:

$$(E_b/N_o)_j = \text{Processing gain} \cdot [S_j/I] \qquad \text{(Equation 12-4)}$$

where I is the total interference and is equal to the total received power at the base station minus S_j (the signal power from user j). This can be rewritten as

$$(E_b/N_o)_j = [C/a_j \cdot R_j][S_j/(S_{total} - S_j)] \qquad \text{(Equation 12-5)}$$

where C is the chip rate, a_j is the activity factor (such as about 65 percent for voice including DPCCH overhead and 100 percent for data), R_j is the user data rate, and S_{total} is the total received signal power at the base station. If we then solve for S_j, we get

$$S_j = S_{total}/[1 + C/(a_j \cdot R_j(E_b/N_o)_j)] \qquad \text{(Equation 12-6)}$$

Substituting from Equation 12-2, we get

$$L_j = 1/[1 + (C/(a_j \cdot R_j)(E_b/N_o)_j)] \qquad \text{(Equation 12-7)}$$

Using Equation 12-1, we now get

$$\text{Load factor} = \sum_{j=1}^{N} L_j = \sum_{j=1}^{N} 1/[1 + (C/(a_j \cdot R_j)(E_b/N_o)_j)] \qquad \text{(Equation 12-8)}$$

In addition to the interference generated by users on the local cell, there will also be interference caused by transmissions from users in nearby cells. If we define the quantity ι to the ratio of nearby cell interference to the interference in our own cell as follows:

$$\iota = [\text{nearby cell interference}]/[\text{local cell interference}] \qquad \text{(Equation 12-9)}$$

then the total load factor for the local cell is

$$\text{Load factor} = (1 + \iota) \cdot \sum_{j=1}^{N} L_j = (1 + \iota) \cdot \sum_{j=1}^{N} 1/[1 + (C/(a_j \cdot R_j(E_b/N_o)_j))]$$

(Equation 12-10)

As the load factor approaches unity, the cell has reached it maximum capacity.

12.3.1.1 Example Uplink Cell Loading for Voice Service In this example, we calculate the cell loading as a function of the number of users assuming that all users are using standard voice service.

Assumptions

$a_j = 0.65$

$R_j = 12.2$ Kbps

$E_b/N_o = 4$ dB(= 2.512) for all users (because all users are voice-only in this example).

$\iota = 50\%$ (that is, of the total interference at the base station, one third is being received from other cells).

Using Equation 12-10, we calculate the uplink load factor for a single user.

$$\text{Load factor for one voice user} = 0.00774 = 0.774\%$$

Thus, for a load factor of 50 percent, we can accommodate approximately 65 simultaneous voice users. For a load factor of 60 percent, we can accommodate approximately 76 simultaneous voice users, and so on. The number of users as a function of load factor (and required interference margin) is shown in Figure 12-4.

Given that the required interference margin (that is, noise rise) means a smaller allowable path loss, it is clear that the cell footprint reduces as the number of users increases. If we consider the link budget shown in Table 12-1 and consider the required interference margin as a function of the number of users, the allowable path loss (which determines the cell size) is as shown in Figure 12-5. We should also note that the maximum path loss shown does not consider building or vehicle penetration losses, which would need to be subtracted from the figures shown.

The calculation we have performed previously provides the loading in terms of number of users. We could easily present the loading results in terms of Kbps. In fact, if we consider that there will be both data and voice usage, then presenting the information in terms of Kbps can be useful as that term will apply both to voice and data.

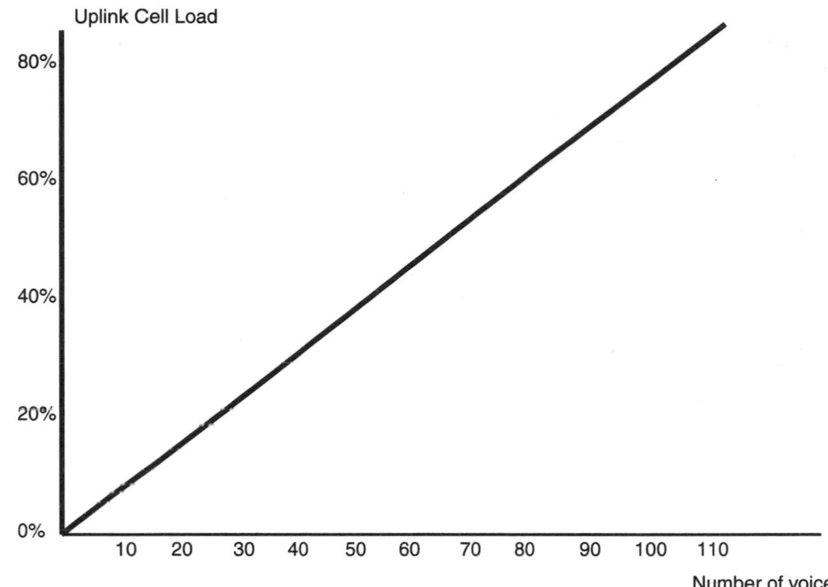

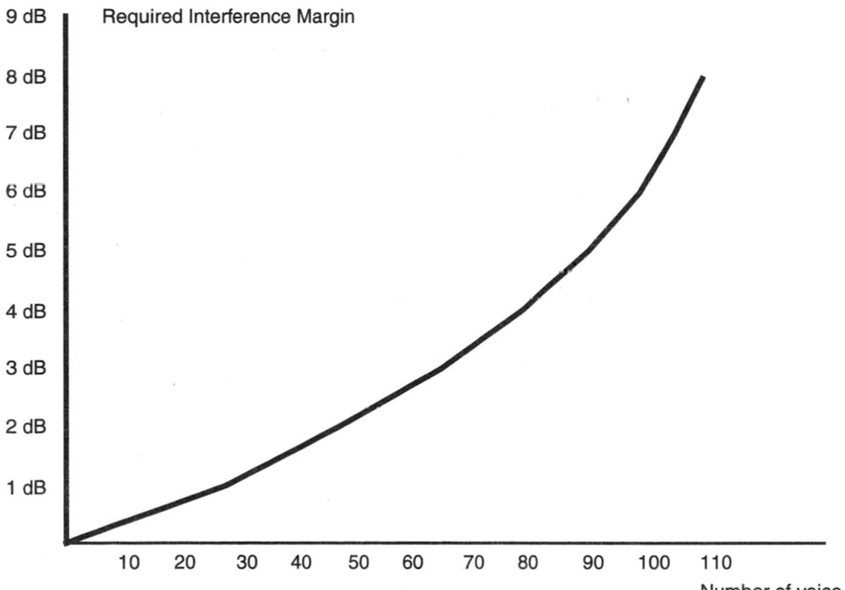

Figure 12-4 Example uplink cell loading and interference as a function of number of users.

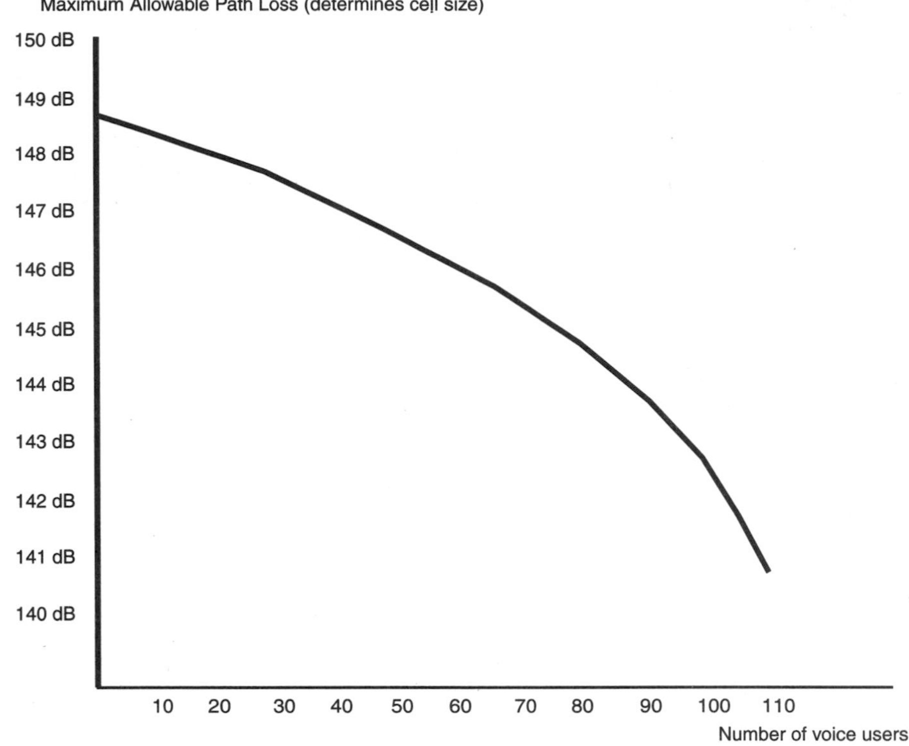

Figure 12-5 Example allowable uplink path loss as a function of number of users.

12.3.1.2 Example Uplink Cell Loading for Data Service In this example, rather than showing the number of users, we show the total throughput (in Kbps) for a given cell loading.

If we look again at Equation 12-10, we note that the equation can be simplified if we assume that all users have the same data rate. The equation is then as follows:

$$\text{Load factor} = N \cdot (1 + \iota)/[1 + (C/(a \cdot R(E_b/N_o)))] \quad \text{(Equation 12-11)}$$

If we note that the term $C/(a \cdot R(E_b/N_o))$ is far greater than 1 for most services, then we can further simply the equation to be

$$\text{Load factor} = N \cdot (1 + \iota)/[C/(a \cdot R(E_b/N_o))] \quad \text{(Equation 12-12)}$$

To get the total throughput (rate times number of users), we rearrange to get

$$\text{Throughput} = R \cdot N = [\text{Load factor } C]/[(E_b/N_o) \cdot (1 + \iota)] \quad \text{(Equation 12-13)}$$

Example assumptions

a = 1.0

E_b/N_o = 1 dB(= 1.259)

ι = 50% (that is, of the total interference at the base station, one third is being received from other cells).

Thus, for a load factor of 50 percent, we have a total throughput of 1,106 Kbps. For a load factor of 60 percent, we have a total throughput of 1,220 Kbps.

12.3.2 Downlink Cell Load

In the downlink, the determination of cell loading uses the same basic approach as for the uplink. The same approach is applicable because the ability of a given mobile to recover a signal that is destined for that mobile is dependent upon how many other signals are being sent to other mobiles in the cell. In other words, for a given user, j, the signals that are being sent from the base station to other users are simply interference. The more such signals, the greater the interference. As is the case for the uplink, the effect of the interference is dependent on the E_b/N_o requirement needed at the mobile. There is also interference caused by downlink common channels and interference caused by other base stations. In the case of interference from other base stations, the amount of interference will depend upon the individual user's location. A user that is close to the serving base station is less likely to experience as much interference from neighboring cells as a user that is near the border between cells.

Finally, we need to factor in orthogonality. In the downlink, for a given scrambling code, transmissions to different users are sent using different channelization codes, which are chosen such that the codes are orthogonal. If the transmission from the base station to a single user arrives over multiple paths, however, and the delay spread across those paths is sufficiently large, the mobile will directly recover only a part of the signal from the base station. The other part of the signal, which arrives over a long delay path, will be seen as interference. This phenomenon needs to be accounted for in our calculation of downlink loading.

Because the same methodology for downlink load factor calculation applies in the uplink, then Equation 12-10 still applies, but with some modifications to account for orthogonality and the fact that interference from neighboring cells is different for each user. Thus, the equation for downlink load factor becomes

$$\text{Load factor} = \sum_{j=1}^{N} L_j \cdot (1 - \alpha_j + \iota_j)$$
$$= \sum_{j=1}^{N} (1 - \alpha_j + \iota_j)/[1 + (C/(a_j \cdot R_j(E_b/N_o)_j))] \quad \text{(Equation 12-14)}$$

where α_j is the orthogonality factor related to user j and ι_j is the interference from neighboring cells experienced by user j.

As in the case in the downlink, for most services, the term $C/(a_j \cdot R_j(E_b/N_o)_j)$ is far greater than 1, which means that the equation can be simplified. Moreover, it is not realistic to determine the orthogonality factor for each mobile in the cell as this will depend on the exact user location and multipath profile. Nor is it realistic to determine the intercell interference experienced by each user as that will also depend on the user's exact location. Thus, we need to consider average values of orthogonality (α) and intercell interference (ι). A typical value for α is 0.4 and a typical value for ι is 0.5.

Including these considerations, the load equation becomes

$$\text{Load factor} = (1 - \alpha + \iota) \cdot \sum_{j=1}^{N} L_j = (1 - \alpha + \iota) \cdot$$
$$\sum_{j=1}^{N} 1/[C/(a_j \cdot R_j(E_b/N_o)_j))] \quad \text{(Equation 12-15)}$$

12.3.2.1 Example Downlink Cell Loading for Voice Service In this example, we calculate the cell loading as a function of the number of users assuming that all users are using standard voice service.

Assumptions:

$a_j = 0.65$ for all users

$R_j = 12.2$ Kbps for all users

$E_b/N_o = 4$ dB (= 2.512) for all users (because all users are voice-only in this example).

$\alpha = 0.4$

$\iota = 0.5$

Because of the fact that all users in this example have the same characteristics, Equation 12-15 becomes

$$\text{Load factor} = N \cdot (1 - \alpha + \iota)/[C/(a \cdot R(E_b/N_o))] \quad \text{Equation 12-16)}$$

Using the previous assumptions, the load factor for one user is

$$(1 - 0.4 + 0.5)/[3{,}840{,}000/(0.65 \cdot 12{,}200 \cdot 2.512)] = 0.0057 = 0.57\%.$$

Thus, for a downlink load factor of 50 percent, we can accommodate approximately 88 simultaneous voice users. For a load factor of 60 percent, we can accommodate approximately 105 simultaneous voice users, and so on.

As is the case for the uplink, the downlink link budget needs to include an interference margin equivalent to the noise rise. The required interference margin is a function of the cell load factor, and the same figures as in Table 12-4 apply. In other words, for a 50 percent load factor, we need a 3 dB interference margin in the downlink.

Assume, for example, a downlink link budget as shown in Table 12-5, where there is a base station transmitter output power of 10 W.

This link budget does not show an interference margin. Such a margin must be included, however. The exact value of the interference will equate to the noise rise, which increases with increasing cell load—that is, throughput. Using the example assumptions outlined previously, Figure 12-6 shows the cell load as a function of the number of users and also the noise rise/required interference margin as a function of the number of users. Figure 12-7 shows the allowable downlink path loss as a function of the number of users. If we compare Figure 12-7 with Figure 12-5, we can determine whether the system is uplink limited or downlink limited for a given number of voice users.

The foregoing examples show how cell loading, in terms of numbers of voice users, can impact uplink and downlink coverage. Using voice service is a convenient example to show how the calculations can be performed. In reality, however, we can expect a significant mix of services—with some subscribers using voice service and some subscribers using data services of one kind or another. Thus, the calculations should be performed individually for each type of service.

While, for a service like voice, the coverage is likely to be uplink limited rather than downlink limited, the same might not apply for data service. With UMTS, data services can be asymmetric—that is, different date rates in the uplink compared to the downlink. Moreover, for many data services (such as Web browsing), we will find that the downlink data rate is far greater than the uplink data rate. Consequently, the effect of interference in the downlink may well be greater than in the uplink, which means that the downlink load may become the limiting factor.

Table 12-5

Example Downlink Link Budget for Speech, Outdoor Pedestrian Service

Transmitter (mobile)		
Base station TX power (dBm)	40	Equal to 10 W
Base station antenna gain (dBi)	18	
Base station feeder and connector losses (dB)	2	
EIRP (dBm)	56	Equivalent Isotropic Radiated Power
Receiver (mobile)		
Thermal noise density (dBm/Hz)	−174.0	Note 1
Receiver noise figure (dB)	5.0	
Receiver noise power (dBm), calculated for 3.84 Mcps	−103.2	= Thermal noise density + receiver noise figure + 10log(3.84×10^6)
Interference margin (dB)	0	This must be set according to expected cell load.
Total noise + interference (dBm)	−103.2	
Processing gain (dB)	25.0	=10 log(3,840,000/12,200)
Required E_b/N_o (dB)	4	Service-dependent
Effective receiver sensitivity (dBm)	−124.2	= Total noise + interference − processing gain + E_b/N_o
Fast fading margin (dB)	4	Enables room for downlink power control
Log normal fade margin (dB)	7.5	Enables for greater cell-edge reliability
Building penetration loss (dB)	0	Typical value for suburban building
Body loss (dB)	3.0	
Soft handover gain (dB)	2	
Maximum allowable path loss (dB)	167.7	

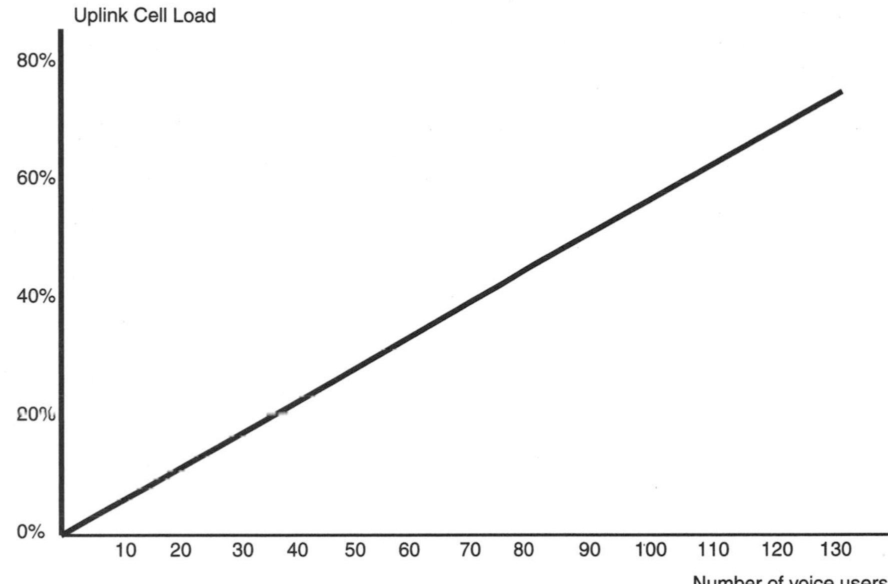

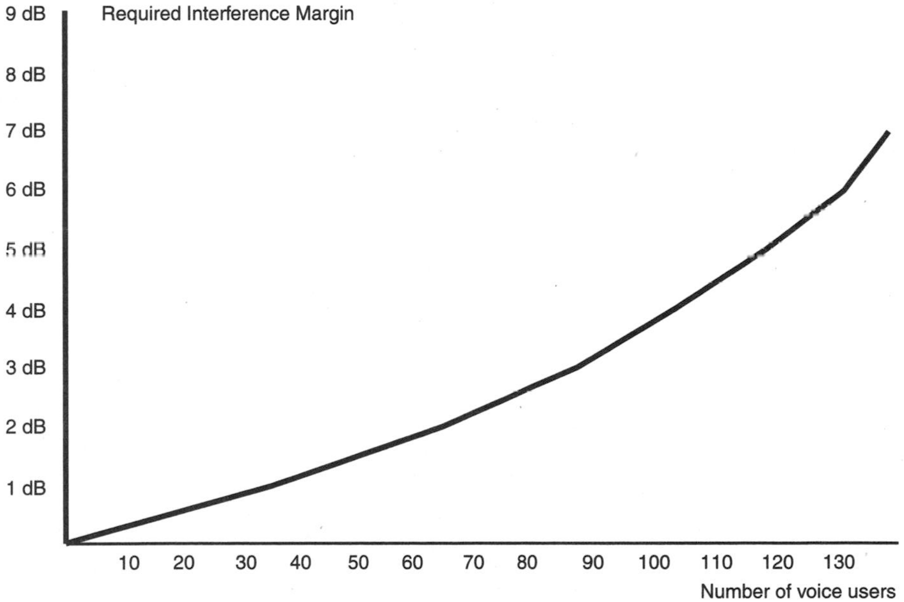

Figure 12-6 Example downlink cell loading and interference as a function of number of users.

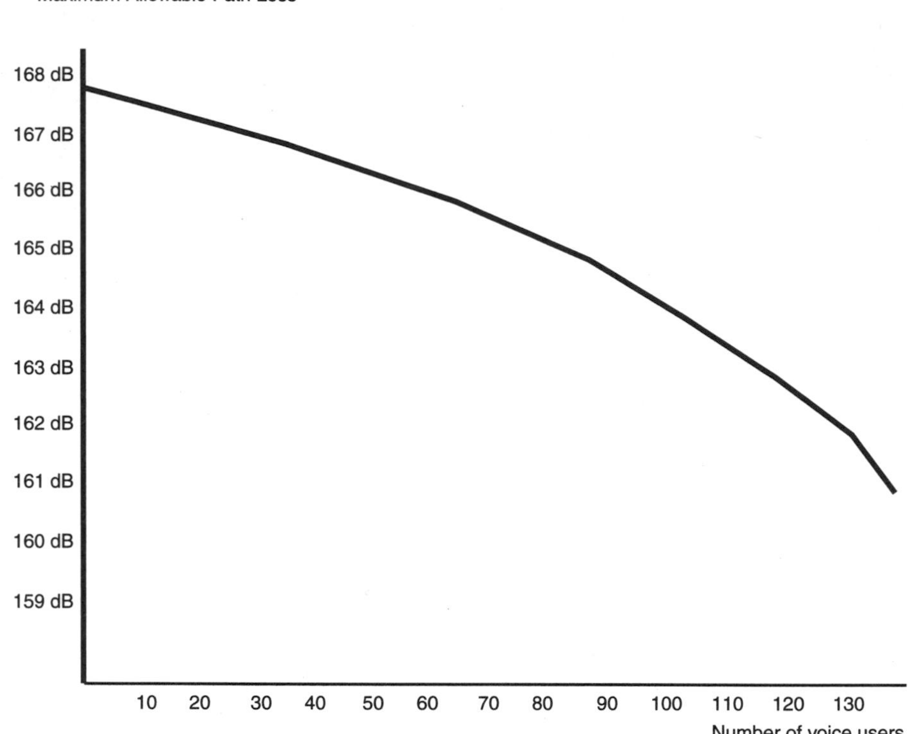

Figure 12-7 Example allowable downlink path loss as a function of number of users.

If we find that we are downlink limited, then we may be able to increase the base station output power and/or add an additional RF carriers subject to spectrum availability. As mentioned in Chapter 6, "Universal Mobile Telecommunications Service (UMTS)," however, the addition of a second carrier will mean that compressed mode must be used (where the MS can tune to other carriers for potential hard handover). Compressed mode means an aggregate lower throughput per carrier, so that, although a second carrier does provide significant additional capacity, it does not mean a capacity increase of 100 percent.

Another downlink limiting factor for a single carrier base station is the availability of downlink channelization (spreading) codes. Recall from Chapter 6 that channelization codes are chosen from a code tree. Recall also that the use of a particular for a channelization code can pre-empt the use of other channelization codes on the same branch of the code tree. For example, consider the channelization code $C_{ch,4,0}$. This code is simply the sequence 1,1,1,1 repeated over and over. Consider the channelization code $C_{ch,8,0}$. This is simply the sequence 1,1,1,1,1,1,1,1 repeated over and over. Clearly, if the base station is using either of these codes in a transmission to a particular mobile, then it cannot use the other code (or any other code that is a series of all ones) in transmission to any other mobile. One way to overcome this limitation, however, is for the base station to use multiple scrambling codes. A given cell can use up to 16 downlink scrambling codes.

12.3.3 Load Sharing

As described in the preceding discussions, inter-cell interference plays a role in the capacity of a given cell. In both the uplink and downlink, higher interference from nearby cells means a lower capacity and possible smaller footprint in the cell of interest. Conversely, lower interference from nearby cells means that the cell of interest can have higher capacity or larger footprint. This means that one cell can effectively "borrow" capacity from one or more nearby cells that is less loaded.

Consider Figure 12-8 for example. Some subscribers move from Cell A to Cell B. Thus, Cell A becomes less loaded and Cell B becomes more loaded. If Cell B were already heavily loaded, then the existing inter-cell interference could have meant that it might not have been possible to accommodate more users in Cell B. However, the fact that Cell A now has fewer users means that it is generating less inter-cell interference in Cell B. Thus, it may well be possible to accommodate the additional load on Cell B. This example shows that the capacity of a cell is not static and it varies with the load on nearby cells.

The foregoing discussions regarding uplink and downlink capacity and their effect on coverage emphasize the fact that coverage and capacity are interrelated. Because we need to develop an RF design that supports both coverage and capacity requirements and because capacity affects coverage, the development of the RF design is an iterative process. We start with an initial coverage-based design, and we check that design against the expected demand. We then modify the design to allow for additional capacity where needed. As the implementation phase proceeds, we may find that we need to deal with other constraints, such as the inability to acquire a cell

Figure 12-8 Example of load sharing.

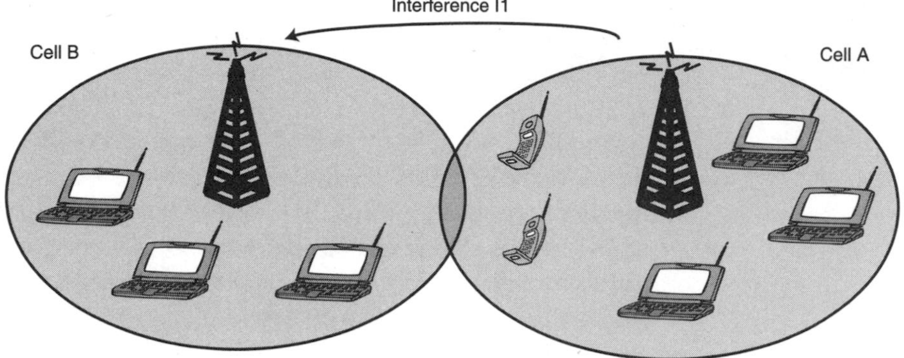

Because of interference (I1) Cell B has only enough capacity to handle one more subscriber

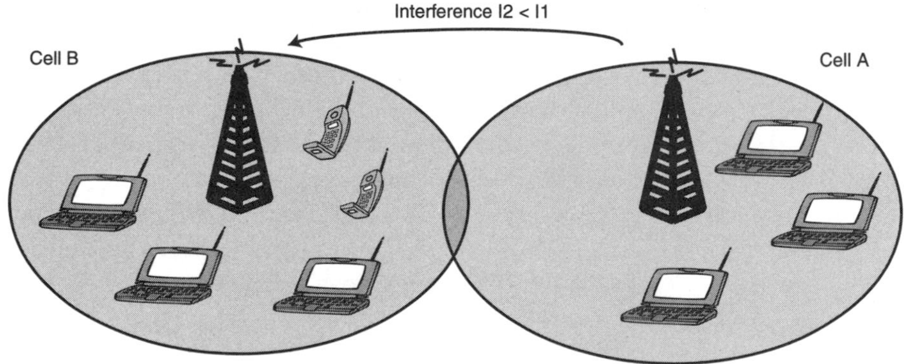

Because of less interference from Cell A, Cell B was able to accomodate two subscribers moving from Cell A

site in the ideal location or drive test results that do not match expectations. In such cases, we will need to change the design to account for different cell site locations, different correction factors, and so on. Several iterations of design may be required until we converge to a point where we can provide both the coverage and capacity required.

12.4 Design of the Radio Access Network

Once an RF plan has been developed, the next step in the design effort is to design a network that will connect the various base stations to their *Radio Network Controllers* (RNCs). This means that we must determine the number of RNCs required, we must determine a suitable placement for the RNCs, and we must design a transmission network from RNCs to the various base stations and between RNCs (for inter-RNC soft handover).

For many GSM *Base Station Controllers* (BSCs), the main capacity limitations are in the numbers of base stations, cells, or transceivers that can be supported. In some cases, there are limits in terms of Erlangs, but such capacity limits are rarely encountered in real networks. With UMTS, however, the capacity of most RNCs is more tightly linked to the traffic mix. While one still finds limitations in terms of total base stations, cells, or RF carriers, the traffic handling limitations play a major role. Traffic limits typically include, total throughput, total Iub interface capacity, and total *busy hour call attempts* (BHCA) for voice calls. Therefore, when determining the number of RNCs required, we need to make sure that none of these limits is exceeded. This means that the RNC network design must be done in close cooperation with the RF network design. To make things more complicated, there is often a trade off between one limit and another. For example, if fewer voice Erlangs are used, then the RNC is likely to be able to support a greater data traffic demand.

In order to simplify the dimensioning effort, a good place to start is with dimensioning of the Iub interface. For many vendors, the total Iub interface capacity is likely to be the most constraining factor. Moreover, the Iub interface is common for voice and packet data. Once we have determined the Iub capacity demand from each of the base stations, we can sum that capacity and determine the minimum number of RNCs needed. In practice, we should add an additional 25 to 35 percent to the RNC capacity that we have determined for three reasons. First, when allocating base stations to RNCs, we need to consider location areas/routing areas. It is to common assign such registration areas such that they align with RNC boundaries. This means that we need some flexibility in how we allocate base stations to RNCs. Second, intra-RNC soft handover is preferable to inter-RNC soft handover as it helps to minimize Iub transport requirements and it helps to minimize the total switching demand on the RNC. Thus, we would like to define RNC boundaries such that they do not align with areas of high traffic. This also means that we need flexibility in how base stations are assigned to RNCs. Third, we never want to find ourselves in a situation where the addition of one or two extra sites (or even RF carriers) would require the addition of a new RNC. In other words, we need to leave some room for growth.

12.4.1 Iub Interface Dimensioning

The physical interface to a base station will be such that the Iub capacity from a given base station has some discreet value. For example, a single T1 offers 1.5 Mbps. Typically in North America, we will find that a UMTS base station has some number of T1, DS-3, or OC-3 interfaces. However, while determining that a particular base station needs one T1 or two T1s is important, we need to determine the total Iub load at the RNC. We will not arrive at that total simply by summing the total Iub capacity available at each base station. Imagine, for example, that 100 base stations each have an Iub bandwidth demand of 1.7 Mbps. We could configure each such base station with, say two T1s, equivalent to about 3 Mbps. However, the total load at the RNC will still be 170 Mbps, not 300 Mbps.

As described earlier in this chapter, the RF design is performed in accordance with both the coverage and capacity demand that we expect. Consequently, information will be available as to the traffic (in Kbps) to be carried on the Iub interface from each base station. Unlike other parts of the network, however, the RF design is unlikely to have a very long build-ahead included. While there should be some build-ahead factored into the design, a build-ahead of 9 or 12 months is not pragmatic. This is because the RF network usually represents the greatest component of the total network cost. A large build-ahead could mean a drastic increase in capital expenditure far in advance of when the capacity is needed. If, however, there is a large build-ahead, then we can simply size the Iub interface based upon the expected throughput (including the build-ahead) and with the addition of perhaps 40 percent for overhead. While this approach is less than scientific, the inclusion of a long build-ahead will mean that the interface will have sufficient capacity for some time in the future. During that time, we have the opportunity to observe the increase in demand and make more accurate predictions of future interface capacity needs. If there is only a small build-ahead (such as three to six months), then we need to be more discerning in our determination of Iub capacity.

To determine the actual Iub capacity required, we need to add a certain amount of overhead to the user throughput. This overhead needs to allow for burstiness of traffic, signaling load, and *operation and maintenance* (O&M) load. Moreover, we need to add *asynchronous transfer mode* (ATM) overhead because all of the user traffic, signaling, and O&M is carried in ATM cells.

The amount of burstiness will depend on the mix of traffic. If only voice service is to be offered, then we can assume zero burstiness. On the other hand, an all-data service could require an overhead of up to 40 percent. An allowance of 25 percent would be typical. In addition, we can assume that, for a given throughput, there will be an extra 10 percent required for signaling. We can also assume that we need an additional 10 percent for O&M load. To each of these, we must then add ATM overhead, which will vary according to the service. To begin with, the cell structure of ATM means that there are five octets of overhead for every 48 octets of payload. This alone means an overhead of 10.4 percent. In addition, as described in Chapter 6, we have *ATM adaptation layers* (AALs), which also consume bandwidth. Each AAL consumes some number of octets in each ATM cell, in addition to the five octets of the ATM header. For AAL2, 3 of the 48 payload octets are consumed by AAL2 information. Thus, for AAL2, the total ATM overhead is approximately 18 percent. For AAL5, 4 of the 48 payload octets may be consumed, meaning that the total overhead is approximately 20 percent. For signaling the *service-specific connection-oriented protocol* (SSCOP) and *service-specific coordination function* (SSCF), as described in Chapter 6, reside on top of AAL5 and generate even more overhead. In order to make calculations straightforward, however, the SSCOP and SSCF overhead should be included as part of the total signaling overhead.

Based on the foregoing, the total required Iub bandwidth is given by

$$\text{Iub bandwidth} = \text{Expected user traffic} \cdot (1 + \text{burstiness}) \cdot (1 + \text{signaling overhead} + \text{O\&M overhead}) \cdot (1 + \text{ATM overhead}) \qquad \text{(Equation 12-17)}$$

If we take typical examples as described previously, this equation becomes

$$\text{Iub bandwidth} = \text{Expected user traffic} \cdot (1 + 0.25) \cdot (1 + 0.1 + 0.1) \cdot (1 + 0.2)$$

$$\text{Iub bandwidth} = \text{Expected user traffic} \cdot 1.8$$

Thus, because of signaling, O&M, and ATM overhead, the Iub interface should be sized to a bandwidth that is almost twice that of the actual raw user traffic. Of course, the user traffic is likely to be asymmetrical, and we are likely to find that the downlink traffic is greater than the uplink traffic. The actual Iub transmission facilities, however, will be symmetrical. In other words, if there is 2 Mbps capacity on one direction, there is also 2 Mbps in the other direction. Therefore, when dimensioning the Iub, we need only to consider the user traffic in one direction—the direction of greater demand. This will usually be the downlink direction.

12.4.2 Determining the Number of RNCs

As previously mentioned, the capacity of an RNC is typically limited by some or all of the following factors:

- Total Erlangs
- Total BHCA
- Total Iub interface capacity (Mbps)
- Total Iur interface capacity (Mbps)
- Total Iu interface capacity (Mbps)
- Total switching capacity (Mbps)
- Total number of controlled base stations
- Total number of RF carriers

In most cases, one will find that the Iub interface capacity is likely to be the limiting factor. For example, a typical Iub limit for an RNC is between 150 Mbps and 200 Mbps. The same RNC might well have a limit of 500 or more RF carriers (that is, cells if only one carrier per cell). Given that we might expect a cell to support 500 Kbps to 1 Mbps, it is clear that the number of RNCs is likely to be driven by the total Iub interface bandwidth than the other factors. Of course, once we determine the number of RNCs based on the Iub bandwidth required, we need to validate that no other RNC dimensioning limits have been exceeded. If they have been exceeded, then additional RNC capacity needs to be added according to the most constraining factor. That, however, would be an uncommon situation.

12.4.3 Designing The UTRAN Transmission Network

Once we have determined the number of required RNCs, based upon Iub bandwidth requirements, we need to develop a homing plan that specifies which base stations are to be controlled by which RNCs. This will define the RNC borders. Analysis of the cells at or near those borders will then allow us to estimate the amount of inter-RNC handover traffic we can expect, so that we can determine the Iur connections required and the bandwidth needed for those connections.

The exact amount of inter-RNC handover will depend on the RF environment near the RNC borders. A reasonable approach, however, is to assume that 50 percent of traffic in border cells is being served by two base stations on different RNCs. This would be a conservative estimate that would allow for additional inter-RNC soft handover involving cells that are not defined on the border. Imagine, for example, a user near the top of a tall building. That user might be served by a cell that is not on the border between RNCs, but because of the user's location, the user might also be able to hear and be heard by a Node B on another RNC. Of course, the exact soft handovers to be allowed in the network will be specified as datafill within the RNCs. But, at the point in the design effort where Node B homing and transmission network design are being performed, that datafill may not yet be defined.

Given that the Iur acts in many ways as a conduit for Iub traffic from a mobile to its controlling RNC, the basic assumptions for determining the Iub bandwidth can be applied to determining the Iur bandwidth. For example, if we assume that the Iub bandwidth needs to be approximately twice the user throughput, then the Iur bandwidth should be close to twice the user throughput for that portion of the traffic that is in inter-RNC soft handover.

Now that we have established the Node B to RNC homing plan and we know the Iub and Iur interface requirements, we need to design a transport network to support all of the necessary connections between Node Bs and RNCs, between RNCs and between RNCs and *Service GPRS Support Nodes* (SGSNs) and *Mobile Switching Centers* (MSCs). Given that all of the interfaces in question are ATM interfaces, we are effectively talking about designing an ATM network.

In the example of Figure 12-9, we have three RNCs, each controlling a number of Node Bs and with each RNC connected to each of the other two RNCs. All three RNCs are connected back to a single SGSN and a single MSC. In this example, all three RNCs are in different locations and all are remote from the MSC and SGSN. This is a somewhat unrealistic situation, but we use this example in order to show complexity and how that complexity could be managed. Figure 12-9 shows the logical connections between the various nodes. To implement each of those interfaces individually, however, would be impractical. Rather, one would like to implement a cost effective transport arrangement that will support each of the logical interfaces. One way to do this could be through the use of a ring arrangement as also shown in Figure 12-9. Basically, each of the locations in question would become nodes on the ring, which might be an OC-12, or perhaps an OC-48 ring or even have a higher capacity depending on demand.

In many cases, however, the distances between nodes could mean that the cost of such a ring could be prohibitive. In that case, one might want to employ one of the configurations shown in Figure 12-10. In the first case, we deploy a separate ATM switching layer that takes care of switching of the various ATM paths between the various nodes. By deploying such a layer, we can reduce the overall transmission cost. Of course, there is a capital cost that must be paid, plus the operational cost of deploying new equipment. In the second configuration of Figure 12-10, we use one of the RNCs as an ATM switch. Back at the MSC site, we may have the possibility to use an SGSN or an RNC at that site as an ATM switch. This option is possible for some equipment vendors because an RNC is fundamentally an ATM switch with additional UMTS-specific functionality. It is not uncommon to find that the total switching capacity of an RNC is several gigabits per second, while the Iub interface capacity may be limited to perhaps 200 Mbps. Thus, we are likely to find that the RNC can switch more ATM traffic than would be required of it as a pure RNC. We can take advantage of this extra switching capacity and reduce overall transmission cost without having to deploy a separate ATM switching network.

The design and cost of the *UMTS Terrestrial Radio Acess Network* (UTRAN) transmission network is interwoven with the placement of the RNCs. There may be multiple options for placement of RNCs. We may choose to place all RNCs at the MSC location, all remotely, or some mix of remote and local RNCs. The placement of the RNCs will be related to the capacity of an RNC, the cost of the RNC, the availability of suitable locations, and the cost of transmission. The final solution must aim for a network topology that strikes a balance between capital cost, operational cost, and network reliability.

Figure 12-9 RNC connectivity—logical connections and possible ring transport.

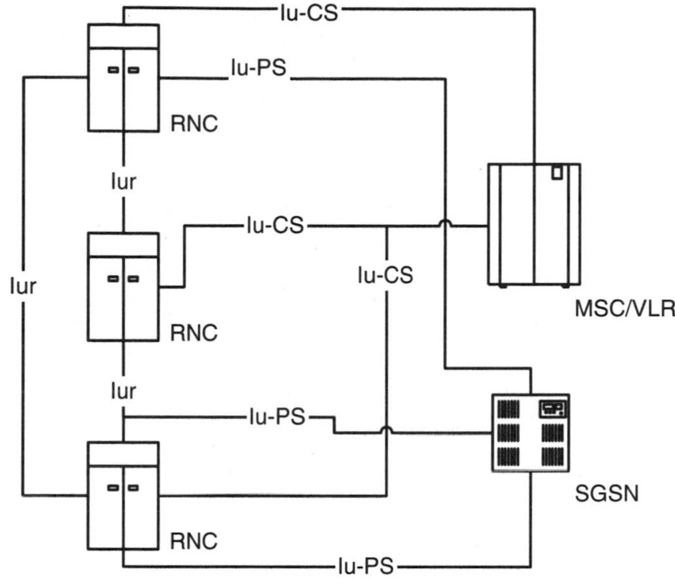

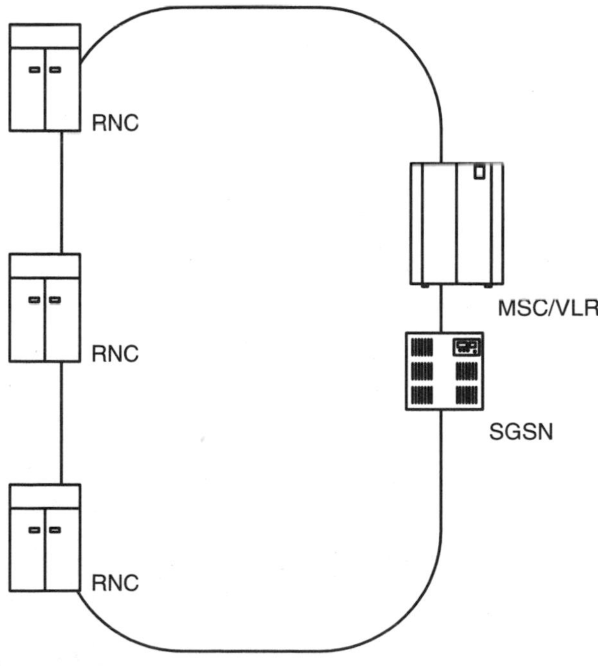

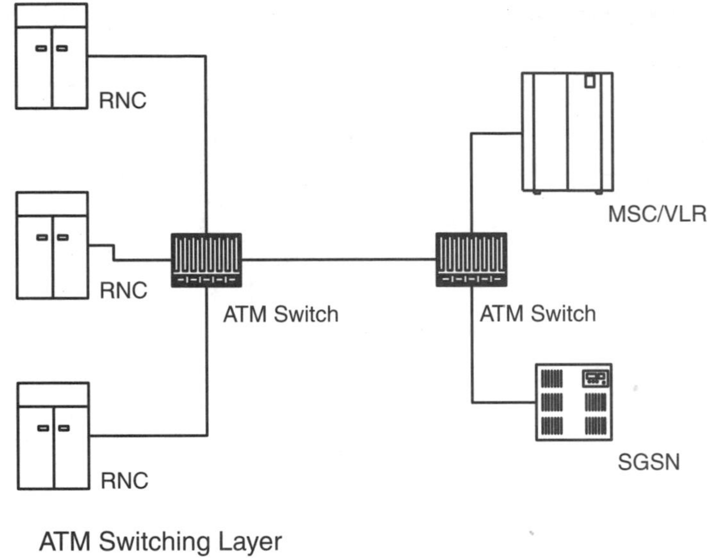

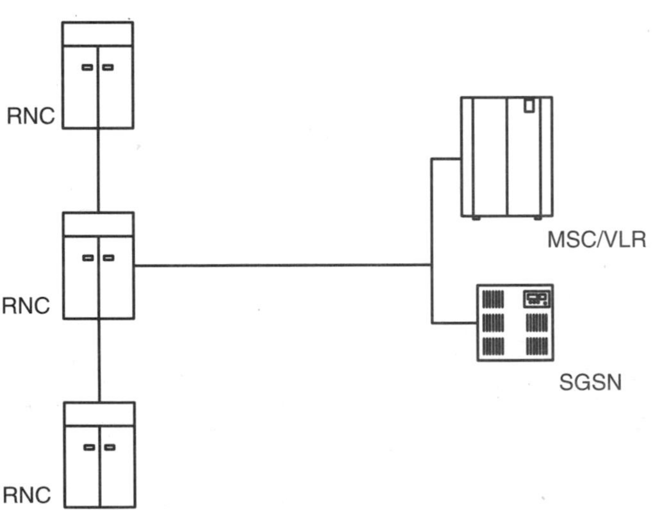

Figure 12-10 Separate ATM switching layer or use of RNC and/or SGSN for ATM switching.

12.5 UMTS Overlaid on GSM

Some network operators will deploy a green-field UMTS network. For many, however, UMTS will be deployed alongside an existing GSM network. Those operators will wish to reuse the components of the GSM network to the greatest extent possible. There is a desire to reuse everything from cell site locations to MSCs, SGSNs, and *home location registers* (HLRs). Because of the fact that the core network of UMTS is essentially the same core network as is used for GSM/GPRS, there is a significant opportunity to reuse existing equipment. For example, a GSM MSC can be upgraded to simultaneously support both GSM and UMTS. Similarly, SGSNs and *Gateway GPRS Support Node* (GGSNs) can be upgraded to simultaneously support both UMTS and *General Packet Radio Service* (GPRS).

In the radio access network, there is also some opportunity for reuse. For most vendors, it will not be possible to upgrade a BSC to simultaneously function as a BSC and an RNC. For base stations, however, several vendors support both GSM and UMTS within the same base station cabinet. In such a situation, it is possible for the GSM and UMTS transceivers to use the same antennas. Even if a given vendor does not support both UMTS and GSM transceivers within the same cabinet, or if the UMTS and GSM systems are provided by different vendors, there may still be the opportunity to co-locate a UMTS base station cabinet with a GSM base station cabinet. This can reduce site acquisition costs and some construction costs.

For GSM systems operating at 1800 MHz or 1900 MHz, the footprint of a UMTS cell and a GSM cell are very similar. In fact, for a cell loading factor of up to approximately 65 percent, the footprint of a UMTS cell for voice service is slightly greater than the equivalent footprint of a GSM cell. For GSM900, however, the difference in frequencies is such that the GSM signal propagates a great deal further, which means that the coverage of a UMTS cell will be less than that of the GSM cell. Thus, when deploying UMTS over an existing GSM900 network, extra cell sites will be required for UMTS. In urban areas, the number of extra UMTS cell sites is likely to be quite limited as the GSM sites will have been deployed in a more dense arrangement for capacity reasons rather than just for coverage reasons. In rural/highway areas, however, there will need to be many more UMTS sites than GSM900 sites, simply because of less attenuation for the lower frequency GSM900 signal.

In the case where a GSM base station and UMTS base station are co-located, or even share the same cabinet, then they can also share the transmission facilities back towards the BSC and RNC. Figure 12-11 shows an

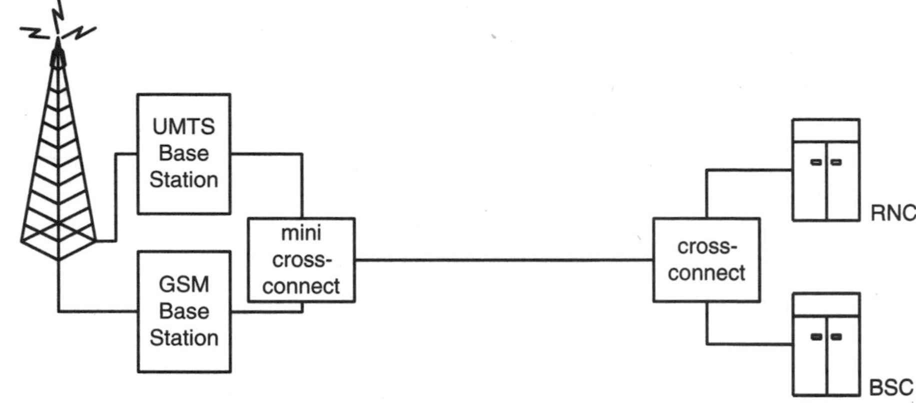

Figure 12-11 Sharing Iub and Abis transport.

example of how this can be done. In that example, a UMTS cabinet is co-located with an existing GSM cabinet. The GSM cabinet already has a T1 connection back to the BSC. Given that a GSM base station requires between two and three DS0s per transceiver, it is quite possible that the T1 is not fully used. In fact, less than half of the T1 might be used, as would be the case for, say, a three transceiver GSM BTS. Provided that the expected Iub bandwidth requirement will consume less than the remaining bandwidth, then we can use that fractional T1 capacity for the Iub interface. In other words, we carry ATM on a fractional T1. Back at the BSC/RNC location, we need to have a cross connect that can perform DS0-level grooming. That cross-connect strips out that part of the T1 that is used by the GSM BTS and sends it to the BSC. The part of the T1 that carries the Iub interface is sent to the RNC. Of course, we also need a mini cross-connect at the base station site. For many GSM base stations, such one-card devices are available as it is not uncommon in many countries to daisy-chain GSM base stations in order to reduce transmission cost.

Of course, we are likely to find that this type of transport sharing will be possible only for those sites that expect relatively low demand. If we are co-siting a UMTS base station with a GSM base station in an urban area, for example, we may find that the GSM base station is already consuming more than half of a T1 (as would be the case for a six-transceiver base station). We are likely to find that the UMTS base station will also require more than half of a T1, particularly when we consider the overhead that the

Iub interface needs to include. In that case, we have little choice but to increase the transport bandwidth to the site.

References

3GPP TS 25.101 UE Radio Transmission and Reception (FDD)

3GPP TS 25.104 UTRA (BS) FDD; Radio Transmission and Reception

3GPP TR 25.942 RF System Scenarios

3GPP TS 23.002 Network Architecture (Release 1999)

3GPP TS 23.101 General UMTS Architecture (Release 1999)

3GPP TS 25.401 UTRAN overall description

CHAPTER 13

CDMA2000 System Design

The system design associated with *code division multiple access* (CDMA2000) systems has multiple factors that are interwoven with each other, making the design aspects a technical challenge. The CDMA2000 designer must not only account for the introduction of packet data services into the radio and fixed network access system but also the legacy systems, variants to 1xRTT, and eventual introduction of 3xRTT. This chapter will attempt to quantify some of the more salient aspects with CDAM2000 system design looking at several key scenarios that or issues that need to be addressed when considering or expanding CDMA2000 compatible infrastructure within a wireless system.

This chapter will cover

- Design criteria
- Traffic assumptions
- Link budgets
- Deployment issues
- Network node dimensioning

for the following three general types of systems:

- CDMA2000-1x (green field)
- IS-95 to CDMA2000-1x
- CDMA2000-1x to 3x

It will be assumed throughout the entire chapter that migration of legacy systems has been done successfully because it is not the intention of this chapter to cover the design aspects of the various legacy wireless systems. The reference section associated with this chapter, however, has several excellent sources for obtaining legacy system design guidance and examples.

With the previous said, the key factor that needs to be addressed, but often one of the most difficult, is what do you want to do? It is a simple question, but one that has profound implications with (all too often) no real answer in response to the question. The questions posed are more a marketing and business decision that are intertwined with the technical platforms that exist for a wireless system. Therefore through the examples listed next, it is hoped that the technical issues associated with the decisions made for service and network deployments can be better weighed, enabling for a better implementation of this exciting technology platform.

13.1 Design Methodology

The methodology for the network and *radio frequency* (RF) design for CDMA2000 needs to be established at the beginning of the process. The establishment of the methodology utilized for the formulation of the report is essential in the beginning stages to ensure the proper baseline assumptions are used to facilitate flexibility in the design and implementation. Flexibility is needed in the design and implementation to address many of the future issues that are really unknown and therefore cannot be properly foreseen at the onset of the design process itself.

Some of the issues that need to be identified at the beginning of study are

- Time frames for the growth plan
- Subscriber growth projections
- Services offered with take rates
- Design criteria
- Baseline system numbers for building on the growth study
- Construction expectations
- Legacy and future technology systems

The time frame for the growth plan is essential to determine at the beginning. The time frames will define what the baseline, foundation, and how much of a future look the plan will present. Therefore the baseline month or time frame associated with an existing system that the data used for generating the plan is critical because the wrong baseline dates will alter the outcome of the report.

The amount of time the plan projection is to take into account is also critical for the analysis. The decision to project one year, two years, five years, or even ten years has a dramatic effect on the final outcome and accuracy of the forecast. In addition to the projection time frame, it is important to establish the granularity of the reporting period—monthly, quarterly, biannually, yearly—or some perturbation of them all.

The particular marketing plans also need to be factored into the report itself. The marketing department's plans are the leading element in any network and RF growth study. The basic input parameters to the network and RF growth plan provided by the marketing department is listed in the following:

- Projected subscriber growth for the system over the time frame
- Projected mErlangs per subscriber expected at discrete time intervals

- Projected Mbps per subscriber expected at discrete time intervals
- Dilution rates for the subscriber voice usage over the time frame
- Dilution rates for legacy subscriber equipment over the time period
- Types of subscriber equipment used in the network and percent distribution of CPE projections
- Special promotion plans over the time frame of the report, such as free Internet access
- The projected amount of mobile data users over the time frame of the study

There are a multitude of other items needed from the marketing department for determining network and RF growth. However, if you obtain the information on the basic eight topics listed previously from the marketing department, it will be enough to adequately start the RF, fixed network, voice, and packet design.

The design, whether it is for new or expansion systems, needs to factor in the following key elements along with the expected traffic loading forecast:

- Spectrum available for use
- Spectrum required and methods for achieving the required bandwidth
- Flexibility to meet the ever changing market conditions
- Cost effective use of the existing and future capital infrastructure
- Standardized systems that enable backward as well as forward compatibility with other networks and data platforms
- QOS/GOS for each type of service offering
- Coverage requirements either new or enhancements to existing coverage

With the previous said, the next step is to establish some guidelines that are specific to CDMA2000 systems.

13.2 Deployment Guidelines

The deployment of CDMA2000 can and does have different faces presented to the designer depending on the situation they are trying to solve. If the design is for a new system, green field the deployment is driven by coverage and then by capacity. On the other hand if the design is for integrating

CDMA2000 into an existing system, like IS-95, then the design is more focused on capacity and possible inclusion of packet data services.

Because CDMA2000-1X occupies the same bandwidth as IS-95A/B, this obviously facilitates the introduction of this platform into that type of system. CDMA2000 can be deployed as a distinct carrier or shared carrier with IS-95 systems leading to many possibilities that a design engineer can possibly utilize to achieve the desired design requirement. Some of the design options for integrating CDMA2000-1x into an existing IS-95 system are shown in the Table 13-1.

For cellular, the F1 is the primary channel for the hunt, although for *personal communications services* (PCS), it is the first channel in the channel selection sequence provided by the operator. The CDMA2000 channel assignment scheme shown earlier is revisited in the Table 13-2 for cellular systems and Table 13-3 for US PCS systems. The asterisk in both Tables 13-2 and 13-3 represents the preferred 3X F1 carrier recommendation at this time of the design cycle. It is important to factor in the possible inclusion or exclusion of 3X in the initial system design and of course the relative location of the particular 1X carriers envisioned for inclusion in a 3X system.

Table 13-1

CDMA2000 1X Deployment Schemes

Option	Method	Advantages	Disadvantages
1	Deploy 1xRTT across all F1 channels.	Maximizes service footprint for packet data services. Seamless 1xRTT service.	Method is capital-intensive and depends on the penetration of CDMA2000 handsets limits full use of services.
2	Deploy 1xRTT on any cdma channel other then F1.	Focused on high capacity locations.	Limited service area for high-speed packet data. Hard handoff with IS-95.
3	Deploy 1xRTT in own spectrum.	Provides additional capacity without impacting IS-95.	Spectrum clearing requirements. Possible reallocation of traffic.
4	Deploy 1xRTT on all CDMA channels.	Full potential for packet and voice services realized.	Extreme capital intensive, except when only 1 CDMA channel is operational.

Table 13-2

Cellular CDMA2000-1X Carrier Assignment Scheme

Cellular System Carrier	Sequence	A	B
1	F1	283	384
2	F2	242*	425*
3	F3	201	466
4	F4	160	507
5	F5	119	548
6	F6	78	589
7	F7	37	630
8	F8 (Not advised)	691	777

Table 13-3

PCS CDMA2000-1X Carrier Assignment Scheme

PCS System Carrier	A	B	C	D	E	F
1	25	425	925	325	725	825
2	50	450	950	350*	750*	850*
3	75*	475*	975*	375	775	875
4	100	500	1000	NA	NA	NA
5	125	525	1025	NA	NA	NA
6	150*	550*	1050*	NA	NA	NA
7	175	575	1075	NA	NA	NA
8	200	600	1100	NA	NA	NA
9	225*	625*	1125*	NA	NA	NA
10	250	650	1150	NA	NA	NA
11	275	675	1175	NA	NA	NA

When introducing CMDA2000 into an existing IS-95 system, there is a need to upgrade different network elements depending on the radio infrastructure supplier that is utilized by the operator. The upgrade from IS-95 to CDMA2000 requires new network elements, namely the *packet data service node* (PDSN) and *Authentication Authorization Accounting* (AAA). However, regardless of which radio vendor chosen or used, all the existing CDMA *Base Transceiver Stations* (BTSs) need some type of modification or upgrade. The specifics for upgrading a base station from any of the 1X platforms to a 3X platform is envisioned to be primarily resident to the radio itself.

With CDMA2000-1X, due to enhancements in modulation schemes, as well as vocoders, it is anticipated to have a net voice capacity gain of 1.5 in the reverse link and 2 times in the forward link than that of 8Kb EVRC. Not only is the 1X platform meant for improvements in overall voice system capacity but as mentioned many times, the introduction of packet data is the driving force for CDMA2000 to be deployed in an existing network.

Packet data usage utilizing 1X has many estimations that are attributed to it. However the average packet data user is expected to use the service for the following services:

- E-mail—65 percent
- Web browsing—30 percent
- Extension of company network (LAN)—27 percent
- Address book/calendar functions—27 percent

The expected migration path either for a new CDMA200 system, an upgrade from IS-95, or one system that chooses to bifurcate their network is to first deploy CDMA2000-1X and then overlay a 3X system on top of it. However there are several versions of CDMA2000-1X:

- CDMA2000-1X
- CDMA2000-1XEV-DO
- CDMA2000-1XEV-DV

The 1XEV versions of CDMA2000-1X are currently under development at this time. CDMA2000-1xEV-DO is a data-only service and is envisioned to begin deployment in 2002, whereas 1XEV-DV is a data and voice offering that enables for higher throughput for data services while sharing resources for voice services and is envisioned for commercial deployment after 1xEV-DO is commercially available. As of this writing, 1XEV-DO has recently been an approval standard.

CDMA3X can be overlaid on top of a CDMA2000 1X system; it was specifically designed in the specification to enable a 3X system to be also overlaid on existing IS-95 systems. To achieve an overlay system, the 3X forward link breaks up the data into three carriers, each of which is spread 1.2288 Mcps, hence the term MC (*multi-carrier*). The reverse link in 3X uses three aggregated 1x carriers which have a combined carrier spread of 3.6864 Mcps.

13.2.1 1x

This in the initial deployment for CDMA2000 involving a single carrier and is typically referred to as CDMA2000 phase 1. The 1x system introduces the use of packet data services for wireless operators. The 1x system utilizes a SR1 and will transport both voice and packet data over the same physical resources.

The data rates envisioned for 1x are listed in Table 13-4.

13.2.2 1xEV-DO

1xEV-DO is the terminology used to describe non-real time, high packet data services that will be offered on a SR1 channel transporting packet data only, hence the name DO. The objective behind deploying a 1xEV-DO service is to enable a higher number of users of the system to utilize packet data services. By separating voice users from data users onto two carriers, it will result in higher data rates for users as well as a higher throughput per carrier. The 1xEV-DO is designed to be directly scalable to a 3X platform.

The data rates envisioned for 1xEV-DO are listed in Table 13-5.

13.2.3 1xEV-DV

1xEV-DV is the last evolution expected for a CDMA2000-1X platform. The 1xEV-DV system will enable both voice and packet data services to share

Table 13-4
1X

	Downlink	Uplink
Building	2 Mbps	144 Kbps
Pedestrian	2 Mbps	144 Kbps
Vehicular	384 Kbps	144 Kbps

Table 13-5
EV-DO

	Downlink	Uplink
Building	2.4 Mbps	144 Kbps
Pedestrian	2.4 Mbps	144 Kbps
Vehicular	600 Kbps	144 Kbps
Vehicular (peak)	1.20 Mbps	144 Kbps

the same resource FA; similar to 1xEV systems, but they have the packet data throughput associated with 1xEV-DO systems.

Effectively, they are the introduction of 1xEV-DV is envisioned to meet or exceed the capacity envisioned for 3X platforms and may have itself precluded the purpose of implementing a 3X system. Data rates are envisioned for 1xEV-DV to reach a peak rate of 5 Mbps with an average throughput of 1.2 Mbps. The current effort to develop 1xEV-DV has begun to question the need for a 3X platform. However, that decision point has not arrived.

13.3 System Traffic Estimation

The traffic estimation for the CDMA2000 system is directly dependant upon the type and quality of services that will be offered and how they will be transported. The traffic estimation process involves not only the radio link, but also the other fixed facilities that comprise the network.

The process and methodology for conducting system traffic engineering, that is, determining the amount of physical and logical resources that need to be in place at different points and nodes within the network to support the current and future traffic. The determination of existing traffic loads is rather more straight forward in that you have existing information from which to make decisions upon. For future forecasts, the level of uncertainty grows exponentially the farther the forecast or planning takes you into the future. However many elements in the network require long lead times, ranging from three weeks to over one year to implement. Obviously, the goal of traffic engineering is to design the network and its sub-components to not only meet the design criteria, which should be driven by both technical, marketing, and sales, but also be done so that it is achieved in a cost-effective manner. It is not uncommon to have conflicting objectives within a design, that is, to ensure that the customers have the highest QOS/GOS for both voice and packet data, but yet have a limited amount of capital from which to achieve this goal. Therefore it is important to define at the onset of the design process, and have some interim decision points where the design process can be reviewed and altered, if required, either by increasing the capital budget, revisiting the forecast input, or altering the QOS/GOS expectations.

Because there will be different variants to circuit- and packets-switched services offered, the variations will be vast. However, there are some commonalties that can be drawn upon.

There are several methods that can be used for calculating the required or estimated traffic for the network.

It is essential to note that there are several key points within the network where the traffic engineering calculations need to be applied:

- BTS-to-subscriber terminal
- BTS to BSC
- BSC-to-packet network
- BSC-to-voice network

There are several situations and an unknown level of perturbations that can occur in the estimation of traffic for a system. In an ideal world, the traffic forecast would be projected by integrating the marketing plan with the business plan, and coupled with the products that should be integral to both the marketing and business plan. However, reality is much harsher, and usually very little information is obtainable by the technical team from which to dimension a network with. Therefore the following is meant to help steer the new system planners in determining their traffic-transport forecast.

Initially, packet data traffic is expected to be low. The higher data speed is a result of the data not being as time-sensitive as voice. Also, packet data services are an enabler for more services offered by the operator.

The forecast would be much more simplified if the system were operational because there would be real traffic information as well as a minimal set of products from which to utilize. The forecast, or growth, could be extrapolated from the business plan or simplified marketing plans, which would specify a specific growth-level desired.

The equation to follow for an existing system would be

$$\text{Total traffic} = \text{existing traffic} + \text{new traffic expected}$$

The new traffic expected could be a simple multiplication of the existing traffic load. For instance, if the traffic is 25 Mbps and the plan is to increase the traffic by 25 percent over the next year, then the traffic forecast for the one-year forecast would be altered by increasing the current traffic load at each node by that amount and determining the requisite amount of logical and physical elements needed in addition to any load sharing that might be achievable.

If the traffic forecast is available only on a country-wide or market level for a new system, then the traffic needs to be distributed in a weighted proportion to each of the markets being designed for the system or homogeneously distributed for a given market.

The forecasting for voice traffic is well documented and will only get a superficial treatment here. However, the real issue with traffic dimensioning lies in the ability to forecast both the circuit switched as well as the new packet services that will be used by the customers of the wireless operator.

The ultimate question that the designer must answer is, "how do you plan on supporting the traffic with their prescribed services?"

Because there are numerous types of services available for both circuit switched as well as packet, some generalizations need to be made in order to have a chance at arriving at some conclusions necessary for input into the design phase. Therefore the symbols defined in Table 13-6 will be used to help define the different classifications of transport services required.

For a new or existing system, the issue of where to begin is always the hardest part. However, one of the key parameters you need to obtain from marketing and/or sales is the penetration rate, take rate for each of the

Table 13-6

Circuit and Packet Data

Symbol	Service Type	Transport Method
S	Voice	Circuit Switch
SM	Short Message	Packet
SD	Switched Data	Circuit Switch
MMM	Medium Multimedia	Packet
HMM	High Multimedia	Packet
HIMM	High Interactive Multimedia	Packet

services types offered. This can be achieved via several methods, such as a general approach where a standard percentage, percent, is used for say packet services. Or you could base the amount of packet data subscribers from the number of handsets expected to be purchased for resale in the market.

Regardless, the first step in any traffic study is to determine the population density for a given market; in the case of an existing system, the population density and primary penetration rates are already built into the system due to known loading issues. However, especially for new services, like packet data, the process of determining the population density for a given area followed by the multiplication of this by the penetration rate will greatly help in the determination of the expected traffic load from which to design the system.

- **Population density** This is a measure of the quantity of people that could possibly utilize the service for a given geographic area. When determining population density, it is important to note that for the same geographic area, there could be different population densities. For example, an area could have 100,000 pedestrians per km^2 but only a vehicle density of $3,000/km^2$.
- **Penetration rate** This is a measure or an estimate of the amount, or rather, percentage of the amount of people in the population density that will utilize a particular service. In the instance of the 100,000 possible users, only 5 percent may want a particular service. Therefore, the possible usage may only be experienced by 5,000 people for that service offering. An important issue is that each service offering will likely have a different penetration and that it is very possible that based on the amount of services offered the total penetration rate could exceed 100 percent because of the various service offerings.
- **Cell site area** The geographic area that a cell site or its sector will cover is determined either via computer simulation or, for a rough estimate, by a two dimensional approach. The equations, or rather formulas for determining the area for a cell, is shown in Table 13-7. The radius for the cell site is determined from the link budget and is dependent upon numerous issues. However the use of a standard cell radius for a given morphology is recommended to be used for the initial design phase. The cell coverage area for a later phase in the design process can be determined through use of computer simulations that should factor in the cell breathing issues that are evident in CDMA systems.

Table 13-7
Cell Area Equations

Cell Type	Cell Area	Sector Area (3-sector cell)
Circular	πR^2	$\pi R^2 /3$
Hexagonal	2.598 R^2	2.598 $R^2 /3$

- **QoS** *Quality of service* is a term used by many and has also the same amount of meanings. For this discussion, QoS is a description of the bear channels capability for delivering a particular *grade of service*, GoS. The GoS is typically defined as a blocking criteria, and for circuit switched data, it is defined by Erlang B, Erlang C, or Poisson equations. For packet data, the relationship for QoS/GoS is blocking, Erlang C, and delay, to mention three of the key attributes.

With the introduction of packet data with CDMA2000, the traffic modeling for packet-switched data involves the interaction of the following items:

- Number of packet bursts per packet session
- Size of packets
- Arrival time of packet burst within a packet sessions
- Arrival times for different packet sessions

The packet usage is relatively an unknown area for wireless mobility systems on a mass-market basis. The issue of where, when, and how much do you dimension a system for packet data will always be a debate between marketing and technical teams. However, in light of the fact that packet data usage is at its infancy, there is little guidance from which to go forth and design the network from. However, ITU-R M.1390, which gives a methodology for calculating the spectrum requirements for IMT-2000, has some guidelines for data dimensioning and the following tables are extracted from that specification. The values in the tables should be used as a guide to establish packet loading for dimensioning when market specific data is not available for a numerous amount of reasons.

It is important to note that all the services defined previously in Table 13-6 and elaborated on in Table 13-8, are either symmetrical or asymmetrical. Of the services listed previously, only MMM and HMM are asymmetrical; the rest are symmetrical service offerings.

Note that the penetration rates shown in Table 13-9 for the services are the same and come to more than 100 percent for any location. It is also important to note that the previous numbers indicate that 73 percent of the system usage is expected to be voice-oriented; 10.8 percent is for SM, 3.51 percent for both SD, MMM, and HMM, whereas 6.75 percent for HIMM. However, the following tables will help provide additional insight into possible traffic dimensioning requirements. (See Tables 13-8, 13-9, 13-10, 13-11, and 13-12.)

The next step is to determine the traffic forecast of user by service type. The method for achieving this value is determined by the equation for each of the service types and locations defined, building/pedestrian/vehicular.

Traffic/user = BHCA $\times$ call duration $\times$ activity factor = 0.9 $\times$ 120 $\times$ 0.5 = 54 call sec during the system busy hour for downlink or uplink voice service for a building environment.

The amount of circuits required for circuit switched voice, switched data, and HIMM services is determined via Erlang B, although the remaining packet data services are determined via Erlang C.

Now the next question is to define the next set of variables that need to be established to help dimension the rest of the packet network. For symmetrical services, the dimensioning is straight forward, well as straight as it can get. However, for asymmetrical service, a few more details are required that are used for the selection and performance of the PDSN:

Transmission time (s) = NPCPS $\times$ NPPPC $\times$ NBPP $\times$ 8 bits per byte/1024 Kpbs

Total session time = packet transmission + [(PCIT $\times$ (NPCPS-1)] + [PIT $\times$ (NPPPC-1)]

Activity factor = packet transmission time/total session time

The data used for uplink and downlink calculations is extracted from Table 13-13. The results are then entered into the traffic calculation section discussed later.

13.4 Radio Elements

Because CDMA2000 is a radio access platform, it leads to reason that the driving force for dimensioning the network to meet the customer demands is to ensure that the radio system is dimensioned accordingly. The radio elements

Table 13-8
Net User Bit Rate

Service Type	Net User Bit Rate Downlink (Kbps)	Uplink (Kbps)
S	16	16
SM	14	14
SD	64	64
MMM	384	64
HMM	2000	128
HIMM	128	128

Table 13-9
Penetration Rates

Penetration Rates (%)			
Service Type	Building	Pedestrian	Vehicular
S	73	73	73
SM	40	40	40
SD	13	13	13
MMM	15	15	15
HMM	15	15	15
HIMM	25	25	25

Table 13-10
BHCA

Busy Hour Call Attempts (BHCA)			
Service Type	Building	Pedestrian	Vehicular
S	0.9	0.8	0.4
SM	0.06	0.03	0.02
SD	0.2	0.2	0.02
MMM	0.5	0.4	0.008
HMM	0.15	0.06	0.008
HIMM	0.1	0.05	0.008

Table 13-11

Call and Session Duration

Call Duration (sec)			
Service Type	**Building**	**Pedestrian**	**Vehicular**
S	180	120	120
SM	3	3	3
SD	156	156	156
MMM	3000	3000	3000
HMM	3000	3000	3000
HIMM	120	120	120

Table 13-12

Activity Factor

Activity Factor		
Service Type	**Downlink**	**Uplink**
S	0.5	0.5
SM	1	1
SD	1	1
MMM	0.015	0.00285
HMM	0.015	0.00285
HIMM	1	1

Table 13-13

Packet Data

MMM/HMM			
Type	**Description**	**Downlink**	**Uplink**
NPCPS	# Packet calls per session	5	5
NPPPC	# Packets per packet call	25	25
NBPP	# Bytes per packet	480	90
PCIT	Packet call inter arrival time	120	120
PIT	Packet inter arrive time	0.01	0.01

that include the base radius, BTS, channel elements, and BSCs, directly influence the circuit switched and packet data network requirements.

There are a few key elements that are associated with the radio dimensioning for a CDMA2000 system whether it is any of the 1X variants or even a 3X system. The key elements for the dimensioning are

- Spectrum
- Channel assignment scheme
- Site configuration (antennas)
- Channel elements
- Link budget

The spectrum requirements for a CDMA2000-1X or 3X network are, of course, directly dependant upon the amount of channels required to be deployed to meet the current or expected demand. In addition, the channel assignment scheme that is utilized will have a direct impact on the possible inclusion of 3X in the future as well as optimal spectrum management of the existing system when using CDMA2000-1x only.

13.4.1 Antenna Configurations

The site configurations for the CDMA2000 sites can and do take advantage of many of the IS-95 lessons learned through the deployment phases. Taking a simplistic view of CDMA2000 antenna requirements, a total of two receive antennas (or paths) are needed per sector, as was the case with IS-95 systems. The diagram shown in Figure 13-1 illustrates the requirement for a single CDMA2000-1X Tx channel and that of a single 3X channel.

Figure 13-1 addresses two issues with 1XRTT deployments: to utilize or not to utilize transmit diversity. Figure 13-1 illustrates for a CDMA2000-1X carrier, a single TX antenna is needed; however, Figure 13-1 shows that two antennas are needed for STD TX diversity. The transmit diversity scheme has technical advantages that can be exploited by the system operator, however, at the cost of deploying or using a second antenna for TX diversity. Now this is normally not an issue with a CDMA carrier because there are usually two antennas or a cross pole used, and the duplexers provide the dual path. Where the rub comes is when a second carrier is deployed, and unless the link budget shows the splitting loss that can be accommodated, more antennas will need to be added to the system or sector.

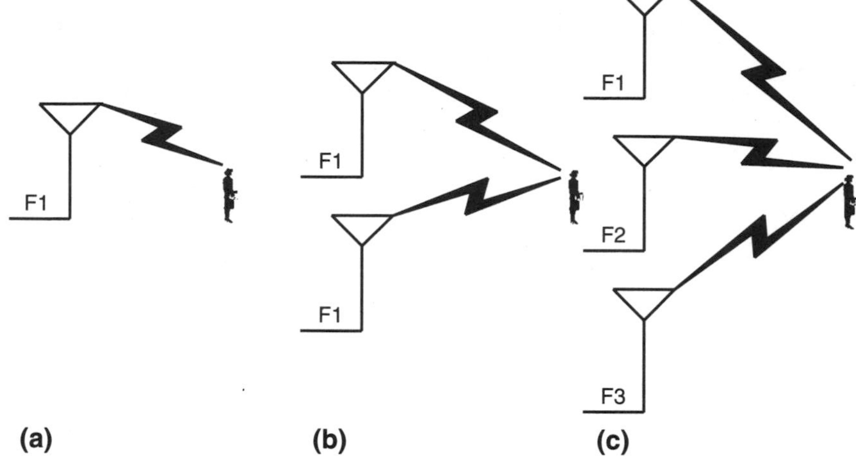

Figure 13-1 CDMA2000 TX configurations: (a) 1×RTT single antenna, (b) 1×RTT TX diversity (STD), and (c) 3×RTT.

TX diversity can also be deployed with a single antenna used for transmit following the *Orthoginal Transmit Diversity* (OTD) method.

13.4.2 BTS

The BTS controls the interface between the CDMA2000 network and the subscriber unit. The BTS controls many aspects of the system that are directly related to the performance of the network. Some of the items the BTS controls are the multiple carriers that operate from the site, the forward power (allocated for traffic, overhead, and soft handoffs), and of course the assignment of the Walsh codes.

With CDMA2000 systems, the use of multiple carriers per sector as with IS-95 systems is possible. Therefore when a new voice or packet session is initiated, the BTS must decide how to best assign the subscriber unit to meet the services being delivered. The BTS in the decision process not only examines the service requested, but it also must consider the radio configuration and the subscriber type, and, of course, whether the service requested is voice or packet. Therefore the resources the BTS has to draw upon can be both physically- and logically-limited depending on the particular situation involved.

The following is a brief summary of some of the physical and logical resources the BTS must allocate when assigning resources to a subscriber:

- **FCH** Number of physical resources available
- **FCH forward power** Power already allocated and that which is available
- **Walsh codes required and those available**
- **Total FCHs used in that sector**

The physical resources the BTS draws upon also involves the management of the channel elements that are required for both voice- and packet-data services. Although discussed in more detail, handoffs are accepted or rejected on the basis of available power only.

Integral to the resource assignment scheme is the Walsh code management, covered in another section in more detail. However, for 1XRTT, whether 1x, 1xDO, or 1xDV, there are a total of 128 Walsh codes to draw upon. However, with the introduction of 3X, the Walsh codes are expanded to a total of 256.

For CDMA20001X, the voice and data distribution is handled by parameters that are set by the operator that involve

- **Data resources** Percent of available resources that includes FCH and SCH
- **FCH resources** Percent of data resources
- **Voice resources** Percent of total available resources

These are best described by a brief example to help facilitate the issue of resource allocation shown in Table 13-14.

Obviously the allocation of data/FCH resources directly controls the amount of simultaneous data users on a particular sector or cell site.

Table 13-14

Carrier Resource Allocation Example

Topic	Percentage	Resources
Total Resources		64
Voice Resources	70%	44
Data Resources	30%	20
FCH Resources	40%	8

13.4.3 Channel Element (CE) Dimensioning

The *Channel Element* (CE) dimensioning will obviously be based on the requirements for both voice- and data-services needs. The total number of channel elements required will be the summation of both the fundamental and supplemental channel elements defined for voice and data services.

The new channel element that is being offered by all the major vendors is compatible with the existing IS-95 system and can be directly substituted for an existing channel element. However, as discussed later, the full replacement of all CEs is not practical; based on the deployment options, used IS-95 legacy systems should be left in place.

For simplification, a *channel element* (CE) is required for each

- Voice cell
- Leg of the soft hand off
- overhead channel
- Data call

The dimension of the channel elements is done in increments of 32 or 64 for CDMA2000-1X-capable CEs. Because CEs typically come in 32/64 cards, it leads to the issue that if 20 CEs are required, a 32-CE card is acquired. Although in the same sense if 33 CEs are needed, a choice needs to be made to either under equip or obtain another CE card to bring the count to 64 when only 33 are needed:

- 32 CE for 13 to 17 percent of cdma-1X full capacity
- 64 CE for 29 to 44 percent of cdma-1X full capacity

The CDMA2000-1X full capacity is derived based on a fixed environment and availability of 128 Walsh codes. Obviously the percentages shown previously depend, of course, on the mix of voice and data traffic within the system as well as mutual interference.

A rule of thumb to follow is that for sites requiring less than 40 CEs, a 32-element card should be used, but for sites requiring more than 40 CEs, then the 64-CE card is expected to be used.

13.4.4 Packet Data Services (RF Environment)

Packet data services have different implications when introduced into a radio environment as compared to a fixed-network environment. More specifically for the radio link, how packet data is handled is dependent on whether the radio link is sharing its resources with voice services or is a data only use. The data-only possibility is available with CDMA2000 1xEV-DO if data services are only permitted on the new channel. However, regardless of this issue, when involved with the wireless link, data services are still a best effort. In addition, signaling traffic has higher priority than voice but voice services and circuit switched, have a higher priority than packet data.

Regarding packet data resource dimension, it is important to remember that the packet session is considered active when data is being transferred. During this process, a dedicated FCH and/or DCCH for traffic signaling and power control exists between the mobile and the network. In addition, the high-speed supplemental channel can be utilized for large data transfers. An important issue that needs to be considered in the allocation of system resources is that while the session is active, channel elements as well as Walsh codes are consumed for use by the subscriber and system independent if data is actually being transferred.

The packet session is alternatively considered dormant when there is no data being transferred, but a PPP link is maintained between the *Packet Data Server Network* (PDSN) and the subscriber. It is important to also note that no system resources are consumed relative to channel elements or Walsh codes while the packet session is considered dormant.

13.5 Fixed Network Design Requirements

The introduction of packet data services of course not only requires the focus on the radio environment for a CDMA2000 system, but also on all the supporting elements that comprise the wireless system. Therefore the following are the major elements that need to be factored into the design of a CDMA2000 system:

- *Mobile Switching Center* (MSC)
- *Baser Station Controller* (BSC)

- *Base Transceiver Station* (BTS)
- *Packet Data Service Node* (PDSN)

In reviewing the previous list, several nodes or elements directly associated with radio elements are listed. The reason the BSC and BTS are listed in the fixed network design requirements lies in the simple concept that connectivity needs to be established between the BTS and the BSC, whether it is via landline services or via a microwave link. The BSCs are listed not only because it routes packet and voice traffic, which requires a certain link dimensioning, but also the BSCs can be local or remote to the MSC depending on the ultimate network configuration deployed.

The fixed network design includes not only element dimensioning but also dimensioning the links that connect the various nodes or elements in order to establish a wireless system. Some of the connectivity requirements involve the following elements listed next. It is important to note that all elements require some level of connectivity whether it is from the *Digital Cross Connect* (DXX), also referred to as DACS, to the MSC, or between the voice mail platform and the switch. However, the list that follows involves elements that usually require an external group to interface with while the internal nodes are more controllable.

- Link between BTS and BSC (usually a leased line)
- Link between BSC and MSC (if remote)
- BTS/BSC concentration method
- Connectivity of the PDSN
  - Router (for internal)
  - Router (for external)
  - AAA
  - HA
  - *Sentence Creation Server* (SCS) and other servers
- Interconnection to the public and/or private data networks

Note as a general practice the routers used for the packet data network for 1XRTT applications should not be utilized for other functions, such as company LAN work.

13.5.1 PDSN

The PDSN needs to have connectivity with the following major nodes:

- CDMA radio network
- Either a public data network, private data network, or both
- AAA server
- DHCP server
- Service creation platform that contains the configuration, policy, profile, sub-provisioning, and monitoring capability

The PDSN is usually connected to the packet data network via an OC3/STM1 or 100baseT connection. The choice of which bandwidth to utilize is determined by not only the proximity of the PDSN to the BSC, but also on traffic requirements. In summary, the PDSN design is based on many factors including the following basic issues:

- Number of BSC locations
- Access type supported (simple IP, mobile IP, and so on)
- Connectivity between nodes
- Network performance requirements

13.5.2 Packet Zone

For the network layout there are several zones that can be assigned within a PDSN network. These zones are referred to as packet zones and should be distributed along the same deployment and logical assignment method used for assigning BSCs. In other words, every BTS connected to a particular BSC should have the same packet zone assigned to it. However, it is possible to have several BSCs residing in the same packet zone, but it is recommended that a separate packet zone be assigned to every BSC.

13.5.3 Design Utilization Rates

The facility utilization goal for the network should be 70 percent of capacity for the line rate over the time period desired. The time period that should be used is a nine month sliding window that needs to be briefly revisited on a monthly performance report, and then during a quarterly design review, following the design review guideline process.

The facility will need to be expanded once it is understood that the 70-percent utilization level will be exhausted within 9 months with continued growth showing exhausting, 100 percent, within 18 months.

For the packet-switched network, the following should be used as the general guidelines:

- **Processor occupancy** 70-percent
- **Switching platform** SCR (25-percent), PCR (90-percent)
- **Port capacity** Design for 70-percent port utilization (growth projection for additional ports based on nine-month forecast)

13.5.4 IP Addressing

The issue of IP addressing is important to a CDMA2000 system design. The introduction of simple IP and mobile IP with and without *Virtual Private Network* (VPN) requires the use of multiple IP addresses for successful transport of the packet services envisioned to be offered. It is therefore imperative that the IP addresses used for the network be approached from the initial design phase to ensure a uniform growth that is logical and easy to maintain over the life cycle of the system.

Not only does the introduction of packet data require an IP address scheme for the mobility portion of the system, each of the new platforms introduced needs to have its own IP address or range of IP addresses. Some of the platforms requiring IP addresses involve

- PDSN
- FA
- HA
- Routers

Some of these new devices require the use of private addresses as well as some public addresses. However, because the range of perturbations for IP address schemes is so vast and requires a specific look at how the existing network is set up and factoring into the mix the desires for the future, a generic discussion on IP address schemes will follow.

The use of IPv4 format is shown in the following. IPv6 or IPng is the next generation, it enables for QoS functionality to be incorporated into the IP offering. However, the discussion will focus on IPv4 because it is the protocol today and has legacy transparency for IPv6.

Every device that wants to communicate using IP needs to have an IP address associated with it. The addresses used for IP communication have the following general format:

Network number	Host number
Network prefix	Host number

There are, of course, public and private IP addresses. The public IP addresses enable devices to communicate using the Internet, although private addresses are used for communication in a LAN/WAN intranet environment. The CDMA2000 system will utilize both public and private addresses. However, the bulk of the IP addresses will be private in nature and depending on the service offering, will be dynamically allocated or static in nature.

Table 13-15 represents the valid range of public and private IP addresses that can be used. The private addresses will not be recognized on the public Internet system and that is why they are used. Also it will be necessary to reuse private addresses within sections of its network, profound as this may sound. Because the packet system is segregated based on the PDSN, each PDSN can be assigned the same range of IP addresses. Additionally based on the port involved with the PDSN, the system can be segregatod into localized nodes, and the segregation enables for the reusing of private IP addresses ensuring a large supply of a seemingly limited resource.

The public addresses are broken down into A, B, and C addresses with their ranges shown in the following.

The private addresses that should be used are shown in Table 13-16.

Table 13-15

Public IP Address

Network Address Class	Range
A (/8 prefix)	1.xxx.xxx.xxx thru 126.xxx.xxx.xxx
B (/16 prefix)	128.0.xxx.xxx thru 191.255.xxx.xxx
C (/24 prefix)	192.0.0.xxx thru 223.255.255.xxx

Table 13-16

Private IP Address

Private Network Address	Range
10/8 prefix	10.0.0.0 thru 10.255.255.255
172.16/16 prefix	172.16.0.0 thru 172.31.255.255
192.168/16 prefix	192.168.0.0 thru 192.168.255.255

Table 13-17

Subnets

Mask	Effective Subnets	Effective Hosts
255.255.255.192	2	62
255.255.255.224	6	30
255.255.255.240	14	14
255.255.255.248	30	6
255.255.255.252	62	2

To facilitate the use IP addressing, the use of a subnet further helps refine the addressing by extending the effective range of the IP address itself. The various subnets are defined in Table 13-17. The IP address and its subnet directly affect the number of subnets that can exist and from those subnets, the amount of hosts that can also be assigned to that subnet.

It is important to note that the IP addresses assigned to a particular subnet include not only the host IP addresses but also the network and broadcast address. For example, the 255.255.255.252 subnet that has two hosts requires a total of four IP addresses to be allocated to the subnet: two for the hosts, one for the network, and the other for the broadcast address. Obviously, as the amount of hosts increases with a valid subnet range, the more efficient the use of IP addresses becomes. For instance, the 255.255.255.192 subnet enables for 62 hosts and utilizes a total of 64 IP addresses.

Therefore you might say, why not use the 255.255.255.255.192 subnet for everything? However, this would not be efficient either, so an IP-address plan needs to be worked out in advance because it is extremely difficult to change once the system is being or has been implemented.

Just what is the procedure for defining the IP addresses and its associated subnet? The following rules apply when developing the IP plan for the system; the same rules are used for any LAN or ISP that is designed. There are four basic questions that help define the requirements:

1. How many subnets are needed presently?
2. How many are needed in the future?
3. How many hosts are on the largest subnet presently?
4. How many hosts are on the largest subnet in the future?

You might be wondering why the use of multiple hosts should be factored into the design phase for CDMA2000. The reason is that it is possible to have several terminals for a fixed application using a single CDMA2000 subscriber unit or fixed unit.

Therefore using the previous methods, an IP plan can be formulated for the wireless company's packet-data platforms. It is important to note that the IP plan is should not only factor into the design the end customers' needs but also the wireless operators' needs.

Specifically, the CDMA2000 operators' needs will involve IP addresses for the following platforms as a minimum. The platforms requiring IP addresses are constantly growing as more and more functionality for the devices is done through SNMP.

- Base stations
- Radio elements
- MicroWave point to point
- Subscriber units
- Routers
- ATM switches
- Work stations
- Servers (AAA, HA, FA, and PDSN)

The list can and will grow when you tally up all the devices within the network both from a hardware and network management aspect. Many of the devices listed previously require multiple IP addresses in order to ensure their functionality of providing connectivity from point A to point B. It is extremely important that the plan follows a logical method. Some

CDMA2000 network equipment may also require an IP plan that incorporates the entire system and not just pieces.

A suggested methodology is to

- List out all the major components that are, will be, or could be used in the network over a 5 to 10 year period.
- Determine the maximum amount of these devices that could be added to the system over 5 to 10 years.
- Determine the maximum amount of packet data users per BSC.
- Determine the maximum amount of packetdata users per PDSN.
- Determine the maximum amount of mobile IP users with and without VPN.
- Determine the maximum amount of simple IP users with and without VPN.

The reason for the focus on the amount of simple- and mobile-IP users lies in the fact that these devices will have the greatest demand for IP addresses due to their sheer volume in the network.

Naturally, each wireless system is unique and will require a different IP address scheme to be implemented. However, the concept presented has been beneficial and should prove useful. If more information is sought on IP address schemes, an excellent source for information is available on the Web at **www.cisco.com**.

13.6 Traffic Model

The capacity for a CDMA2000 cell site is determined through the interaction of several parameters and is driven by the radio access portion of the system, provided the fixed network has the requite number of modules for each platform. The parameters for determining the traffic load at CDMA2000 site are similar to those used for an IS-95 system with the exception that CDMA2000 introduces packet data and the inclusion of 128/256 Walsh code, to mention a few of the previously covered issues.

As with IS-95 base stations, the use of channel element cards is essential for the handling of traffic whether it is for voice or data. The desired result from traffic engineering for a CDMA2000 base station is to be able to determine the amount of channel elements and cards required to support the expected traffic. Another factor that fits into the traffic calculations for the site involves system noise. There is a simple relationship between system noise and the capacity of the cell site. Typically the load of the cell site design is somewhere in the vicinity of 40 to 50 percent of the pole capacity, maximum 75 percent.

The next major element in determining the capacity for a CDMA cell is the soft and softer handoff factor. Because CDMA2000, like IS-95, relies on soft and softer handoffs as part of the fundamental design for the network, this must also be factored into the usable capacity at the site. The reason for factoring soft and softer handoffs into capacity is that if 35 percent of the calls are in a soft handoff mode, then this will require more channel elements to be installed at the neighboring cell sites to keep the capacity at the desired levels.

The pole capacity for CDMA is the theoretical maximum number of simultaneous users that can coexist on a single CDMA carrier. However at the pole, the system will become unstable and therefore, operating at less than 100 percent of the pole capacity is the desired method of operation. Typically the design is for 50 percent of the pole capacity for the site.

However, because soft handoffs are an integral part of CDMA, they need to be also included in the calculation for capacity. In addition for each traffic channel that is assigned for the site, a corresponding piece of hardware is needed at the cell site also.

The actual traffic channels for a cell site is determined using the following equation:

$$\text{Actual traffic channels} = (\text{effective traffic channels} + \text{soft handoff channels})$$

The maximum capacity for a CDMA cell site should be 75 percent of the theoretical limit. Unlike IS-95 systems that were power limited, the CDMA2000 system is anticipated to be Walsh code limited.

13.6.1 Walsh Codes

Reiterating the utilization of Walsh codes has a direct impact upon the radio networks ability to carry and transport the various services. With the introduction of CDMA2000, there are several alterations to the use of Walsh codes that were previously discussed, but only briefly.

With CDMA2000, the Walsh codes now have variable lengths that range from 4 to a total of 256, which is an expansion over IS-95 systems that only had 64 codes. The one effect with utilizing variable length Walsh codes is that if a shorter Walsh code is being used, then it precludes the use of the longer Walsh codes that are derived from it. For instance, if Walsh code 2 is used, then it precludes the use of all the Walsh codes in the code tree that were derived from it.

Table 13-18 helps in establishing the relationship between which Walsh code length is associated with a particular data rate. However how does this play into the use of determining the radio network?

For a SR1 and RC1 there are a maximum number of users that have individual Walsh codes equating to 64, a familiar number from IS-95A. However, if we were to have a R3 capable base radio with a SR1, phase 1 CDMA2000, and we had a total of 12 RC1 and RC2 mobiles under that sector, then this would only allow for three data users at 153.6K, or 6 at 76.8 Kbps, 13 at 38.4, 26 at 19.2, or 104 at 9.6 Kbps. Obviously the negotiated mobile data rate complicates the determination for the total throughput of traffic levels. The real issue behind this is that the type of data that will be enabled to be transported over the network has a direct impact on the available users. If for example, the need were for high-speed data for interactive video. With a R3 capable mobile, 384K of bandwidth, would not be feasible.

Table 13-18
Walsh Codes

		Walsh Codes Tree						
	RC	**256**	**128**	**64**	**32**	**16**	**8**	**4**
SR1	1	Na	Na	9.6	Na	Na	Na	Na
	2	Na	Na	14.4	Na	Na	Na	Na
	3	Na	Na	9.6	19.2	38.4	76.8	153.6
	4	Na	9.6	19.2	38.4	76.8	153.6	307.2
	5	Na	Na	14.4	28.8	57.6	115.2	230.4
SR3	6		9.6	19.2	38.4	76.8	153.6	307.2
	7	9.6	19.2	38.4	76.8	153.6	307.2	614.4
	8		14.4	28.8	57.6	115.2	230.4	460.8
	9	14.4	28.8	57.6	115.2	230.4	460.8	1036.8

At times it is best to see the previous example in a visual format in order to better understand the relationship between the short and long Walsh codes. Table 13-19 shows the relationship, or rather, the Walsh tree, from 4 to 256 Walsh codes and their relative relationship with one another. The relationship is illustrated in Table 13-20.

Table 13-19

Walsh Code Tree

256	128	64	32	16	8	4
0	0	0	0	0	0	0
128						
64	64					
192						
32	32	32				
160						
96	96					
224						
16	16	16	16			
144						
80	80					
208						
48	48	48				
176						
112	112					
240						
8	8	8	8	8		
136						
72	72					
200						
40	40	40				
168						
104	104					
232						
24	24	24	24			
152						
88	88					
216						
56	56	56				
184						
120	120					
248						
4	4	4	4	4	4	
132						
68	68					
196						
36	36	36				
164						
100	100					
228						
20	20	20	20			
148						
84	84					
212						
52	52	52				
180						

Table 13-19 (cont.)

Walsh Code Tree

116	116					
244						
12	12	12	12	12		
140						
76	76					
204						
44	44	44				
172						
108	108					
236						
28	28	28	28			
156						
92	92					
220						
60	60	60				
188						
124	124					
252						
1	1	1	1	1	1	1
129	1					
65	65					
193						
33	33	33				
161						
97	97					
225						
17	17	17	17			
145						
81	81					
209						
49	49	49				
177						
113	113					
241						
9	9	9	9	9		
137						
73	73					
201						
41	41	41				
169						
105	105					
233						
25	25	25	25			
153						
89	89					
217						
57	57	57				
185						
121	121					
249						
5	5	5	5	5	5	
133						
69	69					
197						
37	37	37				
165						
101	101					
229						

Table 13-19 (cont.)

Walsh Code Tree

21	21	21	21			
149						
85	85					
213						
53	53	53				
181						
117	117					
245						
13	13	13	13	13		
141						
77	77					
205						
45	45	45				
173						
109	109	45				
237						
29	29	29	29			
157						
93	93					
221						
61	61	61				
189						
125	125					
253						
2	2	2	2	2	2	2
130						
66	66					
194						
34	34	34				
162						
98	98					
226						
18	18	18	18			
146						
82	82					
210						
50	50	50				
178						
114	114					
242						
10	10	10	10	10		
138						
74	74					
202						
42	42	42				
170						
106	106					
234						
26	26	26	26			
154						
90	90			10		
218						
58	58	58				
186						
122	122					
250						
6	6	6	6	6	6	
134						

Table 13-19 (cont.)

Walsh Code Tree

70	70					
198						
38	38	38				
166						
102	102					
230						
22	22	22	22			
150						
86	86					
214						
54	54	54				
182						
118	118					
246						
14	14	14	14	14		
142						
78	78					
206						
46	46	46				
174						
110	110					
238						
30	30	30	30			
158						
94	94					
222						
62	62	62				
190						
126	126					
254						
3	3	3	3	3	3	3
131						
67	67					
195						
35	35	35				
163						
99	99					
227						
19	19	19	19			
147						
83	83					
211						
51	51	51				
179						
115	115					
243						
11	11	11	11	11		
139						
75	75					
203						
43	43	43				
171						
107	107					
235						
27	27	27	27			
155						
91	91					
219						

Table 13-19 (cont.)

Walsh Code Tree

59	59	59				
187						
123	123					
251						
7	7	7	7	7	7	
135						
71	71					
199						
39	39	39				
167						
103	103					
231						
23	23	23	23			
151						
87	87					
215						
55	55	55				
183						
119	119					
247						
15	15	15	15	15		
143						
79	79					
207						
47	47	47				
175						
111	111					
239						
31	31	31	31			
159						
95	95					
223						
63	63	63				
191						
127	127					
255						

Table 13-20 is an illustration of the interaction of a Walsh code and that of the its higher or lower branches. The Walsh codes that are consumed are depicted in the shaded area. Also the use of Walsh codes for the various channels that are associated with CDMA2000 are not included here because they too draw upon the same Walsh code pool. However, for ease of illustration, they were left out for the example.

In the example shown in Table 13-20, the use of Walsh code 48, which is CDMA2000-capable, and is set up for low-speed packet data and voice applications, precludes the use of high-speed packet data from utilizing this set of Walsh codes, thereby effectively reducing the sites data handling capability by 25 percent with the use of a single voice call. Alternatively the

Table 13-20

Walsh Code Pool Usage Example

256	128	64	32	16	8	4
0	0	0	0	NA	NA	NA
128						
64	64					
192						
32	32	32				
160						
96	96					
224						
16	16	16	NA			
144						
80	80					
208						
48	NA	NA				
176						
112	112					
240						
8	8	8	8	8		
136						
72	72					
200						
40	40	40				
168						
104	104					
232						
24	24	24	24			
152						
88	88					
216						
56	56	56				
184						
120	120					
248						
4	4	4	4	4	4	
132						
68	68					
196						
36	36	36				
164						
100	100					
228						
20	20	20	20			
148						
84	84					
212						
52	52	52				
180						
116	116					
244						
12	12	12	12	12		
140						

Table 13-20 (cont.)

Walsh Code Pool Usage Example

76	76					
204						
44	44	44				
172						
108	108					
236						
28	28	28	28			
156						
92	92					
220						
60	60	60				
188						
124	124					
252						
NA	NA	NA	NA	NA	NA	1
NA						
NA	NA					
NA						
NA	NA	NA				
NA						
NA	NA					
NA						
NA	NA	NA	NA			
NA						
NA	NA					
NA						
NA	NA	NA				
NA						
NA	NA					
NA						
NA	NA	NA	NA	NA		
NA						
NA	NA					
NA						
NA	NA	NA				
NA						
NA	NA					
NA						
NA	NA	NA	NA			
NA						
NA	NA					
NA						
NA	NA	NA				
NA						
NA	NA					
NA						
NA	NA	NA	NA	NA	NA	
NA						
NA	NA					
NA						
NA	NA	NA				

Table 13-20 (cont.)

Walsh Code Pool Usage Example

NA						
NA	NA					
NA						
NA	NA	NA	NA			
NA						
NA	NA					
NA						
NA	NA	NA				
NA						
NA	NA					
NA						
NA	NA	NA	NA	NA		
NA						
NA	NA					
NA						
NA	NA	NA				
NA						
NA	NA					
NA						
NA	NA	NA	NA			
NA						
NA	NA					
NA						
NA	NA	NA				
NA						
NA	NA					
NA						

use of a single high-speed data session using Walsh code 1 eliminates from possible use a total of 64/32 Wash codes. Now there is a difference between both examples; the first is that the data session will end sooner, at least in concept, then the voice call, thereby replenishing the Walsh code pool.

Another very important issue regarding the Walsh code pool is that with 3X channels, the same Walsh code is used for all three carriers associated with the 3X platform.

Therefore, based on the expected traffic mix that is anticipated for the system, the choice of how to deploy the services relative to the carriers is important. To be more blunt, if there is a 50/50 mix between packet and voice traffic and the packet usage may be 70 Kbps or higher, then it is advisable that when deploying CDMA2000, a separate channel is used for packet data only thereby preserving the imbedded voice platforms and of course throughput.

13.6.2 Packet Data Rates

The next part of the puzzle when performing the design aspect is to review the relationship between the data rates and the other components that are affected by the choice of data and its requite speed selected.

The asterisk refers to the fact that there is now a reverse pilot involved with those configurations of CDMA2000.

Looking at Table 13-21, one is drawn to the conclusion or suspicion that there must be some other factor involved with system capacity than Walsh codes as described earlier. As suspected with the differing data rates, there

Table 13-21

Packet Data Rates

Forward

RC	SR	Data Rates	Characteristics
1	1	1200, 2400, 4800, 9600	R=1/2
2	1	1800, 3600, 7200, 14400	R=1/2
3	1	1500, 2700, 4800, 9600, 38400, 76800, 153600	R=1/4
4	1	1500, 2700, 4800, 9600, 38400, 76800, 153600, 307200	R=1/2
5	1	1800, 3600, 7200, 14400, 28800, 57600, 115200, 230400	R=1/4
6	3	1500, 2700, 4800, 9600, 38400, 76800, 153600, 307200	R=1/6
7	3	1500, 2700, 4800, 9600, 38400, 76800, 153600, 307200, 614400	R=1/3
8	3	1800, 3600, 7200, 14400, 28800, 57600, 115200, 230400, 460800	R=1/4 (20ms) R=1/3 (5ms)
9	3	1800, 3600, 7200, 14400, 28800, 57600, 115200, 230400, 460800, 1036800	R=1/2 (20ms) R=1/3 (5ms)

Reverse

RC	SR	Data Rates	Characteristics
1	1	1200, 2400, 4800, 9600	R=1/3
2	1	1800, 3600, 7200, 14400	R=1/2
3*	1	1200, 1350, 1500, 2400, 2700, 4800, 9600, 19200, 38400, 76800, 153600, 307200	R=1/4 R=1/2 for 307200
4*	1	1800, 3600, 7200, 14400, 28800, 57600, 115200, 230400	R=1/4
5*	3	1200, 1350, 1500, 2400, 2700, 4800, 9600, 19200, 38400, 76800, 153600, 307200, 614400	R=1/4 R=1/2 for 307200 and 614400
6*	3	1800, 3600, 7200, 14400, 28800, 57600, 115200, 230400, 460800, 1036800	R=1/4 R=1/2 for 1036800

is a corresponding alteration to the link budget and pole capacity found in the processing gain aspect. Naturally as data rate increases, the processing gain is reduced because the overall spreading rate remains constant.

Table 13-22 shows the relationship between the data rates, defined in Kbps, and the processing gain. It is important to note that the SR and RC are also involved with the decisions, hence their inclusion in the table.

Table 13-22

Data Rate and Processing Gain Interaction

Reverse Link

RC1		**RC2**		**RC3**		**RC4**		**RC5**		**RC6**	
Kbps	**PG**	**Kbps**	**PG**	**Kbps**	**PG**	**Kbps**	**PG**	**Kbps**	**PG**	**Kbps**	**PG**
9.6	128	14.4	85.33	9.6	128	14.4	85.33	9.6	384	14.4	256
				19.2	64	28.1	42.67	19.2	192	28.1	128
				38.4	32	57.6	21.33	38.4	96	57.6	64
				76.8	16	115.2	10.67	76.8	48	115.2	32
				153.6	8	230.4	5.33	153.6	24	230.4	16
				307.2	4			307.2	12	460.8	8
								614.4	6	1036.8	4

Forward Link

RC1		**RC2**		**RC3**		**RC4**		**RC5**		**RC6**	
Kbps	**PG**	**Kbps**	**PG**	**Kbps**	**PG**	**Kbps**	**PG**	**Kbps**	**PG**	**Kbps**	**PG**
9.6	128	14.4	85.33	9.6	128	9.6	128	14.4	85.33	9.6	384
				19.2	64	19.2	64	28.1	42.67	19.2	192
				38.4	32	38.4	32	57.6	21.33	38.4	96
				76.8	16	76.8	16	115.2	10.67	76.8	48
						153.6	8	230.4	5.33	153.6	24
						307.2	4			307.2	12

RC7		**RC8**		**RC9**	
Kbps	**PG**	**Kbps**	**PG**	**Kbps**	**PG**
9.6	384	14.4	256	14.4	256
19.2	192	28.1	128	28.1	128
38.4	96	57.6	64	57.6	64
76.8	48	115.2	32	115.2	32
153.6	24	230.4	16	230.4	16
307.2	12	460.8	8	460.8	8
614.4	6			1036.8	4

What follows next is an example of how to determine the relative number of users that can utilize a single CDMA2000 channel.

$$N = \frac{[W/R]}{\alpha[E_b/N_o][1 + \beta]} + 1$$

where

W/R = Process gain,

α = Activity factor = 0.479 voice and 1.0 data (generally)

$E_b/N_o = 7$

β= 0.6 (omni) and 0.85 (sector)

Examples:

a) RC = 2 and SR = 1

W/R=85.33, $\alpha = 0.479$, $E_b/N_o = 7$ and $\beta = 0.85$ (sector)

N = (85.33)/[(0.479)(7)(1.85)] + 1 = 14.756

Now if $\alpha = 1.0$, then

N = (85.33)/[(1)(7)(1.85)] + 1 = 7.58

b) RC = 3, SR = 1

Data rate = 76.8 Kbps. Therefore W/R = 16, $\alpha = 1.0$, and $\beta = 0.85$ (sector)

N = (16)/[(1)(7)(1.85)] + 1 = 2.235

13.7 Handoffs

CDMA2000 systems utilize several types of handoffs for both voice and packet data. The types of handoffs involve soft, softer, and hard. The difference between the types is dependent upon what is trying to be accomplished. The process for having a call or packet session in handoff for soft, softer, or hard is the same as that used for IS-95. The key exception to this fact is when a packet session is in progress and the subscriber exits the PDSN coverage area, resulting in a termination of the packet session.

There are several user, adjustable parameters that help the handoff process take place. The parameters that need to be determined involve the values to add or remove a pilot channel from the active list, and the search window sizes. There are several values that determine when to add or remove a pilot from consideration. In addition, the size of the search window cannot be too small or too large.

When introducing CDMA2000 into an existing IS-95 system, the choice of how to set up the neighbor list and search windows should mirror the existing system except where there is a transition zone.

13.7.1 Search Window

There are several search windows in CDMA2000 and they are the same as those used for IS-95 facilitating integration and compatibility. As with IS-95 systems, each of the search windows has its own role in the process, and it is not uncommon to have different search window sizes for each of the windows for a particular cell site. Additionally, the search window for each site needs to be set based on actual system conditions. The search window is defined as an amount of time, in terms of chips, that the CDMA subscriber's receiver will hunt for a pilot channel. There is a slight difference in how the receiver hunts for pilots depending on its type.

The search windows needed to be determined for CDMA involve the

- Active
- Neighbor
- Remaining

The method for determining the search window sizes for a CDMA2000 system is the same as that done for IS-95 and covered in Chapter 3, "Second Generation (2G)."

13.7.2 Soft Handoffs

Soft handoffs are an integral part of CDMA. The determination of which pilots will be used in the soft handoff process has a direct impact on the quality of the voice call or packet-data session as well as the capacity for the system. Therefore setting the soft handoff parameters is a key element in the system design for CDMA2000.

The parameters associated with soft handoffs involve the determination of which pilots are in the active, candidate, neighbor, and remaining sets. The list of neighbor pilots is sent to the subscriber unit when it acquires the cell site or is assigned a traffic channel.

A brief description of each type of pilot is the same as that used for IS-95 systems and discussed in Chapter 3; however, it is repeated here for clarity.

The *active set* is the set of pilots associated with the forward traffic channels assigned to the subscriber unit. The active set can contain more than one pilot because a total of three carriers, each with its own pilot, could be involved in a soft handoff process.

The *candidate set* are the pilots that the subscriber unit has reported are of sufficient signal strength to be used. The subscriber unit also promotes the neighbor set and remaining set pilots that meet the criteria to the candidate set.

The *neighbor set* is a list of the pilots that are not currently on the active or candidate pilot list. The neighbor set is identified by the base station via the neighbor list and neighbor list update messages.

The remaining set is the set of all possible pilots in the system that can be possibly used by the subscriber unit. However, the remaining set pilots that the subscriber unit looks for must be a multiple of the Pilot_Inc.

An example of the interaction between active, candidate, neighbor, and remaining sets is shown in Figure 3-30 and the associated description that accompanies the figure.

Several issues need to be addressed regarding soft handoffs with 1xRTT whether it is a 1x, 1xEV-DO, or 1xEV-DV configuration. The issues that need to be factored in are the different radio configurations between all the base stations involved with the soft handoff process. More specifically, the radio configurations involved must be the same. In addition, radio resources must be available for use by the mobile during soft handoff with all involved base stations. The resources available could possibly involve excluding the subscriber unit soft handoff with a target cell due to the lack of resources available.

If the mobile downgrades from one RC, say RC3 to RC2, it cannot upgrade back to RC3 when resources become available.

An equally important issue is that a 2G mobile having RC1 and RC2 capability can be involved with numerous soft handoffs thereby taking resources away from possible 2.5G/3G mobile use.

In addition, when the mobile negotiates a new service option, it can be any one of the available RCs.

13.8 PN Offset Assignment

The assignment of the PN offset for each CDMA2000 channel and/or sector utilizes the same rules that were and are used for IS-95 systems. In CDMA2000, just as with IS-95 systems, the forward pilot channel carries no data but it is used by the subscriber unit to acquire the system and assist in the process of soft handoffs, synchronization, and channel estimation. A separate forward pilot channel is transmitted for each sector of the cell site. The forward pilot channel is uniquely identified by its PN offset, or rather, PN short code that is used. The reverse pilot channel introduced in CDMA2000, however, does not utilize the *Pseudorandom Number* (PN) offset.

The PN sequence has some 32,768 chips that, when divided by 64, results in a total of 512 possible PN codes that are available for potential use. The fact that there are 512 potential PN short codes to pick from almost ensures that there will be no problems associated with the assignment of these PN codes. However, there are some simple rules that must be followed in order to ensure that there are no problems encountered with the selection of the PN codes for the cell and its surrounding cell sites. It is suggested that a reuse pattern be established for allocating the PN codes. The rational behind establishment of a reuse pattern lies in the fact that it will facilitate the operation of the network for maintenance and growth.

Table 13-23 shows what can be used for establishing the PN codes for any cell site in the network. The method that should be used is to determine whether you wish to have a 4, 7, 9, 19, and so on, reuse pattern for the PN codes.

The suggested PN reuse pattern is a N=19 pattern for a new CDMA2000 system. If you are overlaying the CDMA system on to a cellular system, a

Table 13-23 PN Reuse Sequence

Sector	PN Code
Alpha	$3 \times P \times N - 2P$
Beta	$3 \times P \times N$
Gamma	$3 \times P \times N - P$
Omni	$3 \times P \times N$

N = reusing PN cell and P = PN code increment.

N=14 pattern should be used when the analog system utilizes a N = 7 voice channel reuse pattern, or if a PN code scheme has been established for the sector or site, then the same PN code should be used for that sector/cell.

Figure 13-2 is an example of a N = 19 PN Code reuse pattern. Please note that not all the codes have been utilized in the N = 19 pattern. The remaining codes should be left in reserve for use when there is a PN Code problem that arises. In addition, a suggest PN_INC value of 6 is also recommended for use.

The PN short code used by the pilot is an increment of 64 from the other PN codes an offset value is defined. The Pilot_INC is the value that is used to determine the amount of chips, or rather phase shift, one pilot has versed another pilot. The method that is used for calculating the PN offset is shown in Figure 3-1 of Chapter 3 and applies to CDMA2000 as well as IS-95 systems.

Pilot_INC is valid from the range of 0 to 15. Pilot_INC is the PN sequence offset index and is a multiple of 64 chips. The subscriber unit uses the Pilot_INC to determine which are the valid pilots to be scanned. The

Figure 13-2
PN Reuse Pattern.

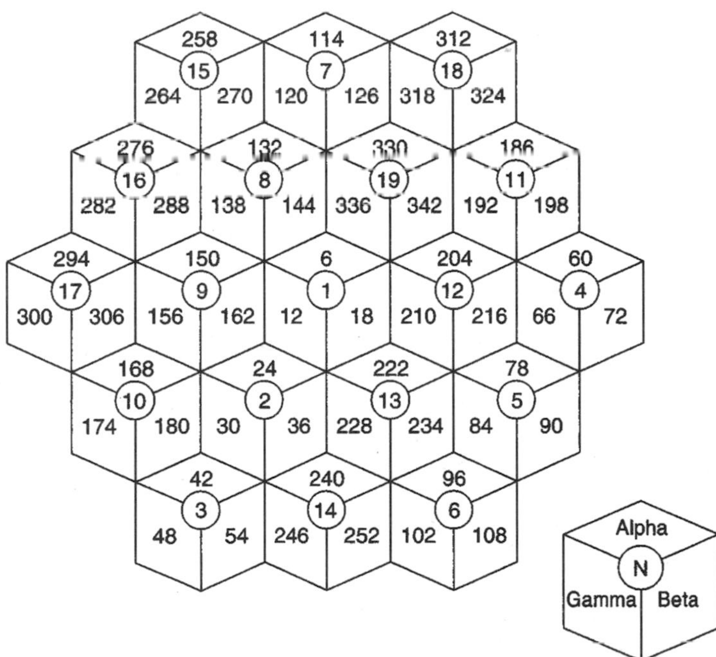

method for calculating the Pilot _INC is the same as that used for IS-95 systems and is a function of the distance between reusing sites.

13.9 Link Budget

The link budget process, as defined in previous sections of this book, is essential for the establishment of a valid RF design to take place. The link budget helps define the cell-size spacing. The cell-site spacing is determined by the link budget using a signal level that is exceeded by 50 percent of the time.

There are two links that need to be determined in the establishment of a link budget: forward and reverse. The forward and reverse links utilize different coding and modulation formats. The first step in the link budget process is to determine the forward before the reverse links maximum path losses. The link budget is defined previously in an earlier chapter.

CDMA2000-1X has a better link budget than IS-95A/B at the same traffic loading therefore offering a high overall capacity at the same traffic load due to vocoder improvements as well as utilizing a coherent demodulation for the reverse link. However, for the link budget that will be used for the design, the link budget parameters primarily associated with IS-95 are utilized due to the prevalence of the RC1 and RC2 subscriber units in the market.

Regarding packet-data services, due to the improved modulation and coding scheme (resulting in a lower target E_b/N_o), the 38.4-Kbps packet data rate for CDMA2000-1x has approximately the same link budget as IS-95 13K voice vocoder, but at higher data rates the service coverage will shrink due to a variety of factors that include process gain as well as power allocation. With 1xRTT, voice is given a priority and therefore data petitions for all available remaining power. Therefore for the design effort put forth a lower data rate of 38.4 Kbps was used per packet data subscriber in the link budget calculations, but 70 Kbps was used for subscriber packet throughput. The disparity was done for ease of discussion.

As stated previously, the link budget calculations utilized directly influence the performance of the CDMA system because it is used to determine power setting and capacity limits for the network. Proper selection of the variables that comprise the link budget is a very obvious issue due to its impact on a successful design.

The following Tables (13-24 and 13-25) represent the link budgets for a CDMA2000 system. Obviously the issue of differing data rates, and subscriber and base radio configurations makes the possible combinations daunting. However, the basic principals that comprise the link budget tables presented in Tables 13-24 and 13-25 can be modified with different process gains as well as a different spreading rate for the uplink path when, and, if a 3X system is deployed.

Table 13-24

Reverse Link Budget

Reverse Link Budget		**Value**		**Comment**
Subscriber Terminal	Tx Power	23	dBm	maximum power per traffic channel
	Cable Loss	2	dB	
	Antenna gain	0	dBd	
	Tx Power per Traffic Channel	**21**	**dBm**	
External Factors	Fade Margin	5	dB	Log Normal
	Penetration Loss	10	dB	(street/vehicle/building)
	External Losses	**−15**	**dB**	
Base Station	Rx Antenna Gain	15	dBd	(approx 17.25 dBi)
	Tower Top Amp Net Gain	0		
	Jumper and Connector Loss	0.25	dB	
	Feedline Loss	1	dB	
	Lightening Arrestor Loss	0.25		
	Jumper and Connector Loss	0.25		
	Duplexer Loss	0.5		
	Receive Configuration Loss	0		
	Handoff Gain	4	dB	
	Rx Diversity Gain	0	dB	
	Rx Noise Figure	5	dB	
	Receiver Interference Margin	3.4	dB	55% pole
	Reciever Noise Density	−174	dBm/Hz	
	Information Rate	41.58	dB	14.4
	Rx Sensitivity	−124.0	dBm	
	Eb/No	7	dB	
	Total Base Station	**−140.77**	**dBm**	
Eb/No	**Eb/No**	**7.00**	**dB**	
	Maximum Path Loss	**139.77**	**dB**	

Table 13-25 Forward Link Budget

Forward Link Budget				
		Value		**Comment**
Tx Power Distribution			units	
	Tx PA Power	39.0	dBm	8 Watts
	Pilot Channel Power	30.8	dBm	15.0% % of Max Power per Channel
	Synch Channel Power	20.8	dBm	10.0% % pilot power
	Paging Channel Power	26.2	dBm	35.1% % pilot power
	Traffic Channel Power	**38.0**	**dBm**	78.2% % of Max Power per channel
	Number Mobiles per Carrier	13		
	Soft/Softer Handoff Traffic	13		1.85 overhead factor
	Maximum # of Active Traffic Channels	26		
	Avg Traffic Channel Pwr	**23.8**	**dBm**	26 Total Traffic Channels
	Voice Activity Factor	0.479		Voice = 0.479, data =1.0
	Peak Traffic Channel Pwr	27.0	dBm	Avg Traffic Ch Pwr/ Voice Activity Factor
Base Station				
	Traffic Channel Tx Pwr	27.0	dBm	
	Duplexer Loss	0.5	dB	
	Jumper and Connector Loss	0.25	dB	
	Lightening Arrestor Loss	0.25	dB	
	Feedline Loss	1	dB	
	Jumper and Connector Loss	0.25	dB	
	Tower Top Amp Loss	0	dB	
	Antenna Gain	15	dBd	
	Net Base Station Tx Pwr	**39.8**	**dBm**	10 Watts ERP per Traffic Channel (voice)
	Total Base Station Tx Power	51.8	dBm	151 Watts ERP per carrier
Environmental	Fade Margin	5	dB	Log Normal
	Penetration Loss	10	dB	(street/vehicle/building)
	Cell Overlap	3	dB	
	External Losses	**−18**	**dB**	
Subscriber	Antenna gain	0	dBd	
	Cable Loss	2	dB	
	Rx Noise Figure	10	dB	
	Reciever Noise Density	−174	dBm/Hz	
	Information Rate	60.90	dB	1230 Kbps
	Rx Sensitivity	−101.1	dBm	
Subscriber Traffic Channel RSSI	Base Tx	39.8		
	Environmental Loss	−18		
	Max Path Loss	139.77		Obtained from Uplink Path Analysis
	RSSI at Sub Antenna	−118.01		

Table 13-25 (cont.)

Forward Link Budget

		Value		Comment
Subscriber Total RSSI	Base Tx	51.8		
	Environmental Loss	−18		
	Max Path Loss	139.77		Obtained from Uplink Path Analysis
	RSSI at Sub Antenna	−105.99		
Interference				
Internal Interference	Orthogonality Factor	−8	dB	0.16 same sector interference
	RSSI at Sub Antenna	−105.99		
	Other User Interference Level	−113.99		Orthoginal Factor *(RSSI total)
	Other Sector Interference	4	dB	
	Interference Density	−109.99		
External Interference	Rx Sensitivity	−101.1		
	external interference	−117	dBm	Depends on local environment
	Total Interfernce on TCH	−100.78		External interference + Rx sensitivity + other user interference
RSSI	Mobile TCH RSSI	−118.01		
	Information Rate	41.58	dB	14.4
	Traffic Channel Eb	−159.59		
	Total RSSI	−105.99		
	Information Rate	60.90	dB	1230
	Traffic Channel No	−166.88		
Eb/No	Traffic Channel Eb	−159.59		
	Traffic Channel No	-166.88		
	Eb/No	7.29		

13.10 Sample Basic Designs

Over the new few pages, three basic designs will be covered. The designs will be rudimentary in nature because the concept of what has to be done needs to be done is stressed, not a particular design for a particular market that will not be relevant for any other system.

- CDMA2000-1X (green field)
- IS-95 to CDMA2000-1X
- CDMA2000-1X to 3X

The exact sequence of migration for any system using CDMA2000 1X to a 3X platform is

- 1X
- 1XEV-DO
- 1xEV-DV
- 3X

The traffic estimate for all three designs will be the same fundamentally with a few variants that are relative to the access platform being deployed. However, a key element to the traffic forecast method is the use of overbooking data services as well as the issue of the volume of CDMA2000 ready subscriber units. One method of determining the number of available subscribers that will be CDMA2000-ready is to obtain the estimate of subscriber handsets that will be procured by the company over the next 6 to 12 months.

13.10.1 CDMA2000-1X

The following is a brief design example that is relevant for a new CDMA2000-1X system being deployed as a green field situation. The design example focuses on the issues that are more relevant to the internal network and does not factor into the mix any possible networking and coordination issues with adjacent systems.

Because this is a new CDMA2000 system, the concerns of legacy equipment are not relevant and it will be assumed that only CDMA2000 capable handsets are used by the system. However in real life, the issue of roaming mobiles into the system that are legacy, IS-95, will need to be factored into the design.

For this design, both CDMA2000-1x and CDMA2000-1xDO channel types will be available for deployment.

The initial design calls for coverage of a selected area within the network. The first step in this case is to determine the desired traffic load for both circuit switched as well as packet data. Utilizing the traffic loading numbers presented earlier Tables 13-26 and 13-27 show the expected traffic

Table 13-26 CDMA2000-1X Greenfield Traffic Forecast

Building			Activity Factor		Call Seconds				
							Penetration	Population	Net Users
	BHCA	Call Duration (s)	UL	DL	UL	DL	35%	50,000	
S	0.9	180	0.5	0.5	81	81	73%	17,500	12,775
SM	0.006	3	1	1	0.018	0.02	40%	17,500	7,000
SD	0.2	156	1	1	31.2	31.2	13%	17,500	2,275
MMM	0.5	3,000	0.15	0.0029	225	4.28	15%	17,500	2,625
HMM	0.15	3,000	0.15	0.0029	67.5	1.28	15%	17,500	2,625
HIMM	0.1	120	1	1	12	12	25%	17,500	4,375

Pedestrian			Activity Factor		Call Seconds				
							Penetration	Population	Net Users
	BHCA	Call Duration (s)	UL	DL	UL	DL	15%	50,000	
S	0.8	120	0.5	0.5	48	48	73%	7,500	5,475
SM	0.03	3	1	1	0.09	0.09	40%	7,500	3,000
SD	0.2	156	1	1	31.2	31.2	13%	7,500	975
MMM	0.4	3,000	0.15	0.0029	180	3.42	15%	7,500	1,125
HMM	0.06	3,000	0.15	0.0029	27	0.51	15%	7,500	1,125
HIMM	0.05	120	1	1	6	6	25%	7,500	1,875

Table 13-26 (cont.) CDMA2000-1X Greenfield Traffic Forecast

Vehicular			Activity Factor		Call Seconds		Penetration	Population	Net Users
	BHCA	Call Duration (s)	UL	DL	UL	DL	70%	50,000	
S	0.4	120	0.5	0.5	24	24	73%	35,000	25,550
SM	0.02	3	1	1	0.06	0.06	40%	35,000	14,000
SD	0.02	156	1	1	3.12	3.12	13%	35,000	4,550
MMM	0.008	3,000	0.15	0.0029	3.6	0.07	15%	35,000	5,250
HMM	0.008	3,000	0.15	0.0029	3.6	0.07	15%	35,000	5,250
HIMM	0.008	120	1	1	0.96	0.96	25%	35,000	8,750

Note: Expect to sell 50K handsets which are CDMA2000 50,000 capable.

Table 13-27 CDMA2000-1X Traffic Forecast (Erlangs and Mbps)

Net User Bit Rates DL Kbps	UL Kbps	Building	Circuit Switch Usage UL	DL		Packet Service Usage UL	DL
16	16		1034775	1034775			
14	14					315	126
64	64		70980	70980			
384	64					3937500	11221.875
2000	128					1181250	3366.5625
128	128					210000	52500
		Total BHCS	1105755.0	1105755.0	**Total BHPS**	5329065.0	67214.4
		Erlang	307.2	307.2	**Mbps**	53.29	0.67
Net User Bit Rates DL Kbps	**UL Kbps**	**Pedestrian**	**Circuit Switch Usage UL**	**DL**		**Packet Service Usage UL**	**DL**
16	16		262800	262800			
14	14					675	270
64	64		30420	30420			
384	64					1350000	3847.5
2000	128					202500	577.125
128	128					45000	11250
		Total BHCS	293220	293220	**Total BHPS**	1598175.0	15944.6
		Erlang	81.45	81.45	**Mbps**	15.98	0.16

Table 13-27 (cont.) CDMA2000-1X Traffic Forecast (Erlangs and Mbps)

Net User Bit Rates DL Kbps	**UL Kbps**	**Vehiclar**	**Circuit Switch Usage UL**	**DL**		**Packet Service Usage UL**	**DL**
16	16		613200	613200			
14	14					2100	840
64	64		14196	14196			
384	64					126000	359.1
2000	128					126000	359.1
128	128					33600	8400
		Total BHCS	627396	627396	**Total BHPS**	287700	9958.2
		Erlang	174.28	174.28	**Mbps**	2.88	0.10
		System Totals					
		Erlang	562.88	562.88	**Mbps**	72.15	0.93

load from a total of 50,000 potential users of the wireless system that sales and marketing expect will use the system. Because the actual throughput is undefined due to the lack of actual traffic data from the network, the design will encompass all the possible traffic loads.

Naturally, if packet data services of only 70 Kbps will be offered, then some of the services included in the example can be eliminated.

Table 13-27 shows the expected load on the overall system in Erlangs and Mbps. The reason for Erlangs is relative for circuit switched data whereas that for packet is in Mbps. In previous comments, if only an estimate from marketing is available regarding packet data usage given in a percentage of voice usage, then the estimation should be done using an Erlang-C model.

Table 13-28 is a summary of the calculations derived for the system traffic load. However some additional information is contained in the table and that is the relative geographic areas associated with each type of traffic. For the purposes of this example, the areas will be considered to be contained adjacent to each other for simplifying the example. However in real life, the areas will be intertwined.

The next step is to determine the number of sites required to support the expected load. An assumption needs to be made at this time and that is all the CDMA2000-1x sites will be sector sites, three sectors per cell. In addition it is assumed that for this design, a total of 8.2 Erlangs per sector can be supported for circuit switch per sector, which is derived from a 2 percent GoS using Erlang B with 14 trunk members. The packet throughput is based on 2.35 trunk members at 76.8 Kbps. Both the packet and circuit switch traffic-handling capacities are very conservative and are driven by the link budget and process gain used.

Table 13-28

Traffic Loading Summary Table

	Region	Area (km^2)	Erlangs	Erlang/km	Mbps	Mbps/km
Building	1	100	307.2	3.0715417	53.29	0.5329065
Pedestrian	2	900	81.45	0.0905	15.98	0.0177575
Vehicular	3	4,000	174.28	0.0435692	2.88	0.0007193
Unserved	4	6,000				
Total		10,000	680.15		72.15	

$$\text{Cell voice Erlangs} = 8.2 \text{ Erlangs/sector} \times 2.64 \text{ (sector gain)}$$
$$= 21.648 \text{ Erlangs per cell}$$

$$\text{Packet throughput} = 2.35 \times 76.8 \text{ Kbps/sector} \times 2.64$$
$$= 453.15 \text{ Kbps per cell}$$

$$N_{\text{Circuit Switched}} = \text{Estimated traffic/cell capacity} = 21.648/680\ .15$$
$$= 32 \text{ cells total for the system}$$

$$\text{Packet data} = \text{(Estimated traffic/overbooking)/cell capacity}$$
$$= (72.15 \text{ Mbps/[10]})/453.15 \text{ Kbps} = 16 \text{ total for the system}$$

The next step is to determine the radius for the site(s) involved with each area. In this example, the same pathloss will be used because it is assumed the same morphology is used for all three areas.

$$PL = 132 + 38\log(r)$$
$$= 132 + 38\log(r)$$

From the link budget PL max = 140 dB.

Therefore radius (r) = 140 = 132 + 38log(r).

$$R = 1.623$$
$$\text{Area of cells} = 8.279 \text{ (building)}$$
$$= 15.18 \text{ (pedestrian PL max} = 145)$$
$$= 41.89 \text{ (vehicular PL max} = 150)$$

Table 13-29

System Sites

	Region	Area (km²)	Coverage	Capacity
Building	1	100	12	
Pedestrian	2	900	60	
Vehicular	3	4,000	96	
Unserved	4	6,000		
Total		11,000	168	48

Obviously from the example, the system is coverage, limited and not capacity-limited. However, in briefly looking at the traffic data, the treatment of one section of the system, building, needs a higher throughput than the vehicular areas, which is obvious. Therefore the deployment recommendation is to have two carriers deployed F1 being 1x while F2 is 1xEV-DO or 1xEV-DV which is assigned for data transport only.

Figure 13-3 represents possible channel deployment schemes that apply to a PCS system operating with 15 MHz of duplexed spectrum. The inclusion of 1x, DO, and DV channels is listed but is really left up to the traffic mix as well as true availability for the technology. A 3X deployment is also included from which to see that the later channels being deployed are positioned correctly with the channel bit map.

Now the next issue is what do you do with this wonderful information. Well you need to lay out a rough system topology from where you can begin to determine if it is valid to centralize or decentralize the BSCs or have intermediate nodes in the network. Typically for a system having 1100 sq km in size, it would be expected to have several MSCs or concentration nodes to reduce the leased-line costs and improve on interconnection transport fees.

It is recommended that the core of the network consisting of the building environment utilize two CDMA-2000 carriers while the pedestrian and vehicular zones use only one carrier. A hard handoff of course will need to take place between the F2 and F1 zone. However, it is recommended that in a situation like this that the BTS F1 carriers process primarily voice traffic while the F2 is more a data only situation. As mentioned earlier, this configuration can be done via software and user-definable parameters.

The various pipe sizes were estimated for the initial concept. From Figure 13-5 it, would be advantageous to collocate BSC 1 with the MSC provided the MSC is located near a tandem. The other BSCs, however, due to their initial traffic load, should be considered to be remotely located provided the operational and support issues can be met. In addition, the BSCs will have on average 15 sites connected to them for the design example with the exception of the core where a total of 12 BTS are associated with the BSC.

The facilities between the BTS and BSC are assumed to be unstructured TDM because this is a more readily-available circuit type. The connectivity to the off-net data networks assumes a 80/20 mix of public verse private networks. The assumption used is that 100 percent of the packet traffic is off-net.

Looking at the BTSs, two different configurations are proposed to help facilitate different areas of the network. The first shown in Figure 13-6 is

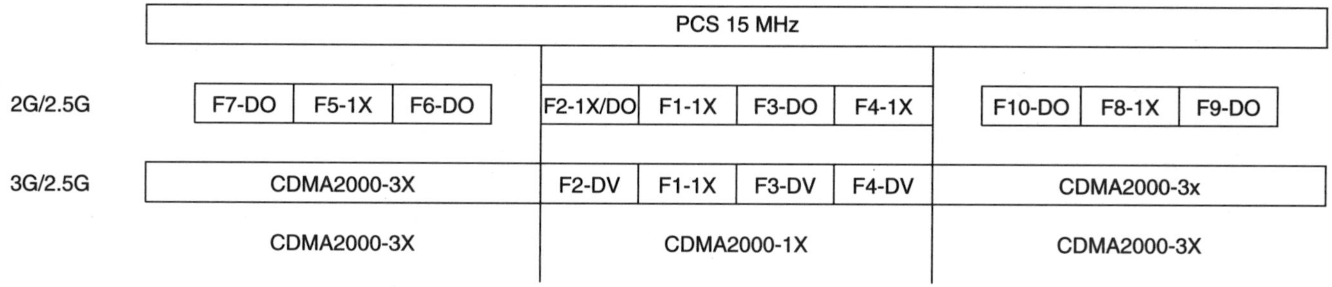

Figure 13-3 CDMA2000 PCS 15 MHz channel deployment scheme.

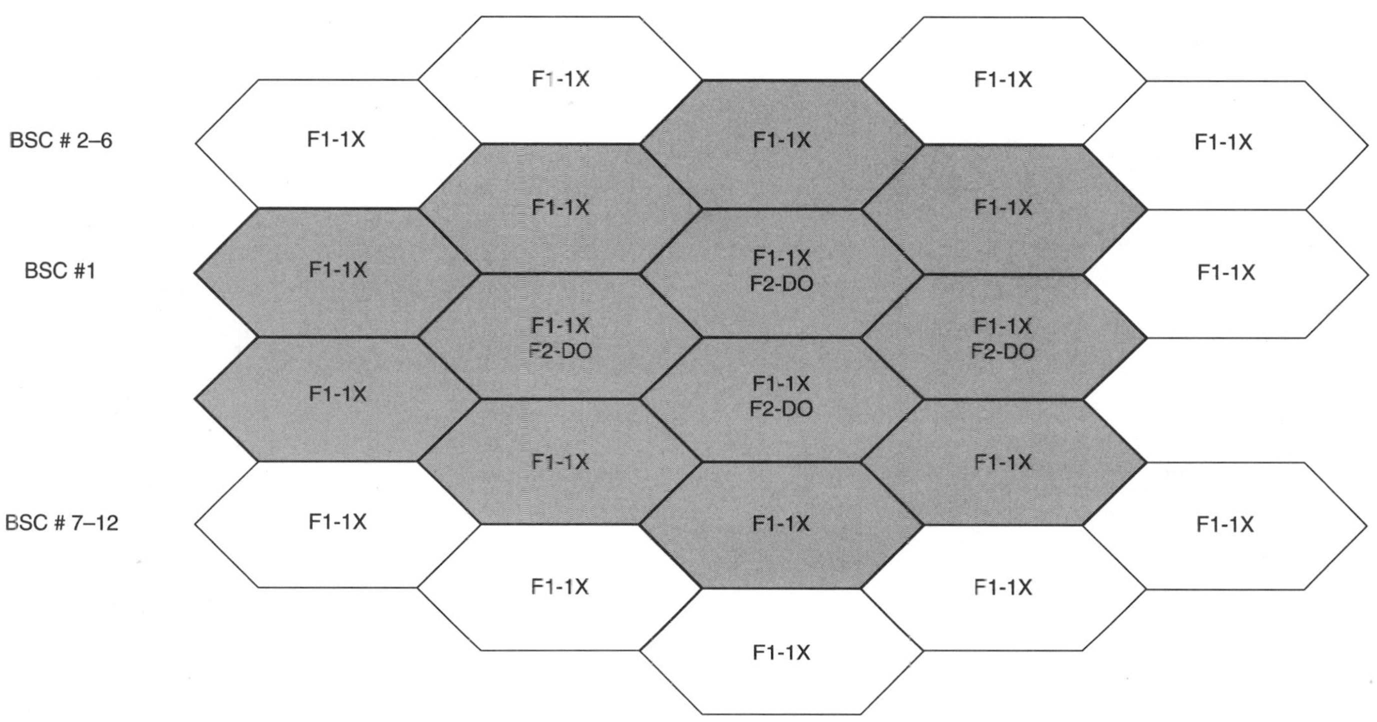

Figure 13-4 CDMA2000-1X carrier deployment scheme within a IS-95 system.

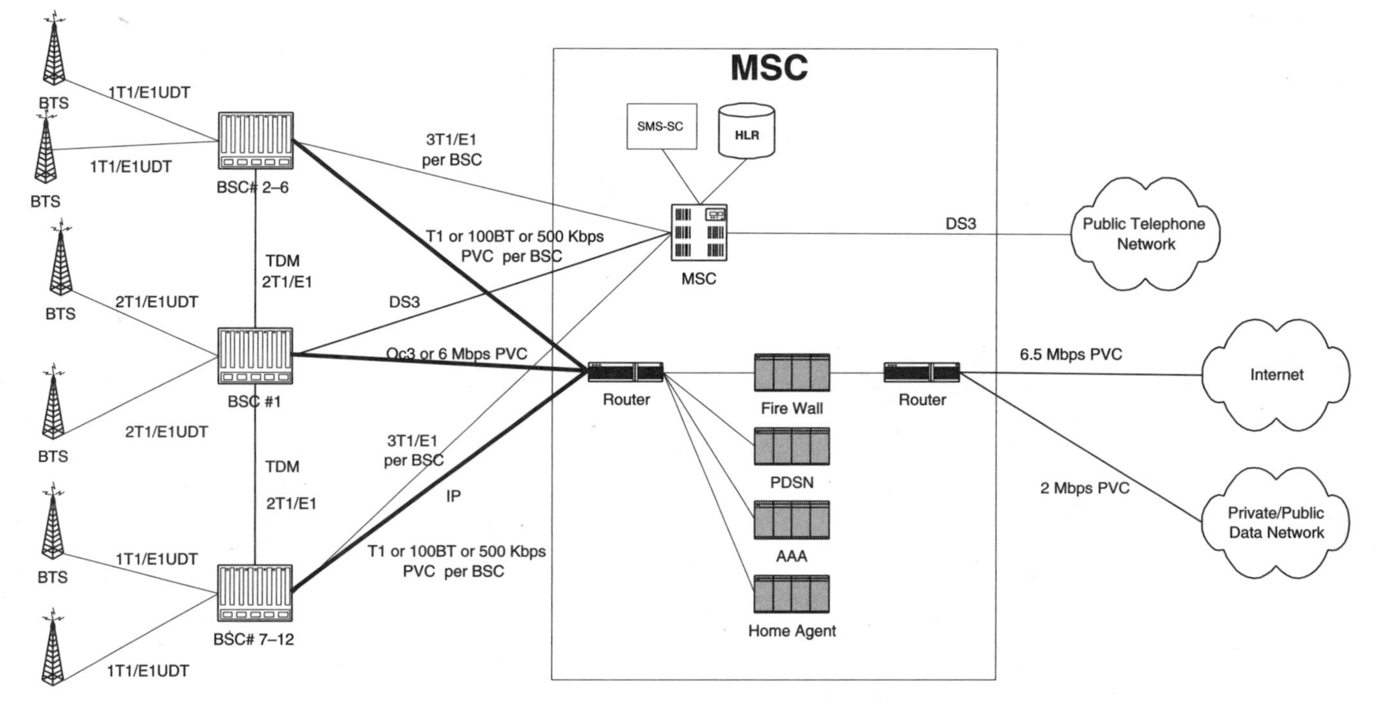

Figure 13-5 Sample CDMA2000-1X system configuration.

for the core area of the network and involves using STD for the transmit diversity scheme because two carriers are initially needed. One could also install more antennas if feasible.

Figure 13-7 shows a configuration recommended for the rest of the network that involves using OTD transmit diversity.

The PN offset assignment scheme that is presented in the earlier part of the chapter should be used for the system design following an N=19 reuse pattern for the PN offsets.

Obviously there are more issues that are involved when designing a CDMA2000 system, but the preceding material should help in the construction of the thought process to achieve the desired goal of supporting the customer requirements for service delivery and transport.

13.10.2 IS-95 to CDMA2000-1X

An all-too-common situation for wireless operators is addressing the issue of how to integrate CDMA2000 into their network. Many operators have devised their own method for implementing CDMA2000 into an existing IS-95 network. However, not all the operators have implemented CDMA2000-1X. Therefore the following will attempt to bring to light

Figure 13-6
Sector STD transmit diversity scheme.

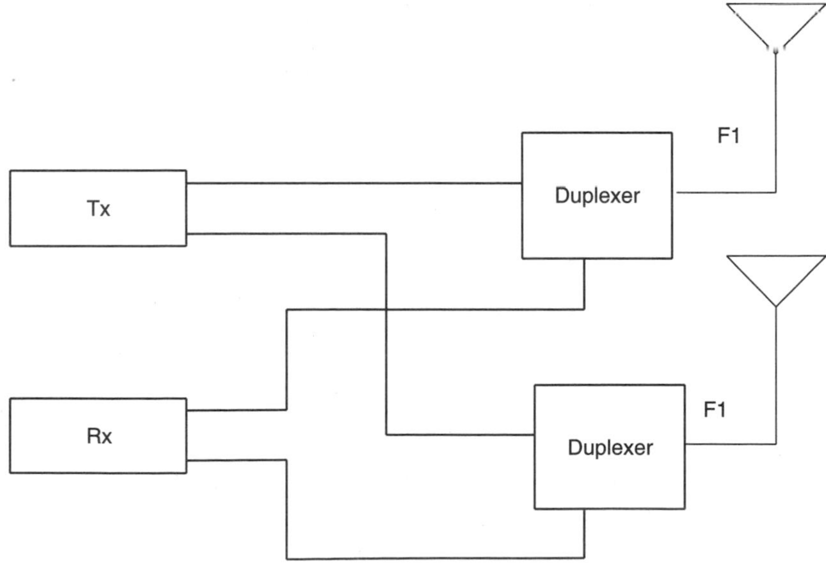

Figure 13-7 Sector OTD transmit diversity scheme.

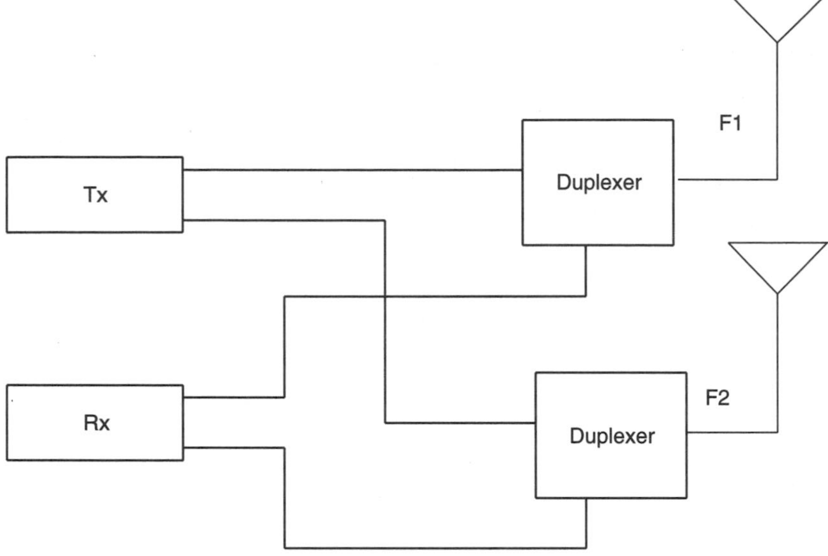

many of the issues associated with integrating a CDMA2000 system with that of a IS-95 system.

Migrating from IS-95 to a CDMA2000-1X platform enables the use of packet data services along with previously stated increases in voice, circuit switched, and carrying capacity. The migration process needs to not only factor in the new services being offered, but also the fundamental problem of still utilizing existing IS-95 equipment.

Figure 13-8 is meant to depict the possible paths that a wireless operator may choose to migrate from an IS-95 system for packet data services. The operator has the choice of waiting for 3X platforms to emerge, but the more rational approach would be to migrate to a 1X platform and then at a future date, when services warrant the move, migrate to a 3X platform.

For this design, both CDMA2000-1xEV-DO and CDMA2000-1xEV-DV channel types will be available for deployment as was the situation with the new CDMA2000-1X system design previously presented. Because the design is a migration to a new technology platform, the system will most likely not be coverage-limited but capacity-driven. Obviously in real life, there are always coverage issues to address, but for the purposes of this example, coverage issues will not be considered.

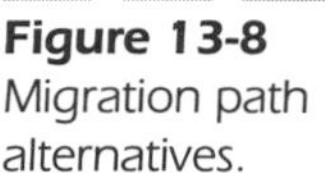

Figure 13-8 Migration path alternatives.

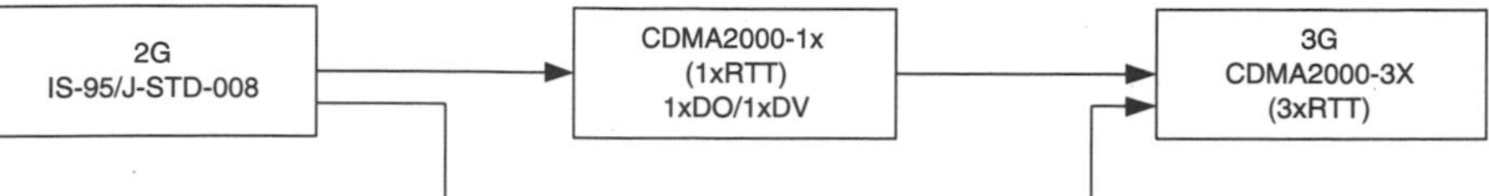

The first step in this case is to determine the desired traffic load for both circuit switched as well as packet data. The circuit switched traffic growth is shown for the system in Table 13-30. The growth represents the increase in circuit switch usage that will be exhibited with CDMA2000-capable handsets.

If the forecasting only involved circuit-switched services, such as voice, then the design process would be straightforward in that a 1:1 replacement of existing IS-95 radios and associated infrastructure would take place only in the areas where capacity was of most concern. Therefore the introduction of CDMA2000-1X would be extremely limited or highly focused into selected areas.

But for this design example, the use of packet data services is included with this design. Therefore utilizing the traffic loading numbers presented earlier Tables 13-31 and 13-32 show the expected traffic load from a total of 17,500 potential users of the wireless system that sales and marketing expect will use the system for packet services. Because the actual throughput is undefined due to the lack of actual traffic data from the network, the design will encompass all the possible traffic loads.

Naturally, if packet data services do not encompass all the speeds possible, then some of the services included in the example can be eliminated.

Table 13-32 shows the expected load on the overall system in Erlangs and Mbps. The reason for Erlangs is relative for circuit-switched data whereas that for packet is in Mbps. In previous comments, if only an estimate from marketing is available regarding packet data usage, given in a percentage of voice usage, then the estimation should be done using an Erlang-C model.

Table 13-33 is a summary of the calculations derived for the system traffic load. However some additional information is contained in the table, and that is the relative geographic areas associated with each type of traffic. For the purposes of this example, the areas will be considered to be contained adjacent to each other for simplifying the example. However in real life, the areas will be intertwined.

Table 13-30

Circuit Switched Usage Forecast

Circuit Switched Usage (Erlangs)	Existing	Growth	IS-95	CDMA2000	Total
Building	199.55	107.45	199.55	107.45	307
Pedestrian	52.9425	28.5075	52.9425	28.5075	81.45
Vehicular	113.282	60.998	113.282	60.998	174.28

The next step is to determine the number of carriers required to support the expected load. Because this is a capacity design, the number of existing sites needs to be identified. The number and relative location within the morphology class is shown in the Table 13-34.

All the BTS sites listed in Table 13-34 are three sector by design.

In addition, it is assumed again that for this design, a total of 8.2 Erlangs per sector can be supported for circuit switch per sector, which is derived from a 2-percent GoS using Erlang B with 14 trunk members. The packet throughput is based on 2.35 trunk members at 76.8 Kbps. Both the packet and circuit-switch traffic-handling capacities are very conservative and driven by the link budget and process gain used.

Cell voice Erlangs = 8.2 Erlangs/sector $\times$ 2.64 (sector gain)
= 21.648 Erlangs per cell (single carrier per sector)

Packet throughput = 2.35 $\times$ 76.8 Kbps/sector $\times$ 2.64
= 453.15 Kbps per cell (single carrier per sector)

$N_{\text{Circuit Switched}}$ = Estimated traffic/cell capacity = 680.15/21.648
= 32 cells total for the system

CDMA2000 $N_{\text{Circuit Switched}}$ = Estimated traffic/cell capacity
= 196.955/21.648 = 9 cells total

Packet data = (Estimated traffic/overbooking)/cell capacity
= (72.15 Mbps/[10])/453.15 Kbps = 16 total for the system

The radius of the particular sites is important to calculate but for this example, the CDMA2000 is a 1:1 overlaid on top of the existing legacy platform.

Table 13-31 Forecasted Packet Data Usage

New Packet Services Building	BHCA	Call Duration(s)	Activity Factor UL	DL	Call Seconds UL	DL
SM	0.006	3	1	1	0.018	0.018
MMM	0.5	3000	0.15	0.00285	225	4.275
HMM	0.15	3000	0.15	0.00285	67.5	1.2825
HIMM	0.1	120	1	1	12	12
Pedestrian	**BHCA**	**Call Duration(s)**	**Activity Factor UL**	**DL**	**Call Seconds UL**	**DL**
SM	0.03	3	1	1	0.09	0.09
MMM	0.4	3000	0.15	0.00285	180	3.42
HMM	0.06	3000	0.15	0.00285	27	0.513
HIMM	0.05	120	1	1	6	6
Vehicular	**BHCA**	**Call Duration(s)**	**Activity Factor UL**	**DL**	**Call Seconds UL**	**DL**
SM	0.02	3	1	1	0.06	0.06
MMM	0.008	3000	0.15	0.00285	3.6	0.0684
HMM	0.008	3000	0.15	0.00285	3.6	0.0684
HIMM	0.008	120	1	1	0.96	0.96

Expect to sell 17.5K handsets which are CDMA2000 capable.

Table 13-32 Forecasted Packet Data Usage

Building							
Penetration	**Population**	**Net Users**	**Net User Bit Rates**			**Packet Service Usage**	
35%	50,000		DL kbps	UL kbps		UL	DL
40%	17,500	7,000	14	14		315	126
15%	17,500	2625	384	64		3937500	11221.88
15%	17,500	2,625	2000	128		1181250	3366.563
25%	17,500	4,375	128	128		210000	52500
					Total BHPS	5329065.0	67214.4
					Mbps	53.29	0.67

Pedestrian							
Penetration	**Population**	**Net Users**	**Net User Bit Rates**			**Packet Service Usage**	
15%	50,000		DL kbps	UL kbps		UL	DL
40%	7,500	3,000	14	14		675	270
15%	7,500	1,125	384	64		1350000	3847.5
15%	7,500	1,125	2000	128		202500	577.125
25%	7,500	1,875	128	128		45000	11250
					Total BHPS	1598175.0	15944.6
					Mbps	15.98	0.16

Table 13-32 (cont.) Forecasted Packet Data Usage

Vehicular							
Penetration	Population	Net Users	Net User Bit Rates			Packet Service Usage	
70%	50,000		DL kbps	UL kbps		UL	DL
40%	35,000	14,000	14	14		2100	840
15%	35,000	5,250	384	64		126000	359.1
15%	35,000	5,250	2000	128		126000	359.1
25%	35,000	8,750	128	128		33600	8400
					Total BHPS	287700	9958.2
					Mbps	2.88	0.10
					System Total Mbps	72.15	0.93

Table 13-33 Traffic Loading Summary Table

	Region	Area (km^2)	Erlangs (IS-95)	Erlangs CDMA2000-1x	Total Erlangs	Erlang/km	Mbps	Mbps/km
Building	1	100	199.55	107.45	307.2	0.5329065	53.29	0.5329065
Pedestrian	2	900	52.9425	28.5075	81.45	0.0177575	15.98	0.0177575
Vehicular	3	4,000	113.282	60.998	174.28	0.0007193	2.88	0.0007193
Unserved	4	6,000						
Total		11,000	365.77	196.955	680.15		72.15	

Table 13-34

Number of Existing BTS Sites

	Region	# BTS Sites
Building	1	12
Pedestrian	2	60
Vehicular	3	96
Unserved	4	0
Total		168

Obviously from the example, the system is coverage-limited and not capacity-limited. However in briefly looking at the traffic data, the treatment of one section of the system, building, needs a higher throughput than the vehicular areas, which is obvious. Therefore the deployment recommendation is to have two carriers deployed F1 being an IS-95 channel or 1x and F2 is 1xEV-DO, or 1x which is assigned for data transport only.

From the previous calculations, a total of 9 sites out of the total 32 are required involved with growth. Because this is an overlay design, the core of the network will be focused on for CDMA2000-1X carrier deployment because the bulk of the growth is coming from the building and pedestrian morphology where in the past, not previously mentioned, the design was for vehicular only.

Figure 13-9 represents a possible channel deployment scheme that applies to a PCS system having operating with 15 MHz of duplexed spectrum. The inclusion of 1x, DO, and DV channels is listed but is really left up to the traffic mix as well as true availability for the technology. However in examining the diagram, the inclusion of a legacy channel is left in place for F1. A 3X deployment is also included from which to see later channels being deployed are positioned correctly with the channel bit map that does require the migration from a legacy channel to that capable of 1X or 3X.

There is of course the cellular band that has many unique issues associated with it when trying to deploy any new technology platform. Figure 13-10 highlights the channel deployment scheme for the A- and B-band cellular operators. The deployment scheme is meant to help transition the new technology but also to address the legacy issues.

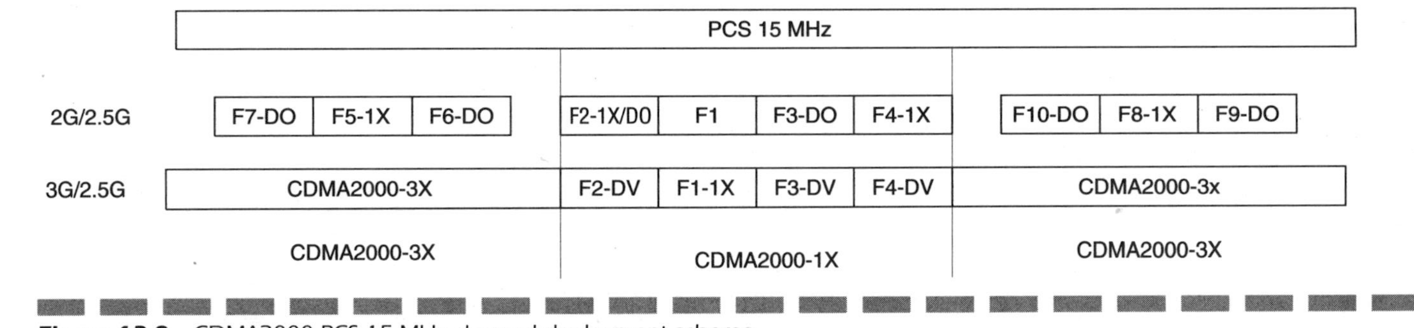

Figure 13-9 CDMA2000 PCS 15-MHz channel deployment scheme.

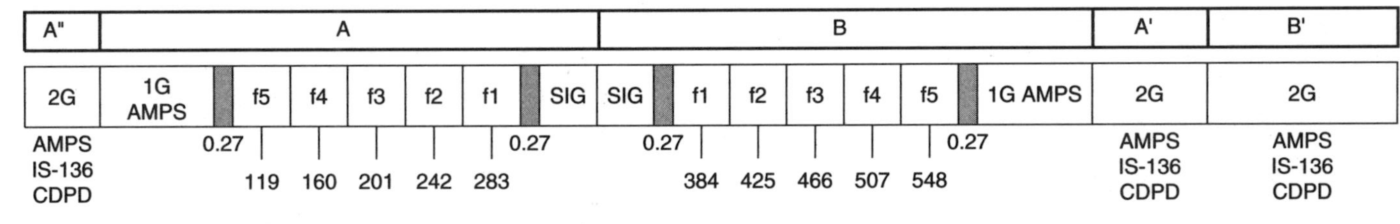

Figure 13-10 Cellular A- and B-band channel deployment scheme.

If a cellular operator, or even a PCS operator has more than one IS-95 channel in operation, the channel deployment schemes can be easily modified by selecting the next channel on the list as the CDMA2000-1x channel of choice. The channel deployment sequence is shown in Table 13-35.

The next step is to deploy the channels in a logical fashion, meeting the individual capacity requirements and maximizing the use of the legacy equipment. The initial system layout is shown in Figure 13-11 and assumes that the BSC 1 is collocated with the MSC. However, the remaining BSCs may be located remotely or also collocated with the MSC. In real life, a system of this size would expect to have more than one MSC or concentration nodes to reduce the leased-line costs.

It is recommended that the core of the network consisting of the building environment utilize two CDMA-2000 carriers while the pedestrian and vehicular zones use only one carrier and that those carriers be a mix between CDMA2000 and IS-95 carriers. A hard handoff, of course, will need to take place between the F2 and F1 zone. However, it is recommended that in a situation like this that the BTS-F1 carriers process primarily voice traffic while the F2 is more of a data only situation. As mentioned earlier, this can be done via software and user-definable parameters.

While poorly represented in the diagram, primarily due to size limitations, Table 13-36 is the breakdown of the carriers by BSC type. The distribution should be based on the individual site loading. The distribution example assumes that packet data services will not be offered throughout the entire footprint of the system. If a true 1:1 overlay was desired, then

Table 13-35
CDMA2000-1X Assignment

	CDMA2000-1X		
Existing IS-95 Carriers	**1X**	**DO**	**Comments**
0	1 (F1)	-	New
0	1 (F1)	1 (F2)	New
1	1 (F1)	-	Overlay
2	2 (F1 and F2)	-	Overlay
2	2 (F1 and F2)	1 (F3)	Overlay and Expansion

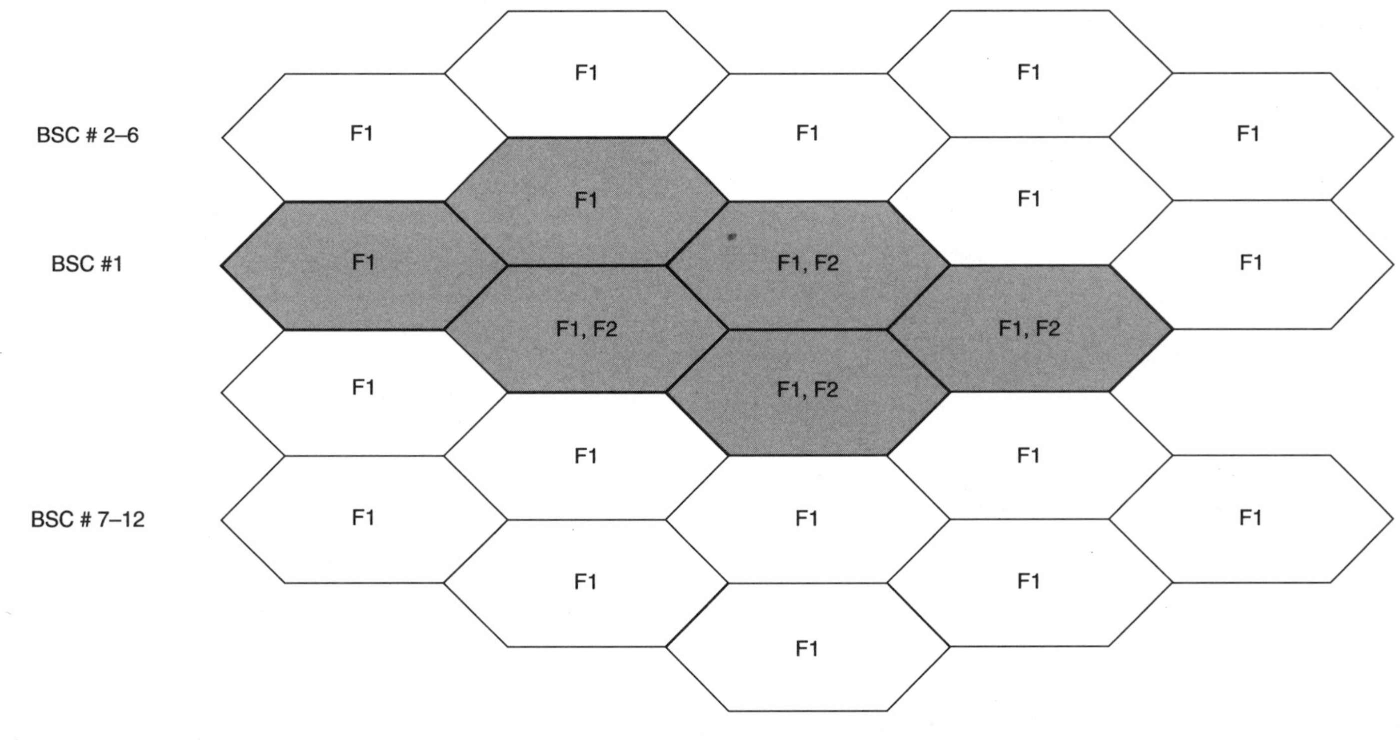

Figure 13-11 IS-95 carrier deployment for sample system.

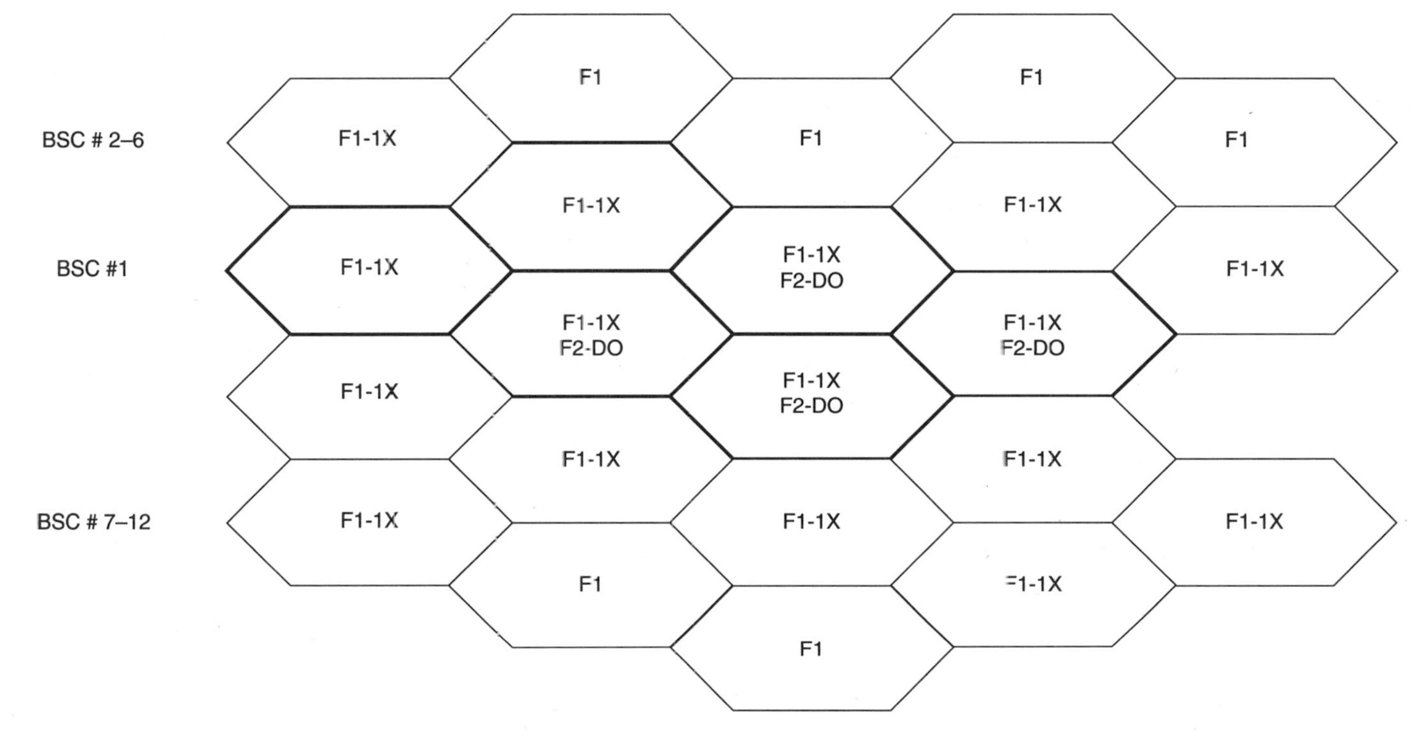

Figure 13-12 CDMA2000-1X carrier deployment scheme within a IS-95 system.

Table 13-36

BTS Type Distribution by BSC

BSC	IS-95 BTS	CDMA2000-1X BTS
1	0	12
2	0	15
3	0	15
4	0	15
5	0	15
6	15	0
7	0	15
8	0	15
9	0	15
10	0	15
11	0	15
12	15	0

the existing legacy equipment should be redeployed or sold on the secondary markets for a more rural application where voice services would only be utilized.

As a brief reminder when handing off from a CDMA2000 channel to a IS-95 system, the loss of packet data services will occur.

Next, the various pipe sizes were estimated for the initial concept. From the diagram it would be advantageous to collocate BSC 1 with the MSC provided the MSC is located near a tandem. The other BSCs, however, due to their initial traffic load, should be considered to be remotely located provided the operational and support issues can be met. In addition, BSC 6 and 12 are considered to be IS-95 only and therefore are not connected to the packet network as depicted in the diagram. While it is possible and advisable to mix the legacy equipment within a BSC, it is not shown in Figure 13-13.

Continuing the facilities between the BTS and BSC are assumed to be unstructured TDM when the BTSs have CDMA2000 channels. The connectivity to the off-net data networks assumes a 80/20 mix of public verse private networks. The assumption used is that 100 percent of the packet traffic

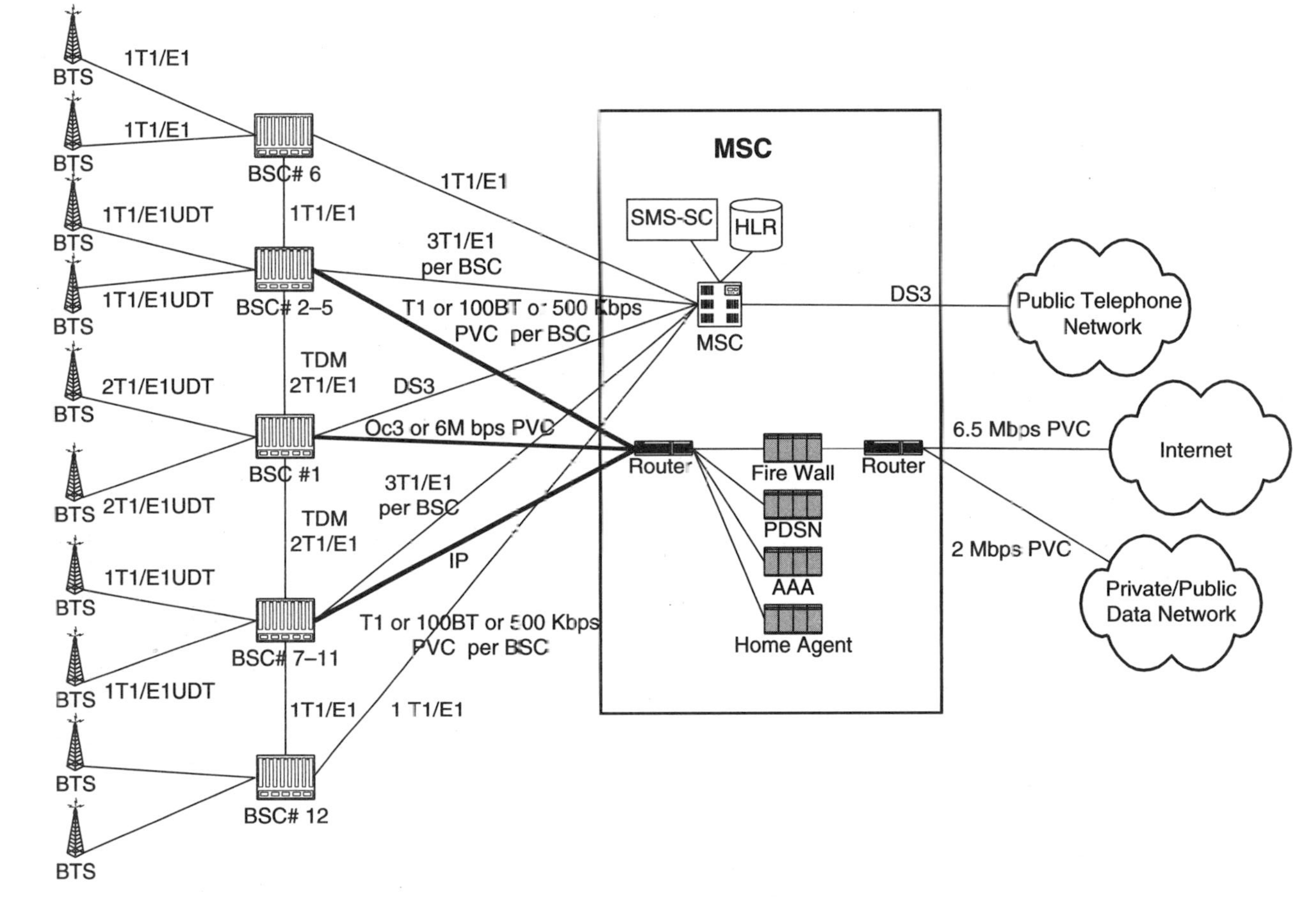

Figure 13-13 Sample CDMA2000-1X system configuration.

is off-net and that mobile to mobile packet sessions will not have a high enough penetration to consider in the design aspect presently.

Looking at the BTS's two different configurations are proposed to help facilitate different areas of the network. The first shown in Figure 13-14 is for the core area of the network and involves using STD for the transmit diversity scheme because two carriers are initially needed. One could also install more antennas if feasible or utilize cross pole antennas.

Regarding the antenna systems, there are some different considerations to take into account when migrating from a IS-95 system to a CDMA2000 system if it is an AMPS or PCS spectrum. Thediversity, and this will be achieved either by a STD or OTD method. However, the STD method is the preferred version. Figure 13-14 (a) shows a STD transmit diversity scheme whereas Figure 13-14 (b) shows an OTD transmit diversity scheme.

Figure 13-14 shows a typical situation where there are two or three antennas per sector available for use. Sometimes there is only one antenna but it is a cross pole antenna, which can be treated as two separate antennas. With an AMPS system as the underlying legacy system, the use of a STD transmit diversity scheme is possible with a configuration shown in Figure 13-14 (a) with the exception that only one carrier is used for CDMA. If a second carrier is added, then OTD diversity is utilized and the configuration shown in (a) is used. Now if the operator has been able to secure more antennas per sector, that is, 5, then the configuration shown in (b) is the desired method where the AMPS and CDMA systems are bifurcated. The use of STD or OTD is again dependant upon the number of carriers required at the site.

The PN offset assignment scheme that is presented in the earlier part of the chapter should be used for the system design following a N=19 reuse pattern for the PN offsets.

Just as with the design example used for a new CDMA2000-1X system, there is a plethora of issues not covered in the example. However, it is believed that the preceding material should help in the construction of the thought process to achieve the desired goal of supporting the customer requirements for service delivery and transport.

13.10.3 CDMA2000-1X to 3X

Migrating from a 1X to a 3X CDMA2000 system is being advertised to be relatively transparent from a radio aspect, provided you have three contiguous 1X channels or you have cleared the spectrum for the new 3X chan-

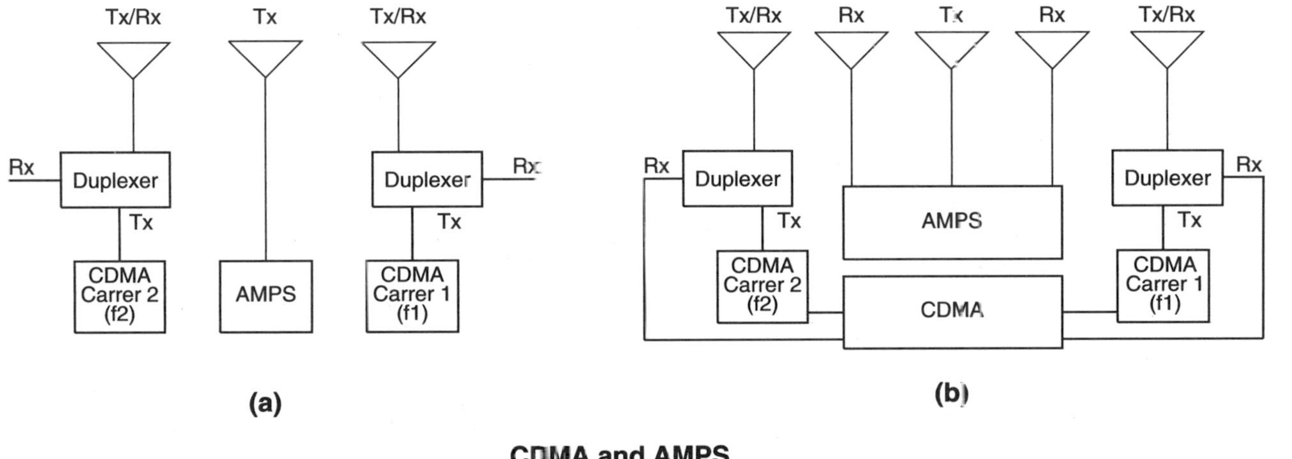

Figure 13-14 CDMA2000/IS-95 and AMPS systems sector antenna configurations.

nel. But in reality, the introduction of a 3X platform will not be transparent due to the variety of operating issues like the real traffic mix.

The fundamental concept behind the migration from 1X to 3X is that the 3X platform comprises of three individual 1X carriers enabling a three fold, with trunking efficiency, in throughput as well as improvements in the modulation scheme and processing gains. The 3X carrier is expected to be overlaid on top of the existing 1X carriers as shown in Figure 13-15.

The various channel schemes that are planned for 3X involve the PCS plans that are shown in Figure 13-16 for a 5-MHz license holder. It is interesting to note that overlaying a 3X platform onto a 1X system needs to be thought out well in advance in order to minimize the impact on traffic loading and carrying. The reason for the traffic concern is that a single 3X mobile will impact all three carriers on a downlink even for a single voice call due to how the Walsh codes are used.

The channel plans shown in Figures 13-17 and 13-18 represent two different alternatives out of the many that are possible. In Figure 13-17, the use of 1X and 3X carriers and their migration paths is shown from a pure 1X environment. It is important to note that the 1X carriers are left for the purpose of supporting circuit switched traffic.

The scenario that Figure 13-18 implies is the possible bifurcation of a 15-MHz PCS license for the purpose of deploying CDMA2000 1X and 3X plus WCDMA.

One can see many possible alternative configurations and options with Figures 13-17 and 13-18. However a very interesting and complex issue arises when focusing on the AMPS band and determining how CDMA2000-3X will be integrated into it. The issue is more complex than just adding a single carrier because a large portion of the spectrum needs to be cleared in order to support the channels introduction. Now the channel associated with 3X may already be operational with CDMA2000-1X carriers, making the transition more efficient. However, if the channels are still in use by 1G systems, then the pain of capacity shifting and migration will need to take place.

Figures 13-19 and 13-20 are examples of how a 3X channel can be deployed into a cellular system. Both figures are slightly different in that Figure 13-19 has two legacy CDMA channels while Figure 13-20 only has one legacy channel. It is assumed that when 3X is introduced to the system that all IS-95 platforms have been retired or moved to voice only areas of the network.

What follows next is an example of traffic calculations associated with the introduction of the 3X platform into the system. For this design, it is assumed that there are no green field applications and that this is a pure integration of an existing CDMA2000-1X system.

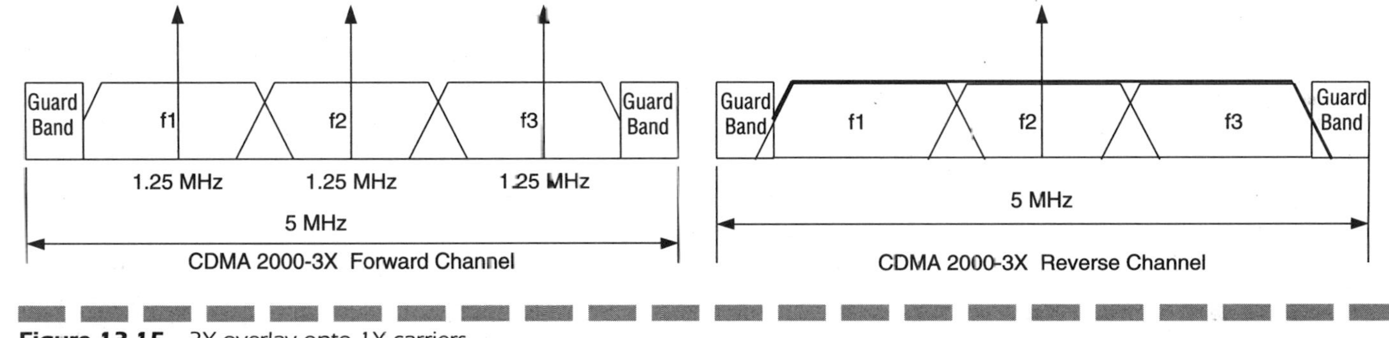

Figure 13-15 3X overlay onto 1X carriers.

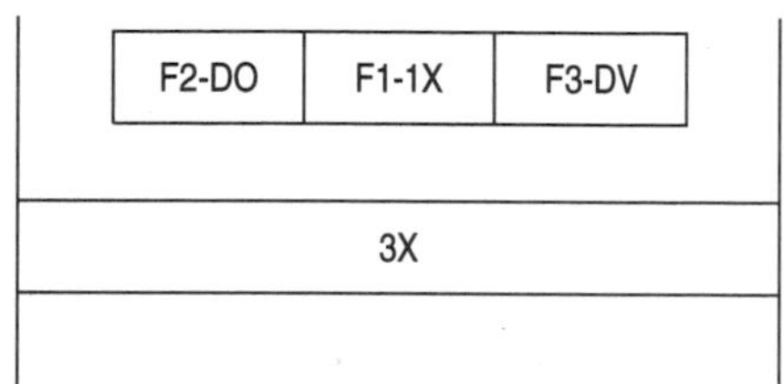

Figure 13-16 PCS 5-MHz channel deployment scheme.

The expansion of the existing system will only take place with CDMA2000-3X-capable handsets. Obviously the mix of handset-compatible units can and will differ depending on the price and delivery factors that have to meet the marketing and sales objectives of the system.

Establishing a simple design for a CDMA2000-3X system calls for a total of 10,000 new subscribers and their relative traffic contributions are shown in the accompanying Tables 13-37, 13-38, 13-39, and 13-40.

Unlike the other designs, the packet data usage is split between the 1X and 3X platforms. Depending on the design objectives defined, the existing packet data users can be rolled up into the new 3X platform. Alternatively, the new packet data users can be allocated to the 3X platform only, and the legacy systems remain in place until the subscribers are migrated over a multi-year process.

If the spectrum is available, then it is recommended to jointly deploy the 1X and 3X platforms. The reason behind this scheme lies in the Walsh code usage because the same Walsh codes are used for all carriers that comprise a 3X radio per sector. Additionally, the 3X platform should be used for packet data only while the legacy systems support voice, circuit switched, until the time that the packet voice is implemented and the legacy subscriber units have been successfully migrated to the new platform.

In examining the traffic defined for the system as a total, which includes existing and new usage, a few issues arise that need to be thought about. With the 1:1 overlay of the 3X system, results in treating new and existing packet data, along with circuit switched data, are being combined.

However, if you were to separate the platforms from a system integration aspect, then 1X could be allocated for circuit switched traffic while 3X is

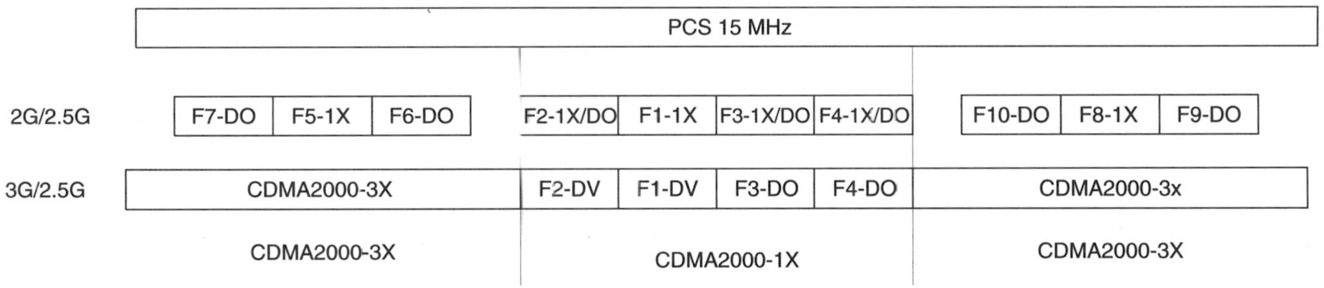

PCS 15 MHz

2G/2.5G
F7-DO | F5-1X | F6-DO
F2-1X/DO | F1-1X | F3-1X/DO | F4-1X/DO
F10-DO | F8-1X | F9-DO

3G/2.5G
CDMA2000-3X
F2-DV | F1-DV | F3-DO | F4-DO
CDMA2000-3x

CDMA2000-3X
CDMA2000-1X
CDMA2000-3X

Figure 13-17 CDMA2000 PCS 15-MHz channel deployment scheme.

PCS 15 MHz

2G/2.5G
F7-DO | F5-1X | F6-DO
F2-1X/DO | F1-1X | F3-1x/DO | F4-1X/DO
GSM

3G/2.5G
CDMA2000-3X
F2-DV | F1-DV | F3-DO | F4-DO
WCDMA

CDMA2000-3X
CDMA2000-1X
WCDMA

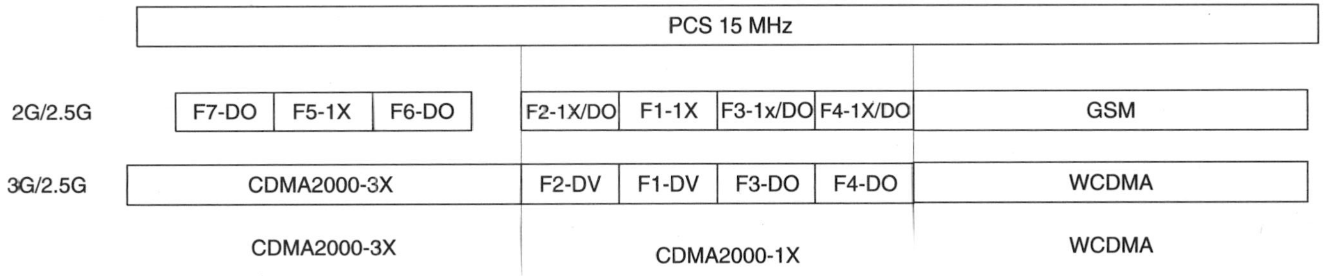

Figure 13-18 CDMA2000 PCS 15-MHz dual system channel deployment scheme.

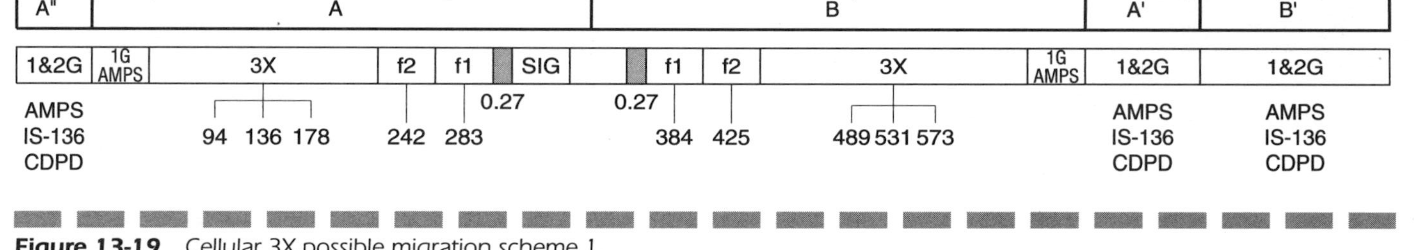

Figure 13-19 Cellular 3X possible migration scheme 1.

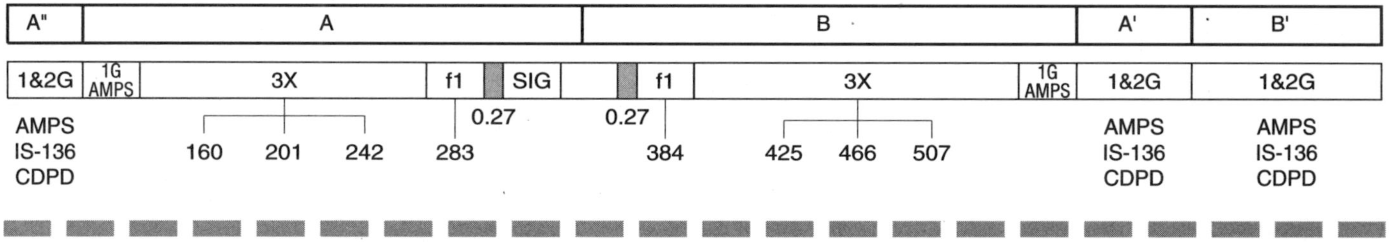

Figure 13-20 Cellular 3X possible migration scheme 2.

Table 13-37 System Growth Forecast

Building			Activity Factor		Call Seconds		Penetration	Population	Net Users
	BHCA	Call Duration (s)	UL	DL	UL	DL			
S	0.9	180	0.5	0.5	81	81	35%	10,000	3,500
SM	0.006	3	[illegible]	1	0.018	0.018	73%	3,500	2,555
SD	0.2	156	[illegible]	1	31.2	31.2	40%	3,500	1,400
MMM	0.5	3000	0.15	0.003	225	4.275	13%	3,500	455
HMM	0.15	3000	0.15	0.003	67.5	1.283	15%	3,500	525
HIMM	0.1	120	[illegible]	1	12	12	15%	3,500	525
							25%	3,500	875

Pedestrian			Activity Factor		Call Seconds		Penetration	Population	Net Users
	BHCA	Call Duration(s)	UL	DL	UL	DL	15%	10,000	1,500
S	0.8	120	0.5	0.5	48	48	73%	1,500	1,095
SM	0.03	3	1	1	0.09	0.09	40%	1,500	600
SD	0.2	156	1	1	31.2	31.2	13%	1,500	195
MMM	0.4	3000	0.15	0.003	180	3.42	15%	1,500	225
HMM	0.06	3000	0.15	0.003	27	0.513	15%	1,500	225
HIMM	0.05	120	1	1	6	6	25%	1,500	375

Vehicular			Activity Factor		Call Seconds		Penetration	Population	Net Users
	BHCA	Call Duration(s)	UL	DL	UL	DL	70%	10,000	7,000
S	0.4	120	0.5	0.5	24	24	73%	7,000	5,110
SM	0.02	3	1	1	0.06	0.06	40%	7,000	2,800
SD	0.02	156	1	1	3.12	3.12	13%	7,000	910
MMM	0.008	3000	0.15	0.003	3.6	0.068	15%	7,000	1,050
HMM	0.008	3000	0.15	0.003	3.6	0.068	15%	7,000	1,050
HIMM	0.008	120	1	1	0.96	0.96	25%	7,000	1,750

Note: Expect to sell 10K handsets which are CDMA2000-3X-capable.

Table 13-38 3X Traffic Forecast

Expansion 3X					
Building	**Circuit Switch Usage**			**Packet Service Usage**	
	UL	**DL**		**UL**	**DL**
	206955	206955			
				63	25.2
	14196	14196			
				787500	2244.375
				236250	673.3125
				42000	10500
Total BHCS	221151.0	221151.0	**Total BHPS**	1065813.0	13442.9
Erlang	61.4	61.4	**Mbps**	10.66	0.13
Pedestrian	**Circuit Switch Usage**			**Packet Service Usage**	
	UL	**DL**		**UL**	**DL**
	52560	52560			
				135	54
	6084	6084			
				270000	769.5
				40500	115.425
				9000	2250
Total BHCS	58644	58644	**Total BHPS**	319635.0	3188.9
Erlang	16.29	16.29	**Mbps**	3.20	0.03
Vehicular	**Circuit Switch Usage**			**Packet Service Usage**	
	UL	**DL**		**UL**	**DL**
	122640	122640			
				420	168
	2839.2	2839.2			
				25200	71.82
				25200	71.82
				6720	1680
Total BHCS	125479.2	125479.2	**Total BHPS**	57540	1991.64
Erlang	34.86	34.86	**Mbps**	0.58	0.02
3X Total	112.58	112.58	**3X Total**	14.43	0.19

Table 13-39 1X Traffic Contribution

Current 1X	Circuit Switch Usage			Packet Service Usage	
Building	**UL**	**DL**		**UL**	**DL**
	1034775	1034775			
				315	126
	70980	70980			
				3937500	11221.875
				1181250	3366.5625
				210000	52500
Total BHCS	1105755.0	1105755.0	**Total BHPS**	5329065.0	67214.4
Erlang	307.2	307.2	**Mbps**	53.29	0.67
	Circuit Switch Usage			**Packet Service Usage**	
Pedestrian	**UL**	**DL**		**UL**	**DL**
	262800	262800			
				675	270
	30420	30420			
				1350000	3847.5
				202500	577.125
				45000	11250
Total BHCS	293220	293220	**Total BHPS**	1598175.0	15944.6
Erlang	81.45	81.45	**Mbps**	15.98	0.16
	Circuit Switch Usage			**Packet Service Usage**	
Vehicular	**UL**	**DL**		**UL**	**DL**
	613200	613200			
				2100	840
	14196	14196			
				126000	359.1
				126000	359.1
				33600	8400
Total BHCS	627396	627396	**Total BHPS**	287700	9958.2
Erlang	174.28	174.28	**Mbps**	2.88	0.10
1X Total	562.88	562.88	**1X Total**	72.15	0.93

Table 13-40

1X and 3X Traffic

	Circuit Switched (Erlang)		Packet (Mbps)	
	DL	UL	DL	UL
1X Total	562.88	562.88	72.15	0.93
3X Total	112.58	112.58	14.43	0.19
Total	675.46	675.46	86.58	1.12

allocated all the packet traffic. Because the radio system is backward compatible, 1X capable mobiles can interact with 3X carriers so the 3X can be used for data only applications.

Another thought comes about for the system layout and that is that the system as it is defined in this example is not capacity-of-coverage driven but rather capability-driven, which is fundamentally different than past designs.

Using the existing sites from the design example done previously for integrating a 1X system into an existing platform, the following underlying numbers will be used to base the 1:1 overlay on shown in Table 13-41.

Taking things just a little further, the basic configuration of the 1X system is shown in Figure 13-21.

The basic configuration shows that parts of the system, in the core region defined as being BSC1, have both 1x and DO channels deployed whereas the rest of the system only has a 1x channel deployed. With the introduction of a 3X platform and the decision to do a 1:1, overlay for the system is shown in Figure 13-22 with the channels associated with the 3X carriers having the legacy 2.5G or 1X configurations for legacy mobiles. The requirement for additional spectrum over the existing 1X system deployment is rather an obvious issue.

Looking at Table 13-42, the issue deployment of 3X into a system is shown in Figure 13-22 but only for the core of the system to facilitate the illustration only.

The next obvious question that needs to be quickly discussed is the issue of what platforms need to be altered in order to support the new 3X system.

Table 13-41

Base System

	Region	Area (km²)	BTS Sites
Building	1	100	12
Pedestrian	2	900	60
Vehicular	3	4,000	96
Unserved	4	6,000	
Total		11,000	168

Table 13-42

3X and 1X Deployment

3X	1X
1	F1–1X
2	F2–1X-EV-DO
3	F3–1X-EV-DV

The change required for migrating from a 1X to a 3X platform is expected to require physical changes to

- BTS radios
- Channel elements

The rest of the PDSN network, as well as the BSC connectivity with the PDSN and circuit-switched networks, should remain the same. The difference would arise if VoIP is deployed but this would impact the BSCs primarily and require the introduction of a VoIP gateway and supporting functions that were covered in detail in Chapter 8, "Voice Over IP Technology."

Therefore the configuration for the previous example, due to the low traffic loading, is the same as shown in Figure 13-5 because the 1X to 3X migration in this example is a capability-driven migration—not capacity-driven.

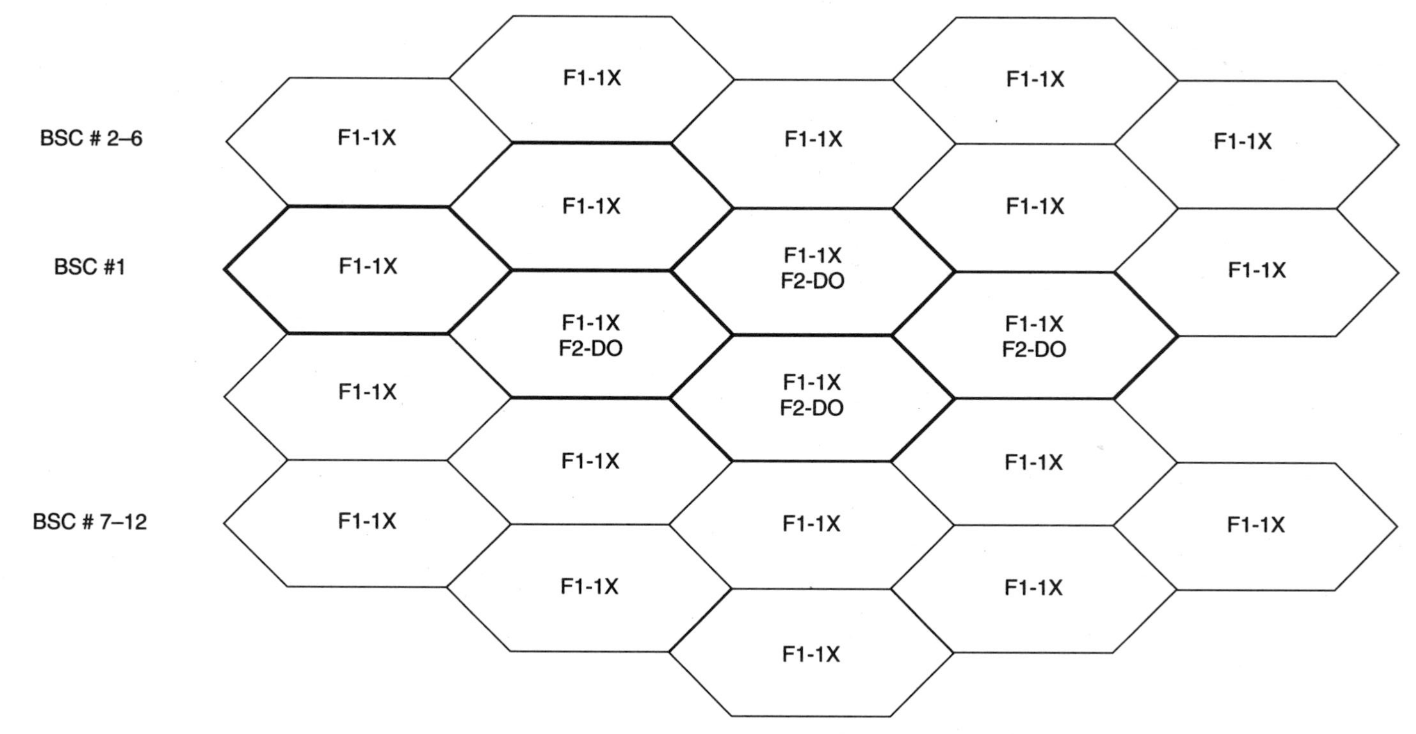

Figure 13-21 1X base system.

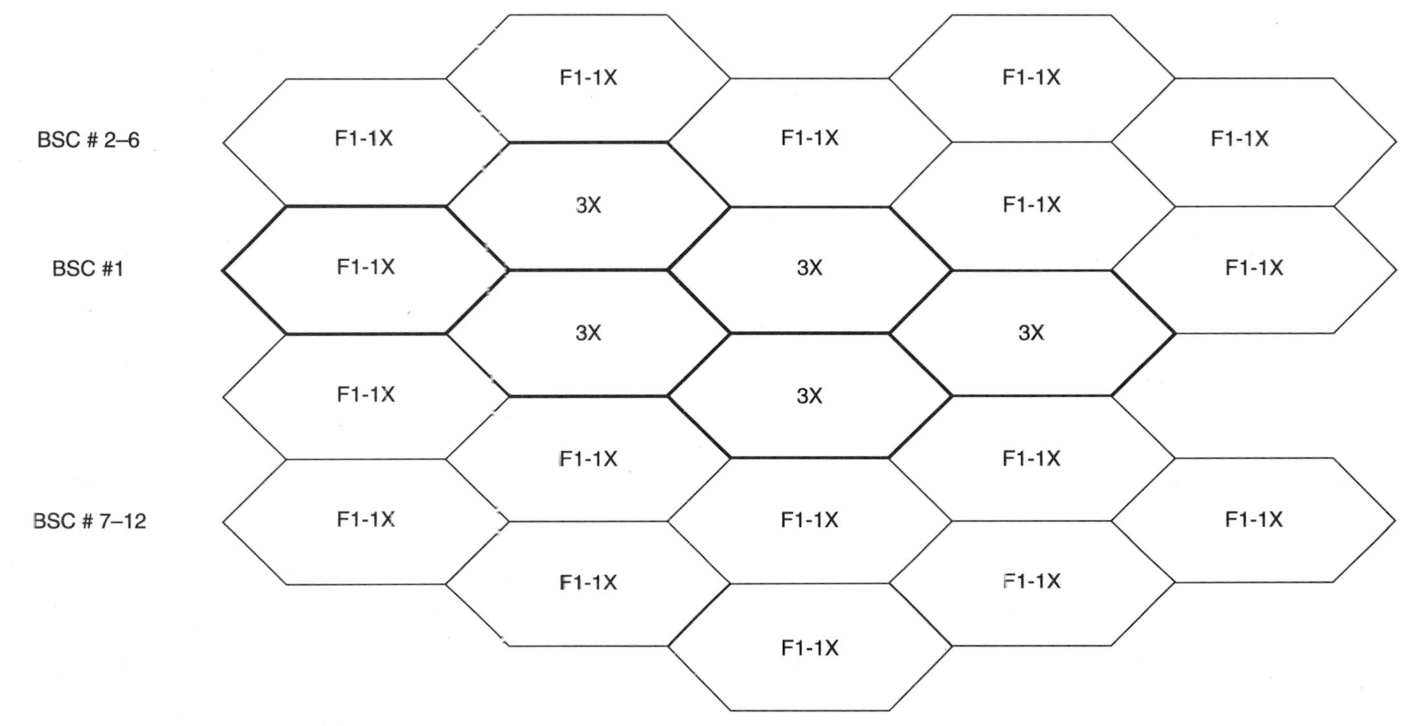

Figure 13-22 3X system deployment—core only.

References

3GPP2 C.S0008-0. *"Multi-carrier Specification for Spread Spectrum Systems on GSM MAP (MC-MAP) (Lower Layers Air Interface),"* June 9, 2000.

Bates, Gregory. *"Voice and Data Communications Handbook,"* Signature Ed., McGraw-Hill, 1998.

Barron, Tim. *"Wireless Links for PCS and Cellular Networks"*, Cellular Integration. Sept., 1995, pgs. 20–23.

Carr, J.J. *"Practical Antenna Handbook,"* Tab Books, McGraw-Hill, Blue Ridge Summit, PA, 1989.

DeRose. *"The Wireless Data Handbook,"* Quantum Publishing, Inc., Mendocino, CA, 1994.

Dixon. *"Spread Spectrum Systems,"* 2nd Ed, John Wiley & Sons, New York, 1984.

Lynch, Dick. *"Developing a Cellular/PCS National Seamless Network,"* Cellular Integration, Sept. 1995, pgs. 24–26.

Jakes W.C. *"Microwave Mobile Communications,"* IEEE Press, New York, 1974.

Molisch, Andreas F. *"Wideband Wireless Digital Communications,"* 2001, Prentice Hall, New Jersey.

Qualcomm. *"An Overview of the Application of Code Division Multiple Access (CDMA) to Digital Cellular Systems and Personal Cellular Networks,"* Qualcomm, San Diego, CA, May 21, 1992.

Salter, Avril. *"W-CDMA Trial&Error,"* Wireless Review, Nov. 1, 1999, pg. 58.

Smith, Gervelis. *"Cellular System Design and Optimization,"* McGraw-Hill, 1996.

Smith, Clint. *"Practical Cellular and PCS Design,"* McGraw-Hill, 1997.

Smith, Clint. *"Wireless Telecom FAQ,"* McGraw-Hill, 2000.

Stimson. *"Introduction to Airborne Radar,"* Hughes Aircraft Company, El Segundo, CA, 1983.

Shank, Keith. *"A Time to Converge,"* Wireless Review, Aug. 1, 1999, pg. 26.

TIA/EIA-98-C. *"Recommended Minimum Performance Standards for Dual-Mode Spread Spectrum Mobile Stations (Revision of TIA/EIA-98-B),"* Nov. 1999.

TIA.EIA IS-2000-1. "*Introduction to cdma2000 Standards for Spread Spectrum Systems,*" June 9, 2000.

TIA/EIR IS-2000-2. "*Physical Layer Standard for cdma2000 Spread Spectrum Systems,*" Sept. 12, 2000.

TIA/EIA IS-2000-3. "*Medium Access Control (MAC) Standard for cdma2000 Spread Spectrum Systems,*" Sept. 12, 2000.

TIA/EIA IS-2000-4. "*Signaling Link Access Control (LAC) Specification for cdma2000 Spread Spectrum Systems,*" August 12, 2000.

Webb, Dr William. "*CDMA for WLL,*" Mobile Communications International, Jan. 1999, pg 61.

Willenegger, Serge. "*cdma2000 Physical Layer: An Overview,*" Qualcomm 5775, San Diego, CA.

CHAPTER 14

Communication Sites

The communication site, usually referred to as the base station, is a critical component in any wireless system. A communication site is a physical location where there is radio equipment that is intended for either receiving, transmitting, or both. With the advent of multiple technology platforms and the need to collocate wireless services on a single structure, there is normally more than one technology and service operator located at any one location. For the design engineer involved with either designing a greenfield or collocation, there is almost an infinite amount of different types of communication site configurations and perturbations that can be considered.

This chapter will cover some of the more salient issues associated with a wireless communication site and the implications that should be considered for installing 2.5G and 3G equipment. Therefore the focus of attention will be directed toward the *radio frequency* (RF) engineer and the issues associated with the design phase. The particulars associated with the operation and construction concerns that are an integral part of the communication sites design criteria will not be covered here because it is outside the scope of this book.

14.1 Communication-Site Types

There are numerous types of communication sites that comprise the 1G, 2G, 2.5G, and future 3G configurations associated with wireless mobility systems. There are also a plethora of other communication sites that the design engineer also may encounter in the design process such as, existing mobility systems, LMDS, PMP, MMDS, SMR, ESMR, paging, broadcast, FM, AM, and so on. Each of these different types of wireless sites, depending on its proximity, may need to be included in the design phase.

The usual co-location considerations are

- **Antenna placement**
- **Frequency of operation** Adjacent channel and co-channel (adjacent market)
- **Intermodulation** Third and fifth order *Intermodulation Distortion* (IMD) products along with spectral regrowth
- **Site maintenance obstructions** Window washing equipment, sand blasting, and so on

The most common types of sites that would be considered for a 2.5G and 3G implementation are

- Macro
  - Omni
  - Sector
- Micro
- Pico

The definition of what macro-, micro-, and pico-cells are is really dependent upon the service area the base station will cover. For instance if the site is to cover 25 square miles, it is considered a macro-cell site. However, if the site is to cover 0.25 miles, it is usually referred to as a micro-cell, whereas a site that is meant to cover a meeting room is often referred to as a pico-cell. Because there is no specification that defines the service area and the name for the particular communication site, the definitions of what constitutes a macro-, micro-, and pico-cell will remain somewhat vague.

A typical cell site, or rather, communication site consists of the following components that are referenced in Figure 14-1. The piece components are the same whether it is for a macro-, micro-, or pico-cell site. The chief difference lies in the form factor that impacts the overall capacity carrying capability for the site and of course power.

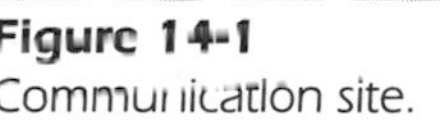

Figure 14-1
Communication site.

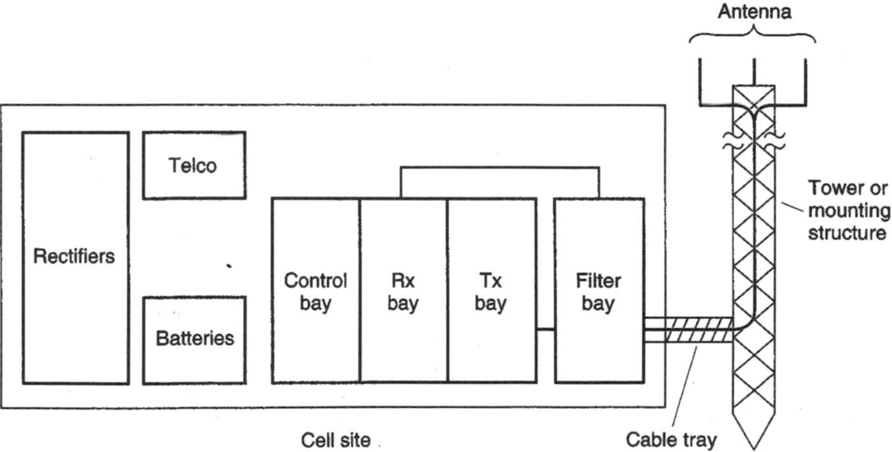

14.1.1 Macro-Cell Site

A macro-cell site is what most people have come to know or expect to see for a communication site for all forms of mobile communication systems. With the advent of *personal communications services* (PCS), the need for macro-cells was initially portrayed as being an item of the past. However, the need to provide coverage to compete with existing wireless operators made the use of macro-cell sites a necessity.

There are multiple configurations associated with each type of technology platform picked for the communication system. For instance AMPS, TACS, GSM, CDMA, NADC (IS-136), and iDEN, to mention a few, can all be configured either as an omni-, bi-directional, or three-sector cell depending on the application at hand.

For the design engineer, the decision of using a macro-cell is driven by multiple reasons. However, there are a few different perturbations with regards to cell sites that need lead to interesting designs. As is often the case in real life, in the city, the amount of green-field locations is not large and in fact, the desire to utilize an existing communication site is receiving much pressure. There are a multitude of reasons why an existing communication site should be used and also why it should not be used.

The reasons for utilizing an existing communication site leads to the issue of community affairs in that there is a strong public awareness of mobile phone systems and the need to limit the amount of towers or any new communication site that are in a community.

There are several types of macro-cell sites that a design engineer considers for possible use depending on the design objectives. The types of macro-cell sites can be classified as either an omni-directional site or a directional site. The omni-cell sites have a coverage pattern of 360 degrees in the horizontal plane while directional sites usually comprise three sectors each covering 120 degrees of the horizontal plane.

14.1.2 Omni-Directional Cell Sites

The omni-directional cell site is used typically in a low capacity area of the network where the system is noise limited and not interference limited. The omni-cell is typically used to cover uniformly in all directions, 360 degrees. Under ideal conditions, the omni-cell site would have a circular pattern when there are no obstructions and the coverage is purely line of sight.

With an omni site there are several methods that can be used for antenna installations. The first is a simple installation on a monopole shown in Figure 14-2. The transmit antenna is highest on the structure with the receive antennas located under the platform, which was the standard installation for 1G and typically is a hold over from that design era. The distance between the receive antennas is usually determined for maximizing spatial diversity so that the mobile signal arriving at both receive antennas are somewhat decorrelated enabling for a diversity gain, or rather, fade margin protection.

There are of course other variants to omni-cell site antenna installations, and they involve installing on a building and when the amount of antennas is limited. Figure 14-3 is an example of an antenna installation that occurs on a building. Please note that the location of antenna needs to meet the required set back rules. If the set back rules cannot be adhered to, then it is possible to install the antennas near the edge of the roof; however, they may become visible to the public at this point, the landlord may not wish this type of installation to take place, or the local ordinances prohibit this from occurring.

Please note that the placement of the receive antennas for rooftop installation should be such that if there is only one major road in the area for the

Figure 14-2
Monopole.

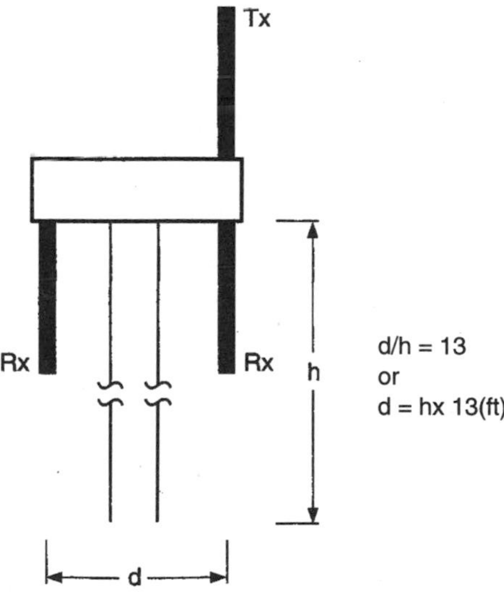

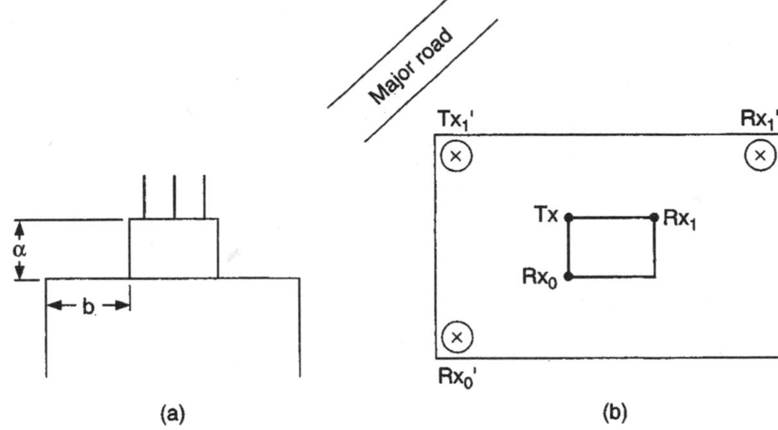

Figure 14-3
Existing rooftop.

cell to cover, then the horizontal diversity placement for the antennas should be such that it is maximized in the direction toward the road. Lastly, if the primary location is not achievable for mounting the antennas, then moving them to the lower level is possible. However, based on the penthouse size, significant blockage may occur in a direction, and this needs to be factored into the design process.

14.1.3 Directional Cell Site

The directional cell-site utilizing three sectors is one of the most popular cell site configurations utilized in the wireless industry, next to the omnicell. The three-sector cell is one that has sectors that cover 120 degrees each, thus having three sectors makes a full circle.

There are a multitude of combinations for transmit receive that can be used for establishing a three sector cell site. However, the following example is the more basic configuration that is used and involves three antennas per sector shown in Figure 14-4.

The configuration shown in Figure 14-4 has a single transmit and two receive antennas per sector. Naturally the amount of receive and transmit antennas can change depending on the technology implemented for 2G, 2.5G, or 3G. For example iDEN 2G systems typically involve three antennas per sector, but all three are usually duplexed to keep the antenna count down while at the same time enabling for three-branch diversity reception.

When designing antenna placements for a site, it is strongly recommended that future configurations be considered at the onset of the design process. For instance implementing 3G services may require the use of a separate Tx antenna and possibly Rx antennas.

14.1.4 Micro-cells

Micro-cells are prevalent in wireless mobility systems as the operators strive to reduce the geographic area each cell site covers thus facilitating more reuse in the network. Micro-cells are also deployed to provide coverage in buildings, subway systems, tunnels, and resolve unique coverage problems. The technology platforms that tend to be referenced as micro-cells involve any communication system that is less than 1/2 kilometer in radius. Typically 4 to 10 micro-cells comprise the footprint that a macro-cell site might be able to perform.

There are currently several types of technology platforms that fall into the general categorization called micro-cells:

- Fiber-fed micro-cell
- T1/E1 micro-cell

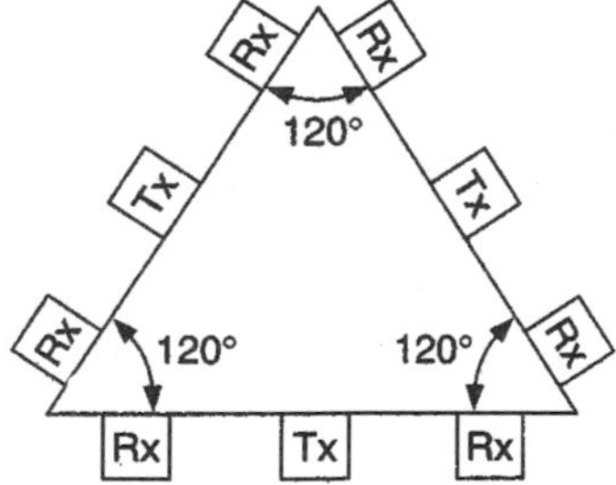

Figure 14-4 Directional cell site.

- Microwave micro-cell
- High-power ReRad
- Low-power ReRad
- Bi-directional amplifier

The choice of which technology platform to utilize is driven by a variety of factors that are unique to that particular situation. One driving factor for the technology platform chosen is the application that is being engineered for, capacity, coverage, or private wireless PBX. Another important factor in the technology platform decision is the configuration options available at that location for providing radio capacity. A third factor in the decision for which technology to utilize is driven by the overall cost of the solution for the network.

A micro-cell typically uses an omni antenna for transmission and reception. The micro-cell also has less Tx power and lower gain receive antennas then does its counterpart the macro-cell site. In addition the micro-cell site typically exhibits a lower elevation then the macro-cell site does helping to contain its coverage area leading to selected trouble spot resolutions. The use of the omni antenna for micro-cells facilitates a smaller, physical appearance leading to installations in more difficult land use areas.

An example of a micro-cell is shown in Figure 14-5. The form factor of the radio hardware is not shown and is assumed to be insignificant or located in the utility box next to the traffic light shown.

14.1.5 Pico-Cell Sites

The use of pico-cell sites by wireless operators is driven by the desire to provide very targeted coverage and capacity to a given area or application. The pico-cell has a very small service area where several pico-cells in concept can cover the same area as a micro-cell.

The pico-cell is a spot coverage and low-capacity site, as compared to a macro-cell site. Pico-cell sites typically have a single omni antenna, as do micro-cells. However, the power and thus the coverage of the pico-cell is less than a micro-cell.

An example of a possible pico-cell is shown in Figure 14-6.

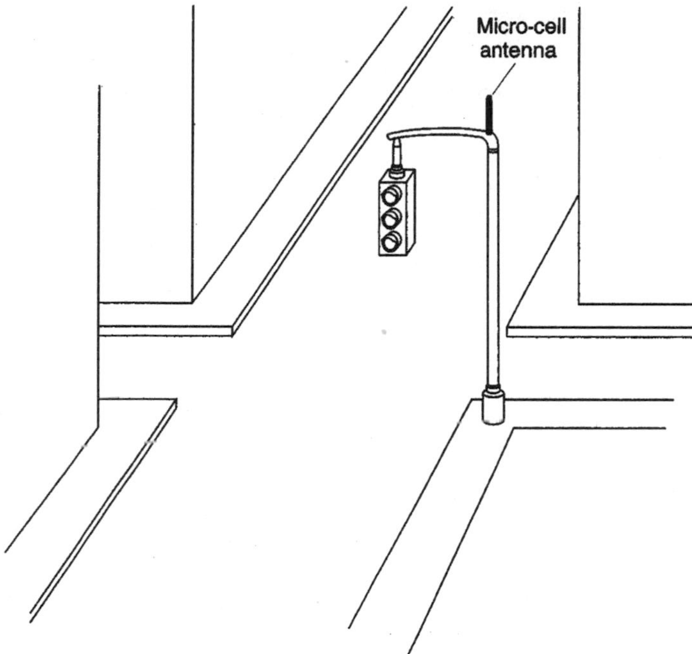

Figure 14-5
Micro-cell.

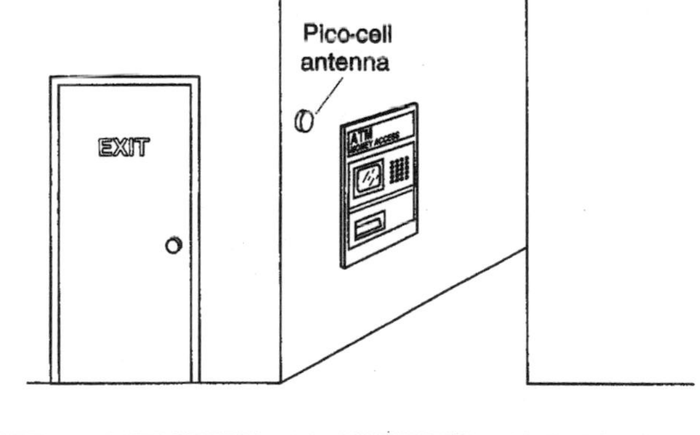

Figure 14-6
Pico-cell.

14.2 Installation

The installation of a cell needs to factor into it many issues that have different elements of importance with them. Some of the issues that need to be factored into the design are the physical placement of the antennas themselves. The physical placement of the antennas depends on some factors that may or may not be in the design engineer's control.

14.2.1 Cable Runs

Some of the physical installation issues that need to be factored into the design involve cable runs from the antenna system, each leg, to the base station equipment. Although this may seem an obvious point, often there are situations where the desired routing of the cables is not practical, making the real installation length much longer than desired. The additional cable run length, when installation reality is factored in, may have made the site nondesirable; however, if this situation occurs too far down the stream of the construction process, it is too late to reject the site or make the appropriate design alterations to correct the situation.

14.2.2 Antenna Mounting

Obviously the mounting of the antennas needs to be taken with extreme care. The following is a brief checklist to ensure for antenna mounting concerns to be checked prior to acceptance of a cell site:

1. What are the number and type of antennas to be installed?
2. What is the maximum cable run allowed?
3. Identify and rank obstructions that would alter the desired coverage.
4. Rx-antenna spacing is adequate; diversity requirements are met.
5. Isolation requirements meet with other services.
6. Antenna-AGL requirements are met.
7. Antenna-mounting parameters are met.
8. Intermodulation analysis is completed.
9. Path clearance analysis is verified (if applicable).

This previous list is just preliminary and can easily be altered based on the situation at hand. However, the list should be modified to meet your particular system-design requirements.

When installing on a tower, the physical spacing or offset from the tower must be selected to the tower's structure which either enhances or does not alter the antenna pattern desired. In addition to the pattern issue, care must be taken in ensuring that there are not degradations caused to the system because of unwanted energy from adjacent systems. It is suggested that an interference analysis be conducted for every site to ensure the proper isolation requirements are met.

When installing on an existing building, the following few items should be considered in the design phase.

14.2.3 Diversity Spacing

The diversity spacing for the receive antennas need to ensure that the proper fade margin protection is designed into the system. Diversity spacing is meant to achieve some de-correlation between the mobile received signal. There have been numerous studies conducted on diversity reception and the system performance improvements associated with the proper implementation of a receive diversity scheme. The diversity scheme is typically achieved through horizontally-placed antennas that are then fed to the radio receiver at the base station.

The base station receiver typically would use either max ratio combing or select diversity as the method of achieving the system performance improvement.

However, for the diversity reception, the antennas for mobile communication for 1G, 2G, 2.5G, and 3G involve horizontal diversity spacing. The initial objective would be to place the receive antennas so that they were as de-correlated. However there is a practical limit: the spacing between the receive antennas when the feedline length between the antennas becomes such that either the feedline loss exceeds the diversity advantage or the signals are completely de-correlated as to eliminate any diversity combining gain possible.

For a micro or pico-cell site, the use of diversity reception is usually a forgone thought due to the antenna configuration—one omni antenna. But when looking at a macro or even micro-cell with multiple antennas for receive, the question arises about what spacing is needed between the antennas. The equation that follows should be used for a two-branch receive system.

Diversity spacing (feet) = [(AGL of antenna (feet)/11) ($835/f_o$)] where f_o is the center receive frequency in MHz.

14.2.4 Roof Mounting

When installing antennas on an existing roof or penthouse, consideration must be taken into account on how high the antenna must be with respect to the roof surface. Obviously the ideal location is to place the antenna right at the roof edge. However, placing the antenna at the roof edge may not be a viable installation design either for aesthetics, local ordinances, or practical mounting issues. When the antennas cannot be placed at the edge of the roof, a relationship between the distance from the edge of the roof and the antenna height exists and needs to be followed.

The basic relationship between the antenna height and the roof edge of the building is depicted in Figure 14-7.

The previous example assumes that there are no additional obstructions between the antenna and the roof edge. If there are obstructions between the antenna and the roof edge, then additional height may be needed. Examples of additional obstructions involve HVAC units and window cleaning apparatus.

Please remember that if there is a desire to implement severe downtilt into the design either at the present or in the future, then the height requirements above the rooftop may need to be increased.

Figure 14-7 Roof mounting.

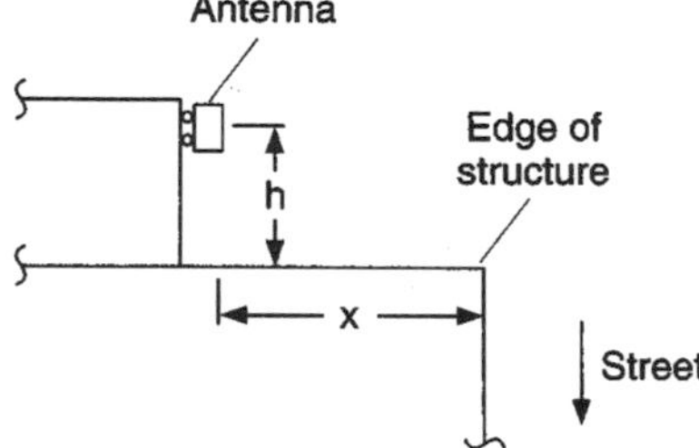

14.2.5 Wall Mounting

For many building installations it may not be possible to install the antennas above the penthouse or other structures for the building. Often it is necessary to install the antennas onto the penthouse or water tank of an existing building. When installing antennas onto an existing structure, rarely has the building architect factored into the potential installation of antennas at the onset of the building design. Therefore as shown in Figure 14-8, the building walls may meet one orientation needed for the system, but rarely all three for a three-sector configuration.

Therefore it is necessary to determine what the offset from the wall of the building structure needs to be. Figure 14-9 illustrates the wall mounting offset that is required to ensure proper orientation for each sector.

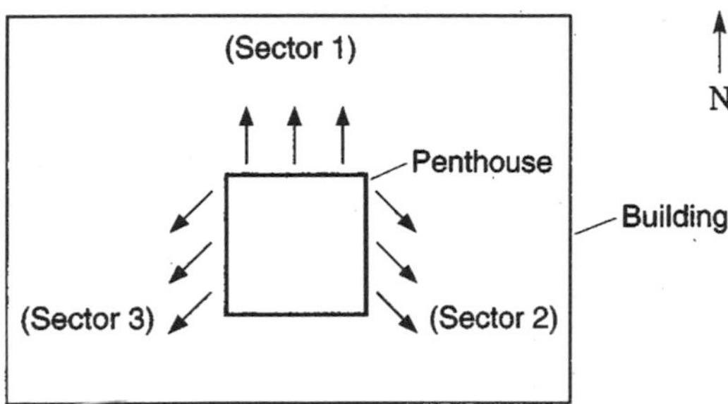

Figure 14-8 Three-sector building configuration.

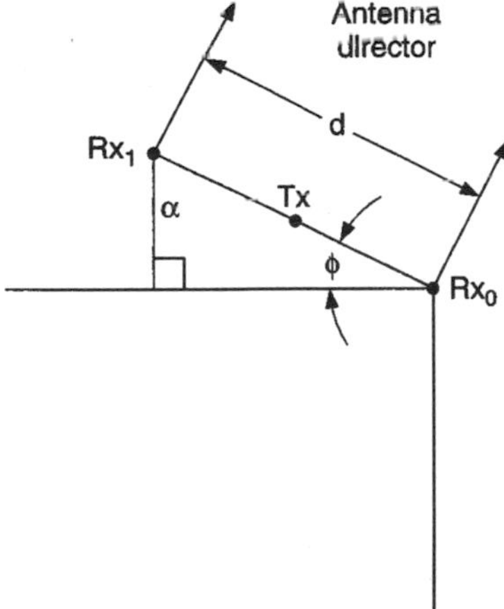

Figure 14-9 Antenna offset mounting.

Obviously, common sense must enter into the situation here when the inclusion of the offset brackets makes the site a metal monster. Tradeoff can be made in the design when the orientation for each sector is within the design tolerance limits for sector orientation. The design tolerance should be within + or − 5 degrees for a three-sector cell.

Lastly, the wall mounting offset must all meet the setback requirements for both antennas and local ordinances.

14.3 Towers

There are numerous types of towers that can and do exist in a wireless network. However, there are three basic types of towers that are more common: self-supporting, guy wire, and monopole. The general configuration for each of the towers is shown in Figures 14-10, 14-11, and 14-12.

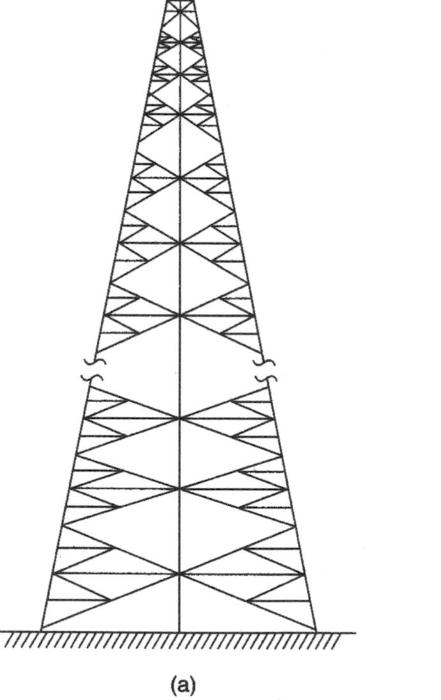

Figure 14-10 Self-supporting tower: (a) Side-view (b) Top-down view

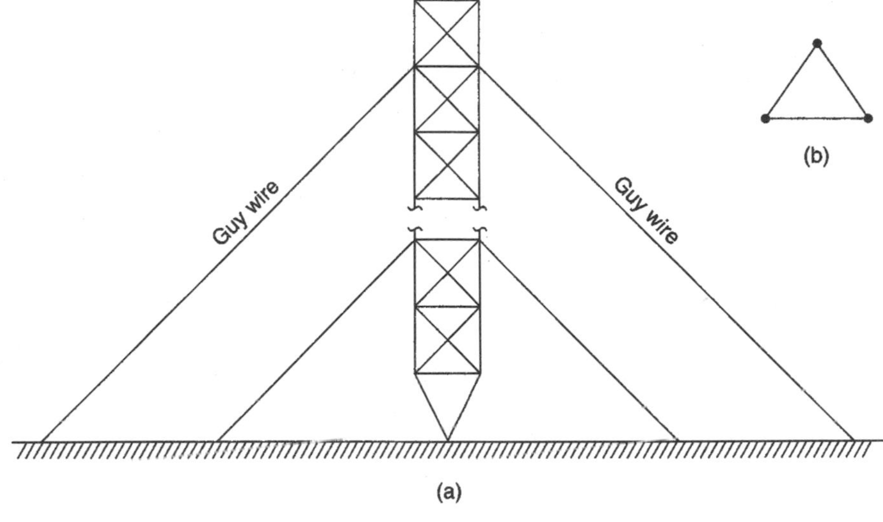

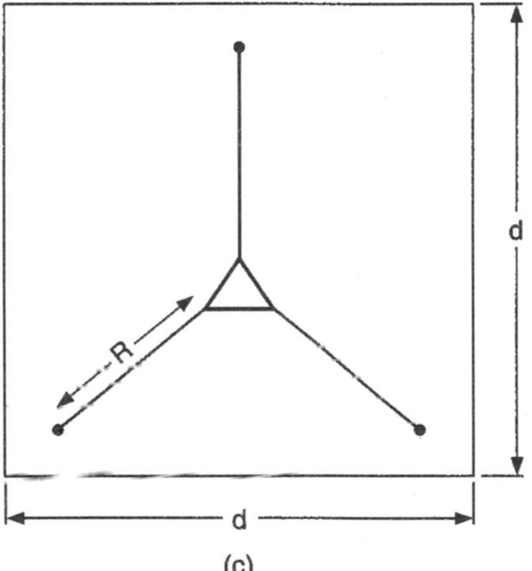

Figure 14-11
Guy-wire tower:
(a) Side-view
(b) Top-down view
(c) Footprint of tower

The cheapest to construct is the guy-wire tower, followed by the monopole, and then the self-supporting. Each has its advantages and disadvantages. The guy-wire tower requires a large amount of room for its guy wires and is shown in Figure 14-11. This can be either relaxed or increased depending on loading and height issues.

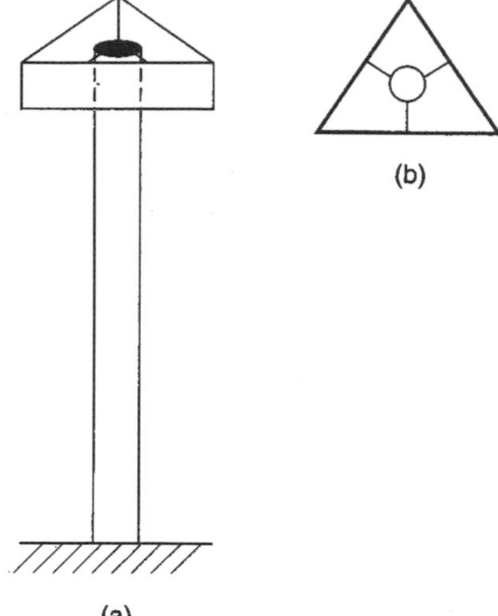

Figure 14-12 Monopole: (a) Side-view (b) Top-down view

The self-supporting tower will enable multiple carriers to entertain operation at the facility whereas the monopole will also accommodate multiple users, although not as many as a similar height self-supporting tower.

14.4 In-building

Wireless systems have numerous applications for in-building applications. The applications include improving coverage for a convention center or large client, disaster recovery, or a wireless PBX to mention a few. With the advent of better transport for data services, the possibility exists that 3G will find more uses for in-building systems.

The propagation of the radio-frequency energy, however, takes on unique characteristics in an in-building application as compared to an outdoor environment. The primary difference in propagation characteristics for in-building versus outdoors is the fading, shadowing, and interference. The fading situation for in-building results in deeper and has spatially closer fades when a system is deployed in an in-building application. Shadowing is also quite different in an in-building application due to the lower antenna heights and excessive losses through floors, walls, and cubicles. The shadowing effects in an in-building application severally limit the effective coverage area to almost *line of site* (LOS) for mobile communications. The interference issue with in-building systems can actually benefit in-building applications because the interference is primarily noise driven and not interference. The reason the in-building systems are primarily noise-driven is due to the attenuation experienced by external cell sites as they transverse into the buildings and various structures.

There are some unique considerations that must be taken into account regarding micro-cell system design for inside a building. Some of the design considerations that need to be factored into an in-building design are

- Base-to-mobile power
- Mobile-to-base power
- Link budget
- Coverage area
- Antenna system type and placement
- Frequency planning

The base-to-mobile power needs to be carefully considered to ensure that the desired coverage is met, deep fades are mitigated in the area of concern, the amplifier is not being over or potentially under drive, and mobile overload does not take place. The desired coverage that the in-building system is to provide might require several transmitters because of the limited output power available from the units themselves. For example if the desired coverage area required 1W ERP to provide the desired result, a 10 W amplifier would not be able to perform the task if you needed to deliver a total of 40 channels to that location, meaning only 25 mW of power per channel was really available. The power limitation can, and often does, makes the limiting path in the communication system for an in-building system the forward link.

The forward link power problem is further complicated by the fact that portable and potential mobile units will be operating in very close proximity to the in-building systems antenna. If the forward energy is not properly set, a subscriber unit could easily go into gain compression causing the radio to be desensitized.

The mobile-to-base power also needs to be factored into the in-building design. If the power windows and dynamic power control are not set properly, then imbalances could exist in the talk out to talk back path. Usually the reverse link in any in-building system is not the limiting factor but the mobile-to-base path should be set so that there is a balanced path between the talkout and talk back paths.

Most in-building systems have the ability to utilize diversity receive but do not utilize it for a variety of reasons. The primary reason for not utilizing diversity receive in an in-building system is the need to place two distinct antenna systems in the same area.

The link budget for the communication system needs to be calculated in advance to ensure that both the forward and reverse links are set properly. The link-budget analysis plays a very important role in determining where to place the antenna system, distributed or leaky feeder, and the amount of micro and pico-cell systems required to meet the coverage requirements.

The antenna system selected for the in-building application is directly related to the uniformity of the coverage and quality of the system. The antenna system, no diversity, primarily provides LOS coverage to most of the areas desired in the defined coverage area. Based on the link budget requirements, the antenna system can either be passive or active. The antenna system for an in-building system may take on the role of having passive and active components indifferent parts of the system to satisfy the design requirement.

Typically a passive antenna system is made up of a single or distributed antenna system; it can also utilize a leaky coaxial system. A leaky coaxial system could also be deployed within the same building to provide coverage for the elevator in the building. The advantage a leaky coaxial system has over a distributed antenna is it provides a more uniform coverage to the same area over a distributed antenna system. However, the leaky coaxial system does not lend itself for an aesthetic installation in a building. The use of a distributed antenna system for providing coverage in an in-building system makes the communication system stealthy.

If the antenna system requires the use of active devices in the communication path, the level of complexity increases. The complexity increases for active devices because they require AC or DC power and introduce another failure point in the communication system. However, the use of active devices in the in-building system can untimely make the system work in a more cost-effective fashion. The most common active device in an in-building antenna system is a bi-direction amplifier.

The frequency planning for an indoor system needs to be coordinated with the external network. Most in-building systems are designed to facilitate hand-offs between the in-building and external cellular system. If the in-building system is utilizing its own dedicated channels assigned to it, then it is imperative that the in-building system be integrated into the external network.

14.5 Intermodulation

Intermodulation is the mixing of two or more signals that produce a third or fourth frequency that is undesired. All radio communication sites produce intermodulation no matter how good the design is. However, the fact that there are intermodulation products produced does not mean there is a problem.

Just what is intermodulation and how does it go about calculating an intermodulation product, IMD? Various intermodulation products are shown in the following for reference. The values used are simplistic in nature so facilitate the examples. In each of the examples, A=880 MHz, B=45 MHz, C=931 MHz, and D is the intermodulation product. The example listed in the following does not represent all the perturbations possible.

Second order:	A + B = D (925 MHz)
	A − B = D (835 MHz)
Third order:	A + 2B = D (970 MHz)
	A − 2B = D (790 MHz)
	A + B + C − D (1856 MHz)
	A − B + C = D (1766 MHz)
Fifth order:	2A − 2B − C = D (739 MHz)

The various products that make up the mixing equation to determine the order of the potential intermodulation. All too often when you conduct an intermodulation study for a cell site, there are numerous potential problems identified in the report. The key concept to remember is that the intermodulation report you are most likely looking at does not take into account power, modulation, or physical separation between the source and the victim, to mention a few. Therefore the intermodulation report should be used as a prerequisite for any site visit so you have some potential candidates to investigate.

Intermodulation can also be caused by your own equipment through bad connectors, antennas, or faulty grounding systems. However, the majority of the intermodulation problems encountered were a result of a problem in the antenna system for the site and well within the control of the operator to fix.

Just how you go about isolating an intermodulation problem is part art and part science. I prefer the scientific approach because it is consistent and methodical in nature.

The biggest step is identifying the actual problem; the rest of the steps will fall in line. Therefore it is recommended that the following procedure be utilized for intermodulation site investigations.

14.5.1 IM Check Procedure

1. Determine if there are any co-located transmitters at this facility.
2. Collect information on each of the following transmitters:
   - Antenna types
   - Emission type
   - Transmit power
   - Location of antennas
   - Operator of equipment
   - FCC license number
3. Conduct an intermodulation study report looking for hits in your own band or in another band based on the nature of the problem.
4. Allocate sufficient time to review the report.
5. Determine if there is a potential problem.
6. Formulate a hypothesis for the cause of the problem and engineer a solution.

Based on the actual problem encountered, the resolution can take on many forms:

- Is the problem identified feasible?
- Can the problem be resolved through isolation alone?
- Is the problem receiver overload-related?

If the intermodulation product is caused by the frequency assignment at the cell site, then it will be necessary to alter the frequency plan for the site, but first remove the offending channels from service.

If the intermodulation problem is due to receiver overload, the situation can be resolved by placing a notch filter in the receive path if it is caused by a discrete frequency. If the overload is caused by cellular mobiles, using a notch filter will not resolve the situation. Instead, mobile overload can be resolved by placing an attenuation in the receive path, prior to the first pre-amp, effectively reducing the sensitivity of the base station receiver.

14.6 Isolation

Isolation is used to describe the amount of attenuation needed between the source, transmitter, and the victim or receiver. All wireless communication systems require some level of isolation between their own transmitters and other transmitters, and their receivers at the base station. The fact that you are using a pico versus a macro-cell site does not mean that more or less isolation is required.

The amount of isolation needed for communication systems is dependent upon a multitude of issues:

- Location of potential offending transmitter to receiver
- Technology platform utilized
- Receiver sensitivity

The methods that follow are based on the simple fact that there is no defective equipment, or there are not out-of-specification transmitters at the location in question. Please keep in mind that the isolation requirements may or may not be directly applicable to the communication facilities that are collocated with you. As often is the case, the offending transmitter is several buildings away.

Isolation can be achieved, once the offender(s) is identified, through antenna placement using both horizontal and vertical separation. Another method could be achieved through more selective filters. Just how much isolation is needed?

An example of how to determine the amount of isolation needed for a communication system is shown in Table 14.1.

$$Tx = 852 \text{ MHz}$$

$$Rx = 849 \text{ MHz}$$

Table 14-1
Isolation

Isolation Requirement		
Tx Power	+50 dBm	
Rx 1-dB compression	−27 dBm	
	77 dB	Isolation needed
Tx Filter Attenuation (in Rx band)		
Tx Filter Attenuation	30 dB	
Vertical Isolation	50 dB (@ 10 ft)	
	80 dB	Isolation

Therefore 80 dB isolation is achieved with 10 feet of vertical separation, which is sufficient to prevent compression. The previous example is for cases where out-of-band emissions are the problem. When the problem is intermodulation-related, it is possible to obtain the necessary isolation to protect the receiver, if the mix is occurring at another location besides in the receive path itself, through simple path loss alone.

Additionally, what is not discussed is the impact of spectral regrowth of the transmitter into the receive band, which can only be resolved through better transmit filtering at the expense of increased insertion loss or size for the base station.

14.7 Communication-Site Check List

Table 14-2 is a brief summary of the major items that need to be checked prior to or during the commissioning of a communication site. The check list that follows is generic and should be tailored for your particular application, that is, add or remove parts where applicable. However, the list that follows is an excellent first step in ensuring that everything is accounted for prior to the communication site going commercial.

Table 14-2

Cell Site Check List

Topic	Received	Open
Site-Location Issues:		
1) 24-Hour Access		
2) Parking		
3) Direction to site		
4) Keys issued		
5) Entry/Access Rerstrictions		
6) Elevator Operation Hours		
7) Copy of Lease		
8) Copy of Building permits		
9) Obtainment of Lean Releases		
10) Certificate of Occupancy		
Utilities		
1) Separate Meter Installed		
2) Auxillary Power (generator)		
3) Rectifiers Installed and Balanced		
4) Batteries Installed		
5) Batteries charged		
6) Safety Gear Installed		
7) Fan/Venting supplied		
Facilities		
1) Copper or Fiber		
2) Power for fiber hookup (if applicable)		
3) POTS lines for Operations		
4) Number of facilities identified by Engineering		
5) Spans shacked and baked		
HVAC		
1) Installation Completed		
2) HVAC tested		
3) HVAC system accepted		
Antenna System		
1) FAA Requirements met		
2) Antennas Mounted correctly		
3) Antenna Azimuth checked		
4) Antenna plumbness check		
5) Antenna inclination verified		
6) SWR Check of antenna system		
7) SWR record given to Ops and Engineering		
8) Feedline connections sealed		
9) Feedline grounds completed		
Operations		
1) User Alarms defined		
Engineering		
1) Site Parameters Defined		
2) Interference check completed		
3) Installation MOP generated		
4) FCC requirements document filled out		
5) Drive Test complete		
6) Optimization complete		
7) Performance Package completed		
Radio Infrastructure		
1) Bays Installed		
2) Equipment installed according to plans		
3) Rx and Tx filters tested		
4) Radio Equipment ATP'd		
5) Tx output measured and correct		
6) Grounding complete		

References

Smith, Clint. "*Practical Cellular and PCS Design*," McGraw-Hill, 1997.

Smith, Clint. "*Wireless Telecom FAQ*," McGraw-Hill, 2000.

Smith, Gervelis. "*Cellular System Design and Optimization*," McGraw-Hill, 1996.

TIA.EIA IS-2000-1. "*Introduction to cdma2000 Standards for Spread Spectrum Systems*," June 9, 2000.

INDEX

Symbols

A

B

C

D

E

F

G

H

I

J–L

Q–R

S

T

U